INTRODUCTION TO MICROFLUIDICS

Introduction to Microfluidics

Patrick Tabeling

*École Supérieure de Physique et de Chimie Industrielles, (ESPCI) Paris,
Paris Sciences et Lettres (PSL) University.*

OXFORD
UNIVERSITY PRESS

Great Clarendon Street, Oxford, OX2 6DP,
United Kingdom

Oxford University Press is a department of the University of Oxford.
It furthers the University's objective of excellence in research, scholarship,
and education by publishing worldwide. Oxford is a registered trade mark of
Oxford University Press in the UK and in certain other countries

First published in hardback in 2005
First published in paperback in 2010

Published in the United States of America by Oxford University Press
198 Madison Avenue, New York, NY 10016, United States of America

British Library Cataloguing in Publication Data
Data available

Library of Congress Control Number: 2023907882

ISBN 978–0–19–284530–6

DOI: 10.1093/oso/9780192845306.001.0001

Printed and bound by
CPI Group (UK) Ltd, Croydon, CR0 4YY

To Isa

Preface

Microfluidics has progressed considerably over the last twenty years, and the time has come to envisage a serious update of the first edition of *Introduction to Microfluidics*, published in 2005. In fact, this second edition is more than an update. Compared to the first one, it keeps the same structure, the same spirit, the same attempts to explain things, on a physical basis, in depth and simply, whenever possible, but it is not reducible to an update. The present edition results from complete rewriting of the first one, nurtured by the considerable amount of information collected in the field over the last two decades. So much information has been gathered in twenty years. So many revisions of the visions of the field have been made. Things that looked impossible in the 1990s gave rise to an important industry, ten years later. This is the case of next generation sequencing (NGS). Things that looked revolutionary turned out to be disappointing. The history of microfluidics is full of dreams that became real and appealing evidence that became wrong. Let us return to the turn of the century. At that time, the microfluidic market (i.e. without ink jet printing) was small, and scepticism was floating around concerning the potential of the technology to find its feet in a market, even though it was regularly announced, here and there, that microfluidics would revolutionarize the twenty-first century. Common sense led to the theory, in fact wrong, that driving flows through tiny channels, at an industrial scale, without leaks, clogging, bubbles, or uncontrolled adsorption, was impossible. The opposite viewpoint believing that it is straightforward to create a complex, functional, microfluidic device was unrealistic. Still, successful microfluidic products emerged, while, in the meantime, the technology penetrated into an increasing number of new domains. The market steadily increased at a two-digit rate, reaching, today, seventeen billions dollars. At the moment, hundreds of millions of devices are sold every year. For example, 1.2 million Illumina microfluidic flow cells, for gene sequencing, are shipped every single year. In the meantime, fundamental phenomena, such as capillarity, wetting, slippage, and nanofluidic transport have been better understood or, in a number of puzzling cases, just understood. Over the years, the early vision of the domain, based on a strict analogy with microelectronics, gradually shifted to a new paradigm, in which the microfluidic tool box is no longer restricted to MOS-FET surrogates, but incorporates a much broader palette of materials and mechanisms.

Since the first edition, twenty books on microfluidics have been published. Many are good or very good. But some look more like galleries of cartoons exhibiting systems supposed to work, with no data showing that they do. This is a style. I tried to avoid this type of presentation. Other books, often engineering-oriented, describe basic phenomena without much depth. In fact, caricaturing subtle mechanisms is a difficult exercise. Sometimes, errors are made. Perhaps, we could content ourselves with this level of description. After all, we can use a computer with no idea how a transistor works. As P. Anderson stated, science is organized in hierarchical levels, built on top of each other, the upper using concepts elaborated by the lower. Both operate independently, elaborating their own laws and paradigms. Then, as just said, computing researchers do not need to know what a transistor is, and there is no necessity for fluid mechanicists to understand kinetic theory of gases. In the 1990s, it was thought that soon, microfluidic researchers would no longer need fluid mechanics. Engineers would design microfluidic machines, without noticing that fluids are running in them. Perhaps, one day, microfluidics will reach this stage. But microfluidics has not reached this level yet. And it is still valuable and probably important, to acquire a minimal knowledge on the subtleties that the field relies upon, at least to know the limitations that they induce. I attempted to provide this type of knowledge throughout this book.

Many calculations developed in the present book are elementary. Still, they allow us to rationalize most of the physics involved in microfluidic flow and transport phenomena. Physico–chemical processes, developing on surfaces or in bulks, so important in microfluidics, are explained on an intuitive basis, without using the powerful machinery of thermodynamics, whose exquisite concepts sometimes obscure the underpinning physics. Emphasis is placed on modern applications, in which biology plays an important part. Also, as in the first edition, a forty-page introduction to microfluidic technologies is given. The book includes enough material to build up courses of various formats, from a few to twenty hours or so.

Fig. 0.1: Ant playing microguitar (From M. Seiffert (2003)).

Acknowledgements

First, I would like to thank C. M. Ho, who hosted me in 2000, at the University of California at Los Angeles (UCLA), at the time when the 'golden years' of microfluidics were just launching. Thanks to him, I discovered the field. On my return to Paris, one year later, P. G. De Gennes offered me to create a microfluidic laboratory at Ecole Supérieure de Chimie Industrielle (ESPCI), which I did with enthusiasm. Since then, for twenty years, I have had the chance to meet extraordinary people, students, postdocs, and colleagues, and to build strong friendships. Writing the second edition is a sort of noetic journey through these exciting years.

Microfluidics is a broad field, and it is impossible to become, all of a sudden, a world class expert on all the subjects it embraces. I was fortunate to receive help from many colleagues, who kindly exchanged correspondance and, in many cases, corrected errors and misconceptions. I acknowledge J. Ottino, who made numerous remarks on the chaotic mixing section. In the strange atmosphere of the COVID confinement days, we had long exchanges over the internet, evoking, sometimes, the years when we pertained to the same community. This left the impression that writing a book is also a look into the past, sometimes exposing the author to vertigo. I had the same feelings with my friends Y. Pomeau, and J.P. Bouchaud, for their remarks concerning Brownian motion and chaos, and K. Moffatt for his comments on Batchelor's approach to fluid mechanics. I had discussions with G. Whitesides and A. Manz concerning microfluidics, its past and future. J. Eijkels and A. van den Berg expressed their enlightening vision of nanofluidics which I refer to in the book. D.Lohse updated the poor knowledge I had on nanobubbles, H. Stone made a detailed reading of my presentation of the Navier-Stokes equations, F. Mugele gave precious remarks on electrowetting and J. Cossy on surfactant chemistry. With my friend S. Quake, I had a long semantic dispute about the word 'vesicle', not ended yet. Z. Z. Li, a former student, now professor at Beijing Institute of Technology, performed experiments to obtain the useful (unpublished) images shown in Chap 4, related to step emulsification technique. I also benefited from great remarks from D. Quere, J. O. Fossum, T. Lecuit, V. Hessel, J. L Viovy, J. Eggers, M. Bazant, R. Ismagilov, P. Doyle, H. Bruus, A. Strooke, F. Wyart, E. Clement, A. Skjeltorp, E. Raphael, M. Tatoulian, Y. Tran, K. Jensen, S. Wereley, J. Bibette, A. Griffiths and D. Weitz . I also acknowledge J. Mouly and C. Midelet, from Yole Development, for their explanations of the microfluidic market. Deep thanks to C. Rollard, for her description of the realm of spiders. I am also indebted to J.O. Fossum, E. Torino, P.A. Netti, for great discussions and for their invitations to sabbatical periods at Nordheim University and the Italian Institute of Technology, in Napoli. I would also like to acknowledge A. Libchaber, from whom I learned a lot. Discussions with my close colleagues, P. Nghe and J. Mc Graw were inspiring and exchanges in my group provided a source of thinking and learning: thanks to P. Garneret, E. Coz, E. Martin, U. Soysal, P. Nieckele, M. Russo, and I. Maimouni. I am also grateful to M. Dhunnoo for her help and L. Dehove, for her appealing cover and the great figures shown in this book.

Contents

1
Introduction

In the 1970s, it became possible to miniaturize electromechanical systems, down to the micrometric scale. This gave rise to a new field, called Micro-ElectroMechanical Systems (MEMS). Later, in the 1990s, the field expanded, creating all sorts of microdevices, in which fluids, driven under control, gave rise to new functionalities. This prompted the birth of a new field – microfluidics – the central subject of this book.

Microfluidics can be defined as the science of manipulation of fluids in systems of micrometric size. Fluids can be gases, liquids, Newtonian or not, mono or multiphasic (e.g. oil and water). Systems can be devices with channels, patterned surfaces, or paper sheets. Micrometric is an order of magnitude. In practice, microfluidic scales range from 100 nm to 1 mm. The definition proposed here is currently used in the field[1] and we will adopt it throughout the book.[2]

1.1 Astonishing microfluidic systems in nature

Obviously, nature manipulates, with exquisite control, fluids at the microscale. Otherwise no life would be possible. The tree is an example. In the tree of Fig. 1.1, tens of thousands of leaves are nourished by a network containing thousands of capillaries of diameters on the order of tens of micrometres (in the trunk and the branches) and billions of pores of several tens of nanometres (in the mesophyll cell system, in the

[1]In an influential paper [1], titled 'The origins and the future of microfluidics', G. Whitesides defined microfluidics as 'the science and technology of systems that process or manipulate small (10^{-9} to 10^{-18} liters) amounts of fluids, using channels with dimensions of tens to hundreds of micrometers'. The definition, although more restrictive about the scales, is essentially is the same as ours.

[2]Two remarks must be made at this level:
- Microfluidics should not be confused with microhydrodynamics. Microhydrodynamics is the study of creeping flows, i.e. flows at low Reynolds numbers. In the definition of microhydrodynamics, 'micro' refers to the Reynolds number, not the system size. In microfluidics, 'micro' refers to the system size, not the Reynolds number. One can have microhydrodynamic flows of decimetric sizes (e.g. honey poured from a spoon) and microfluidic flows operating at substantially high Reynolds numbers (e.g. inertial micromixers).
- 'Microfluid' is a word that sometimes appears in the literature. It is not a physical concept. 'Fluid' refers to a state of matter defined microscopically. In the bulk, gases and liquids circulating in microchannels possess exactly the same microscopic structure as in large containers. There is no new phase. In extremely confined systems, for instance, in carbon nanotubes, the liquid structure can be affected by the walls. However, this concerns only nanometric scales.

Fig. 1.1: This tree possesses a complex network of capillaries (xylem and phloem) that supplies sap homogeneously to the tens of thousands of leaves that it carries on its branches, and redistribute carbohydrates and other organic compounds between leaves, roots, and fruits (credit: Johannes Plenio) [2].

leaf).[3] In the pores, sap evaporates, through a process called transpiration, creating interfaces that pull the sap from the roots to the top [3, 4]. Despite the complexity of the network, the supply of sap is stable in time and homogeneous in space. The hydrodynamics of the tree is extremely subtle. For instance, large trees drive the sap at negative pressures [5,6]. No hydraulic system made by humans functions like this. The reason is that when the pressure is negative, bubbles nucleate and the flow is unstable and becomes out of control. In trees, bubbles grow, but several mechanisms, including mechanical, block their propagation, thereby preventing embolism and death [4].

We will come back to the trees in Chapter 4, when capillary phenomena are discussed. Here, it is enough to observe that the tree provides an example of exquisite microfluidic control.. This control is achieved not for the pleasure of realizing a technological performance, but to ensure survival.

A similar problematic holds for the spider. The spider produces[4] long silken threads, a few dozen micrometres in diameter, forming a complex pattern -the spider web-, each thread developing a resistance to rupture that is twice as great as that of steel [7, 8].

How does the spider manage to produce this material? To make a long story short, [5] the silk solution is contained in the glands, which are highlighted in Fig. 1.2. In most spiders, six glands produce different types of silks. The solution is driven in small capillaries, several tens of μm in diameter. By actuating valves, or selecting the glands,

[3]The mesophyll part is much more resistive than the other parts of the tree. This explains why the tree distributes the sap among the leaves, in a remarkably homogeneous manner.

[4]All spiders produce silk but only a fraction of them produce webs. Those which do not produce webs use the silk wires to build cages protecting their eggs, or to pass from one leaf or one branch to another to move more swiftly.

[5]An excellent, description of the spider silk production is given by the Museum of Australia, in an article named: 'Silk: the spider's success story'.

the spider chooses the silk it wants to produce, for one part of the web or another. Each gland works as part of a pair. In the capillary, one type of silk occupies the central part of the channel, and the other the periphery. Thus, they form a concentric system in which the two silks flow side by side. In their journeys, the silk solutions deshydrate and, combined with the effect of the large deformation rate to which they are subjected,[6] tend to harden. At the end of these journeys, the soft material is extruded at the back of the abdomen, through tens of submicrometric nozzles; it then evaporates and solidifies. The spider thus controls a complex annular flow structure, the dehydratation process along the capillary, and the final extrusion through many nozzles. This technological tour de force allows it to catch insects to eat.

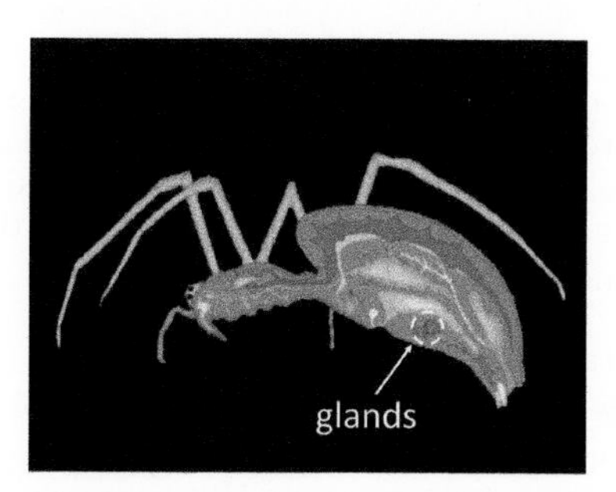

1.2: Two glands produce the silk of the spider web, each gland secreting a particular type of silk. The silk is in a liquid state in the glands. During its journey towards the outlet, the solution hardens under the effect of large deformations, dehydrates, turns into a soft material, exits the spider, evaporates, and solidifies.

1.2 Exquisite microfluidic control in the human body

Humans control flows at the microscale with a high level of precision, and for the same reasons as animals and trees: to be viable. The examples are numerous and they concern almost all parts of the body. Here, just a few are mentioned.

Blood circulation The blood network is formed by vessels of different types: arteries, veins, and capillaries. Each of these plays a specific role in the circulation process. The total length of the vessels is impressive: 100,000 km. Their diameters range from 25 mm (the aorta) to 8 μm, i.e. the size of a red blood cell.[7] Most of the vessels are less than 100 μm in diameter, i.e. they have a microfluidic size. Fig.1.3 (A) shows the vessel network of a lymph node. Its geometrical complexity is evident. Blood itself is a complex fluid: it comprises a plasma, which includes a large variety of solutes, and several types of cells -mostly red blood cells (erythrocytes)-. A number of phenomena, described in the literature, alter the homogeneity of the suspension. The Fahraeus effect [9] whereby, along microchannels whose diameter decreases in the streamwise direction, the cell concentration decreases. The effect is linked to the particular interaction that the cells develop with the walls. In the blood network, different hydrodynamic regimes take place, depending on the ratio of vessel diameter to blood cell diameter. Despite this complexity, the blood circulation is precisely controlled. The averaged flow rate is tuned so as to be

[6]It is considered that the silk solution behaves as a shear-thickening fluid.

[7]Red blood cells are biconcave disks, 8 μm in diameter average

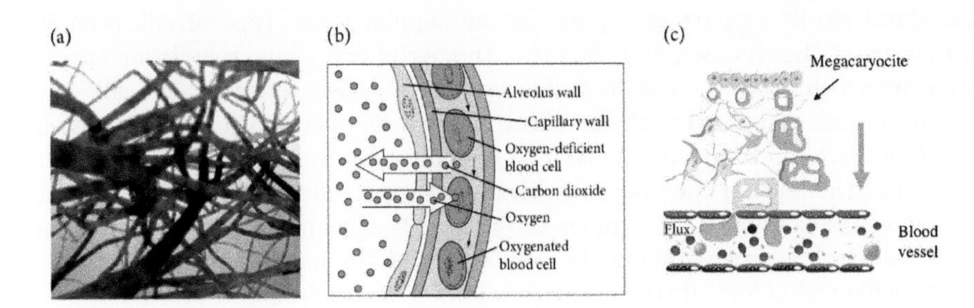

Fig. 1.3: (A) Representation of a network of blood vessels. Gases and nutrients are exchanged between the blood and surrounding tissue through the permeable walls of capillaries, the smallest blood vessels(from Design cell, 2013). (B) Sketch of the gas transfers taking place in the lungs. (C) Migration of a megakaryocyte in the bone tissue, penetration in the blood vessel, and the subsequent platelet formation.

compatible with the metabolism. In the network, no clogging occurs, and the pressure drops are kept small. Driving a suspension, under precise control, in such a complex microfluidic system, represents a formidable challenge that evolution has managed to meet.

Gas exchanges in lungs Lungs take in oxygen and eliminate carbon dioxide. The gas transfer between blood and air, at the alveola level, is schematized in Fig. 1.3 (B). The membrane (made of cells) separating the alveola and the surrounding capillaries is one cell thick, i.e. about 1 μm. With such a small dimension, gases pass quickly through the membrane into the blood. Oxygen-deficient, carbon dioxide-rich blood return to the heart. Although capillaries host only one cell in their cross sections, no clogging occurs. Lungs typically contain 480 million alveoli. The respiratory system provides an example where humans exert an exquisite control of the blood circulation and gas exchanges, at the micron scale.

Platelet production : Platelets are cells measuring 2-3 μm and without nucleus. They prevent haemorrhages. Without them, we would not survive. To carry out this function, platelets, initially spherical, take the form of a star, adhere and aggregate at a wound and trigger thrombin generation and fibrin formation to create a clot, further strengthened by the accumulation of white blood cells [10]. Fig. 1.3(C) sketches the process of formation. Megakariocytes (30 μm large), are produced in the bone marrow, then travel through the bone tissue and enter the blood circulation, by migrating through the vessel wall. In the blood stream, being tethered to the wall, they are stretched out by the shear flow and break up to small vesicular blobs of micrometre size, called platelets [11]. The lifespan of platelets is limited, and we must produce billions of functional platelets every day to renew them. If the blood circulation were too fast, turbulence would develop and the breakup process would become out of control. If the blood circulation were too slow, the megakariocyte would not break up and no platelet would be produced. The process is thus finely controlled.

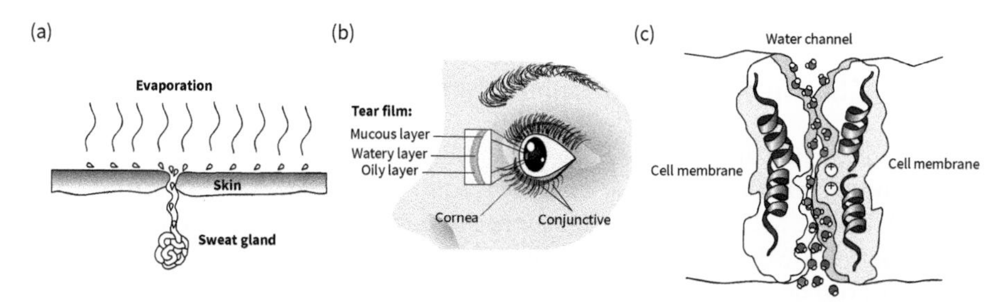

Fig. 1.4: A - Sketch of a skin pore with its eccrine gland; B - Eye with its film: C - Aquaporin, across which water molecules circulate.

Sweating Sweating cools down the body, during exercise, when the temperature is hot or in case of fever [12]. Fig. 1.4 shows an eccrine gland, which produces sweat and drives it through the microfluidic sudoral channel (50 μm in diameter), up to the skin pore. The flow rate is determined by the kinetics of secretion of the gland. Once at the skin level, sweat forms a film or a droplet, which evaporates and, in turn, according to thermodynamics, cools down the blood circulating in the region and thereby the body. Typical energy losses per unit of time are 350 watts, several times the basal metabolism, i.e. the energy, per unit of time, used by a human at rest, as will be seen in Chap. 2. Should the sweat flow be too large, the skin would be covered by a thick film, and, in the opposite case, there would be no thermal regulation. Thereby, again in this case, evolution has led to a fine control of microfluidic sweat flows, which plays an instrumental role in our thermal stability.

Tears Tear film is essential for clear vision and eye health [13]. It protects the ocular surface with moisture, transports waste away, and provides a smooth optical surface. This film has a micrometric thickness. After each blink, it re-forms. Deficiencies in the tear film causes blurred vision, burning, foreign body sensation, and tearing. Flow control is instrumental to avoid these problems.

Aquaporin The preceding examples were concerned with micrometric scales. Humans also control flows at the nanometric scale. An example is aquaporin. The function of this molecule is to transport water across cell membranes in response to osmotic gradients [14]. Without cells could blow, as red blood cells do when they are suddenly immersed in a salty solution. Typically, cells contain up to 30 aquaporins. As illustrated in Fig. 1.4(C), in the presence of an osmotic pressure gradient, water molecules are forced to circulate through the aquaporin, in one direction or the other, to cancel it. Aquaporins are extremely selective: only water molecules pass through them. They never clog, due to the action of an anti-fouling mechanism that is not yet understood. Their permeability is sufficiently high [15] to enable fast equilibration, since if the equilibration process were too slow, cells would be unstable. It has been suggested that the hourglass geometry of the molecule plays a role in this property. Aquaporin illustrates well the exquisite control that humans have developed at the nanoscale to ensure their viability.

1.3 MEMS, the mother of microfluidics

As evoked in the first lines of the introduction, man has begun to create, in the seventies, machines of micrometric sizes. These systems were called Micro ElectroMechanical Systems (MEMS). MEMS sizes currently range between 1 μm and 500 μm. An excellent example of a MEMS, hilglighted in [16], is shown in Fig. 1.5. This MEMS is a microgear whose overall size is 200 μm. It is held by an ant. The figure illustrates the intrusion of a man-made machine into the small animal realm. The entry into micrometric scales is not a new feat, however. Since the invention of the optical microscope in the sixteenth century, the micrometric world has been scrutinized in detail. The microscope has allowed countless scientific discoveries to be made. However, before 1970, humans did not *act* at a micrometric scale, which is precisely what MEMS allows. As a result, when MEMS appeared on the scene, their potential of applications was clear. The first example concerns airbags for cars.

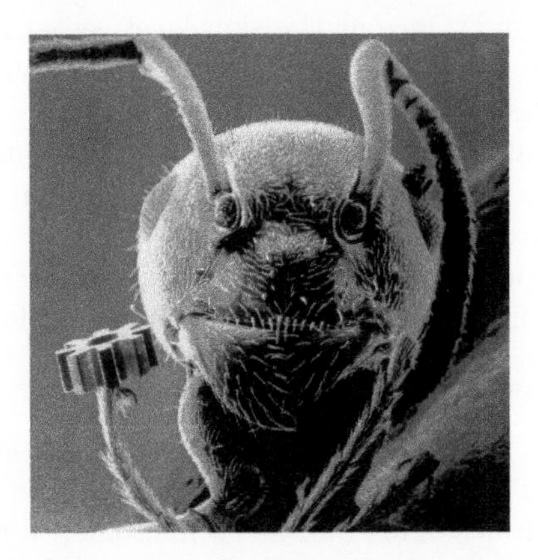

1.5: Ant holding a nickel micro-gear, made by LIGA technology (German for 'lithographie, galvanofomung, abfomung'). This ant was metallized and placed in a vacuum in order to be photographed by electron microscopy. This image was provided by the Karlsruhe group (Germany) [17].

MEMS for airbags, which first appeared in the 1980s, consist of an integrated system on a silicon wafer that is just a few millimetres long. This is shown in Fig. 1.6. The heart of the chip is made of two combs, one fixed and the other mobile; the capacitance of these combs varies under the effect of an impact. As we will see in Chapter 2, the miniaturization of the capacitor element allows the creation of a highly sensitive and rapid detector. The industrial success of MEMS for airbag is not solely due to the improvement in sensor response and sensitivity, but also to the ability to integrate detection, information analysis, and signal processing all on one single chip. Just as with integrated circuits, this chip can be reproduced by the million. The cost, which is critical in the field of automobile manufacturing, becomes very advantageous compared to traditional systems. For this reason, all modern automobiles now use MEMS for their airbags, and tens of millions of these devices are produced each year.

The history of MEMS is interesting. The year 1959 is often considered to be the beginning of the history of micro- and nanotechnology. In December of that year,

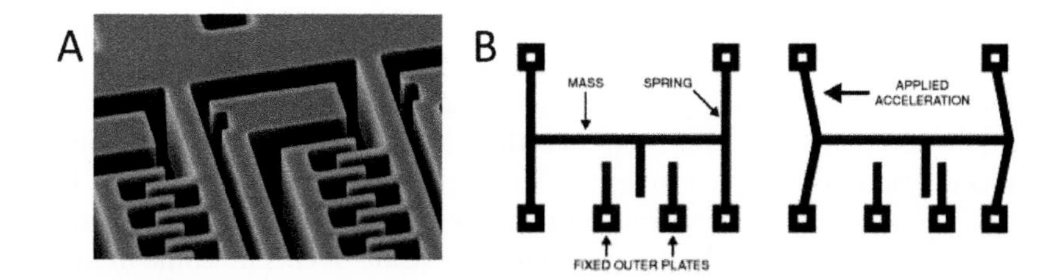

Fig. 1.6: A - SEM image of a part of a MEMS - accelerometer for airbag (Reprinted from Ref. [18], with permission from IOP Publishing, Ltd, Copyright 2022); B principle of functioning of the accelerometer.

a visionary speech was given by Richard P. Feynman during the American Physical Society (APS) meeting at Caltech. This speech was entitled 'There is plenty of room at the bottom' [19]. An early part of the speech is as follows:

I would like to describe a field, in which little has been done, but in which an enormous amount can be done in principle. This field is not quite the same as the others in that it will not tell us much of fundamental physics (in the sense of 'What are the strange particles?') but it is more like solid-state physics in the sense that it might tell us much of great interest about the strange phenomena that occur in complex situations. Furthermore, a point that is most important is that it would have an enormous number of technical applications.

Feynman saw no physical reason why the 50 volumes of *Encyclopedia Britannica* could not be inscribed on the head of a needle. One letter would only need to consist of less than a dozen or so molecules. Confronted with the difficulty of working at micrometric scales, he suggested that we 'train ants how to teach mites' how to construct miniaturized machines'.

How many times when you are working on something frustratingly tiny like your wife's wrist watch, have you said to yourself, 'If I could only train an ant to do this!' What I would like to suggest is the possibility of training an ant to train a mite to do this. What are the possibilities of small but movable machines? They may or may not be useful, but they surely would be fun to make.

These suggestions or predictions did not remain just a fantasy; since three decades later, in 1990, the word 'IBM' was spelled out with 35 xenon atoms [20] (see Fig.1.7).

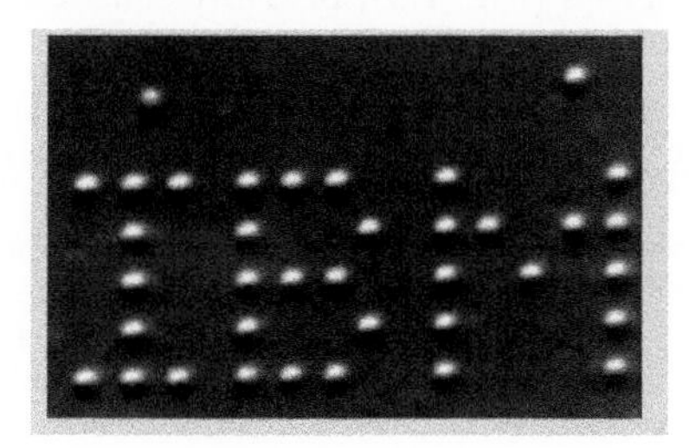

1.7: IBM spelled out with 35 xenon atoms, deposited on a cold (4K) Nickel (110) surface. The atoms were transferred from the sharp Scanning Tunnelling Microscope (STM) tip and imaged with the STM [20]. This image received a large echo.

The first MEMS devices were created two decades after Feynman's speech [21]. The first microbeam was reported in 1982 (Fig. 1.8A), and the first microspring in 1988. The first micromotor was created in 1988 [22,23] (Fig. 1.8B)[8]. It consisted of an electrostatic motor, where the rotating electric field was generated by electrodes evaporated onto polysilicon. One major difficulty was that stiction (the combined phenomena of adhesion and friction), which tended to block the rotor was exacerbated by miniaturization. The solution to this problem consisted of reducing the surface area of the rotor/substrate contact, but this made microfabrication of this machine more difficult.

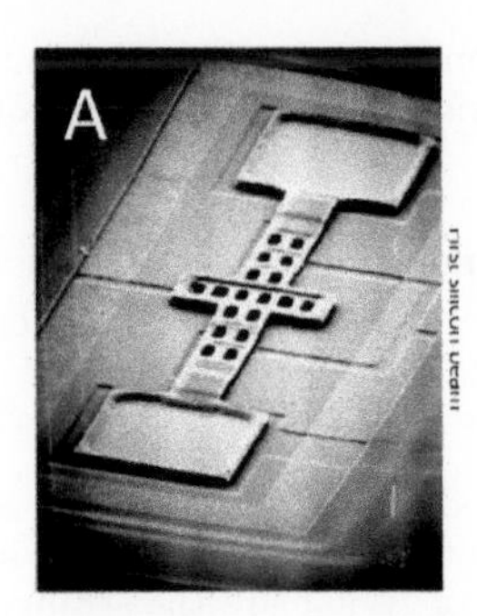

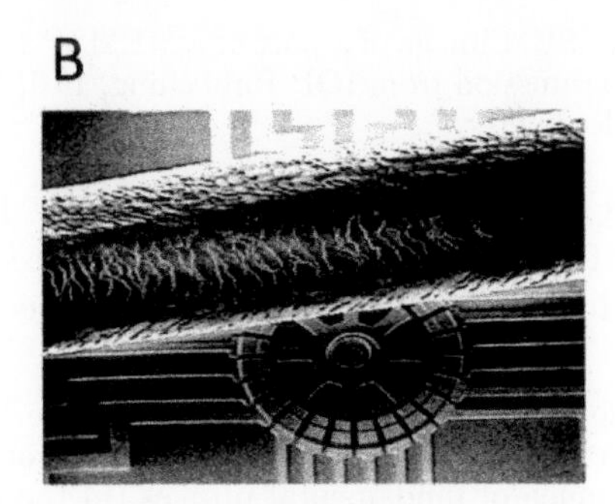

Fig. 1.8: (A) First microbeam (1982). (B) First micromotor, made at UC Berkeley by Tai and Muller in 1989. This motor has been placed next to a human hair whose diameter is on the order of 200 μm. (Courtesy of Professor Richard S. Muller, Berkeley Sensor & Actuator Center, University of California, Berkeley).

Other examples are a microgripper, a hot-wire rake [21], and an astonishing microguitar, shown in Chapter 7, with nanostrings 30 nm in size, vibrating at MHz frequencies.[9] In addition, some unsubstantiated concepts were proposed: for instance, a MEMS consisting of inclinable mirrors that permit communication between the ground and an airborne micro-engine.[10] The project failed, but the concept was appealing. Not all of these objects were practical, but all of them stimulated the imagination.

Today, the MEMS market is estimated to be worth between US 10 and US 13 billions, depending on the agencies. Examples of companies are ARM Holdings. Bosch, Cisco Systems Inc., InvenSense, Knowles Electronics, MediaTek Inc. Microchip Technology Inc., Samsung Developers, STMicroelectronics, and, Texas Instruments. MEMS production includes a variety of products like MEMS for airbags, microgyroscopes for mobile phones, micromirrors for digital projectors (Fig. 1.9), pressure sensors, to men-

[8]We will see in Chapter 2 that this micromotor can comprise the base element of a microturbine that converts chemical energy to electrical energy. It is also interesting to note that microgears, fabricated using MEMS technology, are often used today in clock making.

[9]The guitar, 10 μm long, with six 30 nm wide cords, will be shown in Chapter 7. It was realized at Cornell, in 1997, in the group led by Prof Craighead.

[10]This concerned a project, written by Kristofer S. J. Pister, Joe Kahn, and Bernhard Boser, the University of California, Berkeley, in 1997, whose objective was to build wireless sensor nodes with a volume of one cubic millimeter.

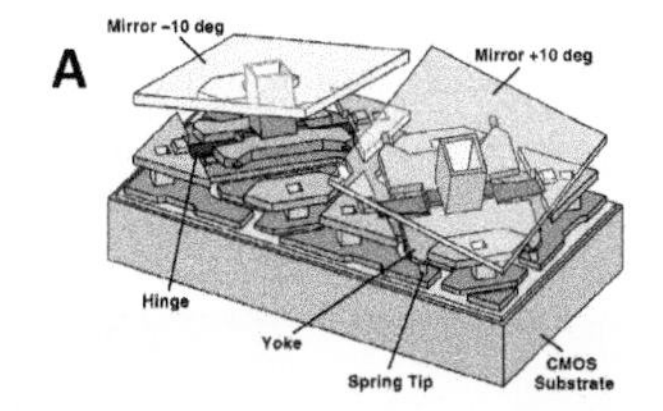

Fig. 1.9: (A) Optical MEMS (MOEMS) made by Texas Instrument, used in digital projectors. (B) An array of micro-mirrors used in a video-projector. The size of each mirror is 10 μm [24](Creative Commons Attribution 4.0 Unported License).

tion a few. The applications are impressive, but there is still room at the bottom. For instance, no MEMS is capable of flying and act, for instance, like a mosquito. Creating a MEMS-mosquiito would necessitate developing an extremely lightweight and powerful energy source, that does not exist yet, and integrating an extraordinarily efficient motor, along with ultraminiaturized micro-pumps. All this, today, looks impossible.

1.4 The birth of microfluidics

We now concentrate on microfluidics. As in the case in many fields, the birth of microfluidics, defined as 'the science of manipulation of fluids at the microscale', as said above, has been preceded by precursors and even precursors of precursors. People have succeeded, in many circumstances, in manipulating fluids at the micrometric scale, in a controlled manner, with only their hands. To do this, they exploited hydrodynamical laws that naturally provide a control of the micrometric scale. An example is painting. Painters deposit micrometric layers on walls, whose thickness is controlled by the speed of the brush. Another example is soap bubbles. The films, stabilized by surfactants, are sub-micrometric. These examples show that without microtechnological toolbox, the micrometric scale, in a number of cases, can be controlled.

Of much greater technological significance is a device producing submillimetric droplets. It was invented in 1964 by R. G. Sweet [25,26] (see Fig. 1.10).

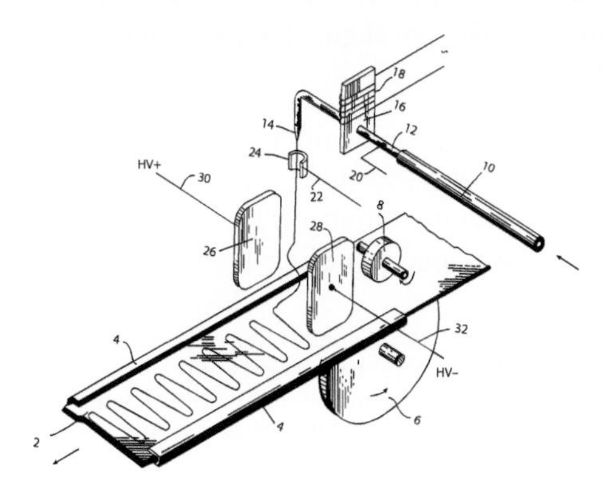

1.10: Sketch of the system used by R. G. Sweet [25].

The invention, with others, played an instrumental role in the field of ink-jet printing, a major application of microfluidics. Here, droplets were emitted and deposited on a moving sheet. A key point was the electrostatic control, which allowed large quantities of nanoliter droplets to be produced, in a reproducible manner, opening an avenue towards the fabrication of ink-jet printers.

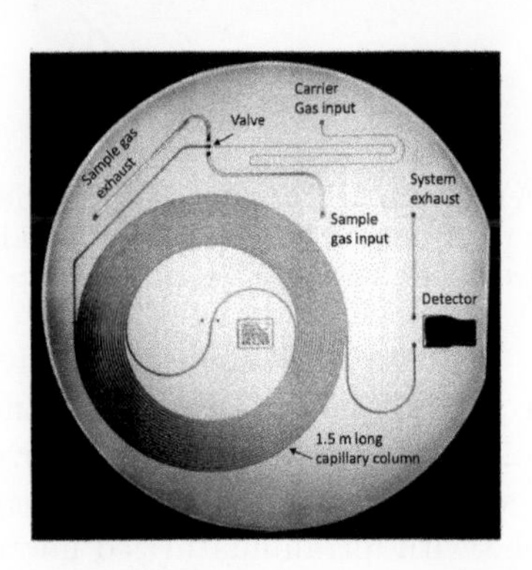

1.11: First microfluidic system (1975) [28]: a miniaturized gas chromatographer, including an injector, a 1.5 m long microchannel and a thermal detector.(Reprinted with the permission of IEEE, copyright 2022.)

In the 1970s, silicon-based MEMS technology was well advanced, many clean rooms were available, and there were no difficulty in etching grooves on silicon wafers to create microchannels. In this context, the first microfluidic device, invented by S.C. Terry [27], appeared on the scene in 1975. Unlike Sweet's invention, the manufacturing and design of the device prefigured the silicon-based microfluidic devices that would later develop, giving birth to microfluidics. The device made by Terry [27, 28] was a miniaturized gas chromatographer. It is shown in Fig. 1.11.

Terry's device circulated gas through a 1.5 m microchannel etched in a silicon wafer, bonded to a glass plate. The system included a miniaturized electromagnetic injector and a thermal detector. Separations of gaseous hydrocarbon mixtures were performed in less than 10 s, without compromising the efficiency, which was an impressive feat at that time.

Nonetheless, although the device was industrialized several years later, the technological jump remained isolated, because, during these years, the separation-science community was not ready to adopt silicon technology [29]. It was only after 1990 that the advantages of miniaturization were thrust into the spotlight, for its application to electrokinetic liquid chromatography [30–32].

A seminal paper appeared in 1990 [30]. By reasoning on scaling laws, A. Manz et al. argued that miniaturization enables the creation of performing separation systems, combining portability, low cost, and high speed, without compromising the efficiency.[11] We will describe these systems in Chapter 6. The paper introduced a new acronym,

[11]The conclusion of the paper was: 'A basic theory of hydrodynamics and diffusion indicates faster and more efficient chromatographic separations, faster electrophoretic separations and shorter transport times for a miniaturized TAS. The consumption of carrier, reagent or mobile phase is dramatically smaller. A multi-channel device would allow the simultaneous performance of a large number of measurements (under the same conditions).'

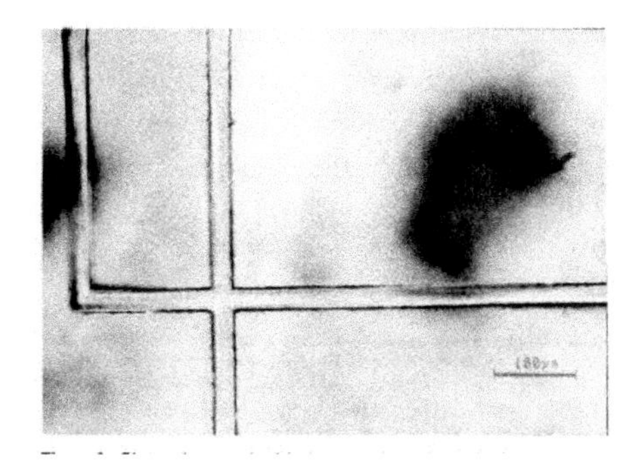

1.12: First electrokinetic separation microsystem (1992). Overall dlmensions are 14.8 cm x 3.9 cm x 1 cm. (Reprinted rom Ref. [31], with permission of Americal Chemical Society. Copyright 2022.)

μTAS, which stood for Micro Total Analysis System. The term looks awkward. What does it mean? A 'Total Analysis System' (TAS) refers to a system in which analytical instruments are transported on a cart, to perform the total analysis of a sample (sampling, sample transport, chemical reactions, detection) in the field. Thanks to miniaturization, no cart was needed, and the equivalent system was called μTAS. The ideas were substantiated two years later, with the realization of the first electrokinetic separation microsystem [30].[12] This demonstrated performances in line with expectations, i.e. high speed of separation, and excellent efficiency, transportability, and low cost. Microfluidics was born.

Later, all sorts of microfluidic systems were fabricated : electrophoretic separation assays [33, 34], electro-osmotic pumps [35], diffusive separation systems [36], micromixers [37–42], DNA amplification systems [43–49], cytometers [50, 51], DNA separation assays [98–102], centrifugal microfluidics (see review [103]) and chemical microreactors [52–56], to cite a few examples. A number of these inventions were to play an important role, a decade later, in the ramping up of the microfluidic market.

1.13: Cover of the proceedings of the first μTAS conference organized by A. van den Berg and P. Bergveld in 1994, in Enschede (the Netherlands). (Reproduced with permission from the Chemical and Biological Microsystems Society (CBMS), copyright 2022 CBMS.)

Up to 1994, the young microfluidic community gathered in different meetings, in particular in MEMS conferences (launched six years before, in 1988), in which one or two sessions were devoted to microfluidics. The situation was not optimal. Controlling

[12] In fact, the device was manufactured in 1988.

fluid flows in microdevices required discussion of subtle hydrodynamic and physicochemical phenomena, and these were far from the focus of these meetings. Therefore, a desire to organize specialized seminars emerged, in order to focus on these phenomena, and, in the meantime, take the specific technological context into account, most often overlooked in traditional academic meetings. The first μTAS conference was organized by A. van den Berg and P. Bergveld in 1994, in Enschede (the Netherlands). The cover of the proceedings is shown in Fig. 1.13. A total of 160 participants attended. Today, μTAS conferences bring together 1,500 participants.[13]

In these early days, a 'paradigm', i.e. a set of concepts and practices that define a scientific discipline at any particular period of time', or, more simply, 'what the members of a scientific community, and they alone, share' [57], emerged. It was thought that the objective of microfluidics was to create basic microfluidic functions, or 'bricks', and assemble them, in a way similar to microelectronics, so as to generate complex functionalities, that could respond to unmet needs in biology and chemistry. The notion of lab-on-a-chip was frequently put forward, and stunning cartoons designed to illustrate the concept, such as those of Fig. 1.14[14], appeared in many broad-readership (for instance Ref. [58]), augmenting the visibility of the field. But why did the community not fabricate complex microfluidic systems, comparable to microelectronic processors, as they envisioned? One bottleneck was valve integration. It appeared impossible, at that time, with silicon, plastics or glass, to integrate more than a few valves on a device. There was no chance to compete with the millions of transistors integrated, at that time, on central processing units (CPUs).

In the same period, hydrodynamic experiments revealed unexpected phenomena; for instance, large slippages, which apparently contradicted the no-slip dogma of hydrodynamics [59, 60]. In the meantime, microfluidic technology revealed the mechanical behavior of single DNA molecules. The experiment was performed by Chu et al. [61] in 1993, in cross-flow microchannels. This work laid down the foundations of a new domain: the study of single molecules. It also showed that microfluidics could provide new tools for investigating fundamental questions. Around that time, a number of fundamental contributions allowed the physical role of the confinement to be clarified. This concerned, for instance, electrohydrodynamics, electrowetting, chaotic mixing, polymer dynamics, gas flows, or fluid interfaces. The list is long and we will return to these subjects later. Molecular dynamics simulations, on the other hand, enlightened

[13] The second meeting, μTAS 96, was held in Basel with 275 participants. The first two meetings were held as informal workshops. By the time of the third workshop, μTAS 1998 (420 participants), held in Banff, the workshop had become a worldwide conference. The number of participants continued to increase in μTAS 2000 (about 500 participants) held in Enschede and μTAS 2001 (about 700 participants) held in Monterey. The number of submitted papers also dramatically increased in this period from 130 in 1998, to 230 in 2000, to nearly 400 in 2001. From 2001, μTAS became an annual symposium (text of presentation of μTAS 2002, held in Nara (Japan), and written by Yoshinobu Baba and Shuichi Shoji).

[14] Fig. 1.14(A) is stunning, but it is misleading in the sense that, in practice, the functionalities shown on the figure are much more difficult to integrate than suggested. Perhaps inspired by this unrealistic illustration, or understanding the analogy between microelectronics and microfluidics in a too literal sense, several companies wasted much time and energy developing complex lab-on-a-chip that had no chance of working

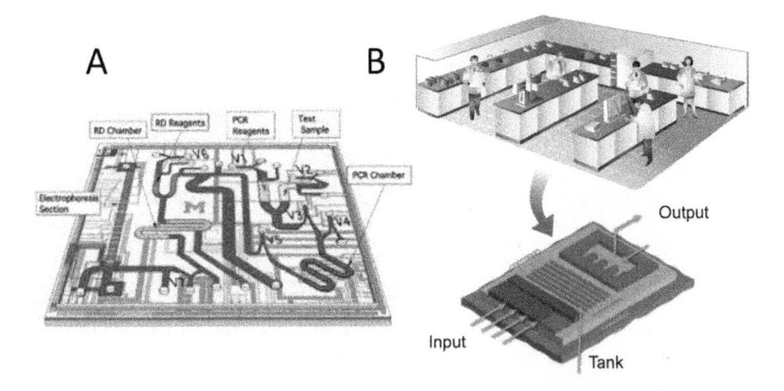

Fig. 1.14: (A) Cartoon published in 1998 [58], showing an imaginary microfluidic device integrating several functionalities: electrophoretic separation, heating, driving, mixing, extraction, and polymerase chain reaction (PCR) amplification. (B) Image elaborated by Caliper, around 2000, that illustrates well the lab-on-a-chip concept.

a number of phenomena, sometimes in conflict with the experiment. This concerned, for instance, slippage, nanobubble dynamics, or electrowetting. Again, we will return to these subjects later.

By the turn of the twenty-first century, many microfluidic devices were created in the labs, some patented (one hundred or so) awaiting commercial developments, many not. Before 2000, the global microfluidic market was a nano-market, worth less than US$100 M.[15] A number of microfluidic products, although elegant, did not meet any market. This was the case of the microfluidic fountain pen [62]. Others did not function with sufficient reliability: this was the case of the glucose watch, dedicated to monitor glucose by sampling interstitial fluids. It will take years before this product finds its place in the market [63]. Other inventions met considerable success: one developed by i-Stat (created in 1983), of extremely simple construction, dedicated to glucose measurements. In the 1990s, i-Stat company shipped several million cartridges every year to hospitals. Two commercial products, whose production volumes were not as important, but which promised an interesting future, must be mentioned: the Bioanalyzer, dedicated to DNA separation, made by Caliper Agilent, and the SmartCycler® System, a Point of Care (POC) molecular system, developed by Cepheid (founded in 1996). The two systems marked the entry of microfluidics in molecular biology. Cepheid would become, years later, a major player in this field. In the meantime, several microfluidic foundries, such as Micronit (founded in 1999), or ChipChop, and later on, Dolomite (2005) for microfabrication, and Fluigent (2006) and Elvesys (2011) for instrumentation, were created. They played an important role in the development of microfluidics.

[15] Although inkjet printing is a microfluidic system, by convention, it is not counted as part of this market.

Who first used the word microfluidics ?

Before 1993, the word microfluidics' was essentially absent from the vocabulary of the microsystem community It did not appear anywhere in the 857 pages of the Microsystem technology 90 Proceedings [67], an important conference of the young MEMS community. Instead, the words 'micro liquid-handling', 'micro-hydraulics', 'micro fluid', with or without hyphen, or micro-liquid flows' were preferred [68]. 'Microfluidics' was also absent from Terry's [28], and Manz's [30,31] papers, published in 1979, 1990, and 1992. Why was this the case ? We may hypothesize that researchers of that time, involved in what we now call 'microfluidics', were reluctant to use a word that looked esoteric. The situation seemed to change in 1993, after the publication, by P.Gravesen, of a survey titled 'Microfluidics: A Review' [65]. The paper acquired a significant visibility. Although no etymologic study has been done, and probably will never be done, one may hypothesize that the review strongly contributed to, if not initiated, the spread of the word. Later on, although invisible in the MicroTas94 Proceedings, the word 'microfluidics' increasingly appeared in the literature. The word was suitable since it represented, as pointed out by R. Zengerle [66], a 'headline' covering all types of actions involving fluids, and performed in microsystems. After 1997, it frequently appeared and eventually acquired a prominent place in the language. Finally, the community working on fluid manipulations at the microscale adopted it to give a name to their field.

1.5 The advent of soft technology

An important step was taken in 1998, with the development, by G. Whitesides, of soft technology [69–71].[16] With soft technology, devices were no longer made in glass or silicon, but in soft materials. From a technological perspective, this represented a considerable shift. Two images, extracted from [69] and [70], are shown in Fig.1.15.

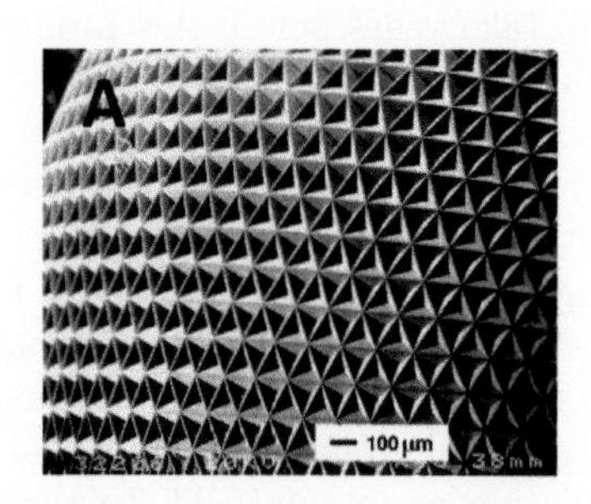

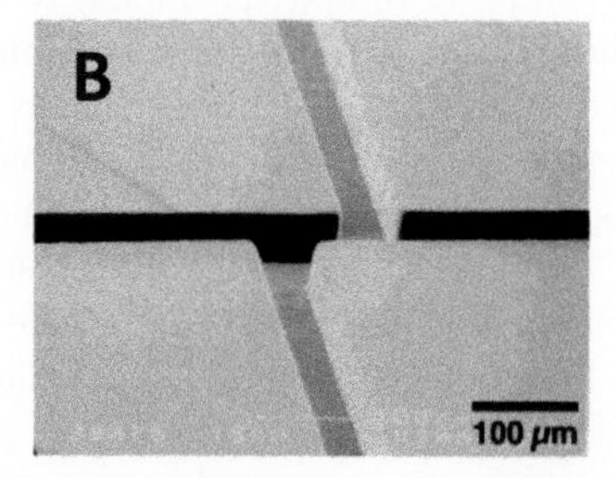

Fig. 1.15: (A) Scanning electron micrographs of a PDMS honeycomb structure, created by molding the polymer against a photoresist mould (Reprinted from Ref. [69] with permission from Wiley and Sons, copyright 2022 Wiley and Sons). (B) Double-T section of the network of channels. The roughness in the side walls of the PDMS channels arises from the limited mask resolution (Reprinted with permission from Ref. [70], copyright 2022 American Chemical Society.).

[16]Whitesides' papers were preceded, in 1997, by two pioneering works [72, 73].

Figure 1.15(A) is taken from Ref. [69]. The paper explained the concept of soft technology, and its two main facets, microprinting and micromoulding, which we will discuss in Chapter 7.

Figure 1.15(B) shows microchannels made in PolyDiMethylSiloxane (PDMS) [70]. We will hear much about PDMS in the book. It is almost a miraculous material. It has properties that no other material possesses: deformable, transparent in the visible range, insulating, hydrophobic, sticking to glass in a reversible manner. We will see, in Chapter 7, that all of these properties make soft lithography possible. In Fig. 1.15(B), the aqueous sample is introduced in the double T, then driven by electroosmotic forces into a long microchannel (not shown), along which electrokinetic separation is performed, in a manner similar to Manz's work [31]. The walls are rough, due to the low mask resolution. The authors achieved ionic separation in the system, with a resolution and a speed comparable to silicon devices. The paper thus suggested, that, even though surface chemistry is not as controlled as in glass or silicon, electrokinetic separation, the major application of microfluidics at that time, is feasible. For chromatographic experts, this was quite a surprise.

Nonetheless, the message received by many researchers was not about separation. It was that PDMS microdevices were easy to create.[17] Once the mould was fabricated, the rest of the technological process could be made outside a clean room, without specialized skills. A master student could learn it in one day. The simplification of the technological process gave rise to a surge of activity. Many laboratories, interested in microfluidics, but with limited access to clean rooms, came to the field and started investigating new directions. One could compare this period to the transition from centralized informatics to laptop computing.

Why did all the community not rush out to use soft technology, abandoning silicon? Two reasons can be provided: the first is that PDMS surfaces are not stable, meaning that performing accurate separations with such a material was impossible. For the community of that time, composed, mostly, of analytical chemists, this represented a serious limitation. Secondly, PDMS devices cannot be produced in large quantities. Scaling up is not possible, while with silicon it is. These two arguments led most of the microfluidic community of that time to keep working with silicon and glass. The arguments were not unreasonable. Today, half of the hundreds of millions of devices produced in the microfluidic market are made with glass and silicon.

Most researchers using soft technology were thus newcomers. In the same period, pressing needs appeared: examples are DNA sequencing, cell sorting, molecule screening, single cell analysis, and proteomics. As mentioned earlier for the particular case of DNA sequencing, all these applications conveyed big numbers: five billion cells to sort for isolating 1-10 circular tumor cells (CTC) [76], hundred of thousands of compounds for drug discovery [80], millions of fluid manipulations for performing gene sequencing [75]. How so many experiments could be carried out rapidly and in parallel, while consuming little reagent [74]? This was the new challenge faced by microfluidics. The

[17]The discovery of SU8, by IBM, in the 1990s, allowed thick moulds to be made in one photolithographic step, facilitating the fabrication of PDMS devices.

challenge was not only intellectual. Today, the aforementioned domains represent the largest share of the microfluidic market.[18]

In this context, a surge of innovations emerged. One landmark was the demonstration, in 2002, that, by exploiting the deformability of PDMS [77], thousands of valves could be integrated on the same device [78]. This is shown in Fig. 1.16. The valve problem, raised in the previous decade, seemed to be solved. The technology gave birth to a company, Fluidigm, whose valuation would soon reach US$1bn. With Cepheid and Fluidigm, two microfluidic unicorns were born in the years 2000-2010.[19]

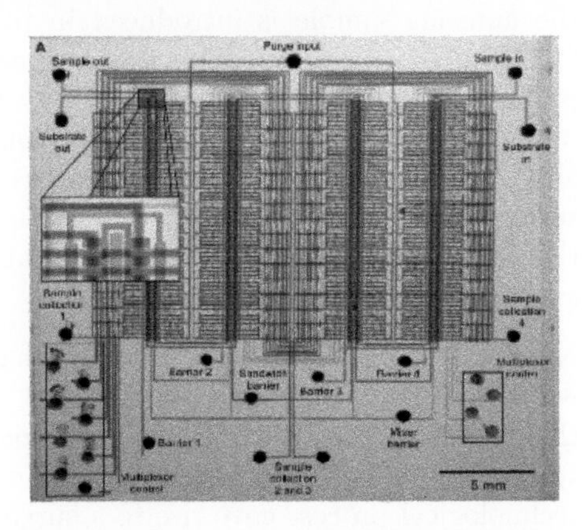

1.16: The device, whose channels are visualized with dyes, contains 2056 microvalves. The system performs distinct assays in 256 nl subreaction chambers. Each of them is individually addressed, using a multiplexed valve system,. With it, only 20 valves are needed to control the assay. (Reprinted from [78], with permission from the American Association for the Advancement of Science. Copyright 2022.)

The device shown in Fig. 1.16 contains 2,056 microvalves. Each of the 256 chambers on the chip is individually addressed. A multiplexer allows for reducing the number of connections, from N (N being the number of chambers) to 2 $log_2 N$ [78]. It is impossible to realize such a system with glass or silicon. For microfluidics, a new period opened.

1.6 Diversification of the technology and broadening of the applications

In the years 2000-08, thanks to soft lithography, the microfluidic community had an easier access to technology. The number of applications, including those involving large numbers, increased, and growing public support raised momentum in universities. In such conditions, the microfluidic community grew substantially, reaching thousands of researchers worldwide. The stimulating atmosphere of that time was perceptible in the μTAS conferences.

Droplet-based microfluidics appeared in the period 2000-2002. We will present the technology in Chapter 4. At first glance, producing microdroplets under control did not appear new. For instance, ink-jet printers and fluorescent active sorters (FACS) already

[18] Before the advent of the COVID-19 pandemic.

[19] At about the same time, Theranos, which promised to perform a hundred different tests with a single drop of blood, reached much higher valuations. Ten years later, the company collapsed. Because of the absence of peer-reviewed publications issued by the company, links between the microfluidic community and this company were inexistent.

performed the task. The new idea was that, by playing with the device, droplets could be filled with reagents, and thus operate as microreactors, being moreover well mixed, thanks to vigorous recirculations developing inside them, as we will see in Chapter 4. Millions of chemical reactions per hour could be run, using minute amounts of reagents (less than one nanolitre or so per droplet). These performances allowed researchers to work with large numbers, responding to the needs appearing at the turn of the century, as mentioned earlier.

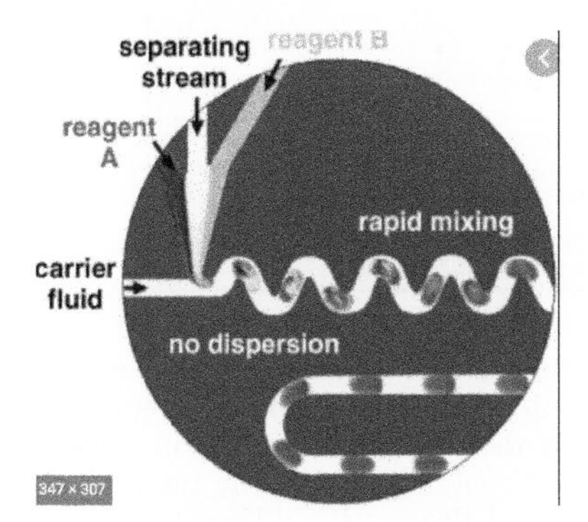

1.17: Two reagents A and B are introduced in a T junction, in a flux of oil, and form droplets. By varying the flow rates, each droplet can acquire different reagent concentrations. This system allows high throughput screening (kHz) of biochemical reactions, using minimal quantities (nl) of reagents. (Reprinted from Ref. [79], with permission of John Wiley and Sons, copyright 2022.)

Fig. 1.17, published in 2003 [79], illustrates well the situation. In short (this will be explained in more detail in Chapter 4), the entry was composed of three channels: one for reagent A, another for reagent B, and, in between a buffer. By varying flow rates, the amount of reagent could be varied in each droplet. Then it was possible to screen a chemical reaction, as a function of the phase ratio. With droplets being emitted at kHz, the screening was orders of magnitude faster than could be achieved by any robot.

In about the same period, it was demonstrated that more complex droplets, in the form of double or multiple emulsions, could also be produced under control. This topics was developed by D.Weitz (Harvard). This was an important step, because it established the link between microfluidics and colloidal and material sciences.

Fig. 1.18 is extracted from a paper published in 2005 [84]. The device consists of three glass tubes, one with a square, the others with circular cross sections, placed inside each other. Along the tubes three liquids, immiscible or immiscible by pairs, are driven, forming droplets inside droplets, i.e. double emulsions. The structures of the emulsions depend on the flow rates, the nature of the phases and the formulation. The device stood against the technological wind of the time: neither soft nor hard technology were used.[20] The work opened a route towards creating objects difficult or impossible to make. Implications in various domains such as cosmetics, oil or food industries were numerous.

[20]Double emulsions have indeed also been created in microchannels, made with soft or hard technologies (see, for instance, [86]). The advantage of glass tube technique is its capacity to obtain well defined hydrophilic and hydrophobic regions. Disadvantages are size limitations (double emulsions below 30 μ are difficult to make), delicateness of the geometrical adjustment of the tube noses, and limited parallelisation.

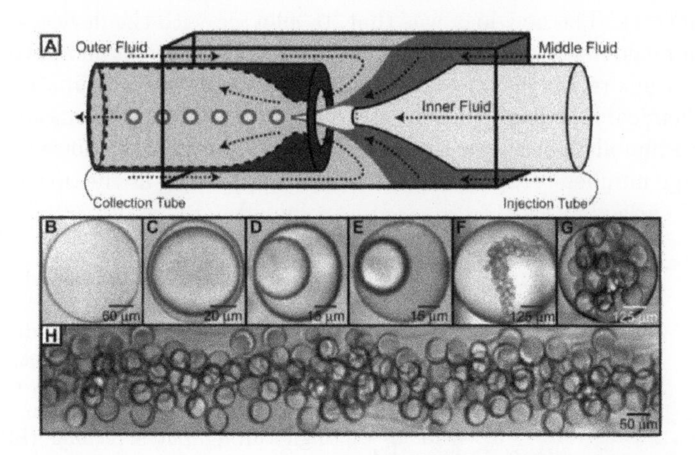

Fig. 1.18: Microcapillary geometry for generating double emulsions from coaxial jets [84]. (A) Schematic of the coaxial microcapillary fluidic device. The geometry requires the outer fluid to be immiscible with the middle fluid and the middle fluid to be in turn immiscible with the inner fluid. (B to E) Double emulsions containing only one internal droplet. (F and G) Double emulsions containing many internal drops with different size and number distributions. (H) Double emulsion drops, each containing a single internal droplet, flowing in the collection tube. The devices used to generate these double emulsions had different geometries. (Reprinted from Ref [84], with permission of the American Association for the Advancement of Science, copyright 2022).

Still in the same period, i.e. around the years 2000, digital microfluidics appeared [87, 88]. The technology exploited a technique, called electroWetting on dielectrics (EWOD), developed first by Berge [89]. EWOD consists of placing a conducting droplet (such as salted water), polarized at some tension, over a thin hydrophobic insulating layer, deposited on a metallic electrode. The electrical tension changes the apparent contact angle. With this, the solid surface can switch, in a fast time (sub-millisecond) from hydrophobic to hydrophilic state. Digital microfluidics exploits this functionality, by displacing droplets over a surface patterned with addressable electrodes. In the period 2000-10, several groups [88, 90–95] demonstrated impressive realizations: rotating droplets, droplets moving on substrates in a complicated manner, interacting together, mixing reagents, initiating chemical reactions and cutting droplets. The technology substantiated well the concept of 'lab-on-a-chip'. Some papers, written by 2010, promised the advent of another revolution [96].

Let us pause for a moment to consider the genome sequencing and its link to microfluidics. Sequencing illustrates two notions: big numbers and complexity without valves. In the years 1990-2005, the human genome programme, in full development, was exclusively based on Sanger sequencing [97]. Massive use was made of restriction enzymes, which fragment the genome into small single-stranded pieces. This is where big numbers came into play. Typically, ten fractionations per kilobase were needed to determine primary sequences [104]. For the human genome, this represented no fewer than ten million separations -a huge number-. The community thus invented new separation technologies that were faster, more integrated, more automatized [58], and more parallelized [98]. Impressive achievements were made. However, the sequencing community did not use this work. The most popular sequencer, ABI Prism 3700, which, historically, sequenced the human genome, used a 96 parallel capillary configuration [104]. There was not a single microfluidic chip in the machine. The penetration of microfluidics came later, with the advent of Next Generation Sequencing (NGS). The first NGS machine, the GS20, appeared in 2005 [104], and it could decipher 20 million bases in 5.5 hours. The success of GS20 was due, in part, to the use of large microfluidic assays. Years later, methods, such as nanopore sequencing (Oxfore Nanopore) or sequencing by synthesis (Illumina) also made full use of micro/nanofluidic technology. This technology has clearly become core to genome sequencing.

These accomplishments demonstrated that a microfluidic device can perform millions of different tasks without using a single valve. How to do that? By relying on molecular actuators: enzymes (polymerases, recombinase, reverse transcriptase, helicase, etc.), molecular motors, selective affinities (such as antibodies), and self assembly processes (which spontaneously pattern molecular assemblies). In NGS microfluidic devices, these processes, working synergistically, allowed extremely complex functionalities to be developed. The same ideas will be used, later, in other fields of application, such as the single cell.

Paper microfluidics came later, in the years 2005-07 [106]. The idea was to replace standard microfluidic substrates (glass, silicon, plastics, or PDMS) with paper, much cheaper (its cost is a few hundredths of a cent per sheet), and easier to source. Being burnable, it reduces the risk of contamination. This material is thus particularly suitable for diagnostic applications in developing countries. Paper had been used for decades for diagnostics, but the novelty brought by G. Whitesides' group was paper patterning [107]. Paper microfluidics appeared as a new branch of microfluidics, holding potential in several areas, while using a cheaper and more available substrate. We will consider this further in Chapters 4 and 7.

1.19: Multiplexed analysis of a model sample of urine (glucose solution) [108]. The white is the paper, the grey is the wax, defining channels. Two colorimetric detections are performed in duplicate. (Reprinted from Ref. [108], with permission from American Chemical Society, copyright 2022)

Later, in the period 2008-21, many inventions and discoveries were made. Thousands of studies were published every year (today, about 5000). Obviously, reviewing them lies beyond the scope of the book. We may nonetheless mention the appearance of new technologies (3D printing, continuous flow lithography, Northland optical adhesive technology (NOA), hydrogel caging, etc.), raising the microfluidic tool box to a level where a considerable number of microfabrication problems can be solved.

Special mention should be made to 3D printing, which has already impacted the way in which microfluidic devices are built. This will be discussed in Chapter 7, but an example is shown in Fig. 1.20.

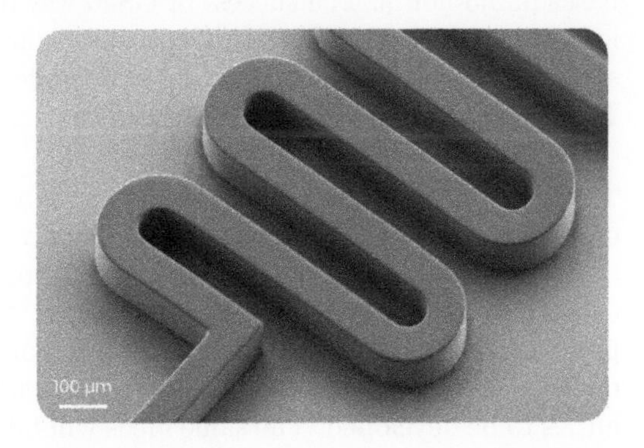

1.20: An example of 3D microprinting. Courtesy of Microlight3D, printed with SmartPrint UV [109].

Another technology worth mentioning is organ on a chip (OOC). OOC creates systems that mimick organs, in the hope of performing drug studies on representative models. It has the potential to impact the pharmaceutical industry, by refining analysis, producing large amounts of pertinent data, speeding up trials, accelerating early phases of regulation processes, and, importantly, suppressing the use of animals. OOC was the missing link between *in vitro* and *in vivo* drug trials. An early example of OOC, mimicking lungs, is shown in Fig. 1.21, and was reported in 2010 [111].

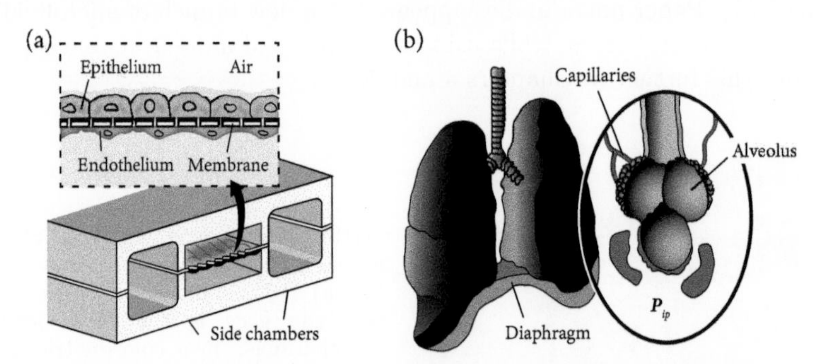

Fig. 1.21: (A) Microfluidic model of a lung [111].(B) Real lung

In Fig. 1.21(A), a membrane, coated with epithelial cells, separates two microchannels: one is for air circulation and the other filled with a liquid. This structure attempts to mimick the real lung, shown in Fig. 1.21 (B). The gap is large, but the hope is that, over the years, it will progressively shrink, reaching levels where drug testing on animals becomes less representative than OOCs. Other OOCs, realized in the last decade are intestines, kidneys, hearts, and vascularization networks [110].

In the meantime, fundamental questions such as slippage were clarified, albeit not solved. After fifteen years of discussion, the nanobubble mystery was resolved in 2015 [85]. Different manners of handling fluids were proposed, sometimes counter-intuitively (such as working at moderately high Reynolds numbers to reorder particles conveyed by a flow, giving rise to 'inertial microfluidics'), and progress in the ability to create complex functionalities has been made (such as handling cells in droplets, barcoding and processing them). The examples given are far from exhaustive.

Looking back on thirty years of microfluidics, it appears that most of the founding microfluidic discoveries have been made in a surprisingly short period of time: between, roughly, 1998 and 2005. One could thus call this period the 'golden years' of microfluidics.

1.7 MicroReaction Technology (MRT)

Microreaction technology (MRT) is part of a general approach called process intensification. The idea is that, as will be seen in Chapter 2, miniaturization enhances heat exchange and facilitates temperature control, which in turn increases the specificity of the reaction, by avoiding the formation of unwanted chemical species [112]. MRT focusses on the millimetric and submillimetric scales, where better flow control can be achieved, thanks to the absence of turbulence. Let us recall that turbulence mixes, but also creates uncontrolled flow heterogeneities: for instance random eddies in the corners, or behind the blades of a mixer. These eddies induce, in turn, heterogeneities in temperature, concentration, and, eventually, product characteristics. The key point of MRT is to operate below turbulence onset to avoid these phenomena[21] and, in the meantime, achieve throughputs of industrial relevance.

The first conference of MRT, international conference on microreaction technology (IMRET), took place in 1997, three years after μTAS 1994. Since then, IMRET has met annually.

From a technological viewpoint, because of high temperatures, and often high pressures, MRT uses hard materials, such as slilicon, glass, ceramics, or metal. High throughput and resistance to clogging impose larger microchannels than in microfluidics, typically in the milimetre range. An example of a MRT device, dedicated to the chemical synthesis of carbamates, is shown in Fig. 1.22 [113].

With MRT, all kinds of synthesis have been performed over the years, starting with quantum dot synthesis, which is better controlled with MRT than traditional techniques [114]. Other applications are combinatorial chemistry and high throughput screening.

[21]The turbulence onset strongly depends on the flow geometry and is characterized by a Reynolds number (see Chapter 3). Because of hysteresis phenomena, turbulence thresholds are difficult to define. In the microfluidic literature, values ranging between 1000 and 3000 are taken without discussion. This range is appropriate for most of the geometries of practical interest, such as straight pipes or channels but does not apply for other geometries, such as converging or diverging flows.

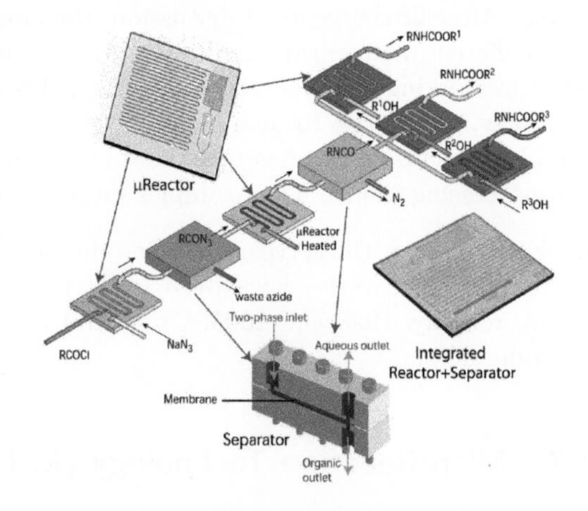

1.22: Multi-step microfluidic chemical synthesis of carbamates starting from aqueous azide and organic acid chloride using the Curtius rearrangement reaction. The scheme involves three reaction steps and two separation steps. Reprinted from Ref. [113], with permission of John Wiley and Sons, copyright 2022.

By working in small rather than large microreactors, efficiency is increased and plant sizes can be reduced. Concerning the throughput, parallelization allows volumes of industrial interest to be reached. In this approach, safety and reliability are improved. The first attempts to substantiate these ideas were made in the years 2005-10. An example is shown in Fig. 1.23. Today, the global MRT market is around US$1bn [116].

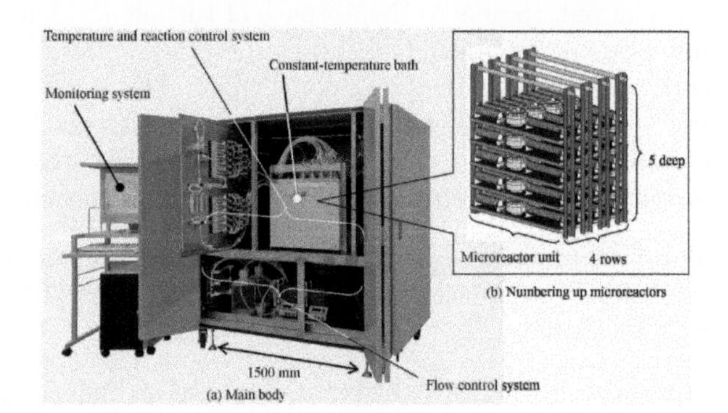

1.23: Is it possible to miniaturize a plant ? An example of a millifluidic plant (Reprinted from [115], with permission of Wiley and Sons, copyright 2022).

1.8 Nanofluidics

As will be discussed in Chapter 3, nanofluidics[22] concerns the 1-100 nm range. Similarly to microfluidics, nanofluidics can be defined as the science of manipulation of fluids at the nanoscale.

Interestingly, humans have manipulated fluids at that scale long before clean rooms were used. For instance, the study of Newton black soap films, 4 nm thick, traces back to 1877 [120]. Ultrafiltration membranes designed by engineers have pore sizes

[22] The word 'nanofluidics' appeared, for the first time, in a review written by H. Craighead [117] in 2000

below 10 nm. Equilibrium films, created by A. Sheludko [121] in the 1960s, for measuring disjoining pressures, had nanometric thicknesses [121, 122]. A landmark in the domain was the invention of the force machine [123], in which fluids were confined in nanometric (open) chambers for studying liquid behavior at the nanometric scale (a detailed description of the instrument will be given in Chapter 2). Other examples can be found in the review by J. Eijkel and A. Van den Berg [126].

In the 1990s, it was possible, with e-beam or optical lithography, to create nanochannels less than 100 nm high, or holes less than 100 nm in diameter. In this context, several contributions were made,[23] such as the realization of a silicon membrane with 50 nm pore sizes in 1990 [118] and a 88 nm nanofluidic field-effect transistor in 1992 [119]. Later, in 1999, a nanofluidic device, built with the intent to develop a chromatographic functionality, based on an original principle, was published by H. Craighead et al [127].[24] It is shown in Fig. 1.24.

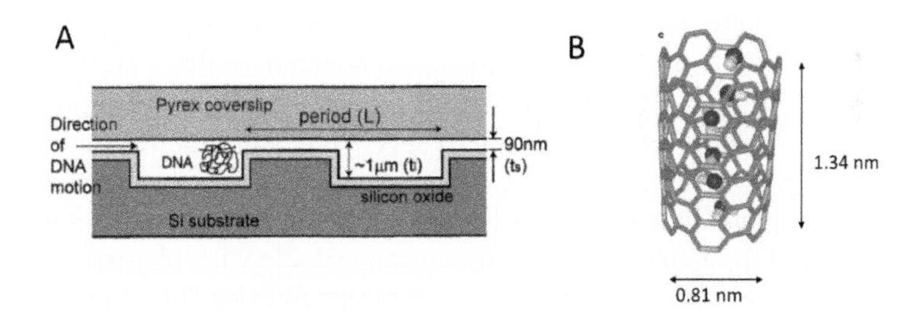

Fig. 1.24: (A) Cell used in Ref. [127] (Reprinted from the preceding reference, with permission of American Physical Society, courtesy of H. Craighead, copyright 2022.) (B) Numerically designed single-walled nanotube 1.34 nm long with a diameter of 0.81 nm , solvated in a water reservoir and simulated for 66 ns. Despite its strongly hydrophobic character, the initially empty central channel of the nanotube is rapidly filled by water from the surrounding reservoir, and remains occupied by about five water molecules during the entire simulation, as shown in the figure [129].

The system was composed of a series of channels of different depths. Large' channels were 1 μm deep, and shallow ones had a depth of 90 nm, i.e. smaller than the gyration radius of λ phage DNAs used in the experiments. Under the action of an applied electric field, DNA strands periodically escape the chambers to migrate through the shallow channels. In this process, they stretch out. Arguments based on entropy show that the probability of escape is an exponential function of the DNA length. This was the efficient size separation mechanism that the authors were seeking. It turned out that the amplitude of the effect was much smaller than expected, for reasons explained

[23] I am indebted to Jan Eijkel for bringing to my attention a number of contributions, along with sharing his vision on the history of nanofluidics, consistent with that exposed here.

[24] In the description we give, one may mention two studies, in which nanofluidic transport phenomena were studied: DNA separation in a nanopillar array in 1998 [124], and mixing in a flow focusing geometry in the same year [125]. These studies did not create nanochannels. They used microfluidic systems to reach, through subtle physical mechanisms, the range of nanoscales

by them.[25] Later, in 2000, the system was modified and efficient DNA separation could be demonstrated [128].

In about the same period (2001), Hummer et al. [129] simulated a single-walled carbon nanotube (CNT), 1.34 nm long, 0.81 nm in internal diameter which was solvated in a water reservoir and tracked for 66 ns. Despite its strongly hydrophobic character, the nanotube was rapidly filled by water.[26] We learned that water molecules circulated in the CNT in single file, developing pulses with peaks of about 30 water molecules per nanosecond.[27] Soon after, CNTs were shown (numerically), to develop large slippages. [132] The effect was directly measured twelve years later [134]. The phenomenon is still not understood today.

The two studies, combined with other experimental [118, 119] (already mentioned) and numerical [133,135,136] contributions, prompted the development of nanofluidics. Soon after, nanofluidics swiftly expanded. An extremely short list of examples, far from exhaustive, includes the observation of polarization concentration induced by the overlapping of Debye layers [137], the invention of nanofluidic diodes and bipolar transistors [138], DNA separation through arrays of nanopillars [139], and negative capillary pressure studies [143]. Fundamental studies dedicated to slippage were also performed, using surface force apparatus (SFA), nano or micro particle image velocimetry (μ PIV), coupled or not to total internal reflection fluorescence (TIRF) or pressure measurements in nanochannels and, more recently measurements through single CNTs [134]. Of particular note is membranes made with aligned CNTs, realized in 2004 [140] (see Fig. 1.25). By offering large permeabilities [141, 142][28] and high selectivities, due to the extremely small pore sizes, they opened a new area in the field of membrane technology.

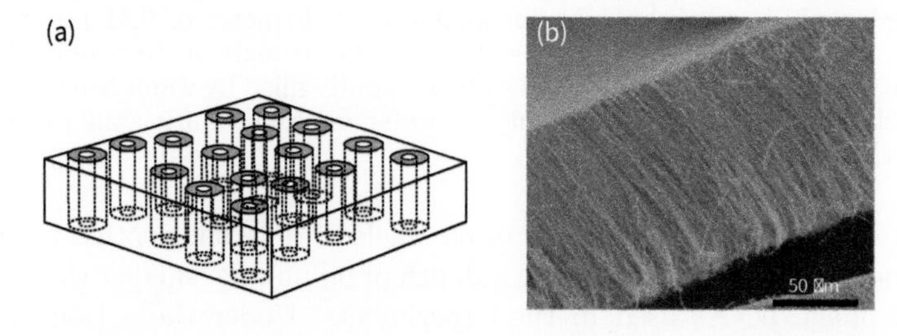

Fig. 1.25: (A) A sketch of a CNT-based membrane structure. A polymer (PS) is embedded between the CNTs; the pore size is their inner-tube diameter [141,142]. (B) Scanning electron microscope (SEM) image of a vertically aligned array of CNTs produced using a Fe-catalysed chemical vapor deposition (CVD) process [141]

[25]The reason was [127]: 'Bigger molecules escape faster simply because more monomers are facing the thin slit, and are able to form a beachhead for escape'. This effect still led to a dependence of the DNA mobility on the chain length, but it was much less strong.

[26]The phenomenon was analyzed in [130]. It is attributed to liquid structuring effects, which reverses the effective wettability of the tube.

[27]Years later, other 'subcontinuum' flow regimes will be discovered [131].

[28]The permeability for the smallest CNTs (1 nm in inner diameter), is four orders of magnitudes larger than no-slip hydrodynamics predicts

In the last few years, a new era came with the advent of novel 2D and 3D materials: functionalized CNTs, graphene, boron nitride (h-BN) and molybdenum disulphide (MoS2) [149–151].

New phenomena were observed in ultra-confined systems (i.e., typically, of sub-2 nm sizes): large slippages (already mentioned), strong decrease of water dielectric permittivity (dead water'), non linear transport along with unexpected large shifts in the freezing transitions [152]. These phenomena may call for a quantum description [149]. If confirmed, this would represent a major conceptual breakthrough. More strongly than in the past, progress made in the domain showed that nanofluidics is not a mere extension of microfluidics, but a distinct field, built on original physical effects and using specific materials and techniques of investigation. The period also inspired new applications, including energy harvesting, high flux membranes, water desalinization, and nanoscale flow sensors.

Performing nanofluidic experiments is challenging, because several forces of various origins, such as electrostatic, van der Waals, entropic or hydrophobic, coupled to the effect of the confinement, come into play together and are uneasy to disentangle [148]. Therefore, when an unexpected phenomenon shows up, it is often difficult to determine its origin. This may explain, for example, why it took fifteen years to understand the origin of nanobubble stability (see Chapter 5). In addition, observation needs sophisticated equipment and the dynamics, often in the nanosecond range, is challenging to track. In this context, it was difficult to envisage the industrialization of a nanofluidic product. This was nonetheless done by Pacific Bio, with zeroth mode cavities, and Oxford Nanopore, with arrrays of nanoholes, both dedicated to sequencing DNA at high speeds and large throughputs (see Fig. 1.26.).

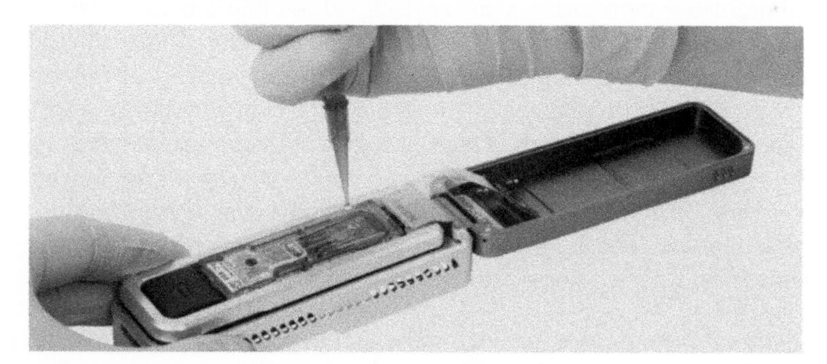

Fig. 1.26: Oxford Nanopore technology. The purified sample is introduced in a chamber containing arrays of nanopores. DNA strands, mobilized by an electric field, cross the arrays and deliver molecular information on the sequence that they contain. The error rate is larger than that of Illumina, but the compacity of Oxford Nanopore product is remarkable. (Courtesy of Oxford Nanopore.)

The Oxford Nanopore system, shown in Fig. 1.26, includes a plate, pierced with nanoholes, across which an electric field is applied. With the help of enzymes, single-strand DNA (ssDNA) pass through the hole and each nucleotide is read, thanks to its electrical signature.

This section mentions a small fraction of the important work done, in nanofluidics, since its birth, i.e. by the turn of the century. The reader may refer to reviews [126,144–149] for detailed information. It should be noted that nanofluidics is part of a broader world which includes, for example drug delivery, colloid science, membrane technology, and nanochemistry. These domains, connected to nanofluidics provide opportunities for developing fruitful synergies.

1.9 The microfluidic market

1.9.1 How to estimate the microfluidic market ?

The delicate task of defining microfluidic products. The microfluidic market integrates microfluidic devices (raw chips), microfluidic products (cartridges with reagents and packagings), and microfluidic instruments. Analysts often distinguish between two segments: instruments and products. We will focus on the second one, which is by far the most important.

How declare that a product pertains or not to the microfluidic market? Let us take the example of dried blood spots (DBS), which store newborn blood samples. Collection is achieved by imbibition and storage by drying. From a physical perspective, imbibition can be viewed as a microfluidic process, because it drives flows, in a controlled manner, through submillimetric pores. Lateral flow tests, such as pregnancy tests, urine dipsticks, and glucose monitoring sensors [162], whose markets are several tens of billions of $, are in a similar situation. Should they be considered microfluidic products? The answer is no. Analysts require a minimal level of technological functionalities, in particular regarding fluid management, in order to decide whether a product pertains or not to the microfluidic market. There is a blurred zone around this requirement, which explains why market estimates may vary from one agency to another. Still, the figures are, within 30%, consistent. Fig. 1.27 illustrates the discussion, for the case of the diagnostic market. The figure is adapted from a recent report published by Yole Development.

The diagnostic market gathers *in vitro* systems enabling the detection of contagious diseases. It includes two types of products: molecular and immunoassay. Each category is in turn divided into centralized (high throughput) and decentralized (point of care (POC) systems). The centralized segment includes large systems, plates, tubes, centrifuge equipment, etc. The machines manipulate μlitre volumes and they use standard assays, which do not incorporate any microfluidic component. It is logical to exclude them from the microfluidic market. By contrast, Cepheid products manipulate fluids in microfluidic devices. The cartridges thereby pertain to the microfluidic market. Immunassays are excluded, for the reasons discussed above. All this leads to Fig. 1.27, in which selected products are enclosed in oblong boxes.

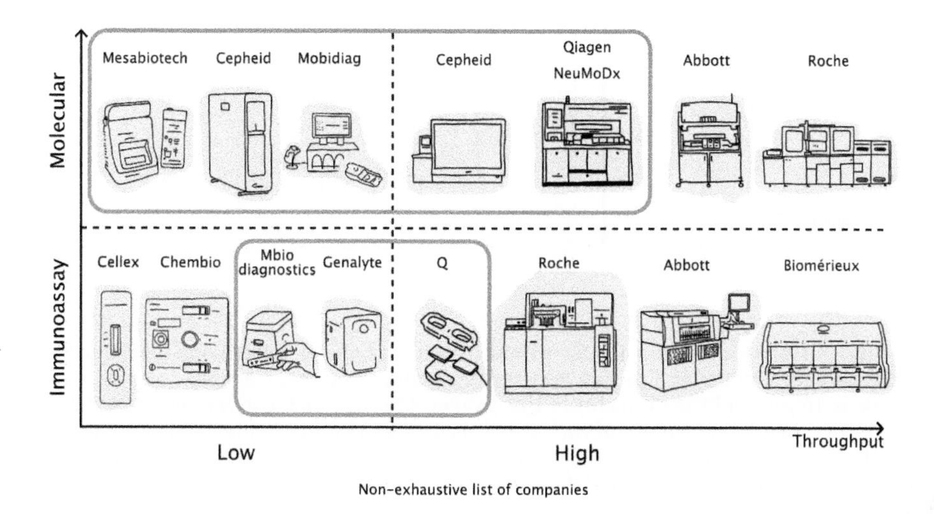

Fig. 1.27: Products of the diagnostic industry, declared belonging (inside oblongs) or not belonging (outside the oblongs) in the microfluidic market. (Adapted from a recent report from Yole Development.)

The case of inkjet printing. Ink-jet printers, for paper, are truly microfluidic systems: droplets emitted by printers are typically 50 μm in diameter. Fig. 1.28 shows the dynamics leading to the formation of droplets, on their way making contact with the paper sheet. Paper inkjets exploit MEMS technology, in both versions (bubble or piezo-injectors [154]) and manipulate microjets with high precision, producing monodisperse droplets, under high throughput conditions. In fact, ink-jet printers represents one the most remarkable achievement of microfluidics. An example is shown in Fig. 1.28.

According to several agencies (Mordor Intelligence, Future Market Insight) (FMI), the overall 2021 market of ink-jet printers is on the order of US$50bn. This is more than twice the rest of the microfluidic market, as will be seen below. Although it could be legitimate to add them up for establishing a 'total' microfluidic market, it is conventional to treat the ink-jet printing market separetly. The microfluidic market that we present here therefore excludes ink-jet printers.

1.9.2 Today's market

As mentioned above, before 2000, the microfluidic market (i.e. without ink-jet printing) was insignificant and scepticism was floating around concerning the potential of the technology to find its feet in a market, even though it was regularly announced, here and there, that microfluidics would revolutionarize the twenty-first century [158]. Common sense led to the belief, in fact wrong, that driving flows through tiny channels, at the industrial scale, without leaks, clogging, bubbles, nor uncontrolled adsorption, was impossible. The opposite viewpoint - believing that it is straightforward to create

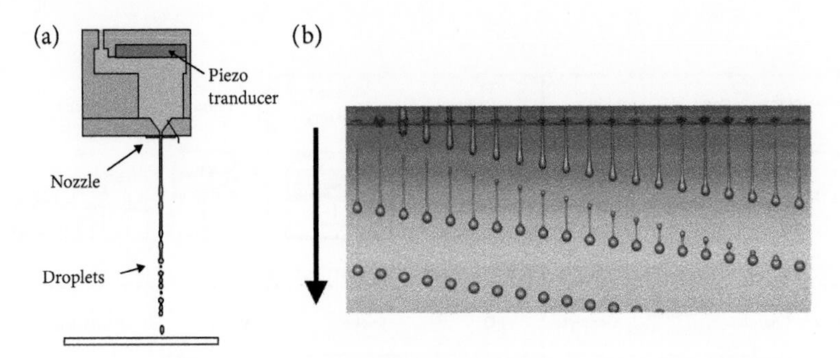

Fig. 1.28: (A) Piezo acoustic inkjet printer of Ref. [156]; (B) Time series of jetted ink droplets, stroboscopically recorded with single-flash photography. (Left to right) Multiple images of single droplets with a delay of 3 μs between the individual droplets. Here the opening radius of the nozzle is 15 μm and the diameter of the droplet 23 μm, which corresponds to a droplet volume of 11 pL. The final velocity of the droplet is 4 m/s. The figure illustrates the imaging quality and the absence of motion blur due to the use of the 8-ns iLIF (illumination by laser-induced fluorescence) technique [155–157].

,

a complex, functional, microfluidic device - was unrealistic. To add to uncertainties, the remark, occasionally made by prominent researchers, that microfluidic devices are small, but, with the tubings and the pumps needed to operate the system, they more often resemble a 'chip in the lab' rather than a 'lab on a chip', casted doubts about the possibility of creating microfluidic devices of industrial relevance.

Despite all this, a number of successful microfluidic products emerged. Technical difficulties could be solved, circumvented, or just found irrelevant. Impressive industrial challenges, previously considered out of reach, could be met, such as driving ssDNA through nanoholes and collect genomic information from it. Finally, over the last two decades, the microfluidic market steadily increased at a two-digit rate. Today, hundreds of millions of devices are sold every year. For example, 400 million Illumina microfluidic flow cells are shipped every year.

Let us analyse three examples, which, today, generate revenues between US$200mn–US$3bn (see Fig. 1.29).

Quidel Triage In Fig. 1.29, the device is a microfluidic adaptation of the Elisa test: the blood sample, in which targeted antigens are fluorescently labelled, is driven, by capillarity, in a straight microchannel, with antibodies immobilized on the walls. As in Elisa tests, antigens are captured by the antibodies, and, after rinsing, fluorescence emission is analysed. With this test, levels of troponin, whose elevation indicates heart muscle damage, can be measured, with a few nanograms per litre sensitivity. This is a major application. Sales are on the order of US$200 mn.

Ilumina New Generation Sequencing (NGS) technology, developed by Illumina, is based, as all NGS methods, on microfluidic technology. The cartridge contains a patterned cell, with wells, called a flow cell, fabricated over a Complementary

Fig. 1.29: Three examples of types of products, based on microfluidics, generating, in 2020, turn overs in the range US$200mn–US$3bn: (A) Triage for cardiac control (Quidel); (B) DNA sequencer (illumina HiSeq 2500); (C) Kindle (e-ink technology)

metal–oxide–semiconductor (CMOS) chip. In the cell, each well faces a photodiode. In a nutshell, sequencing is performed by hybridizing sampled ssDNA on anchored probes, and monitoring, in real time, the light emitted during the process. Microfluidics plays an instrumental role in this domain. Without microfluidics, no NGS based sequencing could exist.[29] Illumina revenues generated by the product reached US$3.2bn in 2020. As noted above, the company ships 400 million flow-cells each year.

e-ink The Pocketbook Inkpad Lite is based on microfluidic droplets (the technology will be described in Chapter 4). Its price ranges between $200 -$300. The display technology, commercialized by E-ink, generated US$530mn in 2021.

Who are the microfluidic players? Fig. 1.30 organizes them in three groups, each corresponding to a range of revenues (the figure is adapted from Yole-Development).

Market estimates are given by several agencies: Yole Development, Market and Markets, Grand View Research, uFluidics, Research & Market, and BCC Research. On average, the microfluidic market is estimated around 17 US$bn with a compound annual growth rate (CAGR) around 15 - 20%. Should 20% be accurate, the market would reach 100 B$ in 2030. A plot showing the evolution of the market over the years, again made by Yole Development, is shown in Fig. 1.31.

Fig. 1.31 moreover provides information on the market segments. In decreasing order (in terms of volume), we have: point of care, tools for pharmateutical and research, clinical diagnostics, industrial diagnostics, optical actuation (liquid lenses), and manufacturing.

[29] NGS, compared to traditional sequencing, based on Sanger's method, allowed to reduce sequencing prices along with increase speeds by orders of magnitude (see Chapter 5)

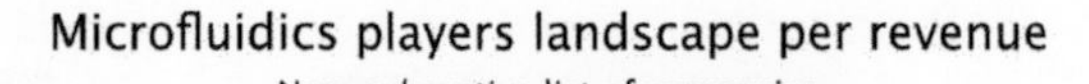

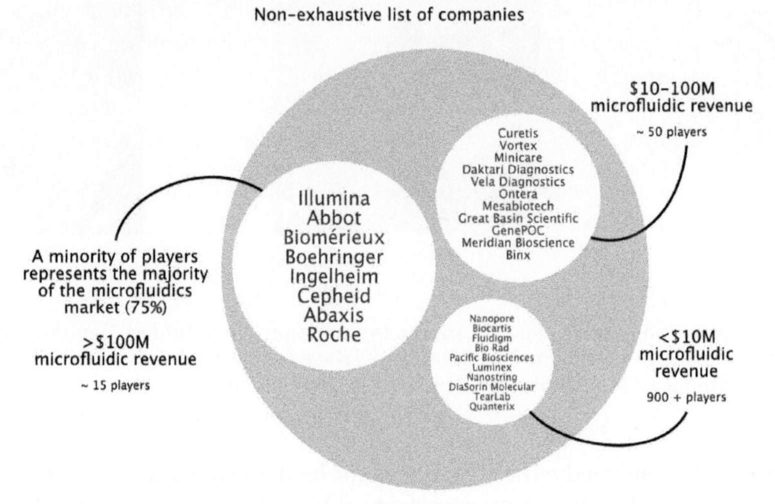

Fig. 1.30: Microfluidic players organized in three groups, each corresponding to a range of revenues [153]. (Adapted from Yole Development.)

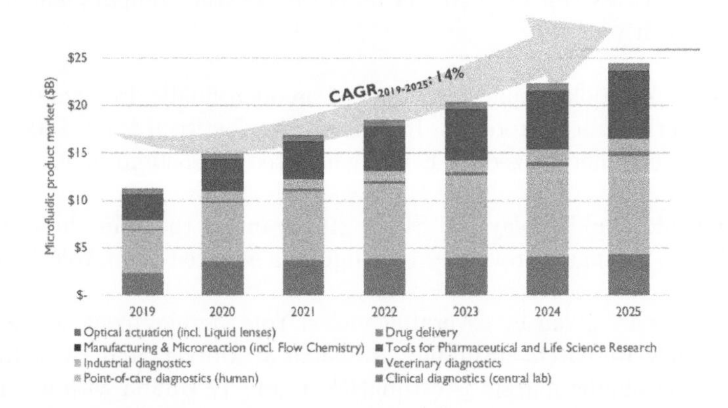

Fig. 1.31: Microfluidic market and its evolution (Courtesy of Yole Development) [153]

1.10 Future of microfluidics

1.10.1 Microfluidics and Pasteur and Bohr quadrants

In 1997, D.Stokes proposed the 'quadrant model [163,164] (see Fig. 1.32). This model classifies research activity into four quadrants. Three of them are defined in the following manner:

Bohr quadrant -pure basic research : N. Bohr's efforts were entirely dedicated to answer fundamental questions of physics. In 1913, he explained the origin of the spectral lines of hydrogen, by representing the atom as a nucleus orbited by an electron [159]. The Bohr atom model, which contributed to laying the foundations of quantum theory, was conceived under the sole momentum of intellectual curiosity, in a manner disconnected from applications. One paradox is that quantum physics led to the invention of the transistor, and further to computing science, internet -all inventions that, without exaggeration, revolutionized the world-.

Edison quadrant -applied research : Edison was a prolific inventor, with more than thousand patents. Examples are the phonograph and the motion picture camera. The electrical bulb lamp was not invented by him, but he managed to increase its lifetime considerably, by using carbon filaments. Edison's approach was empirical, with no serious attempt to understand the physics involved in his inventions.

Pasteur quadrant -use-inspired basic research : Pasteur discovered microbiology by working on diseases of beer and wine, not understood at that time. His discovery of the vaccine for rabies, made in 1885, was stirred up by the bite of a 9 year old child by a rabid dog. Other examples, found when studying his life, illustrate the synergy between societal/industrial inspiration and fundamental discoveries.

The unamed fourth quadrant represents research motivated neither by intellectual curiosity nor by applications. It is difficult to comment on this quadrant.

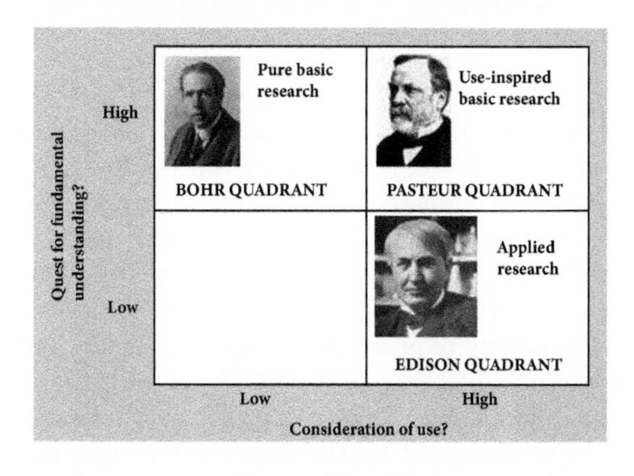

1.32: The four sectors of research: Bohr, Pasteur, Edison, and unnamed [163].

Regarding microfluidics, Pasteur's quadrant is certainly the most relevant: microfluidics offers responses to industrial and societal needs, and, in the meantime, allows us to address basic questions. Since the early days of microfluidics, the community has worked mostly in the Pasteur quadrant.

However, it pays frequent visits to the Bohr quadrant. Often, along the lines of Feynman's talk, discoveries emerged from the excitation of realizing something that has never been done before, or from efforts to understand strange phenomena occurring at the microscale. For example, a precursor work done on microfluidic droplets was motivated solely by curiosity [165–167]. The objective of Ref. [165, 166], was to report the outstanding monodispersivity of droplets produced in microjets, and that of Ref. [167],was to describe droplet patterns observed in microchannels. No application was mentioned. Later, it was realized that droplets could play an important role in colloidal science, chemistry, material science, and biology. Another example is PDMS soft technology, whose precursor studies were mostly motivated by technological challenges [168]. More striking is the case of nanofluidics, regarding which some of the investigators, around the turn of the twenty-first century, had no idea about the applications that the technology could generate. Later, nanofluidics will become a core technology for new genome sequencing. Today, the questions raised in this domain, concerning, for example, the hydrodynamical behavior of carbon nanotubes, clearly pertain to the Bohr quadrant. Other examples can be mentioned (nanobubbles, inertial microfluidics), where research is mainly curiosity driven.

1.10.2 Promising areas of application

The advantages of microfluidics, put forward multiple times over the years, are as follows: small quantities of reagents, small samples, high speeds of analysis, cost reduction, high sensitivity of separation, portability, exquisite flow control, functionality integration, automatization, parallelisation, high level of compartmentalization, and large information throughput. Disadvantages also exist (cleaning constraints for device fabrication, difficulty of producing large volumes of product, no standardization, difficulty of scaling up for PDMS, and, for complex constructions, reliability below industrial standards), but they are outweighted by the advantages offered by the technology.

The general vision is that, with such capabilities, microfluidics can resolve bottlenecks hampering the development, sometimes the existence, of key industrial products (for instance, NGS), or hindering the realization of industrial or research ambitions (for instance, performing in-situ analyses on Mars, retrieving information stored in DNA memories, or building representative models of the kidney). This is what microfluidics has successfully done over three decades. Concerning the forthcoming years, since fluid control at the submillimetric scale is increasingly required in a number of fields, and many domains do not benefit yet from microfluidic technology, or even have no awareness of it, conditions are met for the microfluidic field to keep growing at substantial rate. This linear vision will be disturbed or disrupted by the emergence of new discoveries in microfluidics or in related fields. Examples include the discovery of new enzymes, that could simplify and reduce the cost of biological workflows, and the synthesis of new materials, which could impact microfabrication techniques, as PDMS did in the past. In addition, the priorities of the society, which support a major part of microfluidic activity, may change. Yesterday, a priority was given to the human genome programme. Today, energy production and the environment, are subjects of major importance. All the problems of predicting the future lie in the unpredictable advent of major discoveries and the extreme difficulty to visualize the evolution of societal needs.

During the years 2000-05, institutions attempted to sketch the future. They produced 'roadmaps', or white papers' for microfluidics, to structure the young field and to reassure or stimulate funding agencies and investors, as Moore's law did for microelectronics. Today, on re-reading these roadmaps, the general impression is that they were more concerned with platforms than with the future of the field [169]. This probably shows how difficult it is to predict the future. The scientific literature was more pertinent. Reviews offering a vision of the future of microfluidics, or subfields pertaining to it (see for instance [1, 74, 160, 223]), often pointed in the right directions.[30]

Here, we will look into the crystal ball by listing, without claim for completeness, a number of promising areas of application of microfluidics, in a manner similar to Ref. [161]. This is set out in Table 1.1.

Application	Example
Diagnostics	Perform low-cost SARS COV2 molecular testing with paper microfluidics
Single cell	Establish the transcriptome, at the single cell level, of tumour cancer cells
Screening	Discover antibiotics and enzymes by screening large quantities of biochemical reactions
Delivery	Produce lipid nanoparticles for nucleic acid delivery (for example, for RNA-based vaccines)
Organ on a chip	Create artificial liver, lung or intestine to test drugs
MRT	Produce high-quality materials, benefiting from high throughput and flow control
Cell culturing	Produce, with high throughput and low cost, stem cells of high quality
Material Science	Materials with novel functionalities
Membrane Science	Membranes for energy harvesting or energy production
Cosmetics	Cosmetic products and instrumentation
DNA storage	Retrieve information stored in DNA matrices
Display technology	Display images with microfluidic entities, such as droplets

Table 1.1 : Non-exhaustive list of promising areas of application for microfluidics and nanofluidics

The table deserves few comments

Diagnostics: Today, therapeutic drugs are making rapid progress but common sense says that, in order to cure a disease, one must first identify it. This is where diagnostic comes on stage. Tests pertain to two categories: antigen and molecular. Antigen tests have limited sensitivities, but their costs are low and their usage simple. To-day, molecular tests are mostly based on polymerase chain reaction (PCR). In order to reach high sensitivities, these tests perform an extraction before amplifying the nucleic acids. The technology is thus more sensitive but more

[30] It is amazing to note that, in some cases, authors did not predict the advent of technologies that, later on, they invented [1].

complicated and more costly than antigen tests. Since 2000, simpler methods of amplification have been developed. An example is loop amplification (LAMP), invented in 2000 [171], which operates under isothermal conditions, much simpler to implement. It it is thus regrettable that PCR keeps dominating the molecular market. It has been shown that paper microfluidics, described in Chapters 4 and 7, and based on LAMP, performs molecular diagnostics, with integrated membrane purification, with sensitivities and specificities comparable to those of PCR, much lower cost, and with a minimal logistics [170,172–174]. There is no doubt, that, in the future, paper microfluidics will represent an important technology for molecular point of care (POC) diagnostics. At the moment, the problematic is linked to business, not to technology.

Delivery: Delivery is about creation of particles, to be swallowed, inhaled, injected into the blood circulation, or injected intracutaneously. These particles circulate or diffuse in the body and then, at the end of their journeys, deliver their load, with a controlled kinetics. Words often used are 'targeted drug delivery' or 'controlled drug release'. Delivery primarily concerns human health but also agriculture: for example, pheromones are encapsulated and delivered at certain periods of the year, in order to disrupt insect reproduction. Delivery methods are diverse: physico-chemical (for instance hydrogel dissolution in tissues), acoustic (bubble or droplet bursting) [175], thermal, or chemical. Delivery kinetics ranges from fraction of seconds to months, and the sizes of the particle range from 100 nm to tens of micrometres. There is a natural coupling between delivery and microfluidics [176]. Microfluidics allows, by reducing polydispersivity, improving encapsulaton rates, suppressing of burst phenomena (too much delivery in a short time, which can be lethal) and, more generally, optimizing delivery kinetics.

Let us pause for a moment to consider lipid nanoparticles (LNPs) (see Fig. 1.33(A)). We will see, in Chapter 5, that these objects gain considerably being produced in microfluidic or millifluidic devices. LNPs are complex nanoparticles, 50 to 100 nm in diameter, formed through a self-assembly process that includes, according to Ref. [178], three steps: discoidal cluster formation, aggregation of clusters into larger patches, and vesicle formation.[31] LNPs travel in the blood circulation, attach to a cell thanks to their affinity, and penetrate into it. Having penetrated the cell, LNPs deliver their passengers, designed to interact with the cell machinery (see Fig. 1.33(B)). Molecules carried by LNPs can be DNA, mRNA or siRNA. LNPs have been highlighted during the COVID-19 crisis. They vectorized mRNA into cells and made possible mRNA-based vaccination [177].

There is little doubt that the domain of delivery, coupled to microfluidic technology, will expand in the future.

Organ on a chip (OOC): As noted above, OOCs are biomimetic objects enabling drug testing on representative models. One outcome of the technology is to ac-

[31]LNP formulation includes, to control stability, electrostatic charge, and pH, reagents such as 1,2-dioleoyl-sn-glycero-3-phosphocholine (DOPC), 1,2-dimyristoyl-rac-glycero-3-methoxypolyethylene glycol-2000 (DMG-PEG2000), Citric acid, sodium citrate tribasic dehydrate and cholesterol

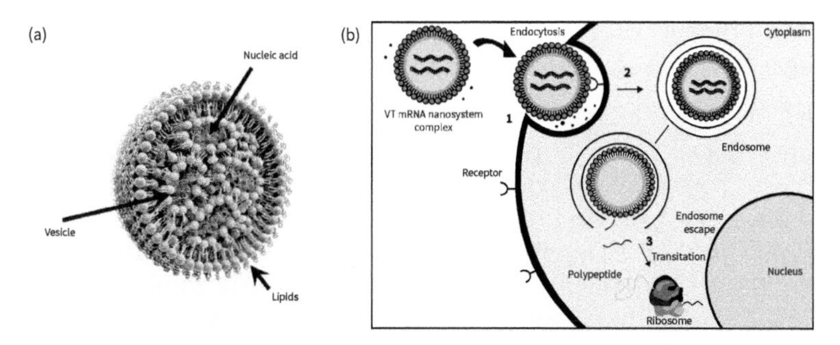

Fig. 1.33: (A) LNP structure: a ionizable lipiidic shell encapsulates vesicles, which encapsulate nucleic acids. (B) Nucleid acid delivery into the cell, including: endocytose, endosome formation, endosome escape, and, for the case of mRNA, protein synthesis in the endoplasmic reticulum (ER) and its finalization, prior to exocytose, in the Golgi apparatus (not shown).

celerate drug discovery, without animal testing. Mimicking *in vivo* conditions is a challenge, for which microfluidics is instrumental. The OOC field is often reviewed (see for instance [110]).

Microreaction technology (MRT): As noted earlier, MRT allows us to produce materials under high throughput conditions with excellent control. It is probable that MRT will progressively replace the traditional techniques, based on large batches, in which turbulence tends to degrade the performances of the product and, in a number of cases, raise safety issues. MRT devices, by working at high throughput but below the turbulence onset, and, by working with smaller, parallelized systems, avoid these problems. Broader discussion on the virtues of MRT can be found in Ref. [187].

Single cell: Single cell domain concerns fundamental research and applications. For more than decade, research has benefited considerably from microfluidic technology [188]. In oncology, the challenge is to capture substantial amounts of cells (typically 10^4-10^5), barcode them and sequence their transcriptome, at the single cell level, i.e. perform single cell RNA sequencing (scRNAseq). With this approach, cell expression heterogeneity can be mapped, guiding therapeutics. This possibility would represent a breakthrough in cancer treatment. The task can be performed in droplets, as shown by 10X Genomics, a fast-growing company [192]. It can also be performed in hydrogel cages [190, 191]. The domain was recently reviewed [189].

Cell culturing: In cell culturing, microfluidics allows one to control or better control the environment, i.e. flow geometry, temperature, gas exchanges and nutrient fluxes (see for instance [193, 194]). The culture of mesenchymal stem cells (MSCs) deserves particular interest. Several regenerative medicine treatments use these cells. The treatments typically use 10^8 MSC cells, whose cost, at the moment, reaching several tens of thousands of dollars per dose, makes regenerative medecine unaffordable [195, 196]. By using microfluidics, it can be argued that costs will seriously significantly, while keeping production levels and quality high.

It is thus probable that, in the future, microfluidics will play an important role in the domain of regenerative medicine, and, more generally in cell therapy.

Materials: Microfluidics can produce all sorts of materials. These include nanoparticles, too small to be directly controlled by the confinement, but still benefiting from better-controlled environments. Micrometric particles gain considerably from being produced in microfluidic systems [84, 179, 182]. Another area is particle assemblies. Examples include clusters formed by hydrodynamic interactions [180, 181], or bubbles assemblies, spontaneously forming crystals, leading, after solidification, to microfluidic foams [183]. These foams may be used as scaffolds for cell culture, but also as complete photonic band gaps materials, [184,185] an area called 'phoamtonics' [186]. This materials are interesting for infrared (IR) communication applications. Other examples are tissues and organoids, obtained by culturing cells.

Membrane technology: CNT-based membranes possess high permeabilities and high surface charges densities. Based on these characteristics, a proposal was made to use these membranes for converting osmotic energy into electrical energy [198]. Membranes separating salt water (from the sea) and fresh water generate osmotic flows, which develop streaming electrical currents (See Chapter 6), and thus electrical energy. In this manner, 'blue' energy can be produced. In the case of a single boron nitride nanotubes (BNNT), performances are impressive [198]: the maximum power density lies in the kW/m^2 range, i.e. three orders of magnitude above that of standard membranes [199]. However, membranes formed with nanotubes are subjected to a number of effects (in particular polarization concentration), which considerably reduce their performances [199, 200]. This domain is the subject of active research.

Cosmetics: There are opportunities to invent new cosmetic creams, perfumes, sprays or nail polish and develop novel instrumentation, based on microfluidics [201,202].

DNA storage: DNA storage is a branch of DNA computing. DNA storage shows impressive performances [203]. An assembly of DNA strands, packed in a container of the size of a shoe box, contains all the internet information circulating in the world. To retrieve the stored information, microfluidics and nanofluidics are obviously needed . At the moment, this domain is still embryonic, but it should grow considerably in the future.

1.10.3 Microfluidics and its interactions with other fields

Although possibly debatable, one may attempt to distinguish between microfluidic technology and interactions between this technology and other fields. This leads to the diagram in Fig. 1.34, which does not claim exhaustivity.

The upper part represents landmarks of the field. There is a range of years, extending roughly, between 1997 and 2003, in which a large part of the foundations of microfluidic science, as it looks today, have been established. These years may be called 'golden years'. They are characterized by a high density of discoveries. The bottom part of

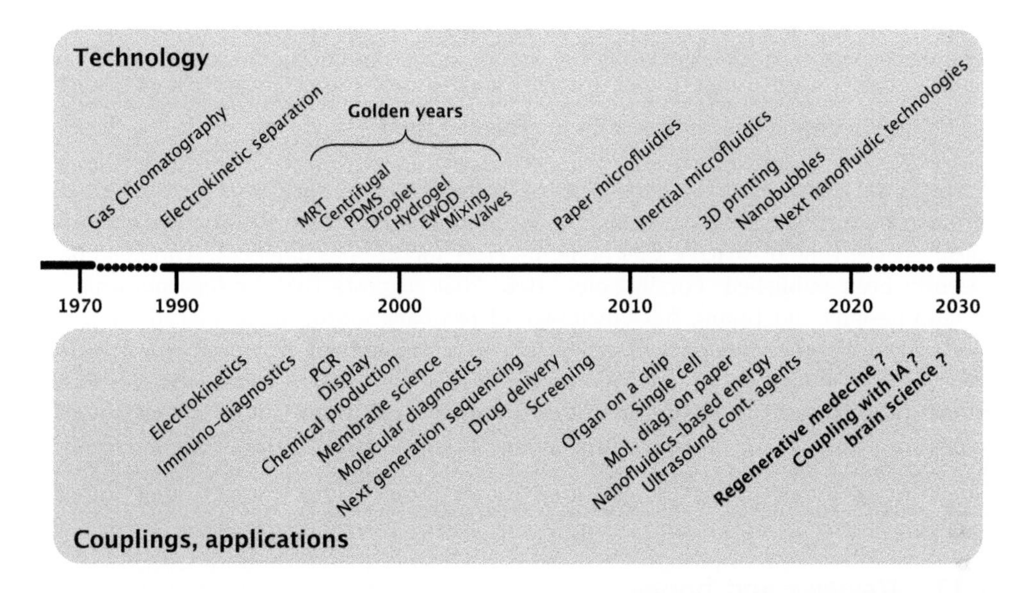

Fig. 1.34: Microfluidics and its coupling with other fields, giving rise to major applications.

the diagram shows successful couplings between microfluidics and other fields, from which, in most cases, interesting applications emerge. They are numerous. Compared to the upper part, the inventions are distributed in a more homogeneous manner. In addition, more strongly than for the upper part, milestones often start off new areas of activity, which expand further, sometimes considerably. This part conveys most of the microfluidic market.

1.10.4 Microfluidics and complexity

In the field of microfluidics, there has been a continuous effort for increasing the complexity of the devices, hoping to develop rich functionalities. For instance, PDMS microvalves have.been used to create microfluidic devices of increasing complexity, that were intended to represent analogs of CPUs. In fact, many mechanisms exist at the bottom, that can be used to actuate the system, without technological effort: examples are self assembly mechanisms [197] for which molecules spontaneously form patterns,[32] amplification processes, selective chemistry, enzymatic activity, molecular motors, and molecular conformational changes, induced by pH or temperature changes. At the larger scale, we have, for instance, convection, Marangoni flows, and capillarity, which can also be exploited to actuate the system. The question is how can we play with all these natural actuators, how can we leverage them, to build up interesting functionalities? How can we use this micro-nanoscopic toolbox? NGS, single cell studies, organ-on-a-chip, and lipid nano-particles [177], provide examples in which the toolbox

[32] A remarkable example, among many others, is DNA origamis.

has been used to perform extremely complex tasks. In the future, microfluidics may take inspiration from these examples to create new outstanding devices.

1.10.5 Microfluidics and Artificial Intelligence

Artificial Intelligence(AI) impacts science. Some domains, such as structural biology or proteomics, will profoundly change their methodologies [204]. What about microfluidics? At the time of writing this book, no clear picture has emerged. Thoughts have recently been published. For instance, Ref. [205] suggests that by incorporating AI in chip designs and taking full advantage of feedback loops, along with data-driven models, one could create systems generating large amounts of data useful for drug discovery, organ modelling, or developmental biology. Thus far, as mentioned in a recent review [71], progress has been slow and commercial adoption has been hampered by issues with reliability, generalizability, and costs. It is probable that, for microfluidics, the importance of AI will grow along the years, facilitating or perhaps enabling the walk towards complexity mentioned in the previous section.

1.11 Reviews and books

1.11.1 Books

Today, around twenty books have been published on microfluidics.

The first book was *Microflows: fundamental and simulation* [206], published in 2002, followed by *Microflows and nanoflows, Fundamentals and Simulation*, appeared in 2005. The book deals with gases, flow modelling and electrokinetics. In the same period (2003), *Fundamental and Application of Microfluidics*, by N.T. Nguyen and S. Wereley, was published [207]. It uncovered a broader part of the subject, mainly from an engineering perspective. The book has been reedited three times. The first edition of my book, in French, appeared in this period (Belin, 2003) [208], followed by the english version, published in 2005, by Oxford University Press [209]. The focus was on the physics of microfluidics. The books of S. Colin [210] (focusing on gases), H. Bruus (*Theoretical Microfluidics* [211], 2007), and B.J. Kirby (*Micro- and nanoscale fluid mechanics : transport in microfluidic devices*, [212], 2010) adopted a similar approach.

The book written by J. Berthier and P. Silberzan in 2009, *Microfluidics for Bioengineering* [214], was instigated by a more engineering perspective. One may also cite (without being exhaustive) *Micro-Drops and Digital Microfluidics* (2008) [215], *Theory and Practice for Beginners* [216], *Introduction to BioMEMS* [217], *Laboratory Methods in Microfluidics* [218], and recently, *Microfluidics and Lab on a Chip*, by Manz et al. [219].

1.11.2 Reviews

Reviews on the research carried out in the field of microfluidics are numerous. They are often well cited, and over the years, they have certainly played a substantial role in the spreading of information. Historically, the first microfluidic review was written by P. Gravesen [65]. As mentioned in this chapter, the review certainly contributed to introduce the word microfluidics' into the community. Later on, in 1999, C.M.Ho and Y. Tai [21] wrote a review on flow control with microfluidics, and M. Gad el Hak [220] on gas dynamics. Reviews written in the same period of time by N. Giordano and J.T. Cheng [221], H. Stone and S. Kim [222], and H. Stone et al [228] ,were centered on the physics and the hydrodynamics of microfluidics. Later, in 2006, Todd et al [74] and A.van den Berg et al in 2010 [229] reviewed the field along similar lines. D. Beebe et al [223] published an early review on biomedical applications, as did G. Sanders and A. Manz [224], E. Verpoorte [225] on biochemical analysis, J. Lichtenberg et al [226] on sample pre-treatment, and P.A. Auroux et al [227] on analytical applications. In 2006, a series of review appeared in a special issue of *Nature*, acquiring a large visibility, in particular G.M. Whitesides's paper 'On the origin and future of microfluidics'.

After 2010, the field was too broad and diverse to review it in the format of an article. Specialized reviews thus appeared, more and more frequently over the years. This seems to be a general evolution. Sometimes, a topics is reviewed several times in a row. These reviews are often of good quality, and are useful to acquire global views on specific topics.

1.11.3 In which journals do microfluidic researchers publish ?

Research articles on microfluidics are dispersed in different journals. Lab-on-a-Chip deserves particular attention because it accounts for about 40% of microfluidic publications. The journal contributed, in synergy with μTAS conferences, to identifying a community. Other journals are *Analytical Chemistry, Sensors and Actuators, Microfluidics and Nanofluidics, Journal of Micromechanical and Microengineering, Biomicrofluidics, PNAS, Nature, Science, Physical Review Letters, Physical Review of Fluids, Journal of Chromatography, Electrophoresis, BioMicrofluidics.*

1.12 Organization of the book

This book consists of six chapters, in addition to this introductory chapter.

- Chapter 2 describes the physics of small systems is described
- Chapter 3 presents low Reynolds number flows.
- Chapter 4 presents droplet microfluidics
- Chapter 5 describes transport phenomena: diffusion, dispersion, mixing, adsorption, thermal exchanges and separation.
- Chapter 6 is about electrohydrodynamics of microsystems; electrophoresis, electroosmosis, and dielectrophoresis.
- Chapter 7 introduces microfabrication techniques : silicon technology, soft technologies, plastic moulding and 3D printing.

References

[1] G. Whitesides, *Nature*, **442**, 368 (2006).
[2] J. Liesche, C. Windt, T. Bohr, A. Schulz, K. H. Jensen, *Tree Physiology*, **35**, 376 (2015).
[3] J. W. C. White, E.R. Cook, J. R. Lawrence, W. S. Broecker, *Geochemica and Cosmochimica Acta*, **49**, 237 (1985).
[4] L. Jingmin, L. Chong, X. Zheng, Z. Kaiping, K. Xue, W. Liding, *Plos One*, **7**, e50320 (2012).
[5] P. F. Scholander, H. T. Hammel, E. D. Bradstreet, E. A. Hemmingsen, *Science*, **148**, 339 (1965).
[6] T. D Wheeler, A. D. Strooke, *Nature*, **455**, 208 (2008).
[7] J. C. A. Zemlin, *A Study of the Mechanical Behavior of Spider Silks*, U. S. Army Natick Technical Report 69-29-CM (1968).
[8] F. K. Ko, J. Jovavicic, *Biomacromolecules*, **5**, 780 (2004).
[9] J. H. Barbee, G. R. Cokelet, *Microvascular Research*, **3**, 6 (1971).
[10] A. T. Nurden, P. Nurden, M. Sanchez, I. Andia, E. Anitua, *Blood*, **114**, 1875 (2009).
[11] C. Dunois-Larde, C. Capron, S. Fichelson, T. Bauer, E. Cramer-Borde, D. Baruch, *Frontiers in Bioscience* , **13**, 3525 (2008).
[12] L . Baker, *Physiology of Sweat Gland Function: The Roles of Sweating and Sweat composition in Human Health, Temperature*, DOI: 10.1080/23328940.2019.1632145 (2019).
[13] C. L. Li, R. J. Braun, K. L. Maki, W. D. Henshaw, P. E. King-Smith, *Physics of Fluids*, **26**, 052101 (2014).
[14] A. S. Verkman, *Curr. Biol.*, **23**, 2 (2013).
[15] S. Gravelle, L. Joly, F. Detcheverry, C. Ybert, C. Cottin-Bizonne, L. Bocquet, *PNAS*, **110**, 16367 (2013).
[16] M. Madou, *Fundamentals of Microfabrication*, CRC Press (2000).
[17] Forschungszentrum Karlsruhe GmbH Technik und Umwelt, Projekt Mikrosystemtechnik (PMT).
[18] S. Renard, *J. Micromech. Microeng.*, **10**, 245 (2000).
[19] R. Feynman, *Engineering and Science*, **23** (5). 22 (1960).
[20] D. M. Eigler, E. K. Schweizer, *Nature*, **344**, 524 (1990).
[21] C. M. Ho, Y. C. Tai, *J. Fluids Engineering*, **118**, 437 (1996).
[22] L. S. Fan, Y. C. Tai, R.S. Muller, *IC-processed Electrostatic Micromotors*, Technical Digest, lEDM, 666 (1988).
[23] L. F. Fan, Y. C. Tai, R. S. Muller, *Sens. Actuators.*, **20**, 49 (1989).

[24] M. Douglass, *Reliability, Testing, and Characterization of MEMS/MOEMS II*, *Proc. SPIE*, 4980 (2003).
[25] R.G. Sweet patent, *High-Frequency Oscillography with Electrostatically Deflected Ink Jets* (March 1964).
[26] R. G. Sweet, *Review of Scientific Instruments*, **36**, 131 (1965).
[27] S. C. Terry, PhD dissertation, *A Gas Chromatography System Fabricated on a Silicon Wafer using Integrated Circuit Technology* (1975).
[28] S.C. Terry, *IEEE Trans. Elect. Devices*, **26**, 1880 (1979).
[29] D. Reyes, D. Iossifidis, P–A. Auroux, A. Manz, *Anal. Chem.*, **74**, 2623 (2002).
[30] A. Manz, N. Graber, H. Widmer, *Sens. Actuators*, **B1**, 244 (1990).
[31] D. Harisson, A. Manz, Z. Fan, H. Ludi, H. M. Widmer, *Anal. Chem.*, **64**, 1926 (1992).
[32] A. Manz, D. Harrison, E. Verpoorte, J. Fettinger, A. Pausus, H. Ludi, H. Widmer, *J. Chromat.*, **593**, 253 (1992).
[33] D. Harrison, K. Fluri, K. Seiler, Z. Fan, C. Effenhauser, A. Manz, *Science*, **261**, 895 (1993).
[34] S. Jacobson, R. Hergenroder, L. Koutny, R. Warmack, J. Ramsey, *Anal. Chem*, **66**, 1107 (1994).
[35] D. Harrison, A. Manz, P. Glavina, *Transducers*, **917**, 92 (1991).
[36] B. Weigl, P. Yager, *Science*, **283**, 346 (1998).
[37] W. Ehrfeld, V. Hessel, H. Mobius, T. Richter, K. Russow, *Dechema*, **132**, 1 (1996).
[38] D. Evans, D. Liepmann, A. P. Pisano, *Proceedings of MEMS 97*, Nagoya, Japan, 96 (1997).
[39] J. Knight, A. Vishwanath, J. Brody, R. Austin, *Phys. Rev. Lett.*, **80**, 3866 (1998).
[40] W. Ehrfeld, V. Hessel, H. Lowe, *Microreactors*, Wiley-VCH, Weinheim (2000).
[41] S. Bohm, K. Greiner, S. Schlautmann, S. de Vries, A. van den Berg, *Proc. μTAS*, 25 (2001).
[42] A. Bertsch, S. Heimgartner, P. Cousseau, P. Renaud, *Proc. MEMS 2001*, Interlaken, 507 (2001).
[43] M. A. Northrup, M. T. Ching, R. M. White, R. T. Watson, *Proceedings of the 7th International Conference on Solid State Sensors and Actuators (Transducers'93)*, Yokohama, Japan, p. 924 (1993).
[44] P. Wilding, M. Schoffner, L. Kricka, *J. Clin. Chem.*, **40**, 1815 (1994).
[45] A. Chaudhari, T. Wounderberg, M. Albin, K. Goodson, *Micro ElectroMechanical Systems*, **7**, 918 (1998).
[46] M. Northrup, B. Benett, D. Hadley, P. Lander, S. Lehew, J. Richards, P. Stratton, *Anal. Chem*, **70**, 918 (1998).
[47] J. Daniel, S. Millington, D. Moore, C. Lowe, D. Leslie, M. Lee, M. Pearce, *Sens. Actuators*, A 81 (1998).
[48] R. Oda, M. Strausbauch, A. Huhmer, N. Borson, S. Jurrens, J. Craighead, P. Wettstein, B. Eckloff, B. Kline, J. Landers, *Anal. Chem.*, **70**, 4361 (1998).
[49] M. Kopp, A. De Mello, A. Manz, *Science*, **280**, 1046 (1998).
[50] E. Altendorf, Zebert, M. Holl, A. Vannelli, C. Wu, C. Schulte, *Proc. μTAS*, 73 (1998).
[51] P. Renaud, U. Seger, S. Gawad, *Proc. Nanotech 2002*, Montreux (2002).

[52] S. Jacobson, R. Hergenroder, A. Moore, J. Ramsey, *Anal. Chem.*, **66**, 4127 (1994).
[53] R. Srinivasan, I. Hsing, P. Berger, K. Jensen, S. Firebaugh, M. Schmidt, M. Harold, J. Lerou, J. Ryley, *AIChE J.*, **43**, 11 (1997).
[54] J. Eijkel, A. Prak, S. Cowen, D. Craston, A. Manz, *J. Chromat. A*, **815**, 265 (1998).
[55] A. Kalmholz, B. Weighl, B. Finlayson, P. Yager, *Anal. Chem*, **23**, 71 (1999).
[56] A. Pabit, S. Hagen, *Biophys. Journal*, **83**, 2872 (2002).
[57] T.S. Kuhn, *Second Thoughts on Paradigms. The Structure of Scientific Theories*, **2**, 459 (1974).
[58] M. Burns, B. Johnson, S. Brahmasandra, K. Handique, J. Webster, M. Krishnan, T. Sammarco, P. Man, D. Jones, D. Heldsinger, C. Mastrangelo, D. Burke, *Science*, **282**, 484 (1998).
[59] J. Pfahler, J. Harley, H. Bau, J. Zemel *Sensors Actuators*, **A21-A23**, 431 (1990).
[60] E. Lauga, M. Brenner, H. A. Stone, *Microfluidics: The No-Slip Boundary Condition.* C. Tropea, A.L. Yarin, J.F. Foss (eds), *Springer Handbook of Experimental Fluid Mechanics*, Springer Handbooks, Springer, Berlin, Heidelberg (2007).
[61] D. Smith, S. Chu, *Science*, **281**, 1335 (1998).
[62] G. Waibel, J. Kohnle, R. Cernosa, M. Storz, M. Schmitt, H. Ernst, H. Sandmaier, R. Zengerle, T. Strobelt, *Sensors and Actuators A*, **103**, 225 (2003).
[63] https://www.pkvitality.com/ktrack-glucose/
[64] R. Karlsson, A. Michaelsson, L. Mattsson *Journal of Immunological Methods*, **145** 229 (1991).
[65] P. Gravesen, J. Branebjerg, O.S. Jensen, *J. Micromech. Microeng.*, **3**, 168 (1993).
[66] R. Zengerle, H. Sandmaier, 'Microfluidics in Europe', published online: 22 Aug. 2012, https://doi.org/10.2514/6.1997-1788 (1997).
[67] 1st International Conference on Micro Electro, Opto, Mechanic Systems and Components Berlin, 10-13 September 1990 (ed. by H. Reichel).
[68] F.C.M. Van de Pol, J. Branebjerg, 'Micro liquid-handling devices -a review', Micro System Technologies 90: 1st International Conference on Micro Electro, Opto, Mechanic Systems and Components Berlin, 10–13 September 1990. Springer Berlin Heidelberg, 1990.
[69] J. McDonald, G. Whitesides, *Angew. Chem. Int. Ed.*, **37** (1998).
[70] D. C. Duffy, J.C. McDonald, O. J. A. Schueller, G. M Whitesides, *Anal. Chem.*, **70**, 4974 (1998).
[71] Y. Xia, G. M. Whitesides, *Angew. Chem. Int. Ed.*, **37**, 550 (1998).
[72] E. Delamarche, H. Schmid, B. Michel, H. Biebuyck, *Adv. Mater.*, **9**, 9, 741 (1997).
[73] C. S. Effenhauser, G. J .M Bruin, A. Paulus, M. Ehrat, *Anal. Chem.*, **69**, 3451 (1997).
[74] T. M. Squires, S. R. Quake, *Rev. Mod. Phys.* **77**, 977 (2005).
[75] R. G. Blazej, P. Kumaresan, R.A. Mathies, *PNAS*, **17**, 175 (2006).
[76] C. Alix-Panabières, K. Pantel, *Nat. Rev. Cancer*, **14**, 623 (2014).
[77] M. Unger, H. P. Chou, T. Thorsen, A. Scherer, S. R. Quake, *Science*, **288**, 113 (2000).
[78] T. Thorsen, S. J. Maerkl, S. R. Quake, *Science*, **298**, 580 (2002).
[79] H. Song, J. D. Tice, R. F. Ismagilov, *Ang.Chem.*, **115**, 792 (2003).

[80] D. Thorpe, *Pharmacogenomics J*, **1**, 229 (2001).
[81] D.J. Beebe, J. S. Moore, J. M. Bauer, Q. Yu, R. H. Liu, C. Devadoss, B.H Jo, *Nature*, **404**, 588 (2000).
[82] A. Richter, G. Paschew, S. Klatt, J. Lienig, K. F. Arndt, H. J. P. Adler, *Sensors*, **8**, 561 (2008).
[83] M. G. Pollack, R. B. Fair, A. D. Shenderov, *Appl. Phys. Lett.*, **77**, 1725 (2000).
[84] A. S. Utada, E. Lorenceau, D.R. Link, P. D. Kaplan, H. A. Stone, D. A. Weitz, *Science*, **308**, 537 (2005).
[85] D. Lohse, X. Wang, *Reviews of Modern Physics*, **87**, 981 (2015).
[86] N. Pannacci, H. Bruus, D. Bartolo, I. Etchart, T. Lockhart, Y. Hennequin, H. Willaime, P. Tabeling, *Phys. Rev. Lett.*, **101** 164502 (2008).
[87] F. Mugele, J.C. Baret, *J. Phys.: Condens. Matter*, **17**, R705 (2005).
[88] W.C. Nelson, C.J. Kim, *Journal of Adhesion Science and Technology*, **26**, 1747 (2012).
[89] B. Berge, *C. R. Acad. Sci. II* , **317**, 157 (1993).
[90] B. Hadwen, G.R. Broder, D. Morganti, A. Jacobs, C. Brown, J. R. Hector, Y. Kubota, H. Morgan, *Lab Chip*, **12**, 3305 (2012).
[91] M. G. Pollack, R.B. Fair, A.D. Shenderov, *Appl. Phys. Lett.*, **77**, 1725 (2000).
[92] K. Choi, A. H.C. Ng, R. Fobel, A.R. Wheeler, *Annual Review of Analytical Chemistry*, **5**, 413 (2012).
[93] Y. Fouillet, D. Jary, A. G. Brachet, J. Berthier, R. Blervaque, et al., *ASME 4th International Conference on Nanochannels, Microchannels, and Minichannels*, Limerick, Ireland (2006).
[94] P. Dubois, G. Marchand, Y. Fouillet, J. Berthier, T. Douki, F. Hassine, S. Gmouh, M. Vaultier, *Anal. Chem.*, **78**, 4909 (2006,).
[95] . S.K . Cho, H. Moon, C. J. Kim, *Microelectromech. Syst.*, **12**, 70 (2003).
[96] M. Abdelgawad, A. R. Wheeler, *Advanced Materials*, **21**, 920 (2009).
[97] F. Sanger, S. Nicklen, A. R. Coulson, *Proc. Natl. Acad. Sci.*, **74**, 5463 (1977).
[98] A.T. Woolley, R. A. Mathies, *Proc. Nati. Acad. Sci.*, **91**, 11348 (1994).
[99] A. W. Moore, Jr., S. C. Jacobson, J.M. Ramsey, *Anal. Chem.*, **67**, 4184 (1995).
[100] F. Von Heeren, E. Verpoorte, A. Manz, W. Thormann, *Anal. Chem.*, **68**, 2044 (1996).
[101] S. Jacobson, C. Culbertson, J. Daler, J. Ramsey, *Anal. Chem.*, **70**, 3476 (1998).
[102] C. T. Culbertson, S. C. Jacobson, J. M. Ramsey, *Anal. Chem.*, **72**, 5814 (2000).
[103] S. Haeberle, R. Zengerle, *Lab on a Chip*, **7**, 1094 (2007).
[104] C. Venter, *Nucleic Acids Res.*, **35**, 6227 (2007).
[105] J. Han, H. G. Craighead, *Science*, **288**,1026 (2000).
[106] A. W. Martinez, S.T. Phillips, J. Manish, G. M. Whitesides, *Ang. Chemie (International Ed.)*, **46**, 1318 (2007).
[107] E. Carrilho, A. W. Martinez, G. M. Whitesides, *Anal. Chem.*, **81**, 7091 (2009).
[108] A. W. Martinez, S.T. Phillips, G. M. Whitesides, *Anal. Chem.* **82**, 3 (2010).
[109] https://www.microlight3d.com/applications/microfluidics
[110] Q. Wu, J. Liu, X. Wang, et al. *BioMed. Eng. OnLine*, **19**, 9 (2020).
[111] D. Huh, B. D. Matthews, A. Mammoto, M. Montoya-Zavala, H. Y. Hsin, D. E. Ingber, *Science*, **328**, 1662 (2010).

[112] K. F. Jensen, *Chemical Engineering Science*, **56**, 293 (2001).
[113] H. R. Sahoo, J. G. Kralj, K. F.Jensen, *Angew. Chem. Int. Ed.*, **46**, 5704 (2007).
[114] B. O. Dabbousi, J. Rodriguez-Viejo, F. V. Mikulec, J. R. Heine, H. Mattoussi, R. Ober, *The Journal of Physical Chemistry B*, **101**, 9463 (1997).
[115] V. Hessel, J. C. Schouten, A. Renken, J. I. Yoshida (eds.), *Micro Process Engineering: A Comprehensive Handbook*, Wiley-VCH (2009).
[116] https://www.magnamrc.com/industry-reports/global-microreactors-technology-market-databank
[117] H. G. Craighead, *Science*, **290**, 1532 (2000).
[118] G. Kit Island, G. Stemme, *Sensors and Actuators*, **A21**, 904 (1990).
[119] S. A. Gajar, M. W. Geis *J. Electrochem. Soc.*, **139**, 2833 (1992).
[120] A. W. Rucker, *Nature*, p. 331 (1877).
[121] A. Shaludko, *Adv. Coll. Interfaces*,**1**, 391 (1967).
[122] P–G. de Gennes, F. Brochard-Wyard, D. Quere, *Capillarity and Wetting Phenomena: Drops, Bubbles, Pearls, Waves*, Springer (2004).
[123] J. Israelachvili, *Intermolecular and Surfaces Forces*, Academic Press, 2nd edn (1991).
[124] W. D. Volkmuth, T. Duke, M. C. Wu, A. Szabo, R. H. Austin *Phys. Rev. Lett.*, **72**, 2117 (1994).
[125] J. B. Knight, A. Vishwanath, J. P. Brody, R.H. Austin *Phys. Rev. Lett.*, **80** , 3863 (1998).
[126] J. C. T. Eijkel, A. van den Berg, *Microfluid Nanofluid*, **1**, 249 (2005).
[127] J. Han, S. W. Turner, H. G. Craighead, *Phys. Rev. Lett.*, **83**, (1999).
[128] J. Han, H. G. Craighead, *Science*, **288**, 1026 (2000).
[129] G. Hummer, J. C. Rasaiah, J. Noworyta, *Nature*, **405**, 188 (2001).
[130] S. Gravelle, C. Ybert, L. Bocquet, L. Joly, *Phys. Rev. E*, **93**, 33123 (2016).
[131] J. A. Thomas, A. J. McGaughey, *Phys. Rev. Lett.*, **102**, 1 (2009).
[132] A. Kalra, S. Garde, *PNAS*, **100** , 10177 (2003).
[133] W. Humphrey, A. Dalke, K. Schulten, *J.Mole.Graph*, **14**, 33 (1996).
[134] E. Secchi, S. Marbach, A. Nigues, D. Stein, A. Siria, L. Bocquet, *Nature*, **537**, 210, (2016).
[135] M. Moseler, U. Landman, *Science*, **269**, 1165 (2000).
[136] J. L . Barrat, L. Bocquet *Phys. Rev. Lett.*, **82**, 4671 (1999).
[137] Q. Pu, J. Yun, H. Temkin, S. Liu, *Nano Letters*, **4**, 1099 (2004).
[138] H. Daiguji, P. Yang, A. Majumdar, *Nano Lett*, **4**, 137 (2004).
[139] N. Kaji, Y. Tezuka, Y. Takamura, M. Ueda, T. Nishimoto, H. Nakanishi, Y. Horiike, Y. Baba, *Anal. Chem*, **76** (2004).
[140] B. J. Hinds, N. Chopra, T. Rantell, R. Andrews, V. Gavalas, L. G. Bachas, *Science*, **303**, 62 (2004).
[141] M. H. O. Rashid, S. F. Ralph, *Nanomaterials*, **7**, 99 (2017).
[142] D. Mattia, H. Leese, K. P. Lee, *Journal of Membrane Science*, **475**, 266 (2015).
[143] N. R. Tas, P. Mela, T. Kramer, J. W. Berenschot, A. van den Berg, *NanoLetters*, **3**, 1537 (2003).
[144] E. Lauga, M. P. Brenner, H. A. Stone, 'Microfluidics: The no-slip boundary condition', arXiv preprint cond-mat/0501557 (2005)

[145] J.O. Tegenfeldt, C. Prinz, H. Cao, R.L. Huang, R. H. Austin, S.Y. Chou, E. C. Cox, J. C. Sturm, *Anal Bioanal Chem*, **378**, 1678 (2004).
[146] L. Bocquet, E. Charlaix, *Chem. Soc. Rev.* **39**, 1073 (2010).
[147] D .G. Haywood, A. Saha-Shah, L. A. Baker, S. C. Jacobson, *Anal.Chem*, **87**, 172 (2015).
[148] L. Bocquet, P. Tabeling, *Lab Chip*, **14**, 3143 (2014).
[149] L. Bocquet, *Nature Materials*, **19**, 254 (2020).
[150] L. Bocquet, *Ann. Rev. Fluid. Mech.*, **53**, 377 (2021).
[151] T. Mouterde, A. Keerthi, A. R. Poggioli, S. A. Dar, A. Siria, A. K. Geim, L. Bocquet, B. Radha *Nature*, **567**, 87 (2019).
[152] K. V. Agrawal, S. Shimizu, L. W. Drahushuk, D. Kilcoyne, M. S. Strano, *Nat Nanotechnol.*, **12**, 267 (2017).
[153] Yole Development, https://www.yole.com.
[154] D. J. Laser, J. G. Santiago, *J. Micromech. Microeng.*, **14**, R35, (2004).
[155] D. Cressey, cover image of the month, *Nature*, 28 March (2014).
[156] A. van der Bos, T. Segers, R. Jeurissen, M. van den Berg, H. Reinten, et al., *J. Appl. Phys.*, **110**, 034503 (2011).
[157] D. Lohse, *Annu. Rev. Fluid Mech.*, **54**, 349 (2022).
[158] MIT Technology Review, https://www.technologyreview.com/10-breakthrough-technologies/2001/ (February 2001).
[159] N. Bohr, *On the Constitution of Atoms and Molecules, The London, Edinburgh, and Dublin Philosophical Magazine and Journal of Science*, **26**, 1, (1913).
[160] 'Microfluidics in Late Adolescence', *Nobel Symposium 162*, Stockholm, Sweden, arXiv:1712.08369v1, (2017);
[161] S. Battat, D. A. Weitz, G .M. Whitesides, *Lab Chip*, **22**, 530 (2022).
[162] N. S. Oliver, C. Toumazou, A. E. G. Cass, D. G. Johnston, *Diabetic Medicine*, **26** ,197 (2009).
[163] D. E. Stokes, *Pasteur's Quadrant -Basic Science and Technological Innovation*, Brookings (1997).
[164] J. M. Dudley, *Nature Photonics*, **7**, 339 (2013).
[165] A. M. Ganan-Calvo, *Phys.Rev.Lett.*, **80**, 285 (1998).
[166] A. M. Ganan-Calvo, J.M. Gordillo, *Phys. Rev. Lett.*, **87**, 274501 (2001).
[167] T. Thorsen, R. W. Roberts, F. H. Arnold, S. R. Quake, *Phys. Rev. Lett.*, **86**, 4163 (2001).
[168] E. Delamarche, A. Bernard, H. Schmid, B. Michel, H. Biebuyck, *Science*, **276**, 779 (1997).
[169] R. Zengerle, J. Ducrée, *The Future of Microfluidics: Low-Cost Technologies and Microfluidic Platforms* (2007).
[170] P. Yager, T. Edwards, E. Fu, K. Helton, K. Nelson, M. R. Tam, B. H. Weigl, *Nature*, **442**, 412 (2006).
[171] T. Notomi, H. Okayama, H. Masubuchi, T. Yonekawa, K. Watanabe, N. Amino, T. Hase, *Nucleic Acids Res*, **15**, E63 (2000).
[172] J. T. Connelly, P. Jason, P. Rolland, G. M. Whitesides. *Anal. Chem.*, **87**, 7595 (2015).
[173] P. Garneret, E. Coz, E. Martin, J–C Manuguerra, E. Brient-Litzler, V. Enouf,

D. Felipe González Obando, J–C. Olivo-Marin, F. Monti, S. van der Werf, J. Vanhomwegen, P. Tabeling, *PLoS One*, **16** e0243712 (2021).
[174] E. Coz, P. Garneret, E. Martin, D. Freitas do Nascimento, A. Vilquin, D. Hoinard, M. Feher, Q. Grassin, J. Vanhomwegen, J-C. Manuguerra, S. Mukherjee, J-C. Olivo-Marin, E. Brient-Litzler, M. Merzoug, P. Tabeling, medRxiv (2021/1/1), https://doi.org/10.1101/2021.10.03.21264480 .
[175] O. Couture, M. Faivre, N. Pannacci, A. Babataheri, V. Servois, P. Tabeling, M. Tanter *Medical Physics*, **38**, 1116 (2011).
[176] I–U. Khan, C. A. Serra, N. Anton, T. Vandamme, *Journal of Controlled Release*, **172**, 1065 (2013).
[177] R. Cross, *Chem. Eng. News*, **99**, 16 (2021).
[178] H. Noguchi, G. Gompper *J. Chem. Phys.*, **125**, 164908 (2006).
[179] L.-Y. Chu, A. S. Utada, R. K. Shah, J.-W. Kim, D. A. Weitz, *Angew. Chem., Int. Ed.*, **46**, 8970 (2007).
[180] B Shen, J Ricouvier, F. Malloggi, P Tabeling *Advanced Science*, **3** (6), 1600012 (2016).
[181] I. Fouxon, B. Rubinstein, Z. Ge, L. Brandt, A. Leshansky, *Physicsl Review Fluids*, **5**, 54101 (2020).
[182] SY Kashani, A. Afzalian, F. Shirinichi, M.K. Moraveji, *RSC Adv.*, **11**, 229 (2021).
[183] S. Andrieux, A. Quell, C. Stubenrauch, W. Drenckhan, *Adv. Colloid Interface Sci.* , **256**, 276 (2018).
[184] J Ricouvier, P Tabeling, P. Yazhgur, *Proceedings of the National Academy of Sciences*, **116**, 9202 (2019).
[185] I. Maimouni, M. Morvaridi, M. Russo, G. Lui, K. Morozov, J. Cossy, M. Florescu, M. Labousse, P. Tabeling, *ACS Applied Materials & Interfaces*, **12**, 32061 (2020).
[186] M. A. Klatt, P. J. Steinhardt, S. Torquato, *Proc. Natl. Acad. Sci.*, **116**, 23480 (2019).
[187] V. Hessel Ed., *Design and Engineering of Microreactor and Smart-Scaled Flow Processes.*, Multidisciplinary Digital Publishing Institute (MDPI) (2015).
[188] G. Velve-Casquillas, M. Le Berr, M. Piel, P. T. Tran, *Nanotoday*, **5**, 28 (2010).
[189] T. Stuart, S. Rahul, *Nature Reviews Genetics*, **20**, 257 (2019).
[190] L. d'Eramo, B. Chollet, M. Leman, E. Martwong, M. Li, H. Geisler, J. Dupire, M.. Kerdraon, C. Vergne, F. Monti, Y. Tran, P. Tabeling, *Microsystems & Nanoengineering*, **4**, 1 (2018).
[191] https://www.espci.psl.eu/fr/innovation/creation-d-entreprises/minos-bioscience.
[192] https://www.10xgenomics.com.
[193] R. Sjoberg, A. A. Leyrat, D. M. Pirone, C. S. Chen, S. R. Quake, *Anal. Chem.*, **79**, 8557 (2007).
[194] A. D. Castiaux, D. M. Spence, R. S. Martin, *Analytical Methods*, **11**, 4220 (2019).
[195] V. Jossen, C. van den Bos, R. Eibl, D. Eibl *Applied Microbiology and Biotechnology*, **102**, 3981 (2018).
[196] R.P. Harrison, N. Medcalf, Q.A .Rafiq, *Regenerative medicine*, **13**, 159 (2018).
[197] G. M. Whitesides, B. Grzybowski, *Science*, **295**, 2418 (2002).

[198] A. Siria, M. L. Bocquet, L. Bocquet, *Nat. Rev. Chem.*, **1**, 0091 (2017).
[199] L. Wang, Z. Wang, S. K. Patel, S. Lin, M. Elimelech, *ACS Nano*, **15**, 4093 (2021).
[200] G. Laucirica, M. E. Toimil-Molares, C.Trautmann, W. Marmisoll, O. Azzaron, *Chem. Sci.*, **12**, 12874 (2021).
[201] https://www.capsum.net/en/
[202] https://www.microfactory.eu
[203] G. M. Church, Y. Gao, S. Kosuri, *Science*, **337**, 1628.(2012).
[204] M. Mann, C. Kumar, W. F Zeng, M. T. Strauss, *Cell Systems*, **12**, 759 (2021).
[205] E. A. Galan, H. Zhao, X. Wang, Q. Dai, W. T. Huck, S. Ma, *Matter*, **3**, 1893 (2020).
[206] G. Karniadakis, A. Beskok, *Micro Flows*, Springer Verlag 2002.
[207] N. T Nguyen, S. Wereley, *Fundamental and Applications of Microfluidics*, Artech House (2003).
[208] P. Tabeling, *Introduction à la microfluidique*, Belin (2003).
[209] P. Tabeling, *Introduction to Microfluidics*, Oxfore University Press (2005).
[210] *Microfluidique*, Traité Egem, Hermès, S. Colin (ed) (2004).
[211] H. Bruus *Theoretical Microfluidics*, Oxford University Press (2007).
[212] B. J Kirby, *Micro and Nanoscale Fluid Mechanics : Transport in Microfluidic Devices*, Cambridge University Press (2010).
[213] *The MEMS Handbook*, 2nd edition, CRC Press (2005).
[214] J. Berthier, P.Silberzan *Microfluidics for Bioengineering* Artech House Publishers (2009).
[215] J. Berthier *Micro-Drops and Digital Microfluidics* (2010).
[216] S. Seiffert, J. Thiele, *Theory and Practice for Beginners*, de Gruyter Textbook (2019).
[217] A. Folch, *Introduction to BioMEMS*, CRC Press (2016).
[218] B. Giri, *Laboratory Methods in Microfluidics* (2017).
[219] A. Manz, P. Neužil, J.S. O'Connor, G. Simone *Microfluidics and Lab On a Chip* (2020).
[220] M. Gad El Hak, *Journal of Fluid Engineering*, **121**, 5 (1999).
[221] N. Giordano, J.T. Cheng, *J. Phys: Condens Matter*, **13**, R271 (2001).
[222] H. Stone, S. Kim, *AiChE Journal*, **41**, 1250 (2001).
[223] D. Beebe, G. Mensing, G. Walker, *Ann. Rev. Biomed. Eng.*, **4**, 261 (2002).
[224] G. Sanders, A. Manz, *Trends Anal. Chem.*, **19**, 364 (2000).
[225] E. Verpoorte, *Electrophoresis*, **23**, 677 (2002).
[226] J. Lichtenberg, N. de Rooif, E. Verpoorte, *Talanta*, **56**, 233 (2002).
[227] P. A. Auroux, D. Iossifids, D. Reyes, A. Manz, *Anal. Chem.*, **74**, 2637 (2002).
[228] H.A. Stone, A.D. Stroock, A. Ajdari, *Ann. Rev. Fluid. Mech.*, **36**, 381 (2004).
[229] W. Sparreboom, A. van den Berg, JCT Eijkel, *New J. Physics*, **12**, 15004 (2010).

2

Physics at the microscale

2.1 The scales of small things

2.1.1 The system of units for small quantities

Let us start with a question of vocabulary. The names used for describing small quantities are shown in the following table:

Name	Scale (m)
Milli	10^{-3}
Micro	10^{-6}
Nano	10^{-9}
Pico	10^{-12}
Femto	10^{-15}
Atto	10^{-18}
Zepto	10^{-21}
Yocto	10^{-24}

Table 2.1 Names of the small units.

These small units are frequently used, a few examples are presented below. Volumes contained in microfluidic systems commonly range from about 10 to a few hundred nanolitres. A biological cell has a size of 10 μm and it includes a volume of a few pL. To take an example pertaining to the early days, Ref. [1] demonstrated a detection threshold of 15 amole (i.e. 15 atto-mole) by coupling a microfluidic system to a mass spectrometer. A zeptomole is around 600 molecules (often employed), and a yoctomole 0.6 molecules (rarely used).

2.1.2 Sizes of ions and small molecules

How can one measure the size of an ion or a small molecule, i.e. a molecule including only a few atoms? Many techniques exist: X-ray, neutron diffraction, gas state equation, gas solubility, viscosity, self-diffusion measurements etc. An important notion is the Van der Waals radius, obtained by using, for gases, the notion of Van der Waals

Ion	Radius (nm)	Ion/molec.	Radius (nm)
Li^+	0.068	F^-	0.136
Na^+	0.095	OH^-	0.176
K^+	0.133	O^{2-}	0.15
Fe^{2+}	0.076	CH_4	0.20
Ca^{2+}	0.099	H_2O	0.17
Ar	0.188	He	0.2

Table 2.2 Van der Waals radii of some ions and molecules

excluded volume. The corresponding Van der Waals sphere is visualized as the smallest sphere encapsulating the molecule. This notion applies to asymmetric molecules, such as water. Some values of the Van der Waals radius are given in Table 2.2. For instance, the Van der Waals radius of helium is around 0.2 nm. Water is 0.28 nm and methane (CH_4) 0.4 nm. Sizes of the simplest molecules are typically a fraction of a nanometre.

2.1.3 Complex molecules, DNA, proteins

Many molecules include a moderate number of atoms, and possess limited functionalities, if none. Examples are alkanes, peptides, sugar, and glucose. Macromolecules, by definition, are much larger: they include between a few hundreds and millions of atoms. Macromolecules are also called 'complex molecules'. Their behaviour is exceedingly rich, and they may possess astonishing functionalities (for instance, aquaporines, that we saw in the introduction, which filter water at the single molecule level, or antibodies which bind to antigens in a highly specific manner). Depending on the solvent in which they are immersed and the temperature, macromolecules can adopt different conformations: helices, globules, sheets, stretched structures, etc. Examples are shown in Fig. 2.1.

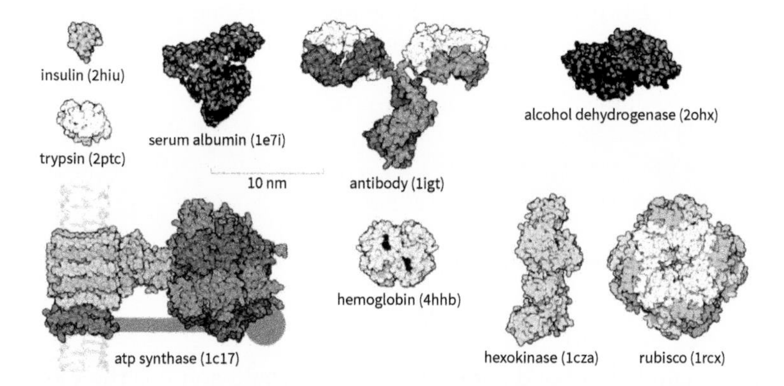

Fig. 2.1: Morphologies and sizes of a selection of proteins: insulin, trypsin, albumin, antibody, deshydrogenase, synthase, haemoglobin, hexokinase, and rubisco. (From David Goodsell, under public licence.)

Different scales must obviously be used to characterize the geometry of these molecules. An important notion is the radius of gyration R_g, defined as the standard type deviation of the atom distances from the molecule centre (defined at the mean position of the atoms). Should atoms be homogeneously distributed inside the volume of a sphere of radius R, R_g would be equal to $\sqrt{(\frac{3}{5})}R$. The radius of gyration thus gives an idea of the size of the molecule, as long as its shape is not too far from a sphere. One advantage of this notion is that it can be directly related to small angle X-ray scattering(SAXS) measurements.

The hydrodynamic radius R_H is another estimate, which, for large molecules, can be inferred from dynamical light scattering measurements. Theory provides the following relation $Rg = \sqrt{(\frac{3}{5})}R_H$. There is a relation between these quantities and the molecular weight M_w. The formula is $R_G \approx 0.7M_w^{0.37}$ (with M_w expressed in kDa [2]. Table 2.3 gives a few gyration radii of proteins, including those of Fig. 2.1.

Molecule	Contour length	r_G
Albumin		2.78 nm
Insulin		1.16 nm
Hemoglobin		23.8 nm
(Bovin) b-Trypsin		1.8 nm
Hexokinase		3.2 nm
ATP synthase		5 nm
Human chromosome	5 cm	1 μm
Human genome	2 m	
E. Coli DNA	1.5 mm	100 nm

Table 2.3 Geometric characteristics of some macromolecules.

For polymers, such as the plasmid shown in Fig. 2.2 [3], another dimension is needed: the contour length L_c. It is the length of the macromolecule measured along its backbone. For a polymer made up of N monomers, each separated by a distance l, we have:

$$L_c = Nl$$

An important relation, established by Flory [4] in 1953, is the following:

$$R_g \sim N^{\nu},$$

where N is the number of bond segments along the chain and ν is the Flory exponent. ν is equal to $1/3$ in poor solvents and $3/5$ in good solvents. Another important length, based on mechanics, is the persistence length, which tells us something about the chain stiffness. For polymers shorter than the persistence length, the molecule behaves like a rigid rod, while for longer polymers, the chain can bend and the molecule looks

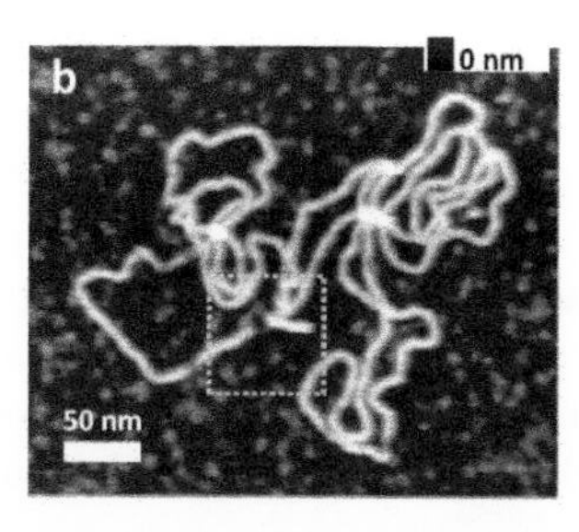

2.2: DNA molecule imaged with AFM. (Reprinted with permission from [3]. Copyright 2022 American Chemical Society.)

like the trajectory of a three-dimensional random walker. To provide an example, a strand of uncooked spaghetti has a persistent length of 10^{18} m [5], a strand of cooked spaghetti 10 cm, and a double helical DNA 39 nm. Another length, the Kuhn length, provides information on the distance within which molecules are correlated. Decades of work have been dedicated to studying polymer conformation, and we will not go further into the subject here.

2.1.4 Visualization of DNA strands in a microfluidic device.

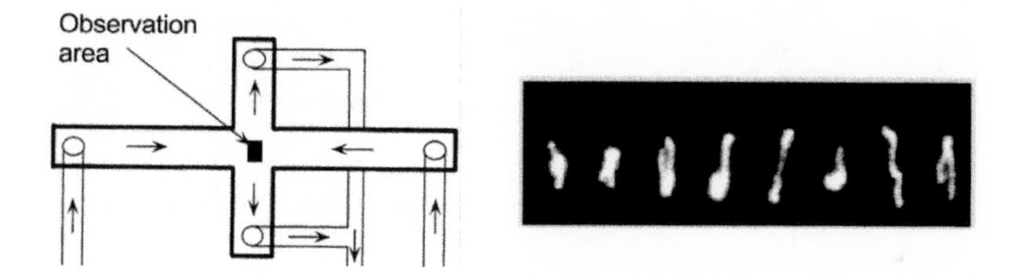

Fig. 2.3: (Left) Microfluidic device used by [6], in which a functionalized PS bead is placed at the centre of a cross-flow, at the stagnation point (black rectangle). (Right) Different conformations of a DNA molecule tethered to the bead, subjected to a strain of 1 s^{-1}. The DNA strand has a contour length of 22 μm. (Photo published with the permission of D.E. Smith et al, *Science*, 283, 1999.) (Copyright 2003. American Association for the Advancement of Science.).

In the 1990s, S. Chu et al [6] imaged the fluorescent labeled DNA of a bacteriophage λ, subjected to the action of a straining field. The work[1] prompted the development of the field of single molecule study. In S. Chu's experiments, the molecule is tethered to a polystyren bead, and the bead is placed at the intersection of two microchannels etched in silicon (Fig. 2.3, left)). Figure 2.3 (right) shows different conformations of the DNA obtained in this type of system. The experiment revealed the considerable richness of the conformational dynamics of DNA, and, in the meantime, demonstrated that with microfluidic technology, it was possible to manipulate isolated molecules.

[1] It was followed by a number of realizations, such as [7].

2.1.5 Viruses, bacteria and cells

Virus sizes are on the order of 100 nm, and (prokaryotic) bacteria 1-2 μm. The sizes of eukaryotic - i.e. possessing a nucleus - animal cells are much larger. They range between 10 and 30 μm. There exist elongated eukaryotic cells (neurons), eukaryotic cells having lost their nucleus (red blood cells), but here, we concentrate on the vast majority of the 3 10^{14} cells forming the human body, i.e. eukaryotic cells of spherical shape, possessing a nucleus. A scheme of the principal cellular elements of these cells is shown in Fig. 2.4 [8].

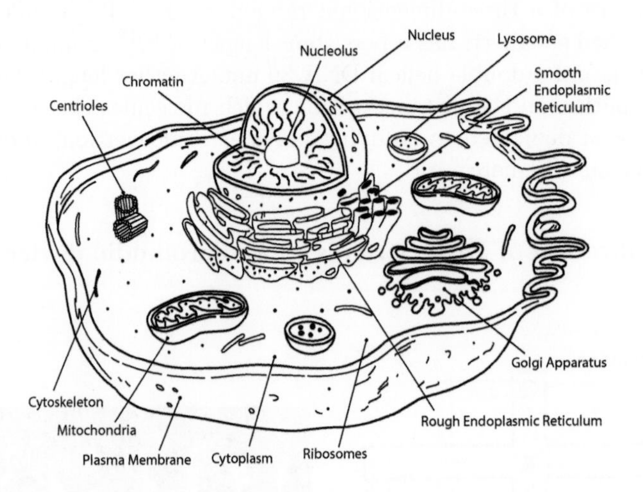

Fig. 2.4: Eukariotic cell, in which several elements are outlined

The cell includes different elements, which would need thick books with titles, such as *The Cell* [8] to describe them. Here we content ourselves to mention basic entities:

- a nucleus of size between 3 and 10 μm. The nucleus itself is a complex object that contains, among other elements, DNA folded into the form of a chromosome. Chromosomes are objects of micrometric size;
- mitochondria, which are nucleated cells of micrometric size. Mitochondria have the cellular function of supplying energy to the cell;[2]
- the endoplasmic reticulum, a collection of vesicles across which proteins are fabricated and transported. The vesicles of the reticulum are micrometric objects;
- the cytoskeleton, containing actin, tubulin, and intermediate filaments. These objects are between 10 and 30 nm in diameter, and a few micrometer in length. They ensure the mechanical cohesion of the cell and
- the membrane, which has a thickness of a few tens of nanometres.

[2]Mitochondria are reservoirs of ADP (acid diphosphate); the transformation of ADP to ATP (acid triphosphate), outside the mitonchondria, supplies the cell with energy.

Table 2.4 presents the size of cellular elements.

Element	Size
Nucleus	4 μm
Chromosome	3 - 5 μm
Ribosome	20 - 30 nm
Membrane thickness	10 nm
Microtubule diameter	25 nm
Lysosome	200 nm
Mitochondria	3 μm

Table 2.4 Sizes of cell components.

2.1.6 Single cells manipulated in microfluidic droplets

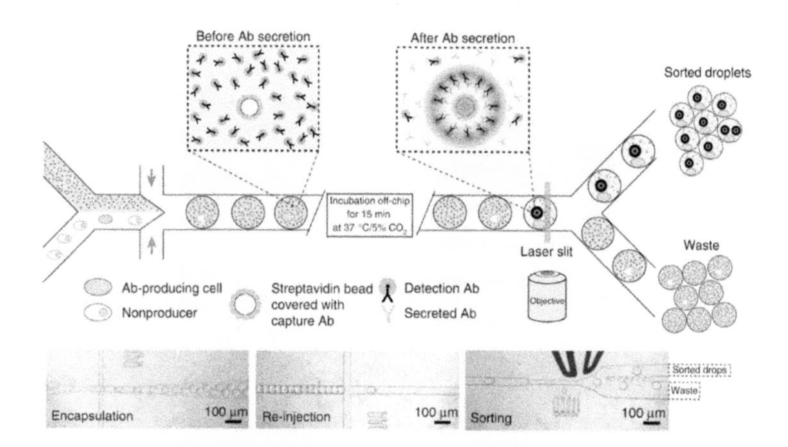

Fig. 2.5: Microfluidic device enabling the production of cells optimized for the secretion of antibodies. Each cell is analysed, processed, selected, at individual level. The device mimicks Darwin selection, but at a much faster rate. (Reprinted with permission from [10]. Copyright 2022. Springer Nature.).

Since the demonstration, in 2002 [9], that clone cells, i.e. possessing the same genome, those can express different proteins, in different amounts and in a random manner, it came as evidence that noise in the expression should be seriously taken into consideration to understand the behaviour of living systems. The paper prompted the emergence of a new domain, called 'single cell', for which microfluidics offered a powerful toolbox. Figure 2.5 shows one of the most advanced systems in the domain [10].

Cells are encapsulated in droplets (see Fig. 2.5, left). The droplet sizes, 100 μm, is suitable for hosting them. Cells, as they travel in their vehicle, may or not secrete antibodies. Depending on their behavior, after incubation, they are sorted, the best performing ones are reinjected for further analysis [10]. The device mimics Darwin selection, but at a much faster rate (hours instead of millions of years).

2.1.7 Multicellular organisms

An example, taken from the billions of multicellular organisms existing on earth, and justified by its importance in neurobiology along with its link with microfluidics, is *C. Elegans*. It is one of the most studied multicellular organisms. It was the first to have its whole genome sequenced. The adult male has exactly 2,015 cells, all documented phenotypically, genetically, transcriptionally, geographically and functionnally. Its basic anatomy includes a mouth, pharynx, intestine, gonad, and collagenous cuticle. It has neither a circulatory nor a respiratory system. The life of the worm does not look exciting: it solely consists of eating, moving, sensing, and reproducing.

The worm is on the order of one millimetre long and two hundred micrometres wide. It can easily be introduced in microfluidic devices to study its behaviour. Several experiments can be performed in parallel, as was done in Refs. [11, 13]. An example, extracted from Refs. [12, 13], is shown in Fig. 2.6.

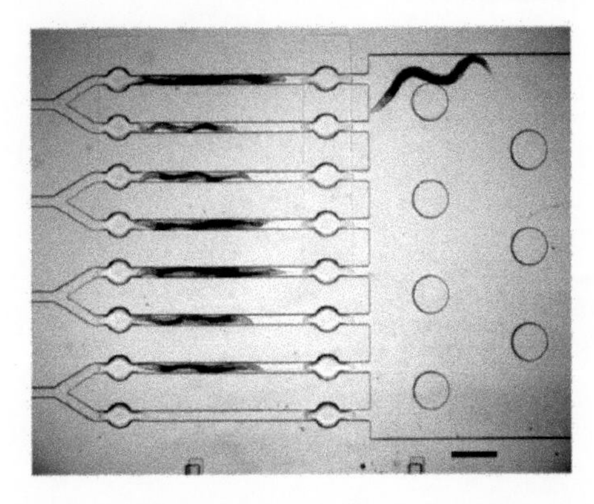

Fig. 2.6: *C. elegans* in a microfluidic device, designed for performing several experiments in parallel. (Reprinted with permission from [13]. Copyright 2022. Elsevier.).

The goal of these experiments was methodological: it offered a platform for screening the response of *C. elegans* to stimuli.

2.1.8 Mean free path in gases.

Let us discuss intermolecular distances in gases. In gases, contrarily to liquids, the intermolecular distances are much larger than molecule sizes. A geometrical relation relates the density n (i.e. the number of molecules per unit of volume) to the intermolecular distance d:

$$d = n^{-1/3}$$

Gas	λ(nm)	τ(ns)
Air	61	0.18
Nitrogen	60	0.17
Argon	64	0.2
Helium	177	0.18

Table 2.5 Mean free paths and collision times in different gases.

Avogadro's law stipulates that, at fixed temperature and pressure, all gases contain the same number of molecules per unit volume At ordinary pressure, and at a temperature of 273 K, this number is $2.69 \times 10^{19}\text{cm}^{-3}$. For nitrogen, taking a molecule diameter a equal to 3 Å, we obtain a value of d equal, approximately, to 3 nm, then ten times a (see Fig. 2.7).

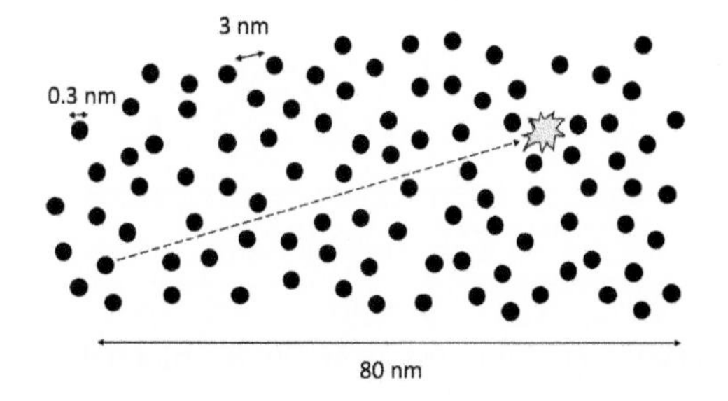

2.7: A schematic view of the mean free path for air in normal conditions. In this case, λ is approximately 60 nm. The molecule travels this distance before hitting a partner

However, for gases, the fundamental scale involved in the establishment of the flow equations is not d, nor a, but the mean free path λ. The reason is that λ is the average distance travelled by a molecule between two successive collisions and, in gases, momentum is exchanged through collisions. Geometrical considerations, coupled to statistics, provide the following expression:

$$\lambda = \frac{1}{\sqrt{2}\pi n a^2} = \frac{kT}{\sqrt{2}\pi p a^2} \tag{2.1}$$

where T is the temperature, k is the Boltzmann constant, and p is the pressure.

Using the formula, and again for nitrogen, and in normal conditions, one finds a mean free path of around 60 nm, This is twenty times the intermolecular distance d. Therefore, a molecule succeeds in avoiding about 95% of their partners before hitting a colleague, like a good rugby player (see Fig2.7.).

Several values of λ along with collision times τ (i.e. the mean duration of the molecule journey between two successive collisions) are given, in normal conditions, in Table 2.5 [14].[3]

[3]The values shown in the table, obtained from viscosity measurements, have been taken from [14]. Note that they vary by 10%, from one publication to the other

Mean free paths for Nitrogen, Argon and air are similar because molecular sizes are close. In the case of Helium, owing to its small size and efficient collision cross-sections, the mean free paths are substantial .

2.1.9 Correlation length in liquids.

Three scales characterize a liquid: the molecular radius, the intermolecular distance and the correlation length (see Fig. 2.8.).

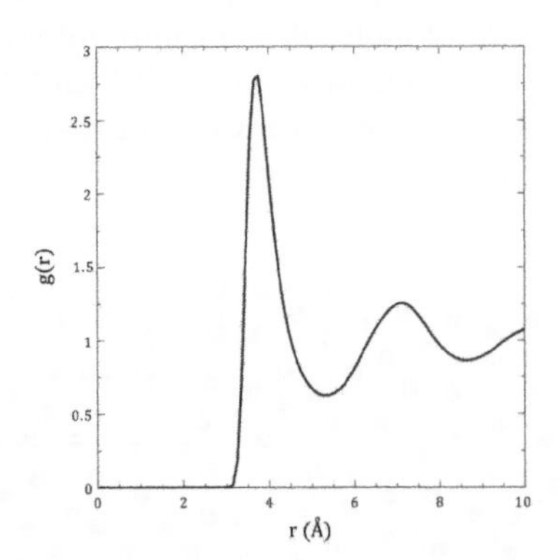

2.8: Correlation function $g(r)$ for liquid argon at 90 K.

Fig. 2.8 shows the pair correlation function $g(r)$, for argon at 90 K, plotted as a function of the distance r from a molecule center, located at $r = 0$. $g(r)$ represents the probability of finding another dense entity at distance r from the centre. The molecule radius is $r \approx 0.2$ nm. Above this value, $g(r)$ increases, up to $r \approx 0.4$ nm, where it reaches a maximum. This maximum marks the averaged position of the nearest neighbor. We can infer that the intermolecular distance is ≈ 0.38 nm. $g(r)$ oscillates, but its amplitude decays with r. It is usually considered that the correlation length l_c corresponds to the value of r for which the decay has reached 10%. Then, in the case of Argon, we find $l_c \approx 1$ nm. A similar order of magnitude hold for current liquids, such as water and oil. The physical interpretation of l_c is the length below which molecules fluctuations remain correlated. This length will play an important role in the next chapter, for the definition the fluid particle size.

2.1.10 Other scales

A few other scales often play an important role in the behaviour of small systems. Some are illustrated in Fig. 2.9.

- **Nucleation length**. In homogeneous nucleation theory, molecules, subjected to Brownian fluctuations, must form a cluster of radius larger than a critical value to initiate crystallization. Nucleation length is twice this radius. The critical radius is on the order of a few nanometres. It results from a competition between surface and bulk energies. An Arrhenius-type statistic describes the nucleation probability. In practice, this statistic holds, but nucleation is most often heterogeneous,

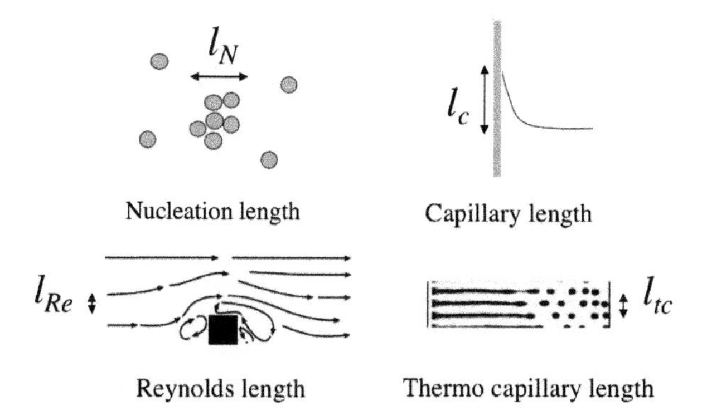

Fig. 2.9: Other scales: nucleation length l_N (twice the nucleation radius), capillary length l_C, Reynolds length l_{Re} and thermo-capillary length l_{CT}.

i.e. controlled by impurities. These features, known for a long time [16], were confirmed in a microfluidic system performing hundreds of nucleation experiments in parallel [17].

As the system size decreases, it becomes less probable to nucleate a supercritical cluster. This explains why, in small systems, water can remain liquid at deeply negative temperature. Altocumulus clouds, a common type of mid-altitude cloud, are mostly composed of water droplets supercooled to a temperature of about -15°C. These droplets often freeze on the airplane wings. In Ref. [18] (1953)), it was found that, in droplets of 10 μm in diameter, the mean freezing temperature of water was -38°C. The phenomenon was unsurprisingly re-observed in microchannels [29]. In extreme confinements, such as sub 2 nm pores, ice never forms [15].

- **Capillary length** Capillary length results from the competition between gravity and capillarity [19]. Its expression is $l_c = \sqrt{\frac{\gamma}{\rho g}}$ and its order of magnitude is one millimetre. As sketched in Fig. 2.9, l_c provides an estimate for the meniscus height developing along a wetting wall. We will return to this subject in Chap. 3. Compared to microchannel dimensions, l_c is large, so that, in microfluidics, gravity is often neglected in favour of capillarity.

- **Reynolds length.** As will be seen in this chapter, and in Chap. 3, the Reynolds number is equal to $\frac{Ul}{\nu}$, where U is a characteristic speed, l is the system size, and ν is the kinematic viscosity. The Reynolds length that we introduce here, is another manner of writing the Reynolds number.[4] It is a manner to keep concentrated on scales. We thus define the Reynolds length as $l_{Re} = \frac{\nu}{U}$. With this definition, $Re =$

[4]The Reynolds length we define here has the same expression as the Oseen length. Oseen length concerns a very specific scale: for flows around obstacles, it tells at which distance Stokes equations become inaccurate. This is not what we are interested in. In order to avoid confusion, we give this scale a different name

$\frac{l}{l_{Re}}$ and $l_{Re} = \frac{l}{Re}$. When the system is much larger than l_{Re}, the Reynolds number is large and turbulence develops. Small l_{Re} (again compared to the system size) correspond to small Reynolds and thus creeping flows. This feature is illustrated in Fig. 2.9. In this case, the Reynolds length is smaller than the obstacle size. Consequently, hydrodynamic instabilities leading to Karman vortices, typical of large Reynolds number behaviour, develop. In terms of orders of magnitude, with U equal to 1 mm/s, and for water, l_{Re} is close to one millimetre.

- **Thermo capillary length**: We will see below that the cost of creating a new interface, per unit of area, is equal to γ, the surface energy. Can we imagine a situation for which this energy becomes comparable to thermal energy? The ratio between them is given by $\frac{\gamma l^2}{kT}$, where l is the system size. This ratio is equal to unity for $l = l_{cT} = \sqrt{\frac{kT}{\gamma}}$. The order of magnitude of l_{cT} is a few nanometers. As shown in Ref. [20], for jets of diameters smaller than l_{cT}, droplets become substantially affected by noise l_{cT} represent the cross-over between the two regimes. It also represents the droplet size at this cross-over, as illustrated in Fig. 2.9. For larger jets, thermal noise can be neglected.[5]

2.2 Intermolecular Forces - basics

2.2.1 Forces between molecules

Non-polar molecules in vacuum: Lennard-Jones potential. Most of the material shown here is extracted from J. Israelachvili's book titled *Intermolecular and Surfaces Forces* [21].

Let us start with the forces existing between two neutral non-polar molecules, i.e. whose barycentres of positive and negative charges overlap. When placed in a vacuum, these molecules develop forces, and thus interaction energies, made up of two contributions.

- The first contribution is called 'short range', because, in all cases, the energy becomes negligible beyond a few Angstroms. This energy has a quantum-mechanical origin that can be either attractive or repulsive. In the attractive case, the force gives rise to the formation of a covalent bond. The force in the repulsive case is known as hard-sphere repulsion, and it ensures the interpenetrability of atoms. We will focus on the latter case.
- The second contribution is Van der Waals attraction. The term is called long range, because, when molecules assemble, for instance, on a surface, the energies of attraction that they develop decay with an exponent smaller than the dimension of space. In such geometries, one can observe the action of the Van der Waals term, at distances on the order of ten nanometres. In the case of a pair of

[5]Not completely: thermal noise is necessary for initiating the instability that breaks the jet into droplets, but does not play a role in the determination of their characteristics, i. their shapes and their sizes

molecules, this 'long range' reduces to one nanometre or so, but this is enough to affect, in a crystal, the second and third nearest neighbours. As the molecule is neutral and non-polar, the existence of a permanent attraction seems strange. So, let us consider the Bohr model of the atom: the pair electron/proton produces an instantaneous dipole. This dipole generates an electric field that deforms the electronic cloud of a neighbouring atom. An induced dipole will thus form. This dipole will in turn produce an electric field affecting the first dipole. This type of interaction is called dipole-induced dipole'. It is attractive. Over the course of time, atoms and their electron clouds fluctuate in position and in orientation, but the attractive interaction remains. The reasoning, held for the Bohr atom, also applies for molecules. The amplitude of the interaction energy is given by:

$$w(r) = -\frac{3\alpha^2 h\nu}{4(4\pi\epsilon_0)^2 r^6} \tag{2.2}$$

in which h is the Planck constant, ν is the orbital frequency of the electron, α is the polarisability, and ϵ_0 is the vacuum permittivity. The Planck constant comes into play though the definition of the electron-nucleus radius. The interaction potential $w(r)$ is called 'London' or dispersion' potential [22]. Its order of magnitude is kT [21].

The two interactions, i.e. Van der Waals and hard sphere, form the basis of the Lennard-Jones potential:

$$V(r) = 4\epsilon\left(\left(\frac{\sigma}{r}\right)^{12} - \left(\frac{\sigma}{r}\right)^{6}\right), \tag{2.3}$$

where ϵ is the cohesion energy and σ the hard sphere diameter'. For example, for argon, $\sigma = 3.045$ Å and $\epsilon = 1.67\ 10^{-21}$ J. σ is not exactly twice the Van der Waals radius (which would lead to 3.76 Å - see Table 2.2 -), but both are comparable.[6] In Eqs. (2.3), the first term is the hard sphere repulsion, associated to the exponent -12. This power law is used for mathematic convenience. The second term is the Van der Waals attraction, of negative sign and associated to the exponent -6.

The Lennard Jones potential $V(r)$ is also called '6-12' potential, or shortened as 'LJ potential'. It is plotted in Fig.2.10. Two initially distant molecules will approach each other and reach the equilibrium point, located at $r_m = 2^{1/6}\sigma$. At the end of their journey, the couple acquires an energy $-\epsilon$. Said differently, they will have reduced their energy by a quantity equal to ϵ. This is Van der Waals bonding energy for neutral molecule in a vacuum. It ensures the cohesion of Van der Waals solids and liquids. As noted above, ϵ is called the cohesion energy. On needs to bring ϵ to the pair to break the bond.

[6]One should in fact, as will be seen later, compare $\frac{1}{2}2^{1/6}\sigma$ to the Van der Waals radius. The difference is 10 %

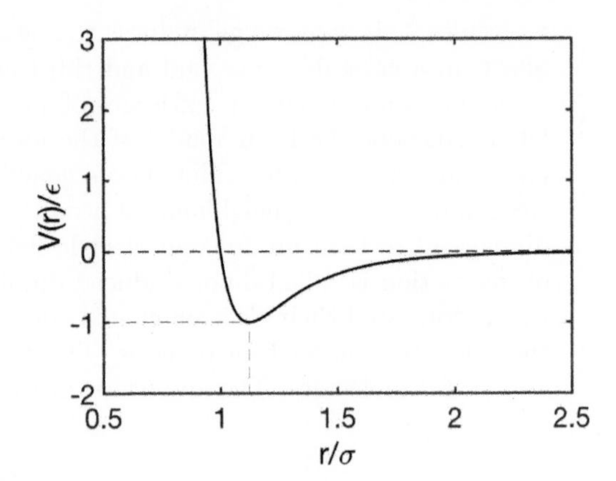

2.10: Lennard Jones potential.

The force between two molecules is the opposite of the gradient of the Lennard Jones potential. We have the following expression:

$$F(r) = -\frac{\mathrm{d}V}{\mathrm{d}r} = 24\epsilon\left(2\left(\frac{\sigma^{12}}{r^{13}}\right) - \left(\frac{\sigma^6}{r^7}\right)\right).$$

The force is called cohesive force. The order of magnitude of this force is roughly, $\frac{\epsilon}{\sigma}$, which leads to $\frac{1.67\ 10^{-21}}{3\ 10^{-10}} \approx$ 5 pN. On must apply this force to break the pair. This force can be converted in pressure by considering a sphere enclosing the molecule. The corresponding cohesion pressure, negative, has the following order of magnitude: $-\frac{\epsilon}{\sigma^3} \sim -$ bars. This negative pressure has a physical significance in solids. It corresponds to the stress engineers must apply to break them. We will return to this question in Chap. 4.

Polar molecules in a vacuum. The above considerations can be extended to polar molecules. Water and carbon dioxide are examples of such molecules. In these situations, there is always a hard-sphere repulsion (which has the same form as in the non-polar case) and a London attraction. However, due to the polar character of the molecules, the attraction term includes supplementary contributions, shown in Fig. 2.11.

So, as a whole, three long-range, attractive, interactions add up when molecules are polar:

- London interaction, already described (see Eq. (2.2)).
- Keesom interaction, due to a dipole-dipole interaction. The corresponding potential is:

$$w(r) = -\frac{u^2 u_2^2}{3(4\pi\epsilon_0)^2 kTr^6}$$

in which u_1 and u_2 are dipolar momentums of molecule 1 and 2 respectively (the other quantities are defined above).

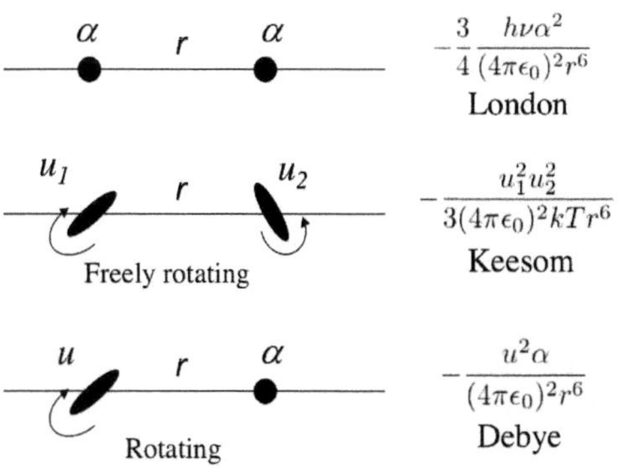

$\alpha \quad r \quad \alpha$ $\quad -\frac{3}{4}\frac{h\nu\alpha^2}{(4\pi\epsilon_0)^2 r^6}$

London

$u_1 \quad r \quad u_2$ $\quad -\frac{u_1^2 u_2^2}{3(4\pi\epsilon_0)^2 kTr^6}$

Freely rotating

Keesom

$u \quad r \quad \alpha$ $\quad -\frac{u^2\alpha}{(4\pi\epsilon_0)^2 r^6}$

Rotating

Debye

Fig. 2.11: The three terms in the Van der Waals forces between polar molecules, in vacuum

- Debye interaction, due to a dipole-induced dipole interaction. The expression is:

$$w(r) = -\frac{u_1^2\alpha}{(4\pi\epsilon_0)^2 r^6}$$

The electrostatic calculations that lead to the above formula are described in Ref. [21]. The important result is that all these contributions decrease with $1/r^6$. We have reached the important conclusion that the Lennard Jones potential can represent not only interactions between neutral molecules, but also those existing between polar molecules in a vacuum. In general, for ordinary molecules, London, Keesom, and Debye interactions have comparable orders of magnitude.

Molecules in a solvent. When two molecules are placed in a solvent, the solvent plays the role of a third participant. Numerous phenomena develop, profoundly modifying the situation described in the previous section. We will not describe them, and instead refer to the work of Israelachvili [21] for a complete presentation of the subject. On important step, that is in no way trivial, is to replace the molecule by a sphere of radius a, characterized by a dielectric constant ϵ and a refractive index n. To make a long story short, the Van der Waals attraction potential $w(r)$, for two identical molecules (medium 1), separated by a solvent 2 , takes the followiing form:

$$w(r) = -\left(3kT(\frac{\epsilon_1 - \epsilon_2}{\epsilon_1 + 2\epsilon_2})^2 + \frac{\sqrt{3}h\nu_e}{4}\frac{(n_1^2 - n_2^2)^2}{(n_1^2 + n_2^2)^{3/2}}\right)\frac{a^6}{r^6} \qquad (2.4)$$

Here, ϵ_1 and ϵ_2 are the dielectric constants of the molecules and the solvent respectively, and n_1, n_2 are the corresponding optical refraction indices. The potential $V(r)$ includes the three contributions (Keesom, London, and Debye), which take a form different from the case where the medium is a vacuum.

What is the order of magnitudes of $w(r)$ when the two molecules are at equilibrium ($r = a(1+2^{1/6})$? Taking, as in the previous example, $h\nu_e$=2.2 10^{-18}J, $kT = 4.11\ 10^{-21}$ at 298 K, and considering n-Dodecan ($\epsilon_1 = 2\epsilon_0$ and $n_1 = 1.41$), with water as the solvent, one finds -0.72 kT for the Keesom and Debye contributions and -1.4 kT for the London contribution. Thus, the order of magnitude of the interaction energy is roughly, in ampliitude, on the order of kT for Van der Waals materials.

Several remarks can be made:

- For two like molecules, i.e for the case we considered, Van der Waals interaction is always attractive. However, when the molecules are different, the interaction can be repulsive. There is no strangeness. Placed in vacuum, each molecule attracts its partner. But when they are placed in a medium, for instance a liquid, the situation changes. In particular, if medium 2 attracts molecule 1 more than molecule 3, molecule 3 will feel a repulsion from its partner. We have here an equivalent of Archimede principle. In a vacuum, a body is attracted by the earth, and falls down. But, as soon as it is plunged in a heavier liquid, it rises, as if it were repelled by the earth.
- Should the materials have the same dielectric constants, and the same refraction index (often the former implies the latter), Van der Waals forces vanish. These forces are thus ubiquitous, but sometimes they are cancelled.

2.2.2 Van der Waals forces between surfaces

We have examined the ranges of forces acting between two molecules, be they polar or non-polar, isolated or placed in a solvent, simple or complex. Now, we consider surfaces.

In this area, the major part of the experimental work has been carried out during the years 1970-1990 [21, 23, 24]. Use was made of a remarkable device called surface force apparatus' (SFA), invented by C. Tabor, R. Winterton and J. Israelachvili [23]. SFA allowed the approach of stages with a precision on the order of just a few angstroms, and to measure the corresponding forces with great precision. One such device is shown in Fig.2.12.

The stage on the device is displaced in an extremely precise manner, thanks to a system of springs with varying stiffnesses, which act as a displacement divider. The space between the stages is generally measured by Fabry Pérot interferometry. Information obtained using this device is presented in the form of a force displacement curve.

In 1986, Atomic Force Microscopy (AFM), and its precursor, Surface Force Microscopy (SFM), were invented [25]. AFM, commercialized in 1989, is now currently used to investigate surfaces at the molecular level, providing remarkable images (an example is given by Fig. 2.2), and enriching the body of information obtained by SFA, by performing, for instance, measurements of the Hamaker constant [26, 27].

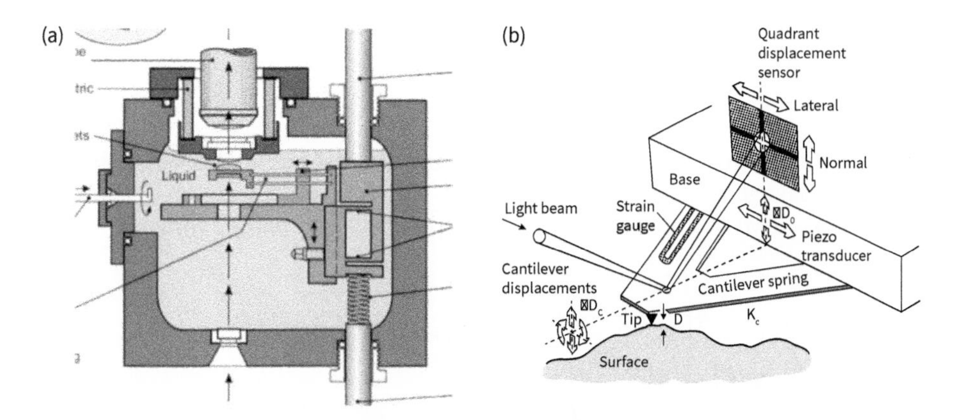

Fig. 2.12: (A) Surface force apparatus (SFA) that enabled, in air and vacuum in 1972 and later, in liquids, the measurement of intermolecular forces near interfaces [21, 23, 24]; (B) Atomic force microscopy (AFM), invented by G. Binnig, C.F. Quate, and Ch. Gerber, in 1986 [25], enabling the analysis of surfaces at the molecular level

Let us return to theory: we saw that the Van der Waals force varies as $1/r^7$ for molecules (where r is the distance between the molecule centres). In order to determine the forces between two plates, a sphere and a plate, or two spheres, we must assume additivity, and sum up the pairwise forces of the molecules composing the *other medium*. In this calculation, r becomes the gap between the objects, for instance the distance between two plates. We do not take the bulk cohesive forces into account, because we do not incorporate, in the summation, the molecule of the *same* medium. So, the forces we calculate are those exerted by the partner, not by the object on itself. It is then a matter of mathematics, well described in textbooks, to obtain that the attractive forces varies as $1/r^2$ for sphere-plane and sphere-sphere[7] systems, and as $1/r^3$ for plane plane systems. The Van der Waals attraction force $P(r)$, per unit of area, between two identical plates along with the interaction energy, $E(r)$ take the following forms:

$$P(r) = -\frac{A}{6\pi r^3} \text{ and } E(r) = -\frac{A}{12\pi r^2}, \tag{2.5}$$

where A is the Hamaker constant. Its expression is:

$$A = \frac{3}{4}kT\left(\frac{\epsilon_1 - \epsilon_2}{\epsilon_1 + \epsilon_2}\right)^2 + \frac{3h\nu_e}{16\sqrt{2}}\frac{(n_1^2 - n_2^2)^2}{(n_1^2 + n_2^2)^{3/2}}, \tag{2.6}$$

in which the involved quantities were defined previously (see Eq. (2.4)). In the above formula, 1 represents the dielectric material forming the plate, and 2 the medium between them. This expression was obtained by Lifschitz and is called Lifschitz theory [28]. Its structure is close to Eq. (2.4), with, again, the first two terms summing up the

[7]for the sphere-sphere case, the relation is valid when the gap r is much smaller than the sphere radii. In the opposite case, we retrieve the r^{-7} law, holding between two molecules

Keesom and Debye contributions, and the third one the London term. The Hamaker constant given by Eq. (2.6), like for Eq. (2.4), is on the order of kT, for the most common materials used in microfluidics.

The theoretical approach was confronted to experiments, performed with sphere plane systems (in this geometry, the expression of Van der Waals forces is $-A/6r^2$). The experimental curve of Fig. 2.13, obtained with mica, confirms the formula.

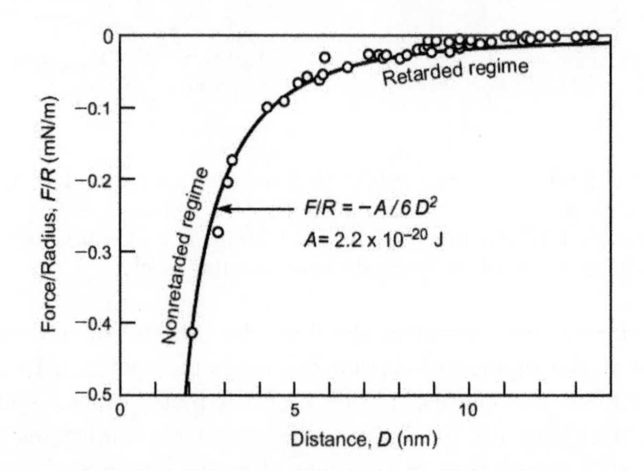

Fig. 2.13: Measurement of the force between two crossed mica cylinders of radius ≈ 1 cm, in water and electrolytes [21]. The solid line if $-A/6D^2$, with D replacing our gap r and $A = 2.2\ 10^{-20}$J. The notion of retarded is related to the fact that interactions between molecules are not instantaneous. Corrections, taking this effect into account, are proposed in the literature. The subject lies outside the scope of the book.

It also allowed us to measure Hamaker constants. Table 2.6 displays some results.

This table calls for several remarks:

- The values in the tables are obtained either from Lifschitz theory, when the experiment is impossible to perform (the case of water/air/water is an example) or when data is not available. In air and in all cases where comparison could be done, theory and experiment were close, within 20 percent or so.
- A is 1-10 kT in air (recall that kT $\approx 4.11\ 10^{-21}$J) and 0.1-1 kT in liquids.
- Hamaker constants in metals are much larger than in Van der Waals solids. This is due to the fact that in metals, due to strong attraction between metal cations and delocalized electrons (metallic bonds), cohesive forces are much stronger.
- Measurements of the Hamaker constant in liquids are delicate and results published in the literature are, sometimes, inconsistent.

Material (1)	Medium (2)	A (10^{-20}J)
Water	Air	3.7
Polystyrene	Air	6.5
PTFE	Air	3.8
Fused quartz	Air	6.5
SiO_2	Air	6.50
Teflon	Air	2.75
Silicon	Air	18.65
Metals	Air	25 -40
Fused quartz	Water	0.63
SiO_2	Water	0.8
PTFE	Water	0.29
Teflon	Water	0.33
Silicon	Water	9.75
Metals	Water	30 -40
Fused quartz	Octane	0.13

Table 2.6 Hamaker constants for different materials.

Microfluidic measurement of the Hamaker constant at high salt concentrations. A microfluidic method for measuring the Hamaker constant at high salinities has recently been reported [30]. The experiment is shown in Fig. 2.14.

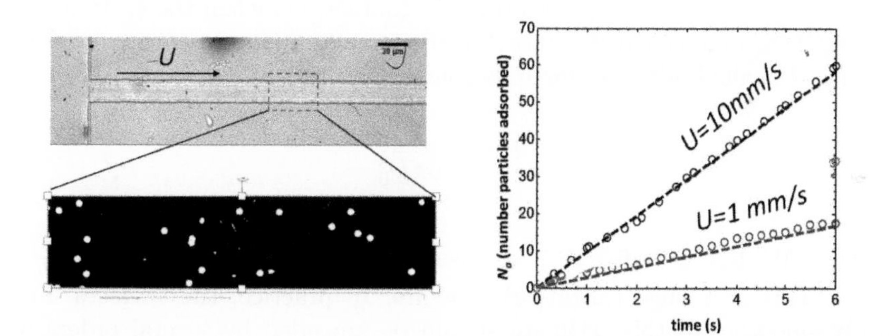

Fig. 2.14: (left): Top: Microfluidic experiment enabling the measurement of Hamaker constant at high salinity. bottom: fluorescent images of the 5 μm particles deposited on the walls, two seconds after they have been injected in the channel.; (right) Evolution of the number $N(t)$ of particles deposited on the microchannel walls as a function of time, for two different flow speeds [30].

A dilute suspension of polystyrene particles, 5 μm in diameter, is injected, in a rectangular, shallow, rectangular microchannel. Most particles circulate through it, from the entry to the outlet, but a fraction of them deposit on the bottom and upper walls. They are somehow captured by the channel. The number of these deposited particles $N(t)$ increases, proportionally to time (Fig. 2.14), right). Theory conducted at high saliniity (where electrostatic forces are screened - see below -) shows that the slope P of these lines is equal to:

$$P = \frac{r}{h/2 - r} \frac{\phi Q}{v_p} \sqrt{S}$$

,

in which ϕ is the mass fraction of the particles, v_P is their volumes, Q is the flow rate, h is the channel height, and S is given by:

$$S = \frac{A}{kT} \xi_L$$

where A is the Hamaker constant, $\xi_L = \frac{LD}{U_r r^2}$ (in which L is the channel length, D the diffusion coefficient of the particles, r their radius, and U_r the flow speed at a distance r from the wall). By measuring the slope, one has access to the Hamaker constant A. The value of A, estimated from Fig. 2.14, is equal to $8 \pm 0.4 \ 10^{-21}$ J, which is close to Lifschitz estimate ($7.6 \ 10^{-21}$ J) for a polystyrene/glass interface with water as a medium.

2.2.3 Macroscopic manifestations of Van der Waals forces

Adhesion forces . When two planes come in contact, i.e. when the surface molecules are separated by a molecular distance D_0, the force of attraction per unit of area, i.e. the pressure P, called adhesion force, becomes:

$$P = -\frac{A}{6\pi D_0^3}$$

Taking $D_0 = 0.3$ nm, $A = 6.5 \ 10^{-20}$J (fused quartz), one obtain $P = \frac{10^{-21}}{6\pi(3 \ 10^{-10})^3} \approx$- 1280 bars. This is a huge (negative) pressure. In practice, the surfaces are rarely atomically smooth and this estimate should be amended by several orders of magnidude. Still, there is space, by optimizing the contact surface, to reach high pressures of adhesion. This is what glues are doing. Glues do not possess adhesive virtues per se, their functionality originates from the fact that they fill the vacant spaces between two surfaces, in order, after solidification, to deploy the Van der Waals forces we just calculated. Along this line of thought, one could imagine assembling plane wings to the fuselage just by gluing them.[8] Taking 10 m^2 as the area of contact between the two pieces, and using the previous Hamaker-based estimates, we would reach an adhesion force of 100 kTons, a force sufficient to prevent the wings to detach during the flight, an event that could scare the passengers. The task of improving the contact area between two solids is well achieved by the gecko. For this purpose, the animal uses a complex

[8]Words taken from Pierre Gilles de Gennes

system of setules, described in the literature [31, 32]. As a result, it suspends its own weight above the floor by applying just one toe on a surface (see Fig. 2.15).

2.15: The gecko suspends his own weight above the floor just by applying one finger on a surface (see the arrow). Each finger possess structures optimizing contact area, so as to take advantage of the Van der Waals attractive forces. (Reprinted from [32] with permission of IOP Publishing, Ltd. Copyright 2022.).

Surface tension. Surface energy γ is defined as the energy required to increase the area of a liquid or a solid by one unit. Surfaces of solids can be increased in many ways, such as by breaking them in two pieces. The corresponding energy cost is equal to the work done by the Van der Waals forces to take the two pieces apart. Its expression is:

$$E = -\int_{D_0}^{\infty} \frac{A}{6\pi r^3} dr = \frac{A}{12\pi D_0^2} = 2\gamma$$

where γ is the surface energy, considering that the cost is shared between the two surfaces. An energy, per unit of area, of 2γ is thus necessary to separate two planes bound by Van der Waals adhesive forces. From this calculation, one concludes that each surface possesses an energy density equal to γ, whose expression is:

$$\gamma = \frac{A}{24\pi D_0^2}$$

The system minimizes it by reshaping the interfaces, when interfaces are deformable. Along this line of thought, a liquid blob will spontaneously become spherical, to minimize its exposed area.

Surface tension is a different concept: it is the tension that the solid or liquid elements are subjected to at the surface. It is expressed as a force per unit of length. We will see, in Chap 4, that in liquids, surface energy and surface tension are equivalent, and they are often legitimately confused, while in solids, in general, they are not. Surface tension gives rise to many spectacular phenomena, such as spontaneous reshaping of liquid blobs into spherical drops, as mentioned above, jet break-up, capillary rising, and imbibition, which will be developed in Chapter 4.

2.2.4 Molecular Dynamic (MD) simulations of interfaces

Today, numerical simulations at the molecular level are important tools for understanding fluid behaviours. Most of them are based on Lennard-Jones potential (Eq. 2.3), to which random noise, mimicking the effect of temperature, is added. Two types of simulations are carried out:

- Monte Carlo (MC) simulations calculate the evolution of the system, based on Newton's law, and reject the new state if it increases the total pairwise energy of the system. MC allows the equilibrium state to be determined in an efficient manner.
- Molecular dynamics (MD) simulations track the evolution of the system, whatever the energetic direction it takes. This approach is often preferred in fluid physics, because, although it takes a longer time to reach the equilibrium, it allows the kinetics to be analyzed. Also, the notion of equilibrium does not apply to all fluidic situations. A typical MD simulations involves thousands of particles. The step is typically one fs, and the accessible range of time several ns, thus millions of steps. In the present state of the technology, it would take more than the age of the universe to calculate the dynamics of a microfluidic system for one second.

Systems computed with LJ potentials exhibit rich phase diagrams, including solid-liquid, liquid-gas and solid-gas transitions, triple point, thus showing similarities with the behaviour of real systems. Fig. 2.16 shows a case where parameters have been tuned so that gas and liquid phases coexist.

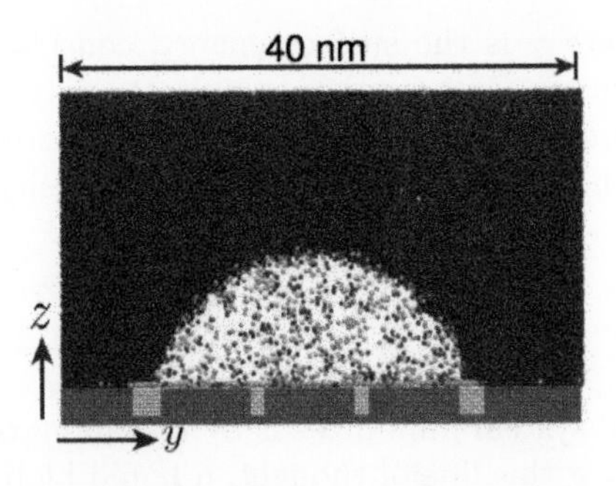

2.16: Lennard-Jones simulation of a nanobubble sitting on a surface patterned with hydrophobic patches. Dark dots are the liquid, grey dots are the gas. The liquid, oversaturated in gas, is held at a pressure of 26 bars. In these conditions, the bubble remains stable for 45 ns, i.e. millions of characteristic molecular times, without any measurable change [33]. In Chapter 5, arguments will indicate that this bubble remains indefinitely stable. (Reprinted with permission from [33]. Copyright 2022. American Chemical Society.).

The simulation shows that, starting with a square distribution of particles, we end up with a semi-spherical interface separating the liquid and the gas, i.e. a bubble. We learn that even though bubbles are nanometric, they possess evident similarities with their micro and millimetric equivalents. The simulation of Ref. [33] moreover showed that, contrarily to naive arguments, nanobubbles pinned on a textured surface do not dissolve in oversaturated liquids (see Fig. 2.16). In fact, the simulation helped to end a long-term controversy (15 years). We will return to this point in Chapter 5. In many cases, simulations provide information on nanosystems, inaccessible to experiment. Because of that, they have contributed considerably to the understanding of nanofluidic phenomena, even though, as mentioned above, they are limited in time to a few nanoseconds, and in size to a few nanometres.

Happy and unhappy molecules. One might ask why the interface found in Fig. 2.16 is semi-spherical. We will develop this point in detail in Chapter 4, but it is worth discussing it here, from a microscopic viewpoint. Molecules in the bulk fluctuate in position, but, on average, they feel the attractive forces exerted by their partners. Let us consider an extremely simplistic model, in which molecules stay close to the equilibrium point, i.e. at a distance equal to $2^{1/6}\sigma$ from their nearest neighbours. This model is that of a solid, and it will be criticized later, when capillary phenomena are discussed in detail (see Chapter 4). Restricting ourselves to cubic structures and immediate environments, the corresponding energetic level, associated to energy plot minima, is -6ϵ. Bulk molecules cannot achieve a better energetic situation, and therefore, they feel happy'. At the interface, the molecules develop nine nearest-neighboring bonds, and, thereby, the corresponding energy level[9] is close to -4.5ϵ. These molecules feel unhappy'. The system, being in a quest for an energy minimum, will minimize the number of unhappy molecules, an objective it meets by adopting a spherical shape, the sphere being the shape that minimizes, for a fixed volume, the exposed area.

2.2.5 Electrostatic forces between surfaces

When a dielectric is immersed in an electrolyte, a surface charge spontaneously appears. For example, a plate of glass immersed in a non-acidic aqueous solution, becomes negatively charged. By preparing a pleasant beverage, such as an orange juice or mint with water, we also cover the internal surface of the glass with charges. The origin of this phenomenon comes from the fact that in aqueous solutions, the silane terminals Si–O–H localized on the glass surface lose hydrogen ions in the presence of the aqueous solution; this protonation leaves Si–O^- terminals on the surface, thus causing the glass exposed to aqueous solution to become negatively charged. The electric potential associated with these charges, for a pH of 7, is on the order of -30 mV. Protonation is not the only cause of spontaneous charging of a surface in contact with water. Different mechanisms exist. We will address this question in Chapter 6. They are described in detail in the literature, for instance, in the text of Cabane and Hénon [34].

These surface charges are equilibrated in the fluid by a double layer of counter-ions, as shown in Fig. 2.17 (A).

The first layer is fixed and is called the 'Stern layer'. We shall discuss its structure in Chapter 6. Although much remains to be learned regarding the Stern layer, its thickness is considered to be on the order of the Bjerrum length, defined by:

[9] Taking the full environment of the molecule into account, one finds -8.38ϵ for molecules in the bulk and -5.88ϵ for molecules at the interface.

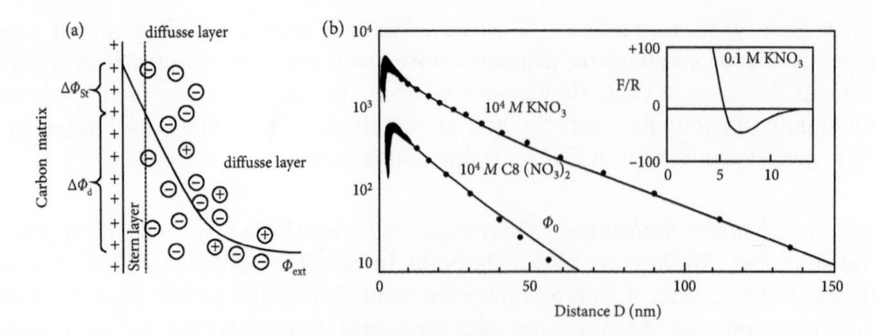

Fig. 2.17: (A) Sketch of the Stern and Debye layers; (B) Measurement of force between two crossed mica cylinders of $R \approx 1$ cm in electrolytes. The forces of the double layer decrease exponentially with distance, and can extend over several tens of nanometres. The solutions used for this experiment are KNO_3, and $CaNO_3$ for the upper and lower curve, respectively. The continuous lines are obtained from the Derjaguin-Landau-Verwey-Overbeck (DLVO) theory, after [21]

$$\lambda_{\mathrm{B}} = \frac{e^2}{4\pi\varepsilon\; kT}$$

where e is the elementary charge. For water at room temperature $\lambda_{\mathrm{B}} \approx 0.7$ nm.

The second layer is diffuse. It results from a statistical equilibrium between thermal agitation (which tends to homogenize the charge distribution) and electric forces (which tend to displace charges of the same sign towards the surface, thus creating a charge profile). We will describe it in detail in Chapter 6. The order of magnitude of the thickness of the double layer, designated by λ_D (the Debye length), is given for a binary symmetric electrolyte, by the following expression:

$$\lambda_D = \sqrt{\frac{\epsilon kT}{2\rho_\infty q}}. \tag{2.7}$$

where ρ_∞ is the ionic concentration (in C/m^3) far from the walls and q is the ion charge ($q = e$ for monovalent ions, such as Na^+ or Cl^-). The others quantities have been defined in the previous sections. Debye layer is around ten nanometres thick for an ionic solution of 1 mM of monovalent salt. As λ_D varies with the inverse of the square root of the charge concentration, it can in principle achieve high values. However, practically speaking, because of the presence of charged impurities, the thickness of double layers does not exceed about a hundred nanometres. This double layer is the origin of electrostatic forces, which had been measured in force machines. Figure 2.17 (B) shows measurements of such forces, between mica cylinders, for two different electrolytes.

The force of the double layer decreases exponentially, as theory predicts (we will return to this point in Chapter 6). Fig. 2.17 (B) also indicates that the fluid feels' the wall at distances of several tens of nanometres.

2.2.6 Other forces.

When a liquid is confined between two surfaces separated by a few nanometers, the liquid tends to structure itself. This structuration effect has important consequences on the way how the fluid behaves, as we mentioned before, and the surfaces interact. The field of forces developing in the tiny space becomes an oscillating function of the distance from the wall [21]. The spatial period is on the order of the intermolecular distance inside the liquid. This oscillatory behaviour plays a considerable role in the behaviour of water in carbon nanotubes: for instance, it leads to a reversal of the wettability or nonlinear behaviors. Other forces arise when surfaces are functionalized. All these forces have given rise to a wealth of studies, well described in textbooks. We will not develop the subject in this book, therefore.

2.3 Nano-Micro and Millifluidics

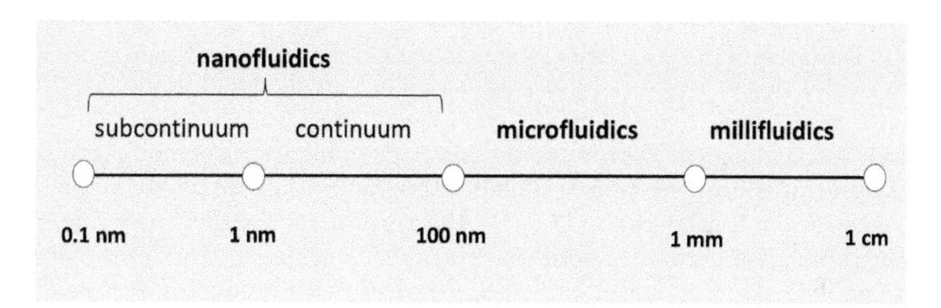

Fig. 2.18: Definition of the three fundamental domains involved in small-scale hydrodynamics: nano-micro and millifluidics. Microfluidics, which covers four orders of magnitude, is the focus of this book.

Fig. 2.18 and 2.19 show the three domains that define what we could call 'small scale hydrodynamics': nanofluidics (between 0.1 and 100 nm), microfluidics (between 100 nm and 1 mm) and millifluidics (between 1 mm and 1 cm).

Nanofluidics Nanofluidics concerns fluidic systems whose sizes pertain to the 0.1 -100 nm range. It extends over three decades. Two subdomains can be defined:

- 0.1 -1 nm (subcontinuum nanofluidics): Molecular scale prevails. Ordinary equations, such as the Navier-Stokes equations, based on a continuum assumption (which we will discuss in Chap 3), are no more valid. This range of scale is called subcontinuum.
- 1 -100 nm (continuum nanofluidics). Continuum hypothesis holds. Many characteristic lengths pertain to the domain: thermo-capillary length, Debye

length, mean free path, Bjerrum length, nucleation radius, slippage length, to mention a few.[10] They are shown in Fig. 2.19. These scales combine with the confinement to give rise to interesting phenomena: unusual regimes of droplet production in jets [35], non-hydrodynamic flow regimes [14], and transistor effect [36]. The list is not exhaustive.

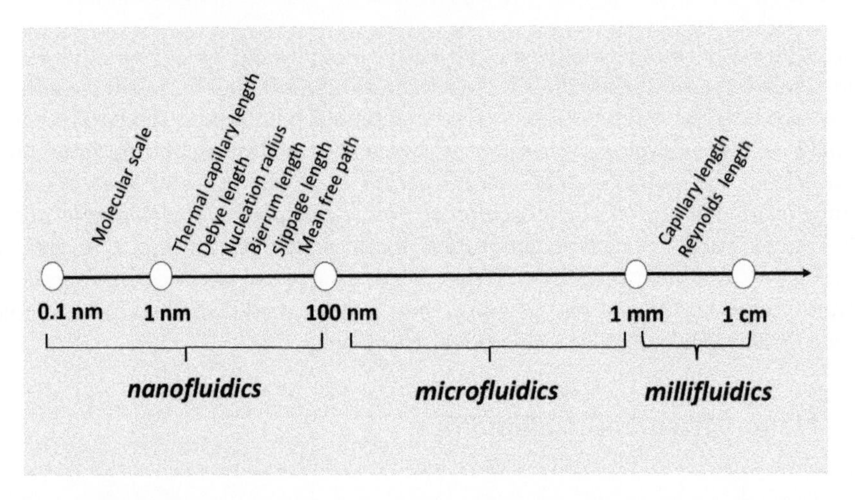

Fig. 2.19: Definition of the three fundamental domains involved in small scale hydrodynamics, in which several characteristic scales are outlined. They are discussed in the text.

From a technological viewpoint, the methods employed in nanofluidics for fabricating devices are based on UV photolithography (limited to sizes above10 nm) or e-beam photolithography [37]. By the turn of the century, new materials appeared, such as carbon nanotubes or graphene surfaces. This fostered the field by enabling the creation of sub-10 nm devices, with unique properties. We have developed the subject in Chapter 1, and will return to it in Chapter 3.

Microfluidics : Microfluidics concerns system sizes pertaining to the 100 nm - 1 mm range. The domain is vast: it extends over four decades[11]. It represents the central subject of this book. In terms of scales, the microfluidic domain looks empty (see Fig. 2.19): not a single characteristic length pertains to it. In this context, microfluidics possesses four outstanding properties:

- The behaviour of a microfluidic system is the same at all sizes. What happens in a device, 100 nm in size, is the same as in a geometrically similar system, ten thousand times larger.

[10]The fact that thermo-capillary, nucleation and Bjerrum lengths are nanometric is true for Van der Waals materials and water. There is a relation between these orders of magnitude and the fact is that solid-liquid and liquid-vapor transitions of these materials take place in the 200-350 K range.

[11]A proposal has been made to single out, within microfluidics, a sub-domain ranging between 100 nm and 1000 nm, called 'extended nanofluidics' [38].

- Microfluidic scales being smaller than capillarity and Reynolds lengths (see Fig. 2.19), gravity and inertia are negligible. By eliminating two parameters, the systems are simpler to analyze and control.
- In the microfluidic domain, all microscopic and mesoscopic phenomena are smoothed out, just as in macroscopic physics. This concerns, for instance, thermal noise and the spatial structure of Van der Waals forces. This feature considerably simplifies the control of microfluidic devices.
- Another important characteristics is geometric: the sizes of a number of interesting objects fall in the microfluidic range. The most important example is the human cell, 10 μm in size, which can be manipulated individually in microfluidic devices. This gives rise to a considerable number of applications: cell sorting, chemotaxis, single cell assays, and single cell -omics - (genomics, transcriptomics, etc..) to mention but a few.

From a technological prospective, microfluidics is the realm of photolithography, moulding, etching and deposition, which will be described in Chapter 7.

Millifluidics Millifluidics concerns fluidic systems whose sizes pertain to the 1mm - 1 cm range. It extends over one decade. This small domain hosts two important characteristic scales: Reynolds and capillary lengths. This implies that in millifluidics, inertia and gravity are generally important. The physics below and above 1 mm is therefore different. Technologywise, creating millifluidic devices requires tools different from nano and microfluidics: for instance, low resolution 3D printing, micromachining and injection moulding. Toolboxes below and above 1 mm differ. We will develop the subject in Chap 7.

2.4 The physics of miniaturization

2.4.1 The Π theorem

The theorem of the product, called Π theorem, allows the number of variables of a system to be reduced. The demonstration of the theorem is an elegant exercise of linear algebra that the reader can find in fluid mechanics textbooks (see, for example, [39]).

The behaviour of a system is expressed by a function f, which relates an observable A to $n-1$ physical quantities, noted $a_1, a_2 ... \ a_{n-1}$.

$$A = f(a_1, a_2, \ldots, a_{n-1})$$

These quantities possess k independent units (for instance three: metre, second, and kilogram). The Π theorem states that the law f can be written in the following form:

$$\Pi = g(\pi_1, \pi_2, \ldots \pi_{n-k-1})$$

where the πs are independent dimensionless products formed from the initial quantities and g, apart from a pre-factor, is equal to f, written in terms of the new dimensionless variables. We now only have a relation between $n - k$, instead of n variables. We thus have reduced the number of variables by k. In many cases, this represents a considerable simplification.

The origin of the Π theorem plunges deep in the history. In the fourth and third millennia BC, the ancient peoples of Egypt, Mesopotamia and the Indus Valley introduced units and homogeneity rules in order to trade. Lengths were measured with the forearm, hand, or finger, and time was measured by the periods of the sun or the moon. These were the units. If the outcome was a price, factors were introduced, such as a price per unit of weight or length, to ensure homogeneity in the calculations.The Π theorem, which formalizes, mathematically, this homogeneity rule, could be viewed as a generalization of these antic practices.

Application 1: Pythagora's theorem. More than one hundred demonstrations of Pythagora's theorem exist. This one is based on the Π theorem.

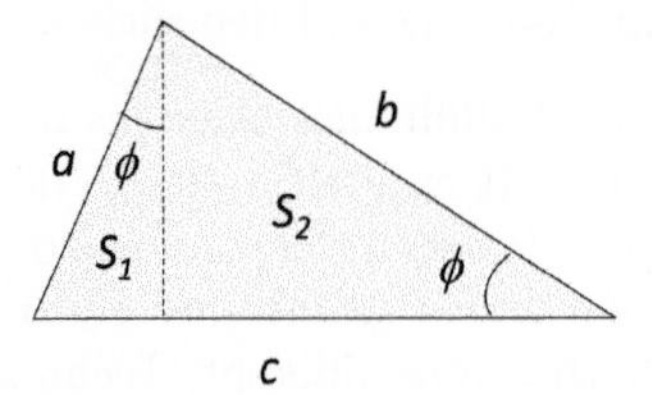

2.20: Triangle used to demonstrate Pythagora's theorem by using the Π theorem.

The surface area S of the grey triangle shown in Fig. 2.20 is a function of ϕ and c, the hypothenuse. It reads:

$$S = f(c, \phi)$$

We have $n = 3$ variables and $k = 1$ dimension, a length. Using the Π theorem, we conclude that the problem is governed by $3 - 1 = 2$ independent dimensionless variables. We choose them equal to $\frac{S}{c^2}$ and ϕ. Therefore, we can write the above relation in the following form:

$$S = c^2 g(\phi)$$

where g is a function.[12] The same law, with the same g and the same ϕ applies for the subsurfaces S_1 and S_2. Consequently, we have:

$$S_1 = a^2 g(\phi).. \text{ and. } S_2 = b^2 g(\phi)$$

By using $S = S_1 + S_2$, and elimination of g, we find:

$$a^2 + b^2 = c^2$$

which is Pythagora's theorem.

[12]Function g is equal to $\frac{1}{4}\sin 2\phi$ but we do not need to know this expression for demonstrating Pythagora's theorem.

Application 2: Reynolds number. Let us consider the incompressible flow around an object of characteristic dimension l (for instance, the sphere radius of Fig. 2.21).

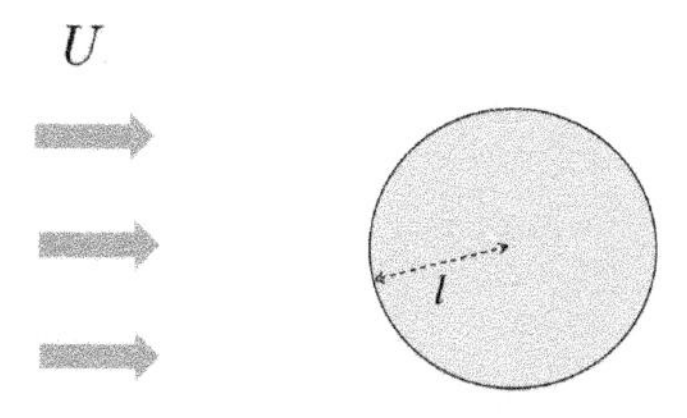

2.21: Flow around an solid sphere.

The local velocity $\mathbf{u}$ (assumed to be stationary) must be written as a function of the following parameters:

$$\boldsymbol{u} = \mathbf{f}(\boldsymbol{x}, l, U, \mu, \rho),$$

where $\boldsymbol{x}$ is the spatial coordinate, l is the sphere radius, U is the upstream speed, μ is the viscosity, and ρ is the volumetric mass of the fluid. We have $k = 3$ and $n = 6$. We deduce that we can form $n - k = 3$ independent dimensionless products, which we choose to be equal to:

$$\frac{\boldsymbol{x}}{l}, \quad \frac{\boldsymbol{u}}{U} \quad \text{and} \quad \frac{Ul}{\nu},$$

where $\nu = \mu/\rho$ is the kinematic viscosity. The product Ul/ν is the Reynolds number *Re*. Thus, the preceding relation can be written in the form:

$$\boldsymbol{u}/U = \mathbf{g}(Re, \boldsymbol{x}/l)$$

where $\mathbf{g}U$ is function $\mathbf{f}$ expressed in terms of the new dimensionless variables. We thus have a much simpler expression than the initial expression.

From this expression, we can conclude that reducing l is equivalent to reducing U or raising the kinematic viscosity ν. Without scrutinizing the flow equations, which we will do in Chapter 3, we can infer that flows at small scales resemble extremely viscous fluids circulating at creeping speeds. An example is the flow of honey around a spoon gently displaced, at constant speed, in a pot. The empirical knowledge of these situations suggests that, in miniaturized devices, there will be no eddy, no hydrodynamic instabilities, and no turbulence.

Comments.

- The two above examples illustrate the force of the Π theorem. By identifying the relevant variables, exploiting the fact that each of them conveys a unit, and units must be homogeneous, the theorem allows insightful conclusions to be drawn *without solving any equation.*

- Π theorem is powerful when the number of variables is small, but it is less operational when many variables come into play. In this case, several dimensionless numbers must be defined, in one way or another, and the final relation, i.e the equivalent of g, includes too many quantities to provide a clear vision of the situation. This often happens in physico-chemistry. In this field, the Π theorem rarely helps in disentangling the inherent complexity of the phenomena.

2.4.2 The concept of scaling law

A 'scaling law' signifies a relation between an observable quantity and a measure of the size of the system. The subject is long-standing. It was discussed by Galileo [40], Sir D'Arcy Thomson [41], and many others. Table 2.7 presents a number of laws applying for physical quantities, and pertinent to microfluidics.

Quantity	Scaling law
Capillary force	l^1
Flow speed at fixed pressure gradient	l^1
Thermal flux at fixed temperature difference	l^1
Viscous force at fixed speed	l^1
Adhesion force	l^2
Electrostatic force at fixed electric field	l^2
Magnetic force at fixed magnetic field	l^2
Diffusion time	l^2
Muscular force	l^2
Mass	l^3
Force of gravity	l^3
Electrical motive power at fixed electric field	l^4
Inertia force at fixed pressure gradient	l^4
Centrifugal force at fixed rotation rate	l^4

Table 2.7 Scaling laws of different quantities.

Table 2.7 deserves a number of explanations and comments.

- For certain quantities, the scaling law is obvious. This is the case for mass where we have the law:

$$M = \rho V \sim l^3$$

where M is the mass of an object of size l, ρ its density and V its volume.

- The exponent of pressure driven flow speed is justified by the following formula, resulting from orders of magnitude arguments made on Stokes equation, i.e in a laminar regime (we will discuss this case in Chap 2):

$$U \sim \frac{h^2 \Delta P}{\mu L}$$

where μ is the fluid viscosity, ΔP is the pressure difference applied along the canal, L is the canal length, and h is its transverse dimension. At fixed pressure gradient $\Delta P/L$, considering there is a single geometric scale in the problem (which is acceptable when the channel is square, the aspect ratio is large, or the cross-section is circular), one obtains $U \sim l$. Therefore, the scaling law exponent is equal to one. Should the pressure gradient be a function of l, the scaling law would be different.

- The scaling law for the electrostatic force F_e is obtained by using the following relation:

$$F_e \sim \epsilon E^2 l^2,$$

where ϵ is the dielectric constant and E is the imposed electric field. With a fixed electric field, the exponent is equal to 2. Similar considerations hold for electromagnetic forces, equal to $\frac{B^2}{2\mu} l^2$, where B is the imposed magnetic field and μ is the magnetic permeability.

- Muscular force: The muscular energy is proportional to the volume of the body, i.e. l^3. This energy is used to perform a work over a distance l (the only scale of the problem). We infer that the muscular force scales as l^2.

In fact, the general rule of thumb is as follows: when two forces are present, it is the force associated with the weaker exponent that becomes dominant in miniaturized systems. Thus, the equilibria we are used to in the human-scale world can be disrupted at the micrometric scale. For example, gravity forces, being associated to a high exponent, are most often negligible in the small world. This situation is the inverse of the human-size world. Scaling laws have far-reaching implications, as will be seen later.

2.5 Scaling laws in nature

2.5.1 Mite clinging to a disk rotating at large speeds

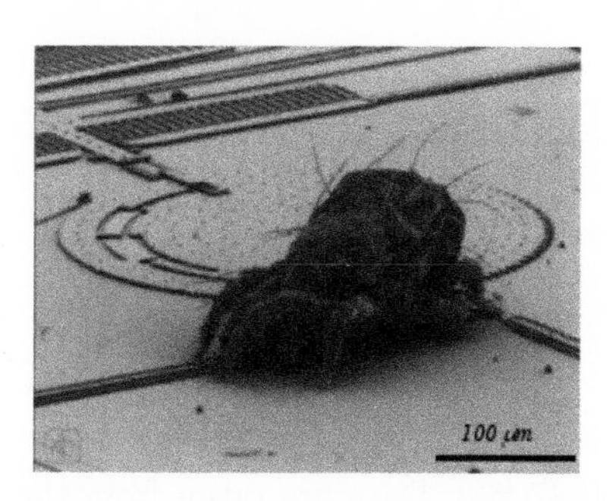

2.22: Mite from New Mexico wandering on a micromotor, before it rotates (From Sandia Laboratories (2005).).

Let us consider an animal of size l and density ρ, placed on a motor rotating at rate Ω (see Fig. 2.22). In these conditions, the centrifugal force the animal is subjected to is:

$$xF_c \sim \rho \Omega^2 l^4$$

We recover exponent 4 of Table 2.7. It is thus not difficult for a mite 100 μm large, to use the adhesion forces (proportional to l^2, as we saw in the previous section and as shown in Table 2.7), to cling to the motor, even though the latter rotates at large speeds, for instance tens of rotations per seconds, as shown in a film by Sandia Laboratories in 2000 (not available today).

2.5.2 Kleiber's law

The basal metabolic rate W_b is a central concept in animal physiology. It is the power needed for an animal to live, when at rest. It involves all what is requested by the organism to survive: breathing, blood circulation, heart beating, food digestion, etc. Two main methods are used for measuring W_b:

- Insert the animal in an ice block and measure the melting rate.
- Place the animal in a confined environment and measure the rate of consumption of the oxygen, assuming that this rate is proportional to W_b.

Both methods were used in 1784 by Lavoisier [42]. The basal metabolic rate W_b in Watt) is represented in Fig. 2.23 in a function of the animal weight M.

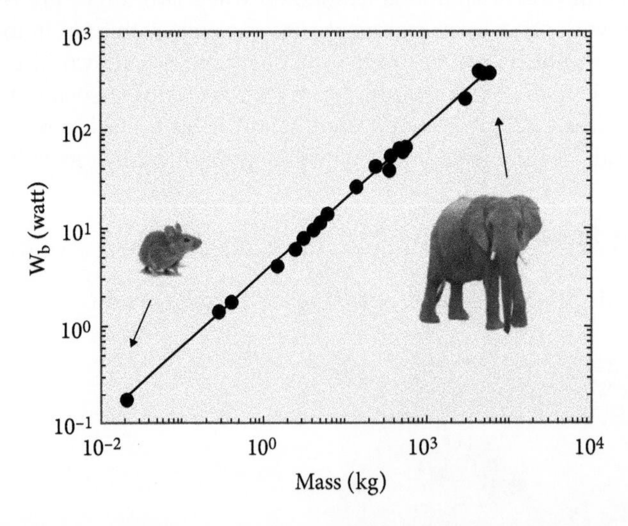

Fig. 2.23: Kleiber's law, relating the basal metabolic rate W_b to the animal mass M. The line corresponds to the formula $W_b = 3.45M^{0.75}$ (best fit obtained excluding the whale, as Kleiber did) [43].

The plot was obtained by Max Kleiber in 1947 [43]. Data are taken from the paper, using modern units. It was a surprise to find that, despite the diversity of animals, W_b follows so neatly, over five decades, a power law. Its expression is :

$$W_b \approx BM^{0.75}$$

where $B \approx 3.45 W/kg^{3/4}$. Before Kleiber, a theory, called 'surface law', proposed an exponent 2/3. The idea was the following: suppose the thermal flux density, across the body surface is the same for all animals, then we would have $W_b = f(M, J)$, which would imply, on using Π theorem, $W_b \sim M^{2/3}$. The surface law, which states that animals produce heat proportionally to their surface areas, remained a dogma for more than a century. It was based on a geometrically similar hypothesis. It was still influential in the 1980s [47], although there was no justification for it. In the first part of the last century, the amount of evidence that nature often obeys allometric laws (i.e. that are not geometrically similar) increased, challenging the law. In fact the exponent found by Kleiber in 1947, was not 2/3, but 0.75 ± 0.03. The difference seems small, but it is significant, and has far-reaching consequences. It leads to different estimates of life span, heart beating frequency, blood pressure, aorta diameter, etc...along with a different vision of the circulation network of air and blood in the body [44].

Kleibler's law remained mysterious for decades. In the nineties, an argument, based on self similarity, emerged [54]. It led to conclude that the exponent should be exactly 3/4. The ingredients of the theory, very well explained in the course of T.Lecuit [45], are sketched in Fig2.24.

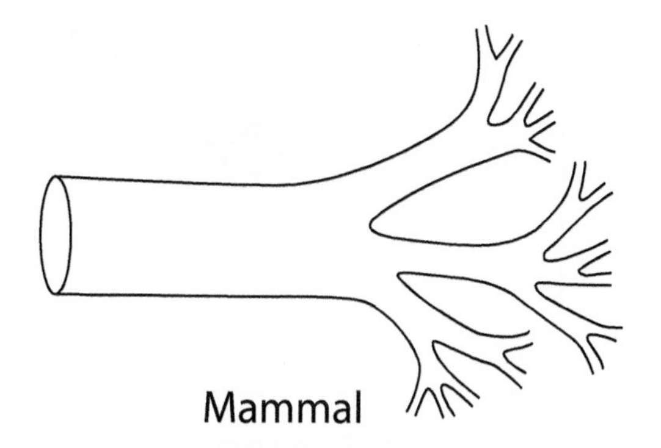

2.24: Kleiber theory: blood and air circulate through a self-similar hierarchical network, with scales decreasing down to a terminal scale, the cell size, the same for all animals. The study of the optimization of this network, from the point of view of the power needed to drive fluids through it, led to Kleiber's law [45].

The system requires efficient delivery to ensure energy conversion at the cell level. The evolution solved the problem by forming self similar structures, sketched in Fig. 2.24. Blood and air circulate through a self-similar hierarchical network, with scales decreasing down to a terminal scale, the cell size, the same for all animals. The network is space filling, because, in the body, cells occupy a substantial, homogeneous fraction of the space. The study of the optimization of this network, from the point of view of the power needed to drive fluids through it, led to Kleiber's law. From the theory, we may propose an estimate for the constant B; in the form $B \sim w_c/m_c^{3/4}$, with w_c the cell energy consumption and m_c its mass. With $w_c \sim 10^{-10} W$ and $m_c \sim 1\, pg$, one finds $B \sim 1 W/kg^{3/4}$, which is consistent with the experimental estimate. The demonstration of Kleiber's law, which is quite elaborate, is well-explained in T. Lecuit's lectures [45]. It stands outside the scope of this book.

2.5.3 The short lifespan of small animals

Kleiber's law has many consequences: for instance, it allows to establish the dependence of the heart beating frequency f with the animal mass M. The argument as follows: the time it takes to bring the oxygen from the lungs to the body is several $1/f$. The transported blood volume, per unit of time, is $l^3 f$. This flow rate, conveying all the oxygen needed to burn the energy used by the animal to stay alive, is proportional to the metabolic rate. Taking Kleiber's law, we conclude that heart beating frequency scales as $l^{-3/4}$.

The calculation of the animal's life span follows. It is based on a mysterious fact that, whatever the animal size, the number of heartbeats during its entire life is the same: about 1.5 billion. It is a sort of credit given to everyone, at its birth. So, based on that, one infers that the lifespan T is:

$$T \sim l^{3/4} \sim M^{1/4}$$

This scaling is compared to the observation in Fig. 2.25 [45, 46].

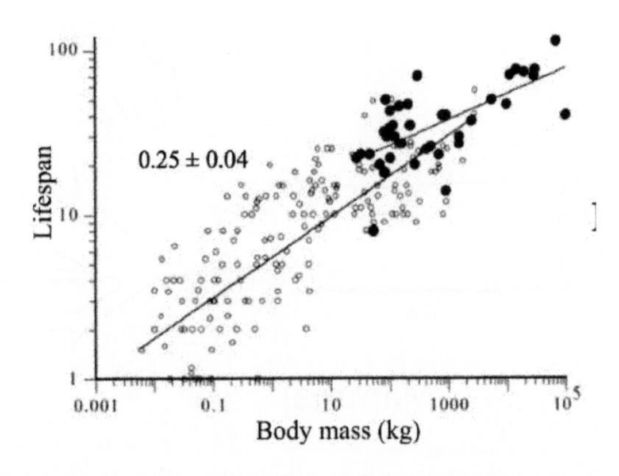

2.25: Lifespans of mammals as a function of their weight [45, 46]. The 1/4 law can be taken to represent the lifespans of all mammals (see the first line, on the left of the graph), even though, for extremely large animals, another power law could be suggested, as is done in [46]. (Reprinted from Ref. [46], with permission from Wiley and Sons. Copyright 2022.).

The spread is large, but the theoretical prediction is consistent with the observations (see the first line, on the left of the graph), even though, for extremely large animals, another power law could be suggested, as is done in [46]. The smallest mammal on earth, the pygmy shrew, a few grams in weight, lives for 12-15 months. The largest animals, the whales, one hundred tons or so, can live for 200 years. A human, with a lifespan of 70 - 80 years, stands in between. To be able to live much longer, as Methuselah (969 years [48]) or his son Lamech (777 years [49]) are said to have done, one would have to drastically reduce one's metabolism rate and, consistently, one's heartbeat frequency. The remaining problem to solve would be avoid falling into sluggishness, which would be quite challenging.

2.5.4 The running of animals

The running of animals, as a function of their size, is a more delicate subject. Naive arguments suggest that the maximum speed is independent of the size. For instance, it can be argued that, in order to acquire a kinetic energy of $\frac{1}{2}MV^2$, the animal must deploy a work equal to $F_m l$, where F_m is the muscular force. F_m scaling as l^2, this would lead to $V \sim l^0 \sim M^0$. The measurements on a large collection of animals, running on firm ground revealed the existence of an exponent, small but not equal to zero. This is shown in Fig. 7.13. There is a large spread but it appears that, below a

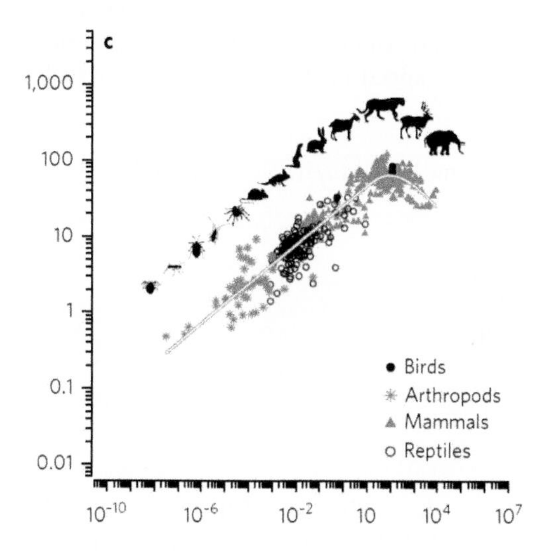

2.26: Maximum speed V (in km/h) of a variety of animals as a function of their body mass M (in kilogram). The line is the formula $V \approx 26\ M^{1/4}$ (Reprinted with permission from [52]. Copyright 2022. Springer Nature).

certain mass (on the order of 100 kg), a power law takes place, over nine decades, with an exponent equal to 0.25 ± 0.1 (see Fig. 7.13). Theoretical arguments, based on an optimization of the total power used by the animal, for struggling against air friction and supplying the body oscillation during the run [53], suggest the following power law :

$$V \sim M^{1/6}$$

which leads an exponent 0.17, consistent with the measurements. In is interesting to note, in the framework of the chapter, that the exponent 1/6 can be explained by using Π theorem: considering that air friction and gravity play roles in the maximum speed estimate, one can write:

$$V = f(\rho_a, \rho, g, l)$$

in which ρ_a is the air friction density, g is the gravity and ρ and l are the body density and animal size. Applying Π theorem, we conclude that only two dimensionless number drive the system. This leads to the following expression:

$$V = g^{1/2} l^{1/2} g(\frac{\rho_a}{\rho}) \sim M^{1/6}$$

The detailed analysis leads to $g(x) = x^{-1/3}$. At the expense of some adaptation, it has been shown that the 1/6 law also applies to birds and fishes [53].

Nonetheless, limits exist. To give a rough idea, should a global power law, such as the one shown in the caption of Fig. 7.13, be true, we would infer that the elephant runs at 211 km/h, which no one has ever seen, even when the animal has an urgency. In fact, as shown in Fig. 7.13, there is a cross-over beyond which the animal speed falls. This cross-over is linked to the kinetics of supply of the muscles in energy [52]. The existence of this cross-over implies, for instance, that the tyrannosaurus and the mammoth, whose masses are around 6 tons, do not run at 228 km/h, as the empirical scaling law of Fig. 7.13 would suggest, but at a speed on the order of 29 km/h [52]. More examples are shown in Table 2.8.

Animal	Mass (kg)	Scaling law (km/h)	Phys. limited
Velociraptor	20	55	39
Allosaurus	1400	159	24
Thyrannosaurus	6000	228	29
Triceraptos	8500	250	26
Apatosaurus	28000	336	12
Braciosaurus	78000	434	18

Table 2.8 Name, mass, scaling law and physiology-limited speeds of several prehistoric animals

For the brachiosaurus, whose weight is 78 tons, the scaling law of Ref. [52], would lead to an impressive 434 km/h, faster than the french TGV.[13] In the same line of thought, is the drawing shown in Fig. 2.27 correct ?

2.27: Is it correct to suggest that the tyrannosaurus will catch these two helpless struthiomimus dinosaurs? (Licence Pixels Library - credit Mohamad Haghani.)

[13]TGV means 'Tres Grande Vitesse', i.e. very fast trains

The flightless birds run, according to their mass (between 30 and 60 kg), at about 50 km/h. They would be easy preys for the tyrannosaurus, if he ran at 229 km/h. However, with the physiological limitation just mentioned, his speed does not exceed 29 km/h. Then according to this estimate, the birds have a good chance to escape the predator. It is interesting to add, in *Jurassic Park*, Spielberg was right to show the tyrannosaurus outpaced by the Jeep, even though the car was not driving as fast as a Ferrari.

2.5.5 The unpleasant life of the pygmy shrew

Before presenting the pygmy shrew, the smallest mammal on the planet, one must discuss the question of thermal regulation.

Mammals are endothermal animals, i.e. they regulate their temperature internally, by adapting the metabolism W, i.e. the production of heat, to the external conditions. W is an adjustment variable, whose lower value is W_b the basal metabolism, and its highest values are 4-5 W_b. W_b follows Kleiber law, as we saw in an earlier section. The equation that expresses the thermal balance of mammals is [50, 51]:

$$W \approx K(T_b - T_e)l \tag{2.8}$$

in which K is the ambient conductivity (for instance, the thermal conductivity of water, for aquatic animals), l is the animal size, and T_e and T_b are the exterior and body temperatures, respectively. For the sake of simplicity, we assume K to be constant. In order to keep the body temperature T_b constant, around 30 -37°C, W must compensate for the variations of the environment. This is done by the hypothalamus, a gland which, based on temperature sensors distributed in the body, regulates the blood flux, the sugar content, among other variables. Fig. 2.28 shows how the animals (e.g. dogs) adapt their metabolism to keep their temperature constant, in thermally changing environments.

Fig. 2.28 is currently used in the field of thermal biology. It simplifies the problem: for instance, there should not be such well-defined plateaus. We will nonetheless take this diagram as a given to discuss the effect of miniaturization.

Let us start with the cold temperatures: on the left of Fig. 2.28(A) (bottom), between 0 and 20 °, the metabolic rate W decreases linearly with the external temperature. The linear decrease is imposed by Eq. (2.8), in which T_b is constant. In this range of temperatures, the animal manages to keep its temperature at 37°. However, below 0°, this is no longer possible and the body temperature drops (see Fig. 2.28 (A), bottom). Depending on the animal, this causes death (in the case of humans), hibernation, or torpo (the case of hummingbirds at night). At higher temperatures, W reaches the basal metabolism W_b. Above this temperature, called the lower critical temperature L_{CT}, W remains constant throughout a range of temperature called temperature neutral zone (TNZ). In this range, the animal feels comfortable. The TNZ marks the range of temperature in which the animal can live, although, in practice, it can endure colder conditions. As the external temperature T_e rises again, above a temperature

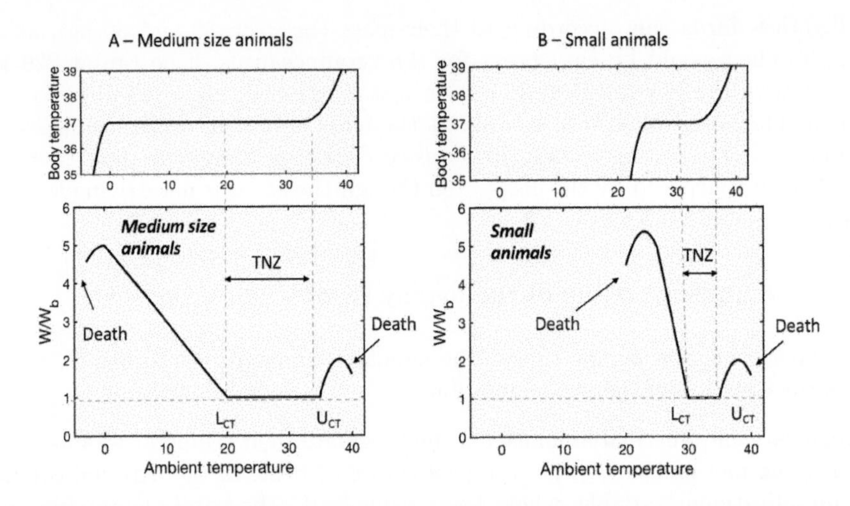

Fig. 2.28: (A) : (Top) Evolution of the body temperature as a function of the external temperature (see text); (bottom)Evolution of the metabolic rate, divided by the basal rate W_b as a function of the ambient temperature. The graph corresponds to a dog. (B): Plots similar to (A), for the case of small animals. (Top): Evolution of the body temperature with the external temperature. (Bottom): Evolution of W/W_b with the exterior temperature.

called upper critical temperature U_{CT}, the plateau ends. The animal has to struggle against heat stresses (by sweating, or lapping). Unfortunately, above a certain temperature, 40°or so, he no longer manages to keep its temperature constant and thus dies. In humid environments, evaporation is slower, and, consequently, body cooling is less efficient.

To discuss the consequences of miniaturization, it is useful to rewrite Kleibler's law in the form: $W_b \sim Al^{9/4}$, where $A \sim \rho^{3/4}B$, with ρ the animal density and $B \sim 1W/kg^{3/4}$ (we work here with orders of magnitude). We thus have:

$$W/W_b \sim KA(T_b - T_e)l^{-5/4} \tag{2.9}$$

The law is sketched in Fig. 2.28 (B). According to Eq. (2.9), with l small, the lower critical temperature, defined by W/W_b=1, cannot be significantly different from the body temperature, T_b. Consequently, the temperature neutral zone collapses. Moreover, because of the steepness of the slope given by Eq. (2.9), it would cost a great deal of energy, for the small animal, to escape the neutral zone. In such circumstances, it becomes difficult to find a place to live. From the plot of Fig. 2.28 (B), one can estimate the sizes of the smallest mammals living on earth and in water. This size is given by the following relation (for the ease of reading, we replace TNZ by ΔT_{NZ}):

$$l \sim (KA\Delta T_{NZ})^{4/5} \tag{2.10}$$

In air, with $\Delta T_{NZ} \sim 1°$, $K \sim 10^{-2} W m^{-1} K^{-1}$ and $A \sim 1 W/m^{9/4}$, one finds 3 cm. In water, where $K \sim 1 W m^{-1} K^{-1}$, one finds 1 m. These estimates are rough, but their orders of magnitude are consistent with the sizes of the smallest mammals living on earth (the pygmy shrew, 2 g weight is 5 cm large (see Fig. 2.29)) and in the sea (the sea otter is one metre big).

The pygmy shrew has an unpleasant life: it must eat all the time to keep the metabolic rate high. During the night, it enters a state of torpor to preserve energy. This reduces the chance to make beautiful dreams. When it awakes, in the morning, it must rush out to eat. The animal hates heat, and prefers to live in the mid-northern part of Europe and America. It also hates cold, and spends a part of its life in niches.

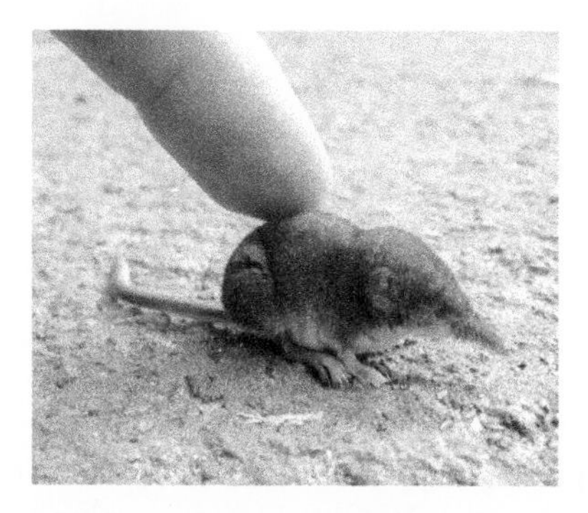

2.29: The pygmy shrew. Its heart beats at 500-800 Hz. Its gestures are fast (a few milliseconds.). The animal ingests its own weight in insects, worms, and spiders on a daily basis. A venom, located in its mouth, paralyses the prey (From Flickr library, Simon Ward.).

Below l, mammals do not survive, but poikilotherm animals, called cold-blooded animals, such as insects, or reptiles, do. For them, temperature is not controlled internally, but fixed by the environment. This looks clever but, on the other hand, as their temperature is out of control, their biological complexity cannot compete with that of the mammals.

2.5.6 Pro and cons being small

There are many advantages to being small:

- Adhesion forces, as we saw, scaling as l and gravity forces as l^3, insects will not feel the effect of gravity as they climb a wall or wander on a ceiling.

- The same argument applies for walking on water, provided that foots are hydrophobic, which is the case.

- Small animals, like ants, must avoid entering bubbles (see Fig. 2.30): muscular forces scaling as l^2 and capillary forces as l, they would be unable to exit, and their colleagues unable to free them.

- Everything goes fast: movements, heartbeats (when there is a heart). We could say that, in the small world, size and time are miniaturized.

- Small animals jump high. The argument is the following: the muscular energy necessary to make a movement of amplitude l is thus proportional to l^3. This quantity must be compared with the gravitational energy of the jump E_p:

$$E_p \sim mgH \sim l^3 H.$$

Comparing the two energies, the height of the jump H scales as l^0. The height attained is thus independent of the size of the animal. This implies that the flea jumps just as high as a horse.[14]

2.30: This drawing, inspired by a famous animated drawing, represents the significant effort it would take for an ant to free a comrade imprisoned in a bubble. At the scale of an ant, capillary forces are very significant with respect to the muscular forces the animal can exert (after [55]).

If we lived in a submillimetric world, the principal phenomenon that would concern us would be capillarity and not gravity or inertia. We would fear getting stuck to the surface of liquids, but we would not risk being hurt by an impact. At this scale, we would undoubtedly have invented (if our intelligence was still intact) numerous machines driven by surface tension. It would seem natural that in our microworld, automobile accidents would not be caused by collisions, but by drops of water left on the road.

2.5.7 Three errors in 'Honey, I Shrunk the Kids' (Walt Disney)

A comic sci-fi film showing life in a miniaturized world was released by Disney in 1989. It met worldwide success. In the scenario, three kids fall into the beam of a shrinking machine built by the dad. They suddenly shrink to a size of 6 mm. The size being less than a pygmy shrew, they need to eat continuously to keep themselves in the neutral temperature range (TNZ), as we saw above. Since the movie director let them starve during a full day, they could not survive. There are several physical errors in the movie, and they are instructive to point out, because they illustrate a number of notions discussed in the book. Three examples are mentioned below:

A The children, who have been miniaturized, are thrown into the back of the garden. They want to come back home. During their journey, the encounter a piece of water, on which a dead insect is floating. The insect is hydrophobic, with a contact

[14]fleas jump 30 cm and horses 1 metre. The jumps are thus comparable, in terms of order of magnitude

angle on the order 100°(this notion will be developed in Chapter 4). Consequently, there should be a meniscus a few millimetres high which, on the scale of the screen, should neatly appear. Instead the water surface is flat.

B The children wander in the night, desperately trying to come back home. The flame they carry less than 1 mm from their fingers, should cause severe burnings. However, the children do not feel any harm. Later in this chapter, we will return to the question of heat conduction in small systems.

C The children are now at home. The dad tries to speak to them. With their small mouths (one hundred microns or so), the children speak with extremely acute voices (in the 100 kHz range), which the dad should not be able to hear. Also, what the dad says should not be captured by the small ears of the children. In the film, they talk without difficulty. Later in this chapter, we will return to the question of resonance frequencies in small systems.

2.6 Miniaturization of electrostatic systems

2.6.1 Dielectric breakdown is retarded in miniaturized systems

The phenomenon of sparks between two electrodes placed in a gas originates from the fact that gas molecules, when subjected to an intense electric field, are ionized, thus forming a plasma. Ionization is propagated through a cascade process, in which a large number of electrons are liberated, and an intense electric current circulates. This current is accompanied by a luminous emission that gives rise to the phenomenon of the electric arc. In air, under normal conditions, the breakdown electric field strength is on the order of 3.4 MV/m (34 kV/cm). This means that, under these conditions, the voltage needed to arc a 1-metre gap is about 3.4 MV.

In miniaturized systems, much higher electric fields can be produced without the generation of an electric arc. In gases, this effect is well-represented by the Paschen curve, depicted in Fig. 2.31.

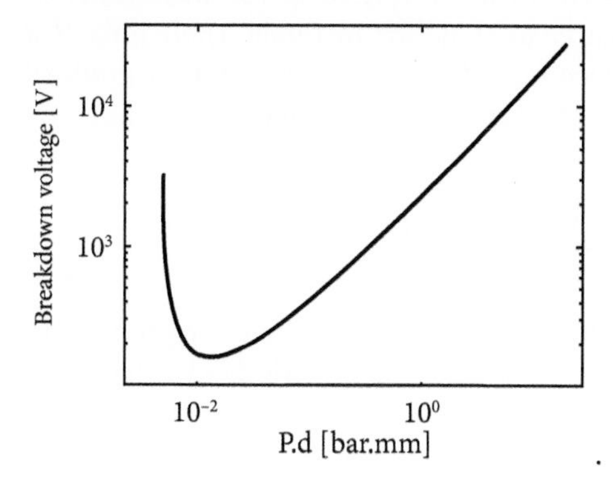

2.31: Evolution of the dielectric breakdown voltage for a parallel-plate capacitor as a function of a variable X equal to pressure times distance between the plates, i.e $X = Pd$; the curve is obtained in air, at 300 K

The Paschen curve represents the dielectric breakdown voltage between two electrodes, measured in air, at 300 K. This curve is traced as a function of $X = Pd$, where P is the pressure and d the distance between the plates. We see that at large values of X, the slope being unity, the breakdown electric field strength E_d is proportional to the pressure but does not depend on d. At one bar, we have:

$$E_\mathrm{d} \approx 3.4\mathrm{MV/m}(= 34\mathrm{kV/cm}).$$

At small values of X the breakdown electric field strength increases considerably The cross-over between the two regimes takes place when $d = \lambda_D$, where λ_D is the mean free path, as we saw in a previous section (see Eq. (2.1)). Physically, when the mean free path becomes comparable to the distance between electrodes, the majority of molecular collisions tends to take place between the gas and the electrodes, and not within the gas itself. This situation inhibits the formation of the cascade process and consequently makes it more difficult for an electric discharge to form in the gas. The effect is strong. For instance, in a 5 μm channel, the breakdown voltage is on the order of 30 MV/m, i.e. ten times the value at large X. In water, the physics of discharge is different. It is suggested that the breakdown is initiated by field emission at the interface of preexisting microbubbles [56]. In practice, the electric breakdown for millimetric gaps is in the 10 -100 MV/m range.

2.6.2 Miniaturization of capacitors

In the field of micro electromechanic systems (MEMS), many devices use miniaturized parallel-plate capacitors as actuators. We will consider the plane capacitor geometry to establish scaling laws. The electrostatic energy W_e stored by a parallel-plate capacitor reads:

$$W_\mathrm{e} = \frac{1}{2}CV^2 = \frac{\epsilon_0 S V^2}{2d},$$

where C is the capacitance, S is the surface area of the plates, ϵ_0 is the dielectric constant, V is the voltage applied on the edges of the capacitor, and d is the distance between plates, whose lateral dimensions are supposed comparable to d. Here we introduce a characteristic scale l, which is on the order of d and, in the meantime, on the order of the lateral dimension of the plate, so that, for instance, their area S is on the order of l^2. The forces used to displace one of the plates result in a gradient of the energy W. For example, a displacement normal to the plane of the capacitor necessitates a force equal to:

$$F = -\frac{\delta W_\mathrm{e}}{\delta z} = \frac{\epsilon_0 S V^2}{2d^2} \sim \epsilon_0 E^2 l^2$$

(z being an axis perpendicular to a plate), where E is the electric field created in the capacitor. In terms of scaling laws, and considering that the electric field is fixed, we find that the energy varies as l^3, and the force as l^2. This is what was noted in Table 2.7. If E is the breakdown field, we have at our disposal an estimation of the *maximum* forces that can release an electrostatic system of a given size. This electrostatic force is dominant over gravitational forces and inertia (which vary as l^3); there is thus no difficulty in exerting rapid accelerations in microsystems. This fast speed is harnessed in car accelerometers.

2.6.3 Electrostatic microactuators

The possibility of creating elevated electric fields has led to the creation of electrostatic microactuators. Figure 2.32 shows an example [57].

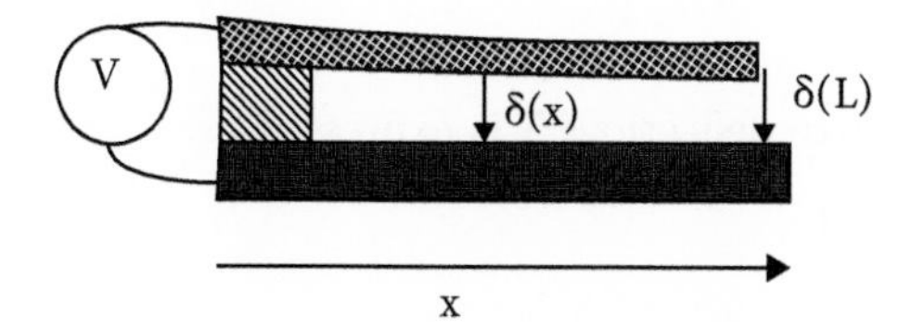

2.32: Cantilever beam, where the beam and the substrate are conductive. The beam, subjected to an electrical potential difference V, deflects and thus can form an actuator.

Here we have a beam and a base subjected to an electrostatic potential, thus forming the two plates of a capacitor. $f(x)$ is designated as the density of electrostatic attraction force (per unit surface) evaluated at the position x.

The expression of the density of force $f(x)$ can be obtained from the electrostatic force given in the previous section. We obtain:

$$f(x) = \frac{\epsilon_0}{2}\left(\frac{V}{d - \delta(x)}\right)^2,$$

where V is the applied voltage, ϵ_0 is the dielectric constant of the vacuum, d is the distance between the beam and the plate for $x = 0$, and $\delta(x)$ the deflection of the beam at the abscissa x (Fig. 2.32).

This force produces a deflection $\delta(x)$, which is controlled by a local elastic equilibrium, given by the relation:

$$\mathrm{d}\delta(x) = \frac{x^2}{6EI}(3L - x)wf(x)\mathrm{d}x$$

where E is the Young's modulus of the beam, I is its moment of inertia, L is its length and w is its width. We recall the expression of the moment of inertia of a beam of width w and thickness h:

$$I = \frac{1}{12}wh^3$$

To determine the deflection at the end of the beam, one must integrate this equation. The profile of the beam is governed by the equation:

$$\delta(x) \approx \left(\frac{x}{L}\right) \delta(L).$$

This permits the determination of the relation between the total force F applied on the beam, and its relative maximum deflection given by:

$$\Delta_b = \delta(L)/d$$

The calculation gives, in the limit of small deflections:

$$F = \frac{\epsilon_0 w L^4 V^2}{2EId^3} \approx \frac{4\Delta_b}{3}$$

In terms of scaling laws, we see that the relative deflection Δ is expressed as:

$$\Delta_b \sim \frac{V^2}{l^2}$$

which shows that it is possible to obtain significant displacements by miniaturizing while maintaining a fixed potential difference (assuming that the breakdown field is not surpassed). This possibility, illustrated in the case of an extremely simple electrostatic actuator, is actually used in a large number of MEMS.

2.6.4 The electrostatic micromotor

One of the first electrostatic micromotors made using MEMS technology is represented in Fig. 2.33.

2.33: First micromotor created in 1988 by Fan et al. [58]. The rotor has a diameter of 200 μm. (Photo courtesy of Richard S. Muller.)

The principle of how this motor functions lies in the fact that the rotor will try to align the region with the strongest dielectric constant with the regions with the strongest electric field. Since the field is rotating, the rotor will tend to follow the maxima of the field, and will thus begin to turn. To analyse the functioning of the motor, the forces in play must be averaged along the angular coordinate, taking into account the fact that the electrostatic field produced by the stator is a rotating field.

Practically speaking, couples are produced on the order of nano- or piconewtonmeters, powers on the order of the microwatt, for thousands or tens of thousands turns per minute. One must envisage rapid rotations in order to produce significant power. Impressive devices are currently being created, including ones that function at several million turns per minute. One example is a micro-turbine developed at MIT which functions at 2.4 million turns/minute and produces 20 W with a yield of a few tens of per cent.A turbine like this, when put into rotation by combustion gases, allows the conversion of chemical energy into mechanical and electric energy. However, in practice, the turbine is complicate to realize.

2.7 Miniaturization of electromagnetic systems

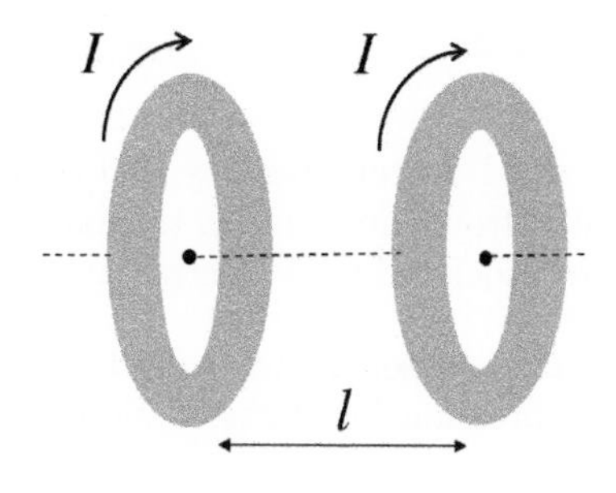

2.34: Two circular coils of diameter l, separated by a distance l, through which an electric current I circulates.

We consider two circular coils of diameter l, separated by a distance l, through which a current I circulates (see Fig. 2.34). We also assume that the cross-section of the wire carrying the electric current is on the order of l^2. This hypothesis may seem astonishing : normally, electric wires have cross-sections much smaller than coil diameters. This hypothesis is not unrealistic, however, for microcoils created with the microfabrication techniques presented in Chapter 6. An example is given in Fig. 2.35.

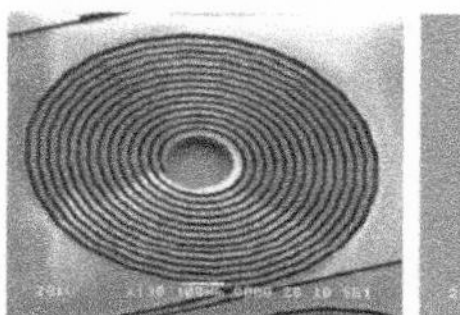

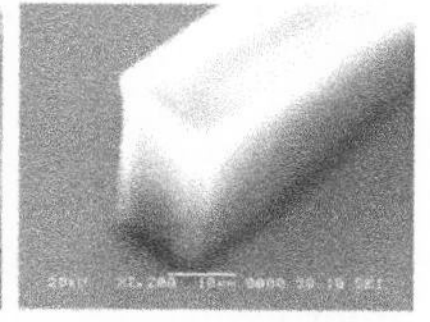

2.35: Microcoil fabricated by H. Fujita et al. [59]. The figure on the right shows a conducting cross-section forming the microcoil.

Thus, the magnetic field B produced by the coil is

$$B \sim \mu_0 jl$$

where j is the current density, which we will consider as fixed. The magnetic energy stored in the coil is:

$$E_{\rm m} \sim \frac{B^2}{\mu_0} l^3 \sim j^2 l^5$$

These expressions show that it is difficult to produce elevated magnetic fields with miniaturizing coils. For a system with a size of 100 μm the maximum magnetic field that can be produced, without excessive heating, is on the order of 100 mT, which in a large number of situations would be considered to be a weak level of induction. Concerning the interacting forces between coils, we have:

$$F \sim jBl^3 \sim j^2 l^4$$

We retrieve the exponent shown in Table 2.7. Consider also the case of a coil immersed in a fixed magnetic field B, produced for example by a large electromagnet. In this case, by using Laplace's law, we obtain:

$$F \sim IBl \sim jBl^3$$

We see here that the force varies as l^3 for a fixed current density j. In all cases, whether using a magnetic field produced locally or from the exterior, the exponents of the electromagnetic forces are higher than those associated with electrostatic forces.[15]

To illustrate these calculations, we proceed to a comparison between electrostatic and electromagnetic microactuators. We recall the expression for electrostatic forces for a parallel-plate capacitor:

$$F \sim \epsilon_0 E^2 l^2$$

where E is the breakdown electric field. For an electric field of 30 kV/cm and a device of 10 μm, we obtain:

$$F \sim 10 \text{ nN}$$

For an electromagnetic system containing N coils in series, we have the following order of magnitude for the force:

$$F \sim \mu_0 N j^2 l^4$$

Taking $N = 100$ coils, $j = 10^7$ A/m^2 and $l = 10$ μm, we obtain the following order of magnitude for the force exerted between two microcoils:

[15] We note that, in principle, it is possible to augment the current density in conductors, by maintaining (as we will see later) intense thermal exchanges in miniaturized systems. A discussion can be found in [57].

$$F \sim 100 \text{ pN}$$

The force produced by the microcoils is less than capacitors. At which scale do electrostatic forces become dominant with respect to electromagnetic forces? We consider two 'prototypic' situations chosen arbitrarily: the case of the parallel-plate capacitor and that of mutually coupled coils. Re-examining the expressions established above, the scale that we find is on the order of a few hundred micrometers. More generally, we allow that ten or so micrometres correspond to the physical limit beyond which electrostatic forces begin to dominate over electromagnetic forces. Ref. [57] presents a detailed discussion of the subject.

2.8 Miniaturization of mechanical systems - the vibrating microbeam

A cantilever beam possesses a resonance frequency determined by the following relation:

$$f \approx \frac{hc}{2\pi L^2},$$

where h is the thickness of the beam, L is its length, and c is the speed of sound in the beam. This expression is obtained by resolving the elasticity equation. In terms of scaling laws, we have the relation $f \sim l^{-1}$. This law can be directly obtained by noting that an acoustic wave associated with the mechanical resonance has a wavelength on the order of the size of the object. We thus have the relation:

$$f \sim \frac{c}{l},$$

which corresponds to the expression deduced from the previous formula.

We find that the resonance frequency is proportional to the inverse of the size of the system. In practice, the law implies that, in the film *Honey, I shrunk the kids*, previously mentioned, the father and the son, contrarily to what the director imagined, could not exchange together, which, in a family, is often problematic. The same scaling law is used to create resonators working in the domain of radiofrequencies. These are called 'RF MEMS' [60]. Figure 2.36 shows an example of one.

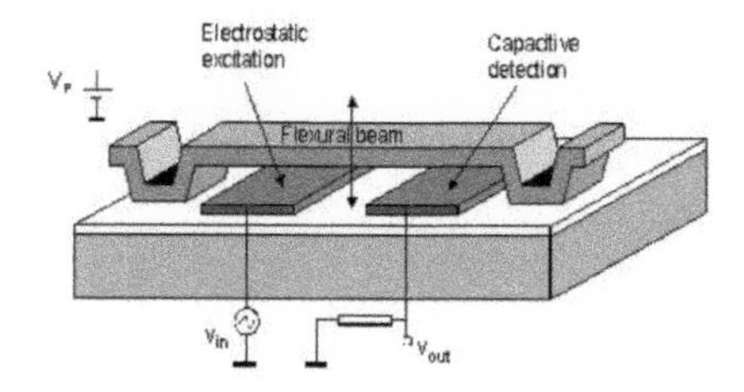

2.36: Sketch of a mechanical resonator excited by an electric field. In practice, the material used is piezoelectric

Figure 2.37 presents a SEM image of a silicon system, made with MEMS technology [61]. In terms of orders of magnitude, we have:

$$c \approx 7470\text{m/s}.$$

Thus, for a beam 1 μm thick and 3 μm long, the formula given above leads to:

$$f \approx 100\text{MHz}.$$

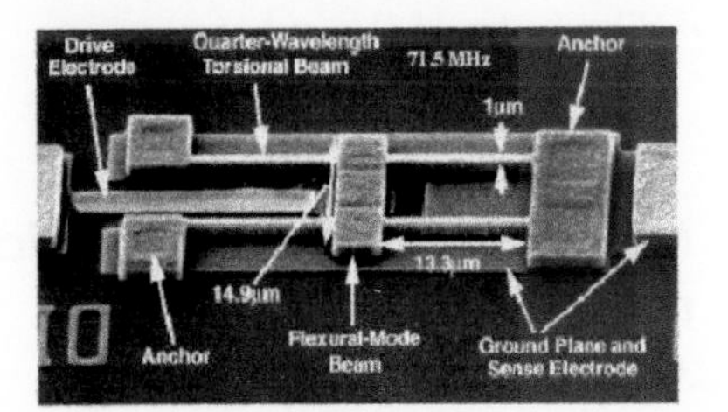

2.37: A resonator functioning at 71.4 MHz, made with silicon at the University of Michigan [61].

The resonance frequency of the sillicon beam of Fig. 2.37, 71.4 MHz, is consistent with the estimate. Resonance falls in the radiofrequency range and therefore the device can be used as an antenna. RF MEMS offer advantages, compared to traditional technologies, in terms of cost, integration, linearity, packaging, power consumption and weight. The domain is vast [60], and its presentation stands beyond the scope of this book.

A note on quality factors. We recall the definition of a quality factor Q, for dampened oscillators. These systems are traditionally modeled by the following equation:

$$m\frac{\mathrm{d}^2 x}{\mathrm{d}t^2} - b\frac{\mathrm{d}x}{\mathrm{d}t} - kx = 0,$$

where m is the mass of the spring, b is the attenuation coefficient, and k is the spring constant. For this type of system, the quality factor is:

$$Q = \omega_0 \frac{m}{b},$$

where ω_0 is the resonance frequency defined by:

$$\omega_0 = \sqrt{\frac{k}{m}}.$$

In the Fourier domain, the resonance curve is given by the expression:

$$A(\omega) = \frac{1}{\sqrt{\left(1 - \frac{\omega^2}{\omega_0^2}\right)^2 + \frac{\omega^2}{Q^2\omega_0^2}}}.$$

in which A is the (normalized) oscillation amplitude and ω the oscillation frequency. The width of the resonance curve is directly related to Q: this width (calculated at mid-height, i.e. for a value of $A(\omega)$ equal to $\frac{1}{2}$), is on the order of the inverse of Q. Thus, the higher the Q, the sharper the resonance. The factor that limits Q is the friction represented by the coefficient b.

For a cantilever beam, the quality factor in a non rarefied gas is estimated by the expression [57]:

$$Q = \frac{\sqrt{E\rho} w h^2}{24 \nu L^2},$$

where ν is the kinematic viscosity of the gas, E is the Young's modulus of the beam material, ρ is its density, w is the width, h is the height, and L is the length. This law indicates that the quality factor decreases with the scale, which may suggest that miniaturization, although interesting from an integration standpoint (all components can be made in silicon, and fabricated with MEMS technology), degrades the sharpness of the resonance. However, in practice, quality factors of current microresonators attain values on the order of1,000. Operating in a vacuum leads to quality factors on the order of 10^4 -10^5.

2.9 Miniaturization of thermal systems

We will treat, in some detail, the miniaturization of thermal systems in Chapter 5. Here, we focus on scaling laws, consistently with Table 2.7. Heat transfers by conduction are governed by the Fourier equation, which is written as follows:

$$q = -K \nabla T$$

where K is the thermal conductivity of the medium, T is the local temperature, and q is the heat flux, i.e. the quantity of heat traversing a surface element per unit surface (the units of q are thus energy per unit time and area, or power per unit of area). The flux density varies as the inverse of the scale, at a fixed temperature difference. A typical situation is a fluid at rest in a cylindrical container, thermally isolated on the sides, and whose ends are in contact with thermostats that impose a temperature difference $\Delta T = T_2 - T_1$ (Fig. 2.38). Under these conditions, the heat flux going through the fluid, along the cylinder axis, is written:

$$Q = \frac{K S \Delta T}{l},$$

where S is the cross-sectional area of the cylinder, and l its length. The following scaling law comes from this:

$$Q \sim K l \Delta T,$$

which governs the heat flux. This law was displayed in Table 2.7, with the assumption that the temperature difference is fixed.

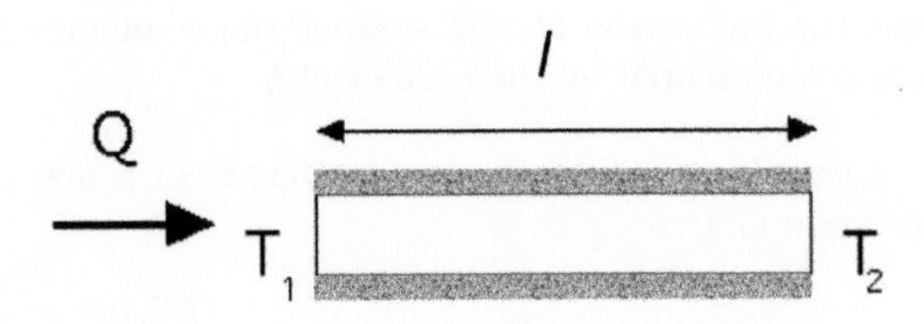

2.38: Thermally isolated bar carrying a heat flux Q, with notations used in the text.

Another quantity of interest is the thermal resistance, defined as the ratio $\Delta T/Q$. According to the previous equation, this ratio is on the order of l. The smaller the size, the smaller the thermal resistance. In the film *Honey, I shrunk the kids*, the thermal resistance being small, children should burn, as in an horror movie. The director preferred an unphysical and more pleasant scenario.

We now suppose that sources of heat exist, in the fluid: Joule heating or heating from a chemical origin (if the fluid is the site of exothermic reactions). In both these cases, we can consider that the heat produced by a volumetric source is:

$$Q_{\mathrm{v}} \sim l^3$$

which represents an exponent higher than that associated with heat evacuated by conduction. We can thus conclude that volumetric heat sources can be easily thermalized in miniaturized systems.

It is interesting to consider this from a dynamic point of view: the heat flux associated with a temporal variation of temperature $\mathrm{d}T$ and a time $\mathrm{d}t$, is governed by the following relation:

$$q = \rho C_{\mathrm{P}} \frac{\mathrm{d}T}{\mathrm{d}t}$$

where ρ is the volumetric mass, C_{P} is the specific heat at constant pressure, and t is the time. Thus the order of magnitude of the time constant of thermalization associated with an object of dimension l is written:

$$\tau \sim \frac{\rho C_{\mathrm{P}}}{K} l^2 \sim \frac{l^2}{\kappa}$$

where κ is the thermal diffusivity of the environment. Consistent with the preceding remarks, this expression shows that miniaturization considerably accelerates the return to thermal equilibrium of a fluid at rest subjected to a sharp temperature change. In other words, miniaturization radically reduces thermal inertia. The preceding logic shows that exo- or endothermic chemical reactions can in principle be finely thermalized in microsystems. This characteristic is very useful for chemical engineering. In general, temperature exchangers currently used by industrial reactors tend to develop unwanted parasite reactions. These reactions limit the selectivity of the process. There are also reactions, impossible to produce industrially in a safe manner, that miniaturization allows to be controlled. For these types of systems, miniaturization offers a source of important improvements. This led to the development, in the 1990s, of the domain of micro reaction technology (MRT), presented in Chapter 1 to the book.

2.10 Sampling and throughput

2.10.1 Microfluidics and low concentration samples

For samples of low concentration, the number of molecules per unit of volume is (by definition) small, and the question is raised whether a given microfluidic device incorporate enough volume to be able to perform a detection. Let us consider a sample of molar concentration C_s. The minimal volume containing at least one molecular entity is given by:

$$v = \frac{1}{C_s N_{\mathrm{A}}}$$

where N_{A} is the Avogadro number. For a cubic microfluidic chamber of side 10 μm, and thus a volume of 1 pL, the minimum concentration C_{min} for which one molecule is present in the chamber is on the order of 10^{-12} mol/L. Should the sample concentration be lower, there would be no molecule in the chamber, and therefore nothing to detect. In the field of diagnostics, a concentration of 1 pM/L is often encountered. For this reason, the sensitivity of pregnancy tests is on the order of fM/L, i.e. three decades below C_{min}. In the molecular diagnostics of SARS COV2, the sensitivities are a few RNA copies per μL. With such low viral loads, there is no chance to capture a single RNA strand in the picoliter chamber.

The solution is obviously to increase the chamber size as much as possible. In practice, typical volumes of detection are several microlitres or tens of microlitres, not picolitres This covers most of the situations of interest. However, in some applications, this is not enough. In the agricultural domain, it is requested to detect one bacteria (for instance salmonella) per 25 ml. This is impossible with microfluidic devices. Then, in this case, a preliminary amplification by culturing, called 'enrichement', is necessary. An early discussion of this type of questions, in the context of the analytical applications of microfluidics, was given by Manz et al [62].

2.10.2 Microfluidics and throughput

In the 1990s, microfluidic devices were considered as suitable for performing analytical studies, but too small for producing materials in quantities of industrial interest. The question of the throughput can be formalized in the following manner: the quantity of volume produced, per unit of time, by a device is:

$$Q \sim Ul^2$$

in which U is the flow speed and l the characteristic cross-sectional size of the channel. With U=1cm/s and $l = 100\ \mu$m, one obtains a production of 10 ml per day. This is far too small for industrial application. Over the years, progress has been made along two lines:

Numbering up : By parallelizing 1000 times, the production reaches 10 litre per day. This number is relevant industrially for high added value products. Numbering up is currently used for producing emulsions under litres per hour throughputs (see, for instance, [63]). An example of a parallelized step emulsion device is shown in Fig. 2.39.

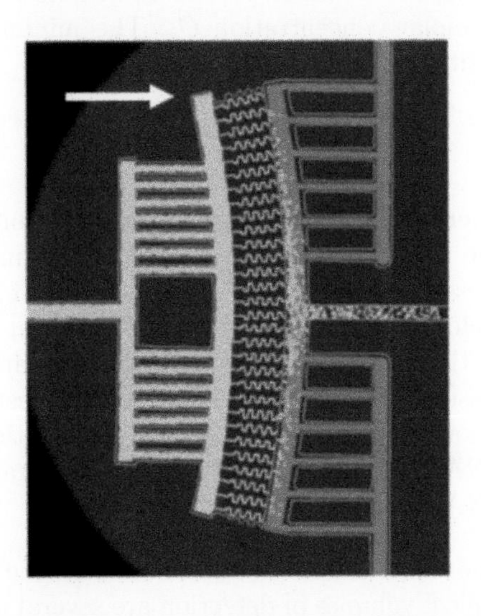

2.39: A 40 fold parallelized device, producing droplets (white dots) in the 30 μm range, at rates on the order of hundreds of μL per minute. (MMN, IPGG, ESPCI.)

The highest parallelisation of droplet emitters, at the time of writing this book, has been reported in Ref. [64] (see Fig. 2.40).

Using a silicon and glass device, 10260 droplet emitters could be placed in parallel, producing trillions of monodisperse droplets per hour. The flow-rate is on the order of liters per hour, and the mass of particles 2.4 ton per year, for a system that could be held in the hand. We may imagine a factory, in which 1000 such systems are put in parallel, leading to 2400 tons of particles produced per year. In many domains (culture cells, functionalized particle,...), contrarily to current thinking, microfluidic technology can respond, on the scale of the planet, to production needs.

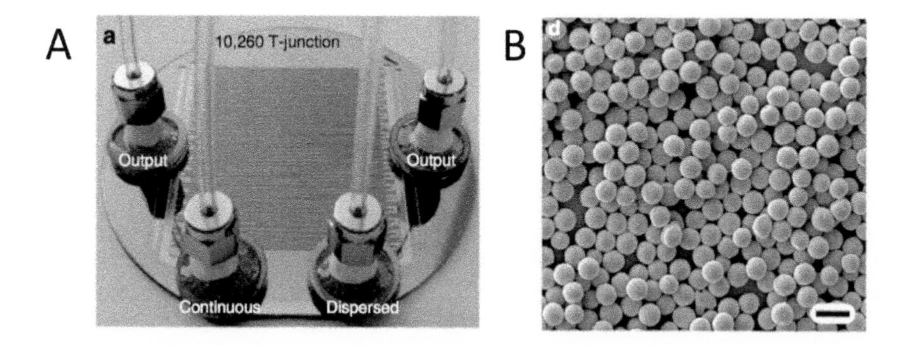

Fig. 2.40: (A) 10260 droplet emitters were placed in parallel, producing trillions of monodisperse droplets per hour. (B) The flow-rate is on the order of liters per hour, and the mass of particles 2.4 tons per year [64]. (Licensed under a Creative Commons Attribution 4.0 International Licence.)

Millifluidics : By raising the dimensions by a factor of 30, we switch to millifluidics, and can reach, by increasing the speed by a factor of 100, 1 m^3 per day, i.e. 360 tons per year (assuming liquid densities equal to 1 ton/m^3). These quantities touch industrial levels. These orders of magnitude are those of micro reaction technology (MRT). This led to realization of miniaturized' industrial plants, as we mentioned in Chapter 1 [65].

Today, one may consider that there is no incompatibility between micro/millifluidics and industrial production of materials.

The question of the information throughput, i.e. the quantity of information produced, per unit of time, by microfluidics, is entirely different. Microfluidics droplets are typically produced at a 1 kHz rate and each droplet conveys one information (for instance, whether an enzyme is produced or not). We thus have the possibility to gather millions of items of information per hour. Because of this large number, microfluidic technology is currently used in domains requesting large screening capacities, such as cancer research, combinatorial chemistry, genome sequencing and drug discovery. An example is shown in Ref. [66].

References

[1] Q. Wang, A. Desai, Y. Tai, L. Licklider, T. Lee, *Proc. MEMS* Orlando, 523 (1999).
[2] D. M. Smilgiesa, E. Folta-Stogniewb, *J. Cryst. Crystal.*, **48**, 1604 (2015).
[3] C. Leung, A. Bestembayeva, R. Thorogate, J. Stinson, A. Pyne, C. Marcovich, J. Yang, U. Drechsler, M. Despont, T. Jankowski, M. Tschope, B. W. Hoogenboom, *Nano Lett.*, **12**, 3846 (2012).
[4] P. J. Flory, *Principles of Polymer Chemistry*, Cornell University Press (1953).
[5] G. V Guinea, F. J Rojo, M. Elices, *Engineering Failure Analysis*, **11**, 705 (2004).
[6] D. Smith, S. Chu, *Science*, **281**, 1335 (1998).
[7] N. Crisona, T. Strick, D. Bensimon, V. Croquette, N. Cozarelli, *Gene Dev.*, **14**, 22, 2881 (2000).
[8] B. Alberts, D. Bray, J Lewis, M. Raff, K Robert, J. D. Watson, *Molecular Biology of the Cell*, Garland Publishing, New York and London (1994).
[9] M. B. Elowitz, A. J. Levine, E.D. Siggia, P. S. Swain, *Science*, **297**, 1183 (2002).
[10] L. Mazutis, J. Gilbert, W. Ung, D.A. Weitz, A.D. Griffiths, J.A. Heyman, *Nature protocols*, **8**, 870 (2013).
[11] S. Elizabeth. H. S. Shevkoplyas, J. Apfeld, W. Fontana, G.M.Whitesides, *Lab On a Chip*, **7**, 1515 (2007).
[12] A. San-Miguel, H. Lu, *Microfluidics as a tool for C. elegans research*, WormBook Ed. (2013).
[13] J.N. Stirman, M. Brauner, A. Gottschalk, H. Lu, H.*J. Neurosci. Methods*, **191**, 90 (2010).
[14] G. Karniadakis, A. Beskok, *Micro Flows*, Springer Verlag (2002).
[15] J. Swenson, S. Cerveny, *J. Phys.: Condens. Matter*, **27**, 033102 (2015).
[16] V. K. La Mer, *JACS.* **74**, 2323 (1952).
[17] P. Laval, A. Crombez, J-B. Salmon, *Langmuir*, **25**, 1836 (2008).
[18] E. K. Bigg, *Proc. Phys. Soc.*, **B 66**, 688 (1953).
[19] P.G. De Gennes, F. Brochard-Wyart, D. Quéré, *Drops, bubbles, pearls, waves*, Springer (2004).
[20] M. Moseler, U. Landman, *Science*, **269** , 1165 (2000).
[21] J. Israelachvili, *Intermolecular and Surfaces Forces*, Academic Press, 2nd edn (1991).
[22] F. London, *Zeitschrift für Physik*, **63** (3–4), 245 (1930).
[23] D. Tabor, R. Winterton, *Proc. R. Soc. Lond. A*, **312**, 435 (1969).
[24] J. Israelachvili, D. Tabor, *Proceedings of the Royal Society of London. A*, **331**, 19 (1972).
[25] G. Binnig, C. F. Quate, C. Gerber, *Phys. Rev. Lett*, **56**, 930 (1986).
[26] C. Argento, R. H. French, *Journal Applied Physics*, **80**, 6081 (1996).

[27] F. L. Leite, C. C. Bueno,, A.L. Da Roz, E. C. Ziemath, O.N. Oliveira Jr, *Int. J. Mol. Sci.*, **13**, 12773 (2012).

[28] E. M. Lifschitz, *Soviet Phys. JETP (Eng. Transl.)*, **2**, 73 (1956).

[29] T. Kitamori et al, *10th International Conference on Miniaturized Systems for Chemistry and Life Sciences*, MicroTAS 2006, Tokyo, Japan (2006).

[30] C. M. Cejas, F Monti, M. Truchet, J.P. Burnouf, P. Tabeling, *Langmuir*, **33**, (26), 6471 (2017).

[31] Y. Tian, N. Pesika, H. Zeng, K. Rosenberg, B. Zhao, P. McGuiggan, K. Autumn, J. Israelachvili, *Proceedings of the National Academy of Sciences*, **103**, 19320 (2006).

[32] K. Autumn, A. Dittmore, D. Santos, M. Spenko, M. Cutkosky *The Journal of Experimental Biology*, **209**, 3569 (2006).

[33] S. Maheshwari, M. van der Hoef, X. Zhang, D. Lohse, *Langmuir* **32**, 43, 11116 (2016).

[34] B. Cabane, S. Hénon, *Liquides. Solutions, Dispersions, Émulsions, Gels*, Belin (2003).

[35] J Eggers, *Physical review letters*, **89**, 084502 (2002).

[36] H. Daiguji, Y. Oka, K. Shirono,*Nano Letters*, **5**, 2274 (2005).

[37] L. Bocquet, P. Tabeling, *Lab on a Chip*, **14**, 3143 (2014).

[38] K. Mawatari, Y Kazoe, H. Shimizu, Y. Pihosh, T. Kitamori *Anal.Chem*, **86**, 4068 (2014).

[39] E. Guyon, J-P. Hulin, L. Petit, *Hydrodynamique Physique*, CNRS Edition, 2nd edn (2001).

[40] Galileo Galilei, *Dialogue Concerning Two New Sciences* (1638).

[41] D'Arcy Wentworth Thompson, *On Growth and Form*, revised version, Cambridge University Press (1942).

[42] M. Karamanou, G. Androutsos *Thorax*, **68**, 978 (2013).

[43] M. Kleibler, *Physio. Rev*, **27**, 511 (1947).

[44] J. B. West, B. H. Brown, *Physics Today*, September (2004).

[45] T. Lecuit, *Course in College de France* (2019).

[46] L. Wittig, *Biological Reviews*, **83**, 259 (2008).

[47] A. C. Economos, *J. Theor. Biol*, **80**, 445 (1979).

[48] Bible, Genesis 5.25.

[49] Bible, Genesis 5.31.

[50] P. F. Scholander, R. Hock, V. Walters, F. Johnson, L. Irving, *Biol. Bull.*, **99**, 237 (1950) .

[51] T. S. Fristoe, J.R. Burger, M. A. Balka, I. Khaliq, C. Hofd, J. H. Brown, *PNAS*, **112**, 15934 (2015).

[52] M R. Hirt, W. Jetz, B. C. Rall, U. Brose, *Nature Ecology*, **1**, 1116 (2017).

[53] A. Bejan, J. H. Marden, *The Journal of Experimental Biology*, **209**, 238 (2006).

[54] G. B. West, J. H. Brown, B. J. Enquist, *Science*,**276**, 122 (1997).

[55] C. J. Kim, *Proc. Symp. Micromachining and Microfabrication*, **4177** (1998).

[56] R. P. Joshia, J. Qian, G. Zhao, J. Kolb, K. H. Schoenbach *Journal of Applied Physics*, **96**, 5129 (2004).

[57] G. Kovacs, *Micromachined Transducers*, WCB, McGraw Hill (1998).

[58] L. C. Fan, Y. C. Tai, R. S. Muller, *IEEE Trans. Electron. Devices*, **ED-35**, 6, 724 (1988).

[59] L. Houlet, G. Reyne, T. Iizuka, T. Bourarina, E. Dufour-Gergam, H. Fujita, *SPIE International Symposium on Microelectronics and Micro-Electromechanical Systems*, Adelaide (Australia), **4592**, 17, (2001).

[60] G. M. Rebeiz, *RF MEMS: theory, design, and technology*, John Wiley & Sons (2004).

[61] A. C. Wang, J.R. Clark, C.T.. Nguyen, *Digest of Technical Papers*, 10th International Conference on Solid-State Sensors and Actuators, Sendai (Japan) (1999).

[62] A. Manz, N. Graber, H. M. Widmer, *Sensors and actuators B: Chemical.*, **1**, 248 (1990).

[63] P. Gelin, I. Bihi, I. Ziemecka, B. Thienpont, J. Christiaens, K. Hellemans, D. Maes, W. De Malsche, *Ind. Eng. Chem. Res.*, **59**, 12784 (2020).

[64] S. Yadavali, H. H. Jeong, D. Lee, D. Issadore, *Nature Com*, **9**, 1922 (2018).

[65] K. Jensen, *Chemical Engineering Science.*, **56**, 293 (2001).

[66] J. J. Agresti, E. Antipov, A.R. Abatea, K. Ahna, A.C. Rowatan, J-C Baret, M. Marquez, A. M. Klibanov, A.D. Griffiths, D. A. Weitz, *PNAS*, **107**, 4004 (2010).

3

Hydrodynamics of microfluidics 1: channels

3.1 The flow equations and the boundary conditions

3.1.1 The notion of fluid particle

How to establish the equations that govern fluid flows ? A natural approach would be analysing the movement of the fluid molecules, and establish equations that describe, from a probabilistic standpoint, their dynamics. This approach led, in 1872, to the Boltzmann equations [1]. These equations are rigorous, but difficult to use in practice. Work has been done, over several decades, to derive approximate forms of these equations. An example is the Burnett equations [2], established for gases in 1936.

In fact, fluid mechanics textbooks do not take the Boltzmann equation as a starting point. Most often, they develop a mathematical approach, by considering an infinitesimal cube, expressing the pressure and inertial forces it is subjected to, and using, for the viscous contribution, a phenomenological law. In this presentation, the cube remains an abstract object. Its size is never given and the role of thermal motion, eliminated without justification, is never discussed. This led G. Batchelor to develop a more physical approach, For this purpose, he introduced the concept of 'fluid particle'. The notion is illustrated in Fig. 3.1. The figure is adapted from *An Introduction to Fluid Dynamics* [3], published in 1967.

Let us imagine a sensor of arbitrary size measuring the *instantaneous* density of a gas. The size of the probing volume is assumed to be comparable to the sensor size. This is a thought experiment, because obviously, no density sensor can respond instantaneously, nor be of arbitrary size. If the sensor size is on the order of the intermolecular scale, the measurement will depend on whether one molecule visits or not, at the time of the measurement, the probing volume. If no molecule shows up, density is zero and if one visitor comes on stage, density suddenly bounces. For sensors of that size, the response will jump from one value to another, as illustrated in Fig. 3.1. With such probes, impossible to measure the density. Let us now consider the opposite situation, where the sensor is so large that its size is comparable to the fluid container. In this case, the sensor response will reflect global features, for instance, the presence of temperature heterogeneities on the scale of the container, which will cause a dependence of the measured density with the probing volume (see the right-hand part of Fig. 3.1). The

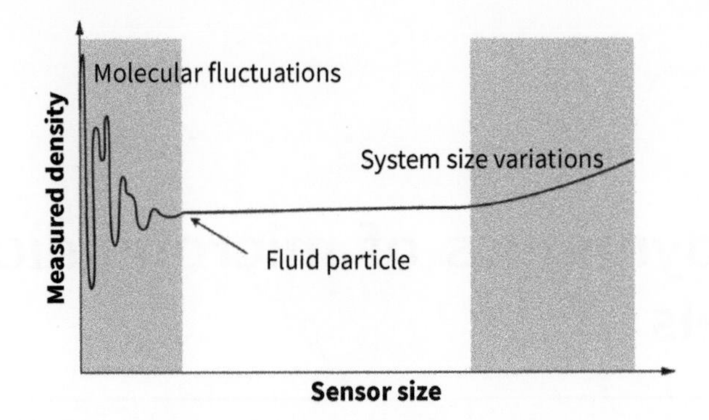

Fig. 3.1: Plot shown in G. Batchelor's textbook [3] representing the variation of the instantaneous measurement of the density made by a probe, as a function of its size. The plot allows to introduce the notion of 'fluid particle' (See text).

measurement performed by the sensor will obviously be non local. In the traditional presentation of fluid mechanics, there is a plateau, extending over several decades, that separates the two limits. In this context, the 'fluid particle' is defined as the smallest probe volume that contains enough fluid molecules to smooth out the effect of molecular fluctuations. Its size – several intermolecular scales – corresponds to the lower edge of the plateau of Fig.3.1. We this notion, all quantities involved in the description of the fluid behaviour, such as density, speed, or viscosity are defined as spatial averages throughout the fluid particle. Gone the thermal fluctuations. They have been eliminated by the averaging process.

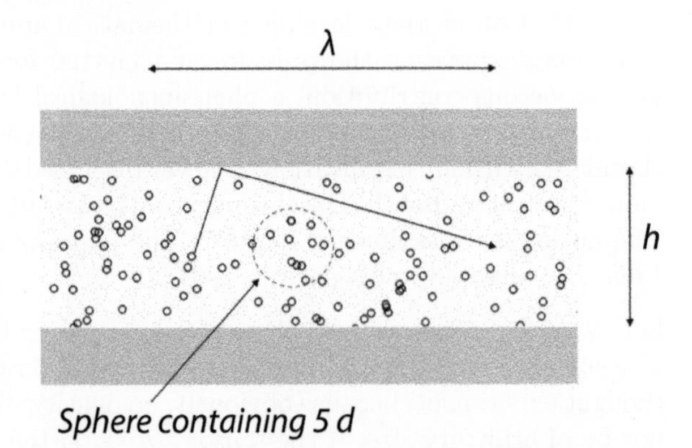

3.2: Gas flowing in a sub-mean free path channel. The sphere represents a Batchelor's fluid particle, i.e. a particle whose size is several d.

However, when small systems are considered – i.e. the subject of the book –, Batchelor's definition of the fluid particle becomes problematic. Fig. 3.2 shows a gas flowing in a small channel, of height h larger than the intermolecular scale d, but smaller than the mean free path λ. In such a situation, by definition of the mean free path, molecules most often, bounce off the walls, and rarely collide with each other. In this system, we can define a fluid particle as Batchelor did (see the dashed circle in Fig. 3.2). However, this particle cannot be representative of the local fluid behaviour, because the molecules it contains mostly interact with the walls. Obviously, there is a problem in the definition.

One way to solve it is to declare that the fluid particle size is not several intermolecular distances, but several *mean free paths*. In such circumstances, molecules forming the fluid particle mostly interact with themselves, and consequently, this particle can be representative of the local fluid behaviour. This is the new definition we propose. The size of the new particle is always larger than Batchelor's one, and, often, much larger: as we saw in Chapter 2, for air, at normal pressure, λ is twenty times larger than d.

Similar arguments can be advanced for liquids. We saw in Chapter 2, that three scales characterize the liquid: the molecular radius, the intermolecular distance and the correlation length. For example, for liquid argon at 90 K, their values are, respectively: 0.2, 0.4 and 1 nm. In a channel, the correlation length is the length below which molecules approaching a wall starts to 'feel' its presence. It can thus be taken as an equivalent to the mean free path. We can therefore propose that, similarly to gases, the liquid particle has a size on the order of the correlation length. A similar criteron was proposed in Refs. [4, 5].

To conclude, it seems pertinent, in the context of small systems, to reconsider the traditional definition of the fluid particle, given by G. Bathelor, and introduce a new one. The size of the new fluid particle should be, in gases, several mean free paths, and, in liquids, several correlation lengths.

3.1.2 The continuum hypothesis

The next step is to consider that the fundamental law of dynamics applies to each fluid particle. A step further is to assume that the variables attached to the particle (density, velocity, pressure) are continuous functions of their position. This assumption is called, in Fluid Mechanics, the 'continuum' hypothesis. On this basis, by using the powerful tools of continuum mechanics, it becomes possible, mathematically, to establish the governing equations of the flow. Being based on the notion of fluid particle, previously defined, these equations cannot describe phenomena on scales smaller than the mean free path in gases or the correlation length in liquids, nor what happens close to the walls, at distances smaller than these scales.

3.1.3 The Navier-Stokes equations

Having defined the continuum hypothesis, we adopt it. Let us consider a fluid element, composed of several fluid particles. Transported by the flow, fluid particles enter and exit the element. When fluid particles enter, the density raises and when they leave, density decreases. The constraint that, throughout the process, mass is conserved is expressed by the mass conservation or continuity equation:

$$\frac{\partial \rho}{\partial t} + \frac{\partial \rho u_i}{\partial x_i} = 0$$

where ρ is the density, x_i, u_i are the i components of position and velocity and t is time. An approximation called 'incompressibility approximation', consists of taking ρ as a constant, i.e. independent of space and time. Subsequently, the mass conservation equation reduces to:

$$\frac{\partial u_i}{\partial x_i} = 0 \tag{3.1}$$

The incompressibility approximation is justified when the fluid velocity is much lower than the speed of sound. The reason is that density may vary from one point to another, likewise pressure and velocity, but, well below the speed of sound, the corresponding variations are so small that they can be neglected [3]. Thereby, density can be taken as constant. As the speed of sound is more than one thousand metres per second in liquids, and hundreds of metres per second in gases, it follows that in the vast majority of microfluidic situations, we can consider the working fluids, whether gas or liquid, as incompressible.

Now we come to dynamical considerations. Applying the fundamental relation of the dynamics of the material point, we obtain the following equations for fluid motion:

$$\rho \frac{Du_i}{Dt} = f_i + \frac{\partial \sigma_{ij}}{\partial x_j} \tag{3.2}$$

where f_i is the volumetric force component the particle is subjected to (for instance, gravity), and σ_{ij} is the stress tensor representing the surface forces applying onto the particle. In this equation, the operator D/Dt is the material derivative, whose expression is written as follows:

$$\frac{D}{Dt} = \frac{\partial}{\partial t} + u_j \frac{\partial}{\partial x_j} \tag{3.3}$$

The left-hand term thus includes two contributions, called local acceleration $\frac{\partial}{\partial t}$ and convective acceleration $u_j \frac{\partial}{\partial x_j}$. In practice, when we refer to inertial terms, we most often include both. As mentioned above, the stress tensor σ_{ij} represents the surface forces the fluid particle is subjected to. Often, to express these forces, we consider a cube , and σ_{ij} represents the stresses applying on the six faces. Mathematically, if σ_{ij} are those forces (per unit of surface), $\frac{\partial \sigma_{ij}}{\partial x_j}$ is the resulting volumetric force.

At this stage, additional assumptions are needed to move forward. A major assumption, made in hydrodynamics, and justified for small deformations, leads to consider that the stress tensor is linearly related to the deformation tensor. Its general form is obtained by taking advantage of the symmetries of the problem and utilizing general results of tensor theory:

$$\sigma_{ij} = -p\delta_{ij} + \mu \left(2e_{ij} - \frac{2}{3}\delta_{ij}e_{mm} \right) + \zeta \delta_{ij} e_{mm} \tag{3.4}$$

where p is the pressure, e_{ij} is the velocity gradient tensor, and δ_{ij} is the Kronecker symbol. The expression of the tensor e_{ij} is written as follows:

$$e_{ij} = \frac{1}{2}\left(\frac{\partial u_j}{\partial x_i} + \frac{\partial u_i}{\partial x_j}\right) \tag{3.5}$$

In Eq. (3.4), μ is the dynamic viscosity and ζ is the *second viscosity* or the *volumetric viscosity*. Expression (3.4) represents a remarkable achievement of tensor theory, The case where μ and ζ coefficients do not depend on the velocity gradient nor the flow history corresponds to 'Newtonian' fluids (see the historical note below). As noted earlier, we will concentrate ourselves on these fluids.

The above relations do not make any assumption about incompressibility. Here, for the sake of simplicity, we will consider the fluid incompressible and, moreover, μ uniform over the fluid. On using Eq. (3.4) and Eq. (3.2), we obtain a simplified form of the Navier-Stokes equation, which will cover all the situations we will analyse in this book:

$$\rho\frac{Du_i}{Dt} = -\frac{\partial p}{\partial x_i} + \mu\frac{\partial^2 u_i}{\partial x_j^2} + f_i \tag{3.6}$$

A vectorial form of the Navier-Stokes equation is often more convenient to use:

$$\rho\frac{D\mathbf{u}}{Dt} = -\nabla p + \mu\Delta\mathbf{u} + \mathbf{f} \tag{3.7}$$

where we recall that $\mathbf{f}$ is the external force per unit volume. In the same notations, the conservation of mass equation can be written as follows:

$$div\mathbf{u} = \mathbf{0} \tag{3.8}$$

History of the Navier-Stokes equations

The first flow equations were obtained by L. Euler in 1757, assuming perfect fluids, i.e. fluids with zero viscosity. The shortcoming of the equation was the absence of dissipation. Without dissipation, fluids subjected to an imposed pressure gradient, for instance, in a channel, or flowing downwards in a small tube under the action of gravity, an example taken by Navier in his article [6], accelerate indefinitely. This was obviously in conflict with the observation, and the solution came by adding a term that dissipates energy, and, in turn allows the flow to reach steady states in the same circumstances. This is what Navier did in 1823 [6]. Amazingly, he used an incorrect phenomenological approach to derive Eq. (3.6), which was correct. The formalization of the viscous term was established in 1843 by Saint-Venant and in 1845, by Stokes, who introduced the concept of viscous stresses. The equation is called the Navier-Stokes equations, in honour of the two scientists but it could just as well have been called the 'Euler Navier-Stokes' equation or even the 'Euler Navier Saint-Venant Stokes' equation.

3.1.4 The concept of viscosity

What is the physical origin of viscosity, which we introduced in a formal manner? For this, we need to return to the microscopic scale. The origin of viscosity is related to the exchange of momentum between molecules moving, in the average, at different speeds. The model was proposed by Maxwell in 1879 [7].

To reach an intuitive understanding of Maxwell's theory, we may imagine, as in Ref. [8]), two trucks, of total mass M (i.e. the vehicle plus the coal sacks, stacked on the tub), circulating at different velocities u_1 and u_2 (see Fig. 3.3).

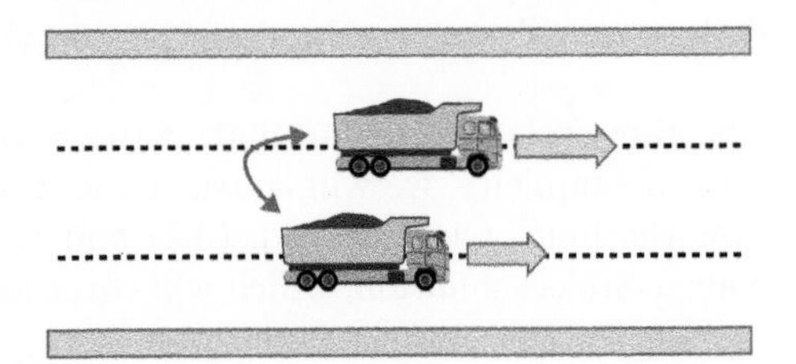

3.3: Two trucks moving at different velocities $u_1 < u_2$ and exchanging charcoal sacks.

The workers at the back of the trucks continually exchange sacks between the vehicles. The new speeds of the trucks, u_1' and u_2', after the sacks have been exchanged is:

$$u_1' = u_1 + \frac{m}{M}\Delta u \text{ and } u_2' = u_2 - \frac{m}{M}\Delta u$$

in which m is the mass of the sacks that were exchanged, and $\Delta u = u_2 - u_1 > 0$, taken as positive for this discussion. We see that, due to sack exchange, truck 1 slows down truck 2, while truck 2 speeds up truck 1, entraining each other, as if they exerted a 'friction' force on one another.

Following this line of thought, let us consider two gas layers, of surfaces S, located at z and $z+\lambda$, in which molecules circulate, in the average, at different speeds, $u(z)$ and $u(z+\lambda)$, as sketched in Fig. 3.4. A molecule situated on the plane z, jumps, at velocity u_{therm}, to the next layer, located at $z+\lambda$, and undergoes a collision. After the collision, the new speed of the jumping molecule should be $u(z+\lambda)$, because of flow stationarity. The corresponding momentum transfer, during this event, is $m(u(z+\lambda)-u(z))$, where m is the mass of the molecule. The time needed to perform the transfer is the collision time τ, i.e. λ/u_{therm}. In a volume $S\lambda$, the total force F associated to this momentum transfer will be $F = \lambda S\rho(u(z+\lambda)-u(z))/\tau \approx S\rho u_{therm}\frac{\partial u}{\partial z}$. The corresponding stress, $\sigma = F/S$ is given by:

$$\sigma \approx \mu\frac{\partial u}{\partial z} \tag{3.9}$$

in which μ is given by:

$$\mu \approx \rho u_{therm}\lambda \tag{3.10}$$

where μ is the viscosity and σ is the viscous stress. Viscous stress acts against the velocity gradient. To maintain the shear, one must apply an external force,. In channel flows, this force is provided by the pressure gradient applied along the channel.

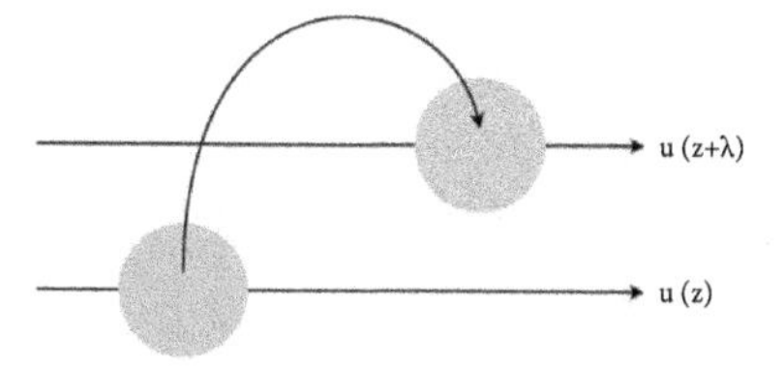

3.4: Two streamlines in a gas, separated by the mean free path λ, and moving at different velocities $u(z)$ and $u(z+\lambda)$.

An important quantity is the kinematic viscosity $\nu \approx \mu/\rho$. An estimate of u_{therm} is c, the sound speed. Therefore, we have $\nu \approx c\lambda$. The three dimensional theory, established by Maxwell [7], incorporates a 1/3 prefactor. In practice, the expression without prefactor provides an acceptable estimate for the kinematic viscosity of gases. For instance, it leads to 1.6 $10^{-5}\text{m}^2\text{s}^{-1}$ for air under normal conditions, while the measured value is 1.5 $10^{-5}\text{m}^2/\text{s}$. Interesting is to note that, according to Maxwell's theory, all diffusivities (momentum, temperature, and mass) are equal.

For liquids, the situation is different and an excellent presentation is given in Ref. [8]. Liquids possess a crystalline order over short distances (a fraction of a nanometre) while still remaining disordered over large distances. Molecules must cross an energy harrier to leave the 'cage' formed by their immediate neighbors, and exchange momentum with a fluid layer moving at a different velocity. Statistical calculation leads to the following form for the viscosity of Van der Waals liquids:

$$\mu = A \exp\left(\frac{E}{kT}\right) \tag{3.11}$$

where A is a constant and E is an activation energy.

Consistently with the previous discussions on the fluid particle, the notion of (bulk) viscosity applies down to nanometric scales in liquids and a few mean free paths in gas [28, 30].

3.1.5 Validity of the Navier-Stokes equations

The case of gases. We saw that in gases, the fluid particle size is on the order of the mean free path. Collision time and mean free paths are linked together by the relation $\lambda \sim \sqrt{\nu\tau}$. It is useful to recall the following formula:

$$\lambda = \frac{1}{\sqrt{2}\pi n a^2} = \frac{kT}{\sqrt{2}\pi p a^2} \tag{3.12}$$

where n is the density (the number of molecules per unit volume), T is the temperature, k is the Boltzmann constant, p is the pressure and a is the size of the molecule. As noted in Chapter 2, at atmospheric pressure, the mean free paths of usual gases, calculated with the above formula, range between 60 and 160 nm. a and λ differ by more two orders of magnitude. The intermolecular distance, of the order of 3 nm at atmospheric pressure, lies in between.

For gases, the question of the validity of the Navier-Stokes equations has been investigated numerically, theoretically and experimentally for several decades (see, for instance, [9–16]). Microfluidic technology has contributed to raise the accuracy of the measurements and, in turn, the sharpness of the conclusions [15,17–19]. It is convenient to define the Knudsen number, as the ratio λ/L, where L is the characteristic length of the system (for instance, the height for a channel). To make a long story short, channels with heights larger than 2λ (i.e. $Kn < 0.5$) are governed by the Navier-Stokes equations, with or without slip at the walls while, above this limit, they enter a new regime, called 'transition regime', where the Navier-Stokes equations no longer apply. A pioneering experiment performed in microchannels [20], comparing Navier-Stokes solution and measurements, supports this conclusion. It is shown in Fig. 3.5 [21].

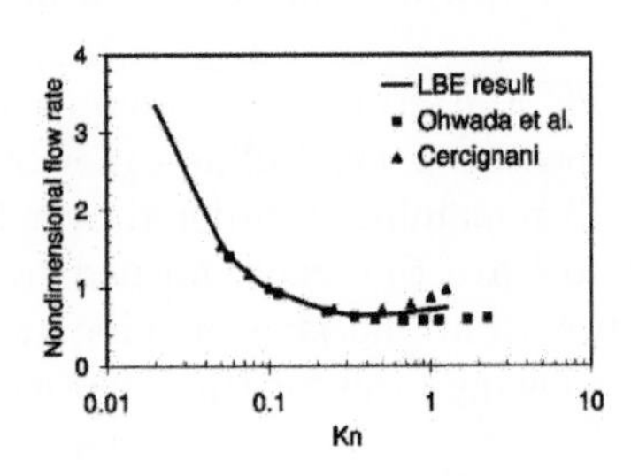

3.5: Nondimensional flow rate as a function of the Knudsen number for rectangular microchannels. Squares are the experiment, triangles the analytical solution of the Navier-Stokes equation. The line is a model of the Boltzmann equations. Deviations between Navier-Stokes and experiment are visible at $Kn > 0.5$. (Reprinted figure with permission from Ref. [21]. Copyright (2022) by the American Physical Society.).

By comparing flow rate measurements with Navier-Stokes predictions (with slippage), one sees that disagreements appear at roughly $Kn > 0.5$ (see Fig. 3.5). λ thus appears, in terms of order of magnitude, as a cross-over for the validity of the Navier-Stokes equations. If we consider that, at this cross-over, fluid particle and system sizes are equal, we may confirm that the order of magnitude of the particle size is λ.

The case of liquids. Fig 3.6 shows examples of submicrometric systems, in which the validity of the Navier-Stokes (NS) equations, for liquids, can be discussed, and, consequently, a practical estimate for the liquid particle size can be obtained, if we assimilate it again to the system size below which NS equations fail. Fig. 3.6(A) shows a 100 nm high channel, fabricated with MEMS technology (described in Chapter 7). In these systems, NS equations provide an accurate description of the flow. Flows in nanocapillaries, such as that of Fig. 3.6(B) [22], are in a similar situation. Large carbon nanotubes (CNT), or boron nitride nanotubes (BNNT), such as the one shown in Fig. 3.6(C) [23], can also be described using NS equations.

In carbon nanotubes larger than 1.4 nm in diameter, and again for water, numerical simulations have shown that the Navier-Stokes equations remain accurate [24, 25]. Nonetheless, when diameters are smaller, complications arise. For CNTs smaller than

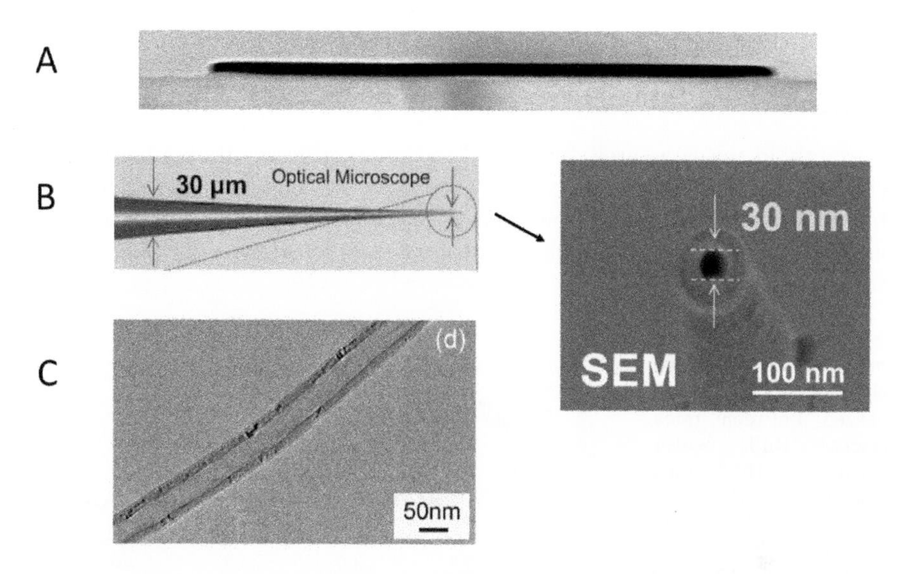

Fig. 3.6: Examples of submicrometric channels or tubes: (A) Submicrometric channel, 100 nm high. (Courtesy of P. Joseph); (B) Nanocapillary, 30 nm in diameter. (Reprinted from [22], with permission from the Royal Society of Chemistry; copyright 2022.); (C) Boron Nitride Nanotube, 30 nm in diameter. (Reprinted from [23], with the permission from the American Chemical Society; Copyright 2022.).

0.83 nm in diameter (2.4 times the water Van der Waals diameter), the liquid molecules flow in single file.[1] The phenomenon is shown in Fig. 3.7.

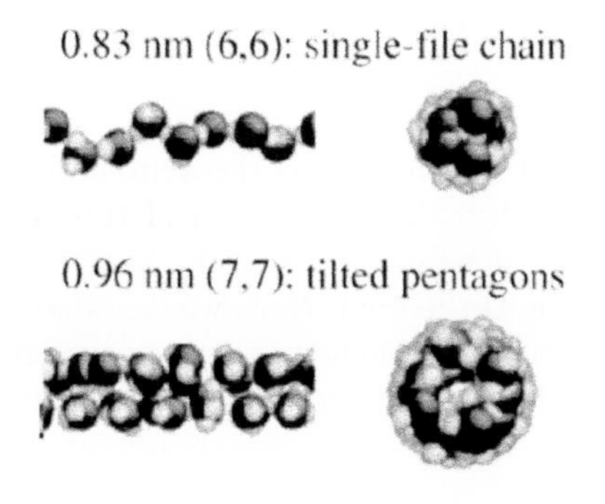

3.7: Molecular Dynamics (MD) simulation snapshots of water structure inside carbon nanotubes of different diameters, adapted from [25]. In these simulations, speeds are on the order of meters per seconds and the pressure gradient is 4 10^{14}Pa/m (Reprinted with permission from Ref [25]. Copyright 2022 by the American Physical Society).

The same simulation showed that when the CNT inner diameter is around 0.96 nm, water molecules cease to structure in a single line; instead, they develop a pentagon pattern. Other structures have been found at larger diameters. Above 1.39 nm, the fluid eventually recovers its ordinary' disordered bulk structure. The domain of diameters below 1.39 nm is called the subcontinuum. It marks the end of Navier-Stokes validity.

[1]This confirmed the pioneering work of Ref. [26].

In this domain, new, still unknown, equations are needed to describe the flow [27]. The fact that in ultra-confined spaces, liquid molecules form structures (leading to force oscillations), was known (see, for instance, [28, 29]). What we learned here was about the dynamics, i.e. how water manages to flow in extremely confined conditions.

Not exactly in the same area, but linked to the question we discuss [5], Fig. 3.8 shows a viscosity study peformed in extremely confined systems [30]. It was found that for sub 2 nm water films squeezed between one cylinder and a flat plate, viscosity departs from the bulk value, by up to five orders of magnitude. Above 2 nm, the bulk viscosity is recovered.

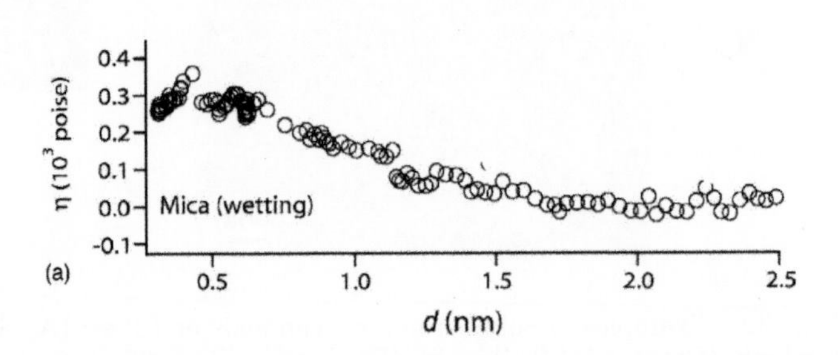

3.8: Viscosity of water versus gap d, measured with SFA, for films squeezed between mica surfaces. Bulk water viscosity is 10^{-3} Pa.s (Reprinted with permission from Ref. [30]. Copyright 2022 by the American Physical Society.)

Although it is difficult to disentangle bulk and surface contributions, one may consider that 1–2 nm, i.e. 1–2 correlation lengths, marks the break-up of the NS equations and therefore provides an estimate for the particle size in liquids. These considerations led the community to distinguish between 'subcontinuum' and continuum' nanofluidics, with a frontier fixed at 1 nm. This distinction, which we showed in Chapter 2 of this book without explanation, is now justified.

3.1.6 The Reynolds number

The Reynolds number was introduced in Chapter 2, using the Π theorem, and we did not need the Navier-Stokes equations to establish it. The definition of this number resulted from dimensional considerations, and there was no way to know what it meant. This the strength and the weakness of the Π theorem. Now, we can answer the question. Let us, thus, recall that stationary incompressible flows characterized by one single spatial scale l are parametrically controlled by a single dimensionless number: the Reynolds number. Its expression is:

$$Re = \frac{Ul}{\nu} \tag{3.13}$$

where U is the characteristic velocity of the fluid and ν is its kinematic viscosity.

For stationary regimes ($\frac{\partial u}{\partial t} = 0$), inertia and viscous forces appearing in Eq. (3.6) can be estimated in the following manner:

$$\text{Inertia forces} \sim \frac{\rho U^2}{l} \tag{3.14}$$

$$\text{Viscous force} \sim \frac{\mu U}{l^2} \tag{3.15}$$

Therefore the Reynolds number can be written as:

$$Re \sim \frac{\text{Inertia forces}}{\text{Viscous forces}} \tag{3.16}$$

The physical interpretation of the Reynolds number is thus the ratio of the inertial to the viscous forces. Large Reynolds number flows are dominated by inertia, and small Reynolds flows by viscous forces.

What is the value of the Reynolds number in microfluidic systems? In such systems, fluid velocities do not exceed 1 cm/s and transverse channel dimensions are on the order of 10 μm so that Reynolds numbers do not exceed 10^{-1}. In fact, in the domain of microfluidics, and in the vast majority of situations, Reynolds number are low or very low.

Still, there exists important exceptions. Devices dedicated to extract the heat generated by microprocessors, a subject that we will present later, need, in order to obtain efficient heat transfer, to operate at substantially high Reynolds numbers. Substantial Reynolds numbers are also involved in inkjet printers. High throughput mixers and reactors, the two basic elements of micro reaction technology (MRT), also operate at moderate or high Reynolds numbers (from ten to several thousands, typically). Last but not least, for the sake of sorting particles or cells, it can be advantageous to work at moderate Reynolds numbers. All these situations pertain to the domains of inertial microfluidics or millifluidics. We will return to these subjects later. For the moment, we assume low Reynolds numbers, which concerns, as mentioned above, the vast majority of microfluidic situations.

3.1.7 The Stokes equation

The field of studying flows at small Reynolds numbers is known as *microhydrodynamics*. As noted in this book, 'micro' refers to small Reynolds numbers, not to small sizes. This domain is well-established in fluid mechanics, and we will review here some of the essential results.

The flows of incompressible Newtonian fluids at small Reynolds numbers are governed by the Stokes equation:

$$-\nabla p + \mu \Delta \mathbf{u} + \mathbf{f} = \mathbf{0} \tag{3.17}$$

To obtain the Stokes equation, we considered the inertial terms to he negligible with respect to the viscous term, owing to the smallness of the Reynolds number. More specifically:

- The first component involved in the inertial force is the acceleration term. It has an order of magnitude of $\rho U/\tau$, where τ is the characteristic time of the variation of velocity. One can imagine situations where the characteristic time of the velocity variations is small (for instance, flow induced by the sudden movement of a plate initially at rest, called the Rayleigh problem), and in this case, the acceleration term cannot he neglected with respect to the viscous term. Here, we consider different types of situations where the flow is quasi-steady. In those cases, which cover a considerable number of practical situations in microfluidics, the acceleration term can be neglected.

- The second component has an order of magnitude is $\rho U^2/l$, where U is a typical velocity and l is the scale of the characteristic variation of the velocity. This term must he compared to the viscous term, which is on the order of $\mu U/l^2$; indeed, the ratio of the first to the second term is precisely $Re \ll 1$, as we showed before.

We will use the Stokes equations to describe, in most cases, microfluidic flows. The domain of inertial microfluidics, which covers cases where the Reynolds reaches important values, a hundred or so, is obviously excluded from this framework. In such cases, we will provide more qualitative descriptions.

3.1.8 The five virtues of Stokes equation

We recall here that in the framework in which we place ourselves, there are no free interfaces and we assume no slippage at the walls. In such conditions, the flows governed by the Stokes equation, which, as discussed above, is valid at small Reynolds numbers, possess the following mathematical properties: linearity, reversibility, instantaneity, uniqueness of solution, reciprocity, and minimum of dissipation.

1. **Linearity**: The Stokes equations are linear. This property permits the superposition of the elementary solutions of the equation to determine a given flow.

2. **Reversibility**: Changing t to $-t$ does not modify the equations. This property implies that, for stationary flows, a fluid particle moving in the forward direction, will move backwards on the same streamline when the flow conditions are reversed. This implies that it is impossible to decide from the sole observation of the streamlines, as an experiment will demonstrate later, in which direction the flow circulates.

3. **Instantaneity**: Since t does not appear explicitly in the Stokes equations, the response of the flow to an external change will be immediate. For instance, if the pressure gradient driving the fluid is suddenly changed, the fluid will react immediately, by setting up a new flow profile.

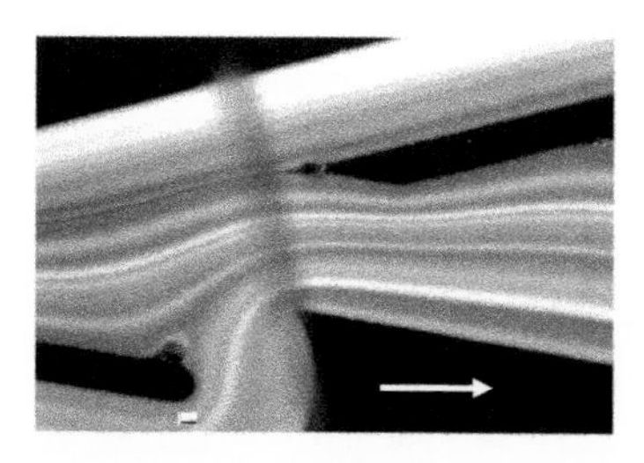
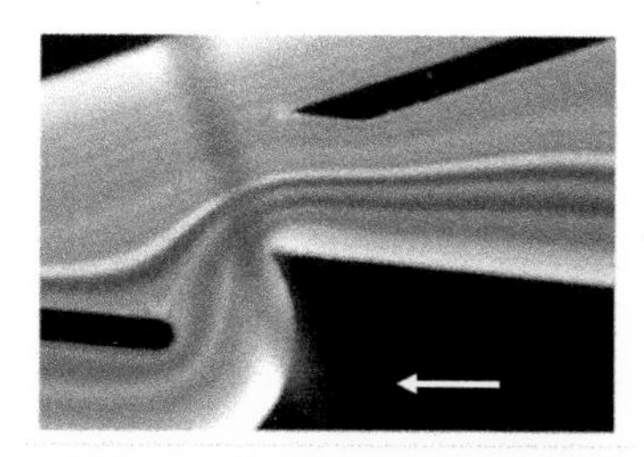

Fig. 3.9: Water flow in a Tesla geometry, in a Silicon glass microsystem. flow rate is 600 μl/min, channel height is 20 μm, channel width 200 μm on average. Speeds are on the order of metres per second In the top photo, the flow is oriented from left to right, and in the bottom, the orientation is inverted. Experiment carried out at the MMN laboratory, at ESPCI (coll. O.Francais [31]).

4. **Minimum of dissipation**: The flow resulting from the solution to the Stokes equation minimizes the kinetic energy dissipation with respect to all kinematically admissible fields compatible with the boundary conditions.

5. **Uniqueness**: The solution to the Stokes equation is unique. This is an important property, because it implies that flow instabilities cannot occur. The reason is that, according to theory, instabilities are associated to bifurcations, i.e. the existence, in the same conditions, of a multiplicity of solutions to the governing equations. This possibility cannot be envisaged with Stokes equations.

Reversibility is illustrated in Fig. 3.9. The geometry of the microchannel is that of the 'Tesla valve' (1920). This valve is made up of a series of hydrodynamic circuits, one unit of which is shown in Fig. 3.9. The flow is driven from left to right in Fig. 3.9 (left) and from right to left in Fig. 3.9 (right). At high Reynolds numbers, the direction of circulation matters, because the positions of the vortices detaching from the boundary layers differ depending on the direction of flow. At low Reynolds numbers, however, the reversibility of the Stokes equation makes the two directions of circulation, with exactly the same streamline pattern, dynamically possible. This is why, from the sole inspection of the streamlines of Fig. 3.9, it is impossible to decide in which direction the liquid circulates.

The reversibility of the Stokes equation has important consequences in the case we examine. It signifies that, unless pushing the fluid at considerable speeds[2] it is not possible to fabricate a 'microfluidic diode', which would present a weak resistance to flow in one direction, and a high resistance in the other direction. This hope, expressed by some investigators in the early days of microfluidics, was rapidly abandoned.

The properties listed above can be viewed as virtues in the sense that they enable to reach an exquisite level of flow control, by inhibiting hydrodynamic instabilities,

[2]We note that in this case, although the speed is several metres per seconds, the Reynolds number, based on the height of the channel, is less than ten. This suggests that the range of validity of Stokes equations may, for some flow geometries, be larger than simple order of magnitude argument may suggest

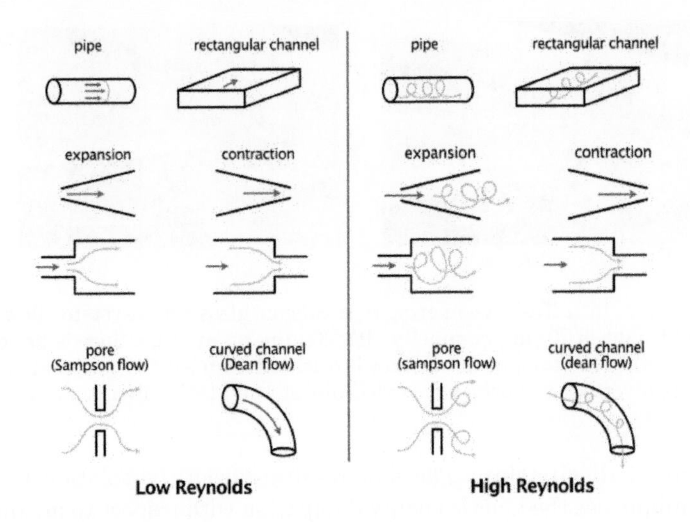

Fig. 3.10: Different flow configurations, at low and high Reynolds numbers. From H. Stone (Cours les Houches, 2009)

responding instantaneously to external changes of flow conditions, adding contributions in a linear manner, which simplifies the design and understanding of the flow behaviours. Also, reversibility can facilitate the operation of microfluidic devices, when rinsing or other operations are required. Figure 3.10 stresses this point. It contrasts the behaviour of low Reynolds number flows, governed by Stokes equations, to that of high Reynolds number flows, governed by the Navier-Stokes equations, for a variety of geometries. In the former case, flows are laminar, reversible, while in the latter, detachment phenomena (which we will present later), instabilities, and turbulence develop, jeopardizing fine control over the flow.

3.1.9 The Navier boundary condition

Figure 3.11 allows us to introduce the Navier boundary condition. Here, a fluid is flowing along a plate at rest, with a speed $u(z)$, z being the coordinate normal to the wall, the system being invariant along the other directions of space.

In the same article in which he derived the Navier-Stokes equations ([6]), Navier stipulated that, at the wall, the speed $u(0)$ adopted by the fluid, i.e. at $z = 0$, is not zero, but is given by:

$$u(0) = b\left(\frac{\partial u}{\partial z}\right) \tag{3.18}$$

where b is the Navier length, or the slip length. In the above expression, the derivative is calculated at $z = 0$.

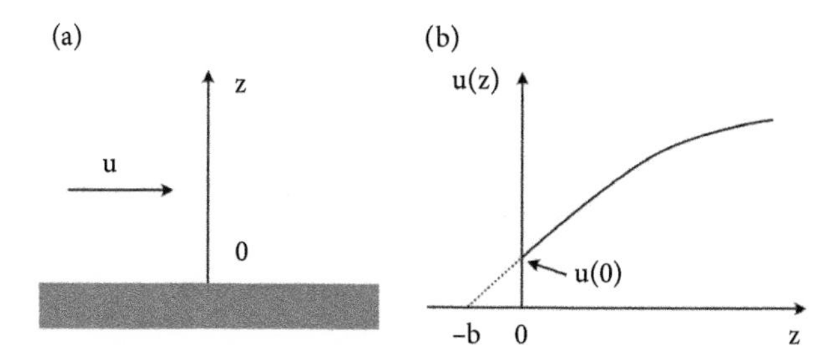

Fig. 3.11: (A) Flow over a plate, with the system of coordinates.(B) Geometrical illustration of the Navier length. The tangent at the origin crosses the z axis at $z = -b$.

A geometrical illustration of the Navier boundary condition, for $b > 0$, is provided in Fig. 3.11. The tangent to the velocity profile at the origin crosses the z axis at $z = -b$. Note that b can be positive or negative. In ordinary fluids, it is positive, but in complex fluids, such as polymer solutions, where adsorption of the chains leads to the build-up of a solid wall inside the fluid, the flow speed can vanish at $z > 0$, yielding a negative value for b.

At the time Navier wrote his' boundary condition, there was no support for it. Sixty years later, Maxwell provided a justification for the case of the gases [7], by showing that b is proportional to λ. The result was published in the same paper as that establishing an expression for the gas viscosity. In short, momentum is exchanged by collision between the walls and molecules located at a distance λ from it. On transposing the relations used in the gas kinetic model for viscosity, to the case of molecules colliding diffusively with a wall, one obtains $b = \lambda$. In fact, in Ref. [7], Maxwell treated the general case where a fraction σ of molecules undergo diffusive reflection, the others reflecting the walls in a specular manner. Calculations lead to $b = C\lambda$, where $C = \frac{2-\sigma}{\sigma}$. For the purely diffusive case, $\sigma = 1$ and thereby $C = 1$ while, for pure specular reflection, $\sigma = 0$ and C is infinite. In this case, the wall behaves as a free surface. For liquids, as will be seen later, things are more complicated.

Decades of controversies on the existence of slippage

The question of whether b is non zero has been much debated in the fluid mechanics community. Navier, Stokes, Rayleigh, Kelvin, and Taylor participated in the discussion [32]. H. Lamb, after having advocated the idea that fluids slip over rigid surfaces [32], later argued that b should be strictly equal to zero. The argument was formulated for the case of rectilinear channel flows. In the continuum framework, if fluids slipped at the wall, the wall shear stress would be infinite, which would contradict the experiment. Taking the words of Huh and Shih, pronounced in another context [33], even Heracles would not be able to push a fluid in a tube. This reasoning, supported by a multitude of experiments carried out over one century, was so well-accepted that textbooks never questioned it. Now, we know that it is physically wrong: liquids, in almost all situations, slip at the wall.

3.2 Slippage in gases

3.2.1 The different gas flow regimes

As the mean free path is a central quantity in gases, it is natural to introduce a dimensionless number defined by:

$$Kn = \frac{\lambda}{l} \tag{3.19}$$

in which l is the system size. This number is called the Knudsen number, in honour of Martin Knudsen [12, 34, 35]. Following Schaaf and Chambre [10], four regimes can be defined. They are shown in Fig. 3.12.

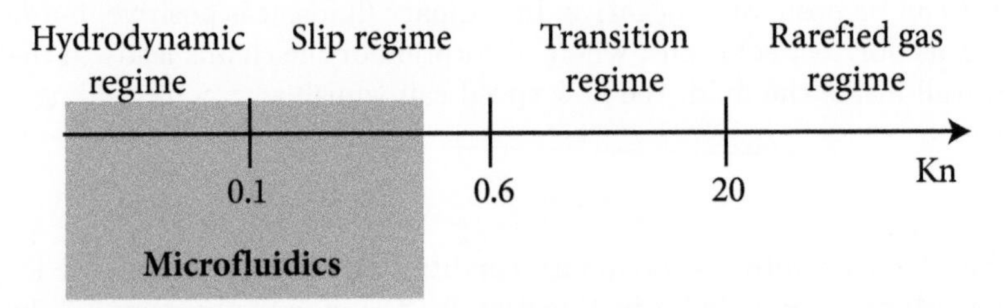

Fig. 3.12: Different gas flow regimes, depending on the Knudsen number. The values indicated on the diagram represent rough estimates of the Knudsen numbers determining the ranges of existence of each regime. The grey zone represents current values of the Knudsen numbers found in microfluidic devices.

The four regimes are:

- For Knudsen numbers smaller than 0.1, the fluid particle is much smaller than the system size, and the flow is described by the Navier–Stokes equations. At the wall, Knudsen layers, of thickness λ, occur, but they can be neglected. Slippage exists, but is negligible. This is the hydrodynamic regime.
- For Knudsen numbers between 0.1 and (roughly) 0.5, especially in the lower range of this domain, Knudsen layers are thin, so that most of the fluid volume is governed by the Navier-Stokes equations. At the wall, slippage is significant. The flow is described by the Navier-Stokes equations with slippage at the wall, which defines the 'slip regime'.
- For Knudsen numbers between 0.5 and 10, the 'intermediate' or transition' regime holds: gas molecules more often undergo collisions with the walls than with their colleagues. Rarefaction effects are significant, both in the bulk and at the wall. Continuum hypothesis breaks down and the Navier-Stokes equations no longer apply. To take rarefaction into account, the Burnett equations, derived from the Boltzmann equations were established in 1936 [2]. In practice, they are rarely used because of their complexity. One sometimes defines an early transition' regime, for

Kn around unity, in which rarefaction effects take place, but with small amplitude [9].

- Finally, for Knudsen numbers greater than 10, the system is in the 'rarefied' regime. The flow is described by the Boltzmann equations.

As indicated by the grey zone in Fig. 3.12, is not rare to find microfluidic flows in the slip regime. Examples are micro- or submicrometric-sized channels, working at atmospheric or sub-atmospheric pressures.

3.2.2 Gas flows in channels in the slip regime

We consider here the case of gas flows in microchannels, where the Knudsen numbers fall in the slip regime. In such conditions, the gas is governed by the Navier-Stokes equations. As mentioned above, we may consider that the Navier boundary condition applies at the wall:

$$u(0) = C\lambda \left(\frac{\partial u}{\partial z} \right) \tag{3.20}$$

in which $u(0)$ is the velocity at the wall and $\left(\frac{\partial u}{\partial z}\right)$ is the deformation rate at the wall.

There is no space here to develop the calculations of the flow in the channel. This was done in a number of papers [17–19], and in the first edition of the present book [36]. The following relation between mass flow rate and pressure, was obtained:

$$Q_m = \frac{\Delta P P_{\mathrm{m}} w h^3}{12\mu RTL}(1 + 6CKn) \tag{3.21}$$

where P_{m} is the average pressure in the channel and ΔP the difference between pressure at the entrance $P(0)$ and the exit $P(L)$, for a channel of length L. Here, the Knudsen number reads $Kn = \frac{\lambda}{h}$, where λ is the mean free path, calculated at P_{m}. A dimensionless quantity facilitating the discussion on the slip is the slip coefficient S, defined by the expression:

$$S = \frac{12\mu P_m Q_v L}{P(0)\Delta P w h^3} \tag{3.22}$$

where Q_v is the volumetric flux at the exit of the canal $x = L$. The calculation shows that S satisfies the following equation:

$$S = 1 + 6CKn \tag{3.23}$$

The evolution of S with the Knudsen number, obtained for nitrogen, in various experiments, is shown in Fig. 3.13.

Should there be no slip, we would have $S = 1$. The curve of Fig. 3.13 shows that slip exists, and that if neglected, the flow rate would be underestimated by one order of magnitude. The effect of slippage, which enhances flow rates, is thus considerable in this case. From the data, by looking at the slope at the origin, one finds $C \approx 0.92$,

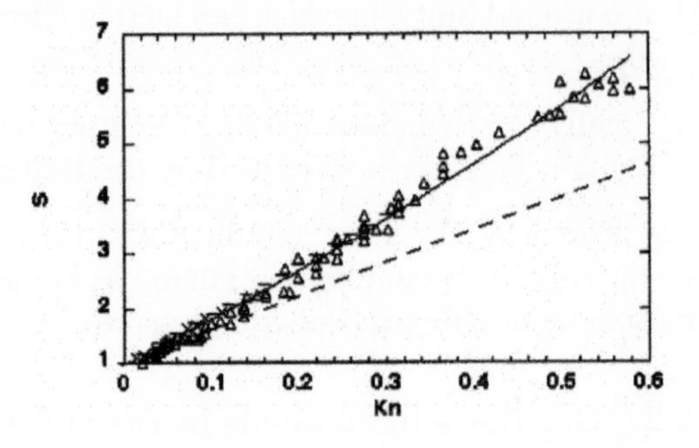

3.13: Evolution of slip coefficient S as a function of the Knudsen number, for a microcanal of rectangular cross-section [19] (empty triangles). The dashed line shows the expression 3.23 with $C = 1$ and the continuous line a polynomial fit of degree 2 .The plot also shows data from [17, 18]. A similar plot, up to $Kn = 0.8$ was obtained with helium.

in agreement with hard sphere model with purely diffusive wall reflection [38]. It is interesting to note the existence of the small positive curvature in the $S - Kn$ plot. This small curvature, called 'second order effects', indicates that as Kn increases above, roughly, 0.3, Navier boundary condition must be modified [18,37,39]. The experiments of Ref. [19], as well as others (see, for instance, [15]) showed that slip regime may be extended, as far as flow ratemeasurements are concerned, to a Knudsen number of 0.8.

What happens at larger Knudsen numbers ?

Microfluidics has contributed to make progress in the field of rarefied gas dynamics, by providing techniques allowing to perform measurements of unprecedented accuracy in a large range of Knudsen numbers. The plot of Fig. 3.14 shows an example. On this plot, the dimensionless mass flow rate G (see [15] for its expression) is plotted against $\delta_m = \frac{\sqrt{\pi}}{2} Kn^{-1}$. The experiments, performed in helium, allowed three orders of magnitude of variation in Knudsen number to be spanned.

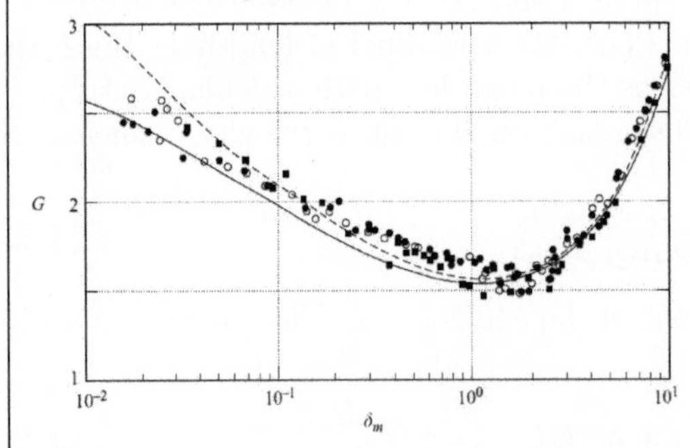

Fig. 3.14: Evolution of the dimensionless mass flow rate G with δ_m, in a channel, for three different pressure ratios (between the entry and the outlet). Dashed and full lines are theoretical curves (From [15], reprinted with permission from Cambridge University Press.)

What we learn from this plot is, among other things, the confirmation of the existence of a minimum (known as the Knudsen paradox), the pertinence of the Boltzmann equations, and that of various models attempting to simplify them. This plot is taken as a reference in the field of rarefied gas dynamics.

3.3 Slippage in liquids

3.3.1 The friction model

The structure of water close to a hydrophobic surface has been calculated by several authors [40–43]. An example is shown in Fig. 3.15 [43].

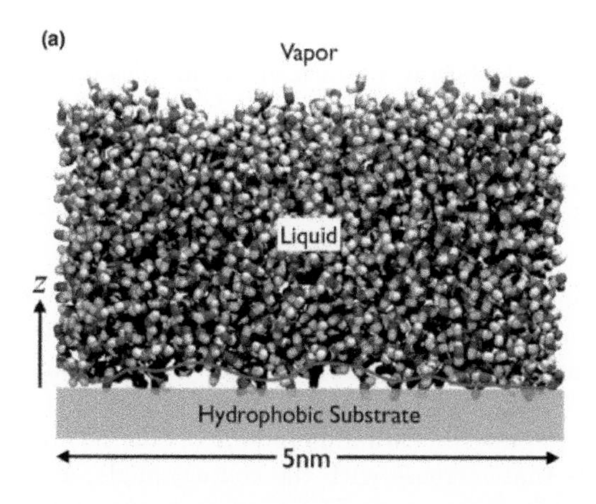

3.15: Molecular simulation of water over an hydrophobic surface (Reprinted by from [43], by permission from AIP. Copyright 2022).

Close to the wall, holes form, depleting the local fluid density. The physical origin of these holes lies in the fact that water molecules are more attracted by the bulk than by the wall. Then, they frequently bounce off the wall to join the bulk. In the hydrophilic case, no such holes form, because molecules, most of the time, stick to the wall [44].

In the presence of a flow, and in the hydrophobic case, it is conceivable that water molecules, being frequently captured by the bulk, are entrained by it. Slippage originates from this effect. In the hydrophilic case, since molecules remain at the wall most of the time, where they are held by Van der Waals forces, there is no slippage.

How can we formulate a boundary condition that could represent the situation? In fact, boundary conditions are like trees hiding forests. In our case, the trees hide a complex physical situation, in which roughness, crystallography, the possible presence of a vapour phase, Van der Waals force field, flow dynamics, and thermal motion presumably play roles. Incorporating these aspects in the expression of the boundary condition is challenging.

According to [45,46], the simplest manner to address the problem, from a phenomenological viewpoint, is to assimilate slippage to a friction process with a friction coefficient independent of the slip velocity. In [45], the following relation between the frictional stress and the hydrodynamic shear stress σ_w (which forces the fluid molecules to slip at the wall), was proposed:

$$\sigma_w = \mu \left(\frac{\partial u}{\partial z} \right) = \lambda_w u(0) \tag{3.24}$$

where μ is the fluid viscosity and λ_w is the friction coefficient.[3] The above relation has exactly the form of the Navier boundary condition, with a slip length defined by:

$$b = \frac{\mu}{\lambda_w} \tag{3.25}$$

What is this λ_w and how large is b? The calculation, performed in [45], lies outside the scope of this book. The following expression was obtained:

$$b \sim \frac{kT\mu D}{C\rho\sigma\epsilon^2} \approx \frac{b_0}{(1+\cos\theta)^2} \tag{3.26}$$

in which D, k, T, C, ρ, σ and ϵ are parameters[4] and whose detailed expressions can be found in Refs. [5, 45, 46]. The formula on the right includes the phenomenological constant b_0. Interestingly, the formula states that, in the case of perfectly smooth walls, slippage is infinite. Molecular roughness, along with crystal anisotropy, neglected in fluid mechanics, plays a crucial role here. In practice, assuming pre-factors on the order of unity, Eq. (3.26) provides slip lengths that increase with the contact angle, up to tens of intermolecular distances, i.e. several nanometres. These values are consistent with the numerical simulations (see Fig. 3.19 in the next section). The slippage we discuss here is often called 'intrinsic slippage'.

3.3.2 Hydrodynamics of liquid flows with slippage

We now analyse the hydrodynamical consequences of the presence of slippage in liquids, using again the Navier boundary condition:

$$u = b \left(\frac{\partial u}{\partial z} \right) \tag{3.27}$$

We assume a stationary pressure-driven flow between two infinite plates, and consider that the flow has only one component, u, directed along the direction x along which the pressure gradient is applied (see Fig. 3.16). With the symmetries, u only depends on the coordinate z, normal to the walls. To calculate the flow profile, we project the Stokes equation onto the x axis. The projection imposes the pressure gradient

[3] Here, the frictional force is proportional to the speed, not to the speed gradient (i.e. the speed divided by an intermolecular distance). This where Lamb's argument is objectionable. In fact, Lamb's argument is based on the hypothesis that liquid/solid interfaces behave as fluid/fluid interfaces. The assumption is acceptable for gases, as we saw, but, for liquids, it has no theoretical justification, and leads to conclusions in conflict with molecular dynamical simulations and experiment.

[4] D is the fluid diffusion constant, k the Boltzmann constant, T the temperature, C a constant characterizing the molecular roughness, ρ the density, σ a microscopic length scale and ϵ a fluid-solid energy.

$G = -\frac{\partial p}{\partial x}$, in which p is the pressure, to be constant, i.e. independent of x and z. With that, the equation to solve is an ordinary differential equation of the second order, coupled to the boundary conditions (3.27). The solution is:

$$u(z) = \frac{Ghb}{2\mu} - \frac{G}{2\mu}\left(z^2 - \frac{h^2}{4}\right) \tag{3.28}$$

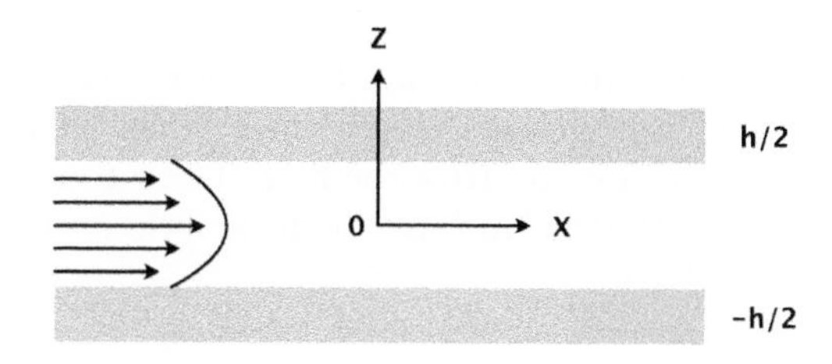

3.16: Flow between two infinite plates, located at $z = \pm\frac{h}{2}$, driven by a pressure gradient applied along x, with slip

The velocity profile with slip is a superposition of a plug flow (uniform velocity of magnitude $\frac{Ghb}{2\mu}$) and a Poiseuille flow. To illustrate this, the velocity profile without slip is represented in Fig. 3.17. Obviously, at fixed ΔP, the presence of slip increases the flow rate. We can describe the profile, as is often done, as a Poiseuille profile, but with a wall located at $z < -h/2$ in the negative part and $z > h/2$ in the positive one. For $b << h$, the (virtual) wall is located at:

$$z = \pm\left(\frac{h}{2} + b\right) \tag{3.29}$$

The pressure drop ΔP can be calculated by integrating the Eq. 3.28 along z:

$$\Delta P = \frac{12\mu LQ}{wh^2(h + 6b)} \tag{3.30}$$

in which w is the channel width (assumed, in practice, much larger than h in order to consider that u only depends on z). The above relation again shows that slippage reduces the pressure drop needed to drive the fluid.

An interesting case occurs when b is much larger than h. This happens in carbon nanotubes a few nanometres in diameter. In this case, slippage dominates the right

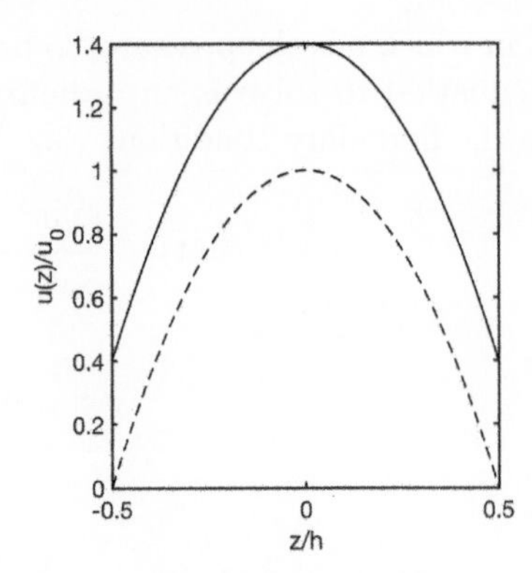

3.17: Velocity profiles, with slippage (full line) and without slippage(dashed line). $u_0 = \frac{Gh^2}{8\mu}$ and, for the slippage case, $b = h/10$.

hand side r.h.s.) of Eq.(3.28) and the flow profile is flat. On the other hand, again in the limit of large b/h, for a given Q, ΔP becomes extremely small. Walls behave as free surfaces and no pump is needed, in principle, to drive the flow. These considerations raised hope, in membrane technology, to develop large permeabilities, which could facilitate energy production (see Chapter 6).

We give here another formula, equivalent to Eq. (3.32), and valid for tubes. It can be obtained by performing the same task as above, in cylindrical coordinates. We find:

$$\Delta P = \frac{128\mu LQ}{\pi d^4(1+8b/D)} \tag{3.31}$$

in which D is the (inner) tube diameter, and L its length.

In microfluidic channels, Q and ΔP can be measured. This led experimentalists to use Eqs. (3.32) or (3.31) for determining slip lengths [47]. For instance, in the rectilinear case, one has:

$$b = \frac{h}{6}\left(\frac{12\mu LQ}{\Delta P w h^3} - 1\right) \tag{3.32}$$

which shows how b can be inferred from pressure and flow rate measurements carried out in a microchannel. Although the method seems straightforward, the main difficulties of the approach are that, since, in microfluidic channels, b lies on the nanometre or decananometre range, great accuracy on flow rate and pressure measurements are needed. Moreover, the presence of minute leakages, difficult to detect, may introduce biases in the flow rate measurements and affect the slip length determination. Nonetheless, a number of experimentalists have resolved the issues and obtained plausible (but moderately accurate) measurements of slip lengths with this method [86].

3.3.3 Slippage measurements in liquids over smooth walls

The measurements of slip lengths, in liquids, over smooth walls[5] have long been controversial (see the review of E. Lauga, M. Brenner and H. Stone [47], who discussed the fifty or so slip measurements performed between 1920 and 2005). By the turn of the century, the confusion increased even more, after three modern' measurements revealed surprisingly large slip lengths [49–51] (on the order of 200 nm -2.5 μm), further unconfirmed by other investigators. In the meantime, numerical studies consistently reported much smaller slip lengths, on the order of a few nanometres, while a second generation of experiments ([52, 53, 84, 86, 87]), using improved instrumentation, and, in the case of the force machine, analysing biases and artefacts in depth ([52])), consistently led to slip lengths at most in the deca-nanometre range.[6] An example of experimental study, performed in 2010, in which the flow profile close to the wall could be resolved down to 100 nm, is shown in Fig. 3.18 [87]. .

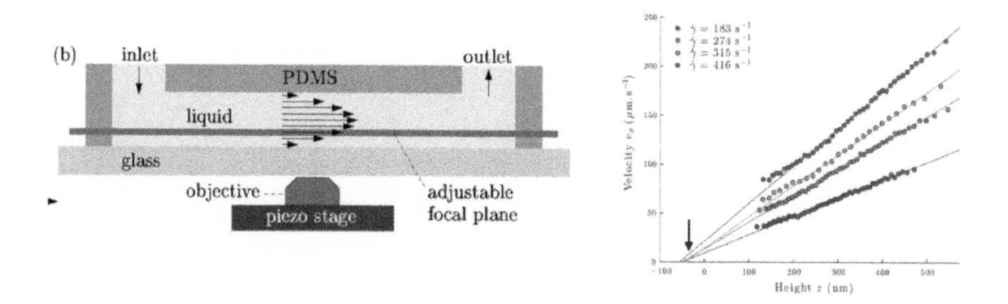

Fig. 3.18: (Left) Experimental setup, showing a microchannel low, with an objective mounted on a piezo-electric transducer, allowing the focal plane to span the near-wall region. Evanescent waves illuminate the fluid and fluorescent nanoparticles seed the flow.(Right) Velocity profiles obtained in this system, for hydrophobic walls, showing, by extrapolation, wall slippage [59, 87].

The ensemble of results, numerical and experimental, were summarized, for water - used in the vast majority of studies - in Ref. [5]. The data are shown in Fig. 3.19. Numerics and experiments show that slip lengths, are indistinguishable from zero when the fluid fully wets the surface. Concerning numerics, slip lengths increase with the contact angle, reaching several nanometres for the largest contact angle, consistently

[5]Smoothness is molecular on bare surfaces of silicon and mica. With hydrophobic layers deposited onto them, the roughness, measured with Atomic Force Microscopy AFM), raises up to a few nanometers. This parameter is systematically measured in the controlled experiments performed on the subject

[6]Stimulated by preliminary observations of nanobubbles at the water/solid interfaces ([57]), it was proposed that a nanometric gas layer, confined between the wall and the fluid, could play the role of a lubrication film, allowing the fluid to slip at substantial speeds and thus develop large slip lengths ([58]). However, series of studies made with X-ray and neutrons, could not confirm the hypothesis.

with intrinsic slippage theory[7]. A similar trend is observed in the experiments. However, a discrepancy exists between the two sets of data (numerical and experimental), unresolved at the time the book was written.

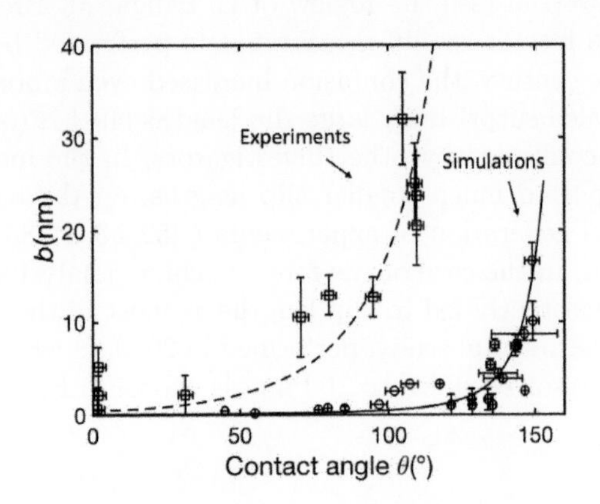

3.19: Experimental and numerical measurements of slip length b as a function of the contact angle θ (adapted from Ref. [5]). The full line on the right is given by $b = 0.3(1 + cos\theta)^{-2}$. The dashed line provides a guide for the eyes; it has no theoretical significance.

[7]Note that one parameter has been tuned in the theory to match the simulations.

Slippage in small carbon canotubes: a puzzling situation

At the moment, regarding the question of slippage in CNTs, the situation is puzzling. The subject stands at the cutting edge of research and, even though important issues are still unresolved, it is interesting, in the context of the book, to provide information on the situation. Fig. 3.20, established in [60], collects data reported in the period 2000- 2017. The reader should pay attention to the fact that a double logarithmic scale is used to represent the data. Water flows extremely fast in CNTs, but the measured flow rates, compared to classical hydrodynamics predictions, are scattered over five orders of magnitude. Slip lengths of 1 to 500,000 nm, resulting in almost zero to 500,000 flow enhancement factors, are reported for water in CNTs with diameters of 0.8 to 10 nm.

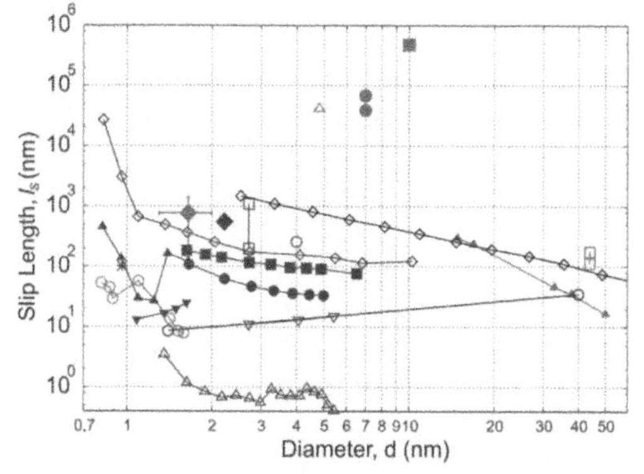

Fig. 3.20: Slip length of water in carbon nanotubes. The graph gathers data obtained from experiments and simulations [60]. Details can be found in the reference. The figure reveals a puzzling situation (Reprinted from Ref. [60] with permission from Materials Research Society. Copyright 2022.).

Why is it so scattered ? There is no response yet and the discussion of this graph stands outside the scope of the book. The reader may refer to recent reviews for more information. See, for instance, [61].

3.4 Microfluidics at small Reynolds numbers

3.4.1 Flows in channels with rectangular cross-sections

We now concentrate on bulk phenomena and proceed to the calculation of unidirectional flows through channels of rectangular cross-section, as sketched in Fig. 3.21, using the Stokes equation, which are valid, as discussed above, at small Reynolds numbers. This geometry is fundamental in microfluidics. Microfabrication techniques naturally create microchannels of this geometry. An example, shown in Fig. 3.21 (right) was taken in Ref. [62]. Much ingenuity is required to build channels of circular cross-sections, and in practice, they are rarely used. This justifies us bringing particular attention to the rectangular case.

Here we assume a slip length $b = 0$ and hypothesize that the flow reduces to a single velocity component along x. This assumption is crucial. By projecting the Stokes

A

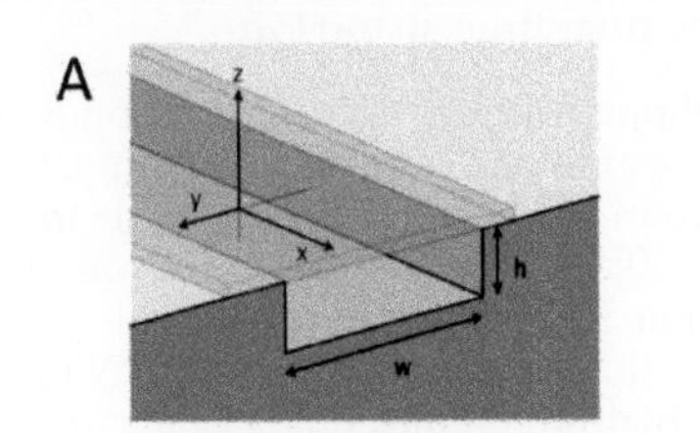

B

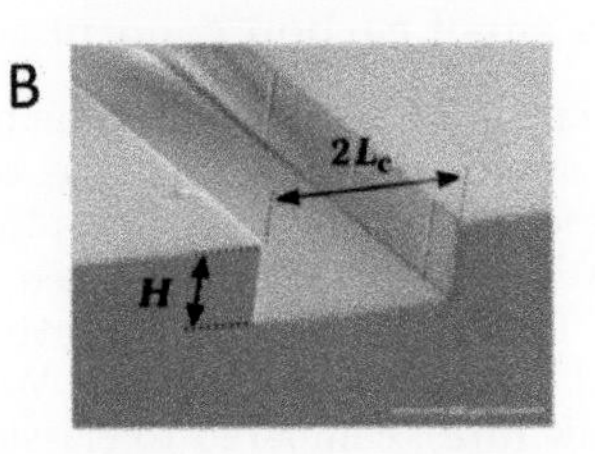

Fig. 3.21: (A) Model: Straight channel of rectangular cross-section, with the system of coordinates.(B) Real PDMS channel. (Reprinted from Ref. [62]).

equation onto the z and y axis, one infers that pressure p depends only on x (otherwise, transverse velocity components would develop). Then, by projecting the Stokes equations along x, we further infer that the pressure gradient $G = -\frac{\partial p}{\partial x}$ should be a constant (otherwise, the flow would depend on x). Using G, the flow equation reads:

$$\Delta u = -\frac{G}{\mu} \tag{3.33}$$

where $\Delta = \frac{\partial^2}{\partial y^2} + \frac{\partial^2}{\partial z^2}$. The boundary conditions are:

$$y = \pm\frac{w}{2} \text{ and } z = \pm\frac{h}{2}, u = 0 \tag{3.34}$$

In the mathematical language, we have a Poisson equation with Dirichlet boundary conditions. Several methods exist to solve it. Fourier series expansion is particularly efficient in this case. In practice, the velocity is expanded into Fourier series expansion along y or z. After some calculations, we obtain the following expression of the velocity $u(y, z)$ (the expansion is made along y):

$$u(y, z) = \frac{4G}{\mu w} \sum_{n=1}^{\infty} \frac{(-1)^{n+1}}{\beta_n^3} \left(1 - \frac{ch\beta_n z}{ch\beta_n \frac{h}{2}}\right) \cos \beta_n y \tag{3.35}$$

where β_n is defined by:

$$\beta_n = (2n-1)\frac{\pi}{w} \tag{3.36}$$

The iso-velocity contours, in the case of a shallow channel $r = h/w = 0.125$ is shown in Fig. 3.22. Except close to the sidewalls $y = \pm w/2$ (i.e. $y/h = \pm 4$), the velocity is independent of y. A misconception of hydrodynamics sometimes leads to think that the velocity profile should be doubly parabolic, i.e. parabolic along y and z. This is not the case.

The volumetric flux Q is defined by:

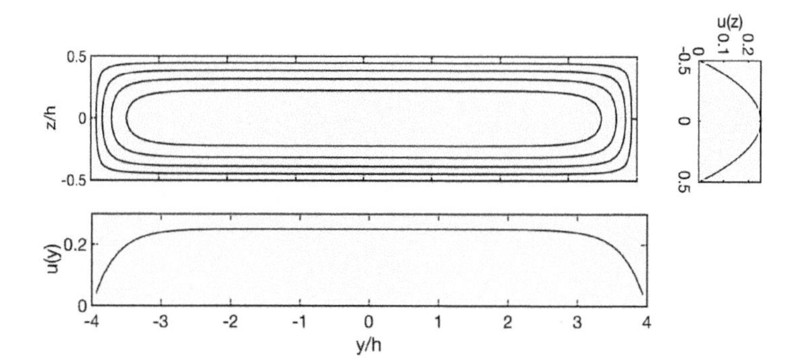

Fig. 3.22: Flow contours and profiles in a rectangular channel of aspect ration $h/w = 0.125$. On the top left, iso-velocity contours. On the right velocity profile $u(z)$, for $y = 0$; at the bottom, velocity profile $u(y)$, for $z = 0$ Except close to the side-walls $y = \pm w/2$, the velocity u depends on z, but not on y. A misconception of hydrodynamics sometimes leads scientists to think that the velocity profile should be parabolic along y and z. This is not the case.

$$Q = \int_{-w/2}^{w/2} \int_{-h/2}^{h/2} u(y, z) \mathrm{d}y \mathrm{d}z \tag{3.37}$$

satisfies the following relation:

$$Q = \frac{8Gh}{\mu w} \sum_{n=1}^{\infty} \frac{1}{\beta_n^4} \left(1 - \frac{2}{\beta_n h} th \beta_n \frac{h}{2}\right) \tag{3.38}$$

A close estimate (up to about 10%) of Q, valid for $h \leq w$, is given by the following expression :

$$Q \approx \frac{wh^3 G}{12\mu} \left(1 - \frac{6 \times 2^5 h}{\pi^5 w}\right) \approx \frac{wh^3 G}{12\mu} \left(1 - 0.63\frac{h}{w}\right) \tag{3.39}$$

In the limiting case $h/w << 1$, the flow rate is:

$$Q_0 = \frac{wh^3 G}{12\mu} \tag{3.40}$$

We plotted Q/Q_0, with Q given by Eq. (3.38), as a function of the aspect ratio $r = h/w$. The result is shown in Fig. 3.23:

For completeness, Table 3.1 displays values of the ratio $Q/Q_0 = f(r)$ as a function of $r = h/w$.

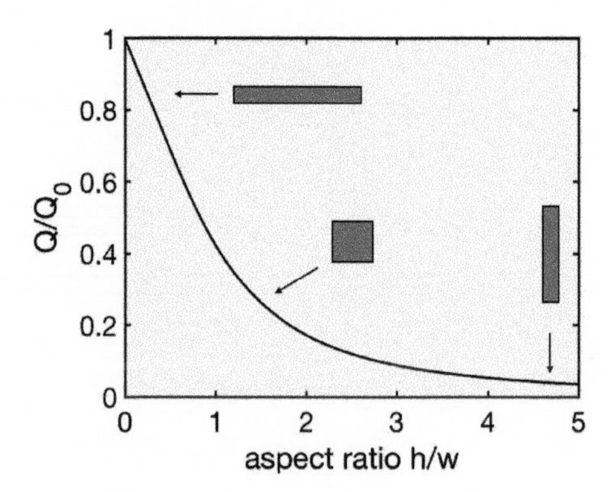

3.23: Evolution of Q/Q_0 (Q is given by Eq. (3.38), and Q_0 by Eq. (3.40) as a function of the aspect ratio $r = h/w$.

$r = h/w$	Q/Q_0
0.01	0.994
0.25	0.843
0.5	0.686
1	0.422
2	0.172
4	0.0527
6	0.0249
10	0.0094

Table 3.1 Various values of $Q/Q_0 = f(r)$ as a function of aspect ratio $r = h/w$. One may check that $f(\frac{1}{r}) = r^2 f(r)$.

3.4.2 Flow around a sphere - the Stokes law

The Stokes law is one of the most remarkable achievement of hydrodynamics. Let us consider Fig. 3.24, in which a sphere of radius a, is held in a homogeneous stream, coming from the left, with speed U_∞ oriented along x.

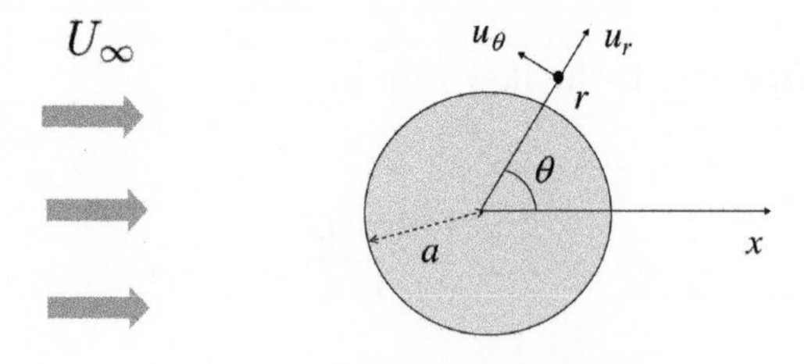

Fig. 3.24: Flow around a sphere.

The flow is perturbed by the sphere. In a spherical system of coordinate, the flow is independent of the azimuthal coordinate. It only depends on r, the distance from the sphere centre, and θ, the polar coordinate. The flow field has two components, u_r and

u_θ. The Stokes equation reads:

$$-\nabla p + \mu \boldsymbol{\Delta} \mathbf{u} = 0 \tag{3.41}$$

in which p is the pressure. Here, we neglect slippage, so that the boundary conditions are:

$$\mathbf{u} = 0 \text{ for } r = a \text{ and } \mathbf{u} = \mathbf{U}_\infty,\ p = P_\infty \text{ for } r = \infty \tag{3.42}$$

in which P_∞ is the (constant) pressure at infinity. To calculate the flow, we need to express Eq. (3.41) in a spherical system of coordinates. The strategy, for figuring out the solution, consists in determining the pressure field $p(\mathbf{r})$, by taking the divergence of Eq. 3.41 and calculate $\mathbf{u}(r,\theta)$ by taking the curl. This is not a straightforward calculation. We recommend Ref. [8], which explains well the steps of the calculation. Here, we skip this technical part and directly provide the solutions. It can found in many fluid mechanics textbooks (see, for instance, [3, 8, 63]):

$$p(r,\theta) = P_\infty - \frac{3\mu a U}{2r^2}\cos\theta \tag{3.43}$$

$$u_r(r,\theta) = U_\infty \cos\theta \left(1 + \frac{a^3}{2r^3} - \frac{3a}{2r}\right) \tag{3.44}$$

$$u_\theta(r,\theta) = U_\infty \sin\theta \left(-1 + \frac{a^3}{4r^3} + \frac{3a}{4r}\right) \tag{3.45}$$

$$\psi(r,\theta) = \frac{U_\infty}{2}\sin^2\theta \left(r^2 - \frac{3a}{2}r + \frac{a^3}{2r}\right) \tag{3.46}$$

The streamfunction ψ and the iso-pressure lines p are plotted in Fig. 3.25.

Pressure develops lobes which are positive on the left-hand side of the figure and negative on the right. There is a streamwise pressure gradient that tends to push the sphere downstream. It is called pressure drag'. Flow streamlines develop around the sphere. Far from the sphere, they stack with an increasing density. This is due to the spherical geometry.

One extremely important outcome of the calculation is the determination of the drag exerted by the flow on the sphere. The drag includes two components: a pressure drag (mentioned above), and a viscous drag, associated to the tangential stress $\tau_{r\theta} = -\frac{3\mu a^3 U_\infty \sin\theta}{2r^4}$. Stresses developed by the two components must be integrated over the sphere surface to determine the force. As expected from symmetry, one finds zero in the direction normal to the upcoming flow. Along x, the calculation leads to:

$$F = 6\mu a U_\infty \tag{3.47}$$

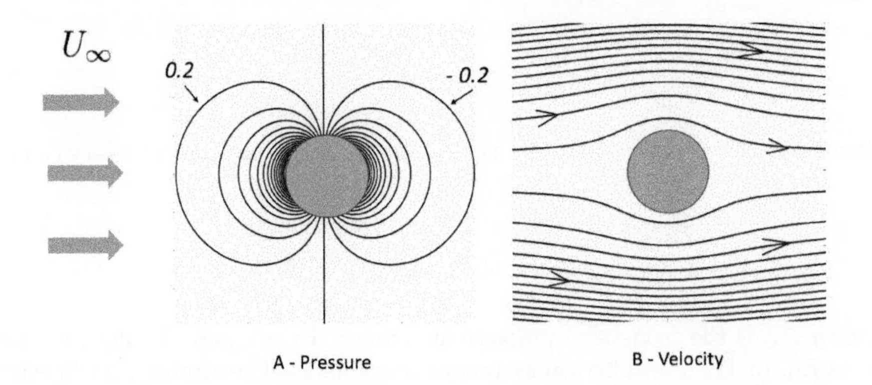

Fig. 3.25: (A) Iso-pressure levels around a sphere, in unit of $\mu U_\infty / a$, given by Stokes's solution (3.43), with P_∞=0. Levels are separated by 0.1.(B) Flow streamlines around a sphere, given by the Stokes solutio (3.46), in unit of $U_\infty a^2$. The minimum level is 0.1; the other levels are separated by a dimensionless quantity equal to 0.5.

This is the famous Stokes law, established in 1851 by George Gabriel Stokes. It is valid at small Reynolds number, for all sphere sizes. We will use the law several times in the book (in Chapter 5, for Brownian motion and particle deposition phenomena, and in Chapter 6 for dielectrophoresis and electrophoresis). Note that in the Stokes law, 2/3 of the force come from the shear stress and 1/3 comes from the pressure difference between the fore and the aft surfaces of the sphere.

3.4.3 Flow around bubbles and droplets

The same approach can be taken for droplets and bubbles. General formula exist, depending on the viscosity ratio between the inner and outer fluids. Let us consider the case of bubbles. In bubbles, the boundary condition, at the liquid-gas interface, is stress-free, i.e., on the liquid side, the tangential stress is equal to zero. There is a non-zero velocity at the interface, but no stress. In such conditions, by applying the same method as before, one obtains the following streamfunctions:

$$r > a : \psi(r, \theta) = U_\infty a r \sin^2\theta (1 - \frac{a}{r}) \tag{3.48}$$

$$r < a : \psi(r, \theta) = \frac{U_\infty r^2}{2a^2} \sin^2\theta (r^2 - a^2) \tag{3.49}$$

These streamfunctions are represented in Fig. 3.26.

It is important to note that two counter-rotating eddies develop inside the bubble. These circulations also exist in droplets. We will see, in Chapter 5 that they allow reactants to mix inside the droplet, which is critical for the biochemical applications

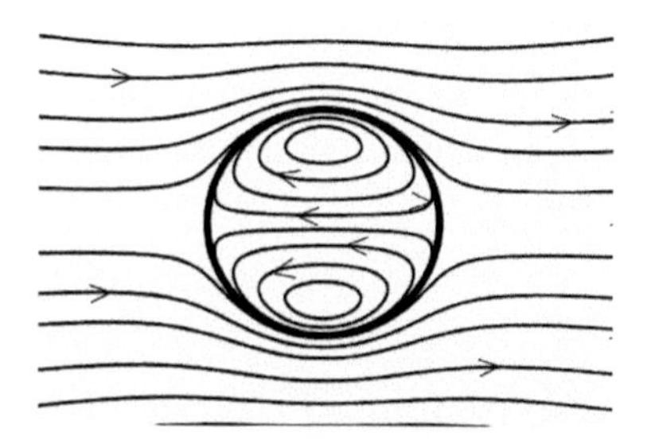

3.26: Flow around and inside a bubble. Two counter-rotating eddies develop in the bubble. A. similar pattern holds in droplets (i.e. for which the fluid inside the sphere is a liquid.)

of microfluidics. Interestingly, the drag exerted by the flow on the bubble is equal to $F = 4\mu a U_\infty$ (the factor 6 has changed).

3.4.4 Flows in Hele–Shaw cells

Returning to flows in channels of rectangular cross-section, the particular case $h << w$ deserves particular attention, because it is frequently encountered in microfluidics.[8] The reason is that, because of the limitation of photosensitive resins, it is uneasy to realize channels with heights above 200 μm. On the other hand, it is often tempting, for throughput reasons, to enlarge the channel, which leads to $h << w$. This type of geometry is called, in the hydrodynamic literature, Hele–Shaw cells, in celebration of Henry-Selby Hele-Shaw, who introduced it in 1898, to investigate the origin of the liquid/solid friction. A Hele-Shaw cell is shown in Fig. 3.27. In general, the channel is not straight; it varies along x, keeping the height h constant, but allowing for slow variations of w. As w is a function of x, velocity $\mathbf{u}$ will also depend on this coordinate. Here we perform the calculation in the case we just mentioned, i.e. the variations of w along x involve a scale much larger than h. Also, as repeatedly said, the Reynolds number is vanishingly small. Despite the apparent complexity of the situation, simple formulas can be obtained.

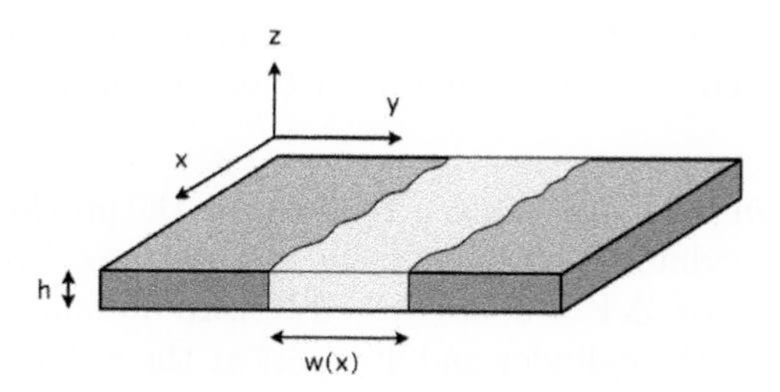

Fig. 3.27: Geometry of Hele–Shaw flows, with the system of co-ordinates.

A fundamental remark is that, in the framework of these approximations, the flow component normal to the plates is negligible. This can be shown by analysing the

[8]We showed an example in Fig. 3.6

mass conservation relation $div\ \mathbf{u} = \mathbf{0}$.[9] This implies, still taking the Stokes equations, that p does not vary with z, i.e. is a constant along the vertical dimension. Based on that, taking again the Stokes equations, and approximating the operator Δ by $\frac{\partial^2}{\partial z^2}$ (because velocity varies only slowly with x and y), $U(x, y, z)$ is found to have the following expression:

$$\mathbf{U}(x, y, z) = \frac{3}{2}(1 - \frac{4z^2}{h^2})\mathbf{V}(x, y) \tag{3.50}$$

where $\mathbf{V}$ is a horizontal vector function of x and y. It should be noted that $\mathbf{V}$ represents the average of $\mathbf{U}$ across the channel height. In most cases, rather than $\mathbf{U}$, $\mathbf{V}$ is used to discuss flow phenomena in Hele-Shaw cells. This is what we shall do here. The relation of $\mathbf{V}$ to pressure is:

$$\mathbf{V}(x, y) = -\frac{h^2}{12\mu}\nabla P \tag{3.51}$$

where P is the pressure and, as in the rest of this chapter, μ is the dynamic viscosity. This analysis is valid as long as the velocity varies slowly with x and y, i.e. far from the vertical boundaries.

Eq. (3.51) is the equivalent of Darcy's law in porous media. The velocity field is potential, implying that vorticity is everywhere zero. There cannot be closed circulations, i.e. eddies, in Hele-Shaw flows. An interesting relation on the pressure, obtained by applying mass conservation $div\ \mathbf{u} = 0$, is the following:

$$\Delta P = 0 \tag{3.52}$$

showing that pressure is an harmonic function. Looking at the function current $\Psi(x, y)$, defined by $\mathbf{curl}(\Psi\ \mathbf{e}_z) = \mathbf{V}(x, y)$, in which $\mathbf{e}_z$ is the unit vector normal to the (x,y) plane, one gets $\Delta\Psi = 0$. These properties characterize Hele-Shaw flows, and in many cases allow the determination of their structure using, for instance codes available in MatLab that are dedicated to electromagnetic or thermal calculations.

Flow around a cylinder in Hele-Shaw cells. To illustrate the previous discussion, let us consider a flow around a cylinder, in a Hele Shaw cell, produced by an upstream speed U_∞. By solving the equation $\Delta\Psi = 0$, associated to zero normal speed component at the surfaces (i.e. $\Psi = 0$ at the cylinder and $\Psi = Cst$ at the walls), one finds that the stream function $\Psi(r, \theta)$ is given, in cylindrical coordinates, by the following expression:

$$\Psi(r, \theta) = U_\infty r \left(1 - \frac{R^2}{r^2}\right) \sin\theta \tag{3.53}$$

[9] If u_z is the vertical component of the velocity, mass conservation equations tells us that $u_z \sim h/wU$, where U is a typical horizontal speed. As h/w is small, u_z can be neglected.

where r is the distance from the center and θ the polar angle. Fig. 3.28 (A) shows the calculated streamlines and Fig. 3.28 (B) those observed in a Hele-Shaw experiment. The patterns agree well.

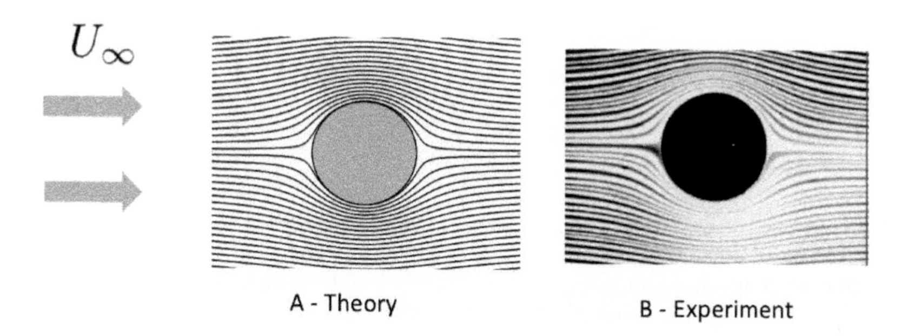

Fig. 3.28: (A) Theoretical stream function.(B) Observed flow pattern, developing around a cylinder, embedded in a Hele-Shaw cell, 1 mm high. The cylinder is several cm in diameter and U_∞ is 1 mm/s(Photo D. H. Peregrine)

Note that solution (3.53) slips at the wall. This is physically possible, independently of the material wall properties, because of the presence of thin fluid boundary layers, of thickness h, in which velocity decreases to zero. These layers are neglected in the Hele-Shaw approximation. Interestingly, the flow streamlines of Hele-Shaw flows, obtained at small Reynolds number, are identical to those of a perfect fluid, i.e. a fluid of zero viscosity slipping over the cylinder surface.

3.4.5 Flow along cavities - Moffatt eddies

Microchannels with cavities or holes are frequently encountered in microfluidics and it is interesting to present the flow patterns developing in such geometries. The streamlines are shown in Fig. 3.29, in a geometry where the flow is two-dimensional, i.e. invariant along the coordinate normal to the plane of the figure. (This geometry is fundamentally different from that of Hele-Shaw, in which the flow is *confined* along the same coordinate.)

Calculations and mathematical analysis show that when the depth of the cavity increases, a larger number of vortices forms inside. This phenomenon was explained by Moffatt in 1964 [64], for wedges of arbitrary angles. In the rectangular cavities we consider, the inner corners correspond to Moffatt wedges of ninety degrees angle. In practice, recirculations inside the cavities are much weaker than the main stream.

3.4.6 Flow over patterned surfaces

To enhance slippage, one idea is to pattern the wall surfaces, by including zones, called no shear zones', that possess infinite slip lengths. In this manner, one may hope that, if the proportion of such zones is important, the walls will expose, globally, a large

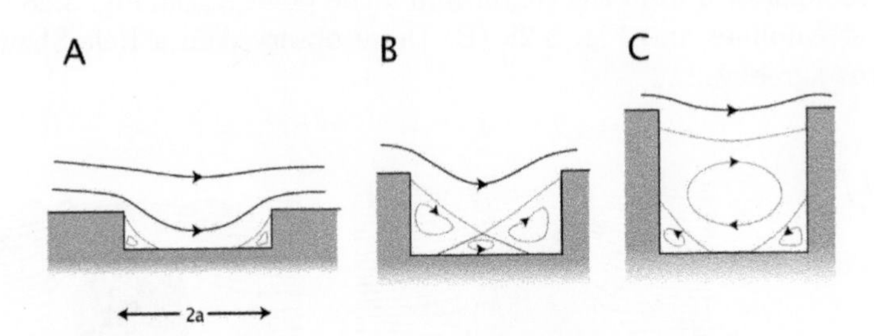

Fig. 3.29: Different flow structures in cavities at small Reynolds numbers. Here, the system is invariant along the direction normal to the plane of the figure. Eddies appear in the cavity. They are due to a mechanism explained by Moffatt [64], which gives rise to the so-called Moffatt eddies. (The figures originate from Ref. [65].)

slip length to the fluid. In practice, the zones with infinite slip lengths are created by cavities trapping gas bubbles, forming a Cassie state, which we will describe in Chapter 4. Two geometries are currently considered [66]. They are shown in Fig. 3.30, for tubes, but equivalents also exists for planes:

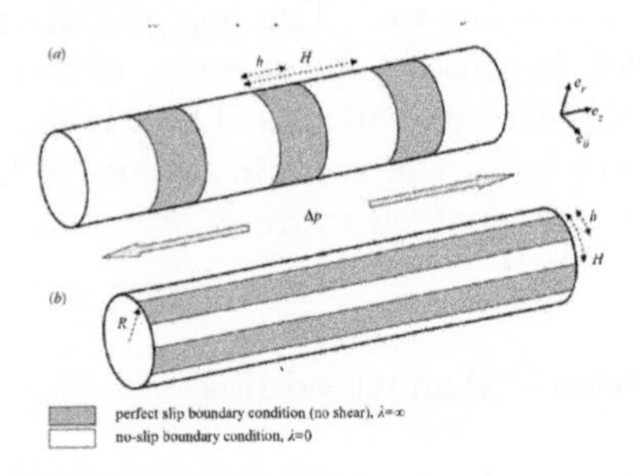

Fig. 3.30: Schematic views of the elementary models used: pressure-driven flow in a pipe of radius R with distribution of no-shear regions of width h and separation H; λ denotes the microscopic slip length.(A) The no-shear regions are transverse to the flow direction.(B) The no-shear regions are parallel to the flow direction. [66]

In Fig. 3.30, the inner tube walls include no slip and infinitely slipping (called 'no-shear') zones. In most cases, in such geometries, whether cylindrical or planar, no exact solution exists. Nonetheless, the case of Fig. 3.30 (b) could be analytically solved by Philip in 1972 [67, 68]. The solution is quite complicated and we do not present it in the book. The reader can refer to Refs [66, 67] for detail. Instead, we focus on the slip length, whose expression is given by the formula:

$$b = \frac{H}{\pi} ln \left(\frac{1}{cos(\frac{h\pi}{2H})} \right)$$

in which the definitions of h and H are the stripe width and inter distance, respectively (see Fig. 3.30). When the area fraction of no-shear bands equals unity, i.e. $\frac{h}{H} = 1$, slip length b becomes infinite. There is thus a manner, in principle, to produce low pressure loss systems, if these no-shear regions could be realized experimentally.

Unlike the case of longitudinal bands, the case of transverse bands, shown in Fig. 3.30 (A) has no exact solution in the general case. Nonetheless, as expected intuitively, as the fraction of no-shear fraction reaches unity, one finds that, in this system, slip also becomes infinite.

Slippage over CNT forests. Since 2003 and later, a number of theoretical, numerical and experimental studies have been carried out on the subject (see for instance [69–74]). Confirmation was given that large effective slip lengths can be obtained when the solid area exposed to the fluid is minimal. Ref. [74] provides an example of an experiment, in which a forest of hydrophobic CNTs covers the wall surface (see Fig. 3.31).

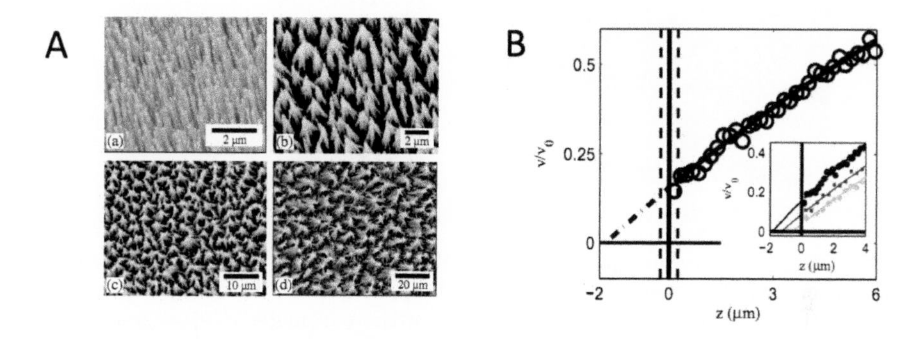

Fig. 3.31: (A)SEM images of superhydrophobic CNT forests, after functionalization with thiols: L: 1.7, 3.5 and 6 μm (pictures b, c and d). B - Velocity profile in the region close to the CNT surfaces, normalized by the velocity in the centre of the channel v0 (v0 = 360 μm/s). Inset: the three different profiles correspond to CNT surfaces with different roughness length scale L. From bottom to top, L = 1.7, 3.5, 6 μm, and v_0 = 550; 760; 360 μm/s [74].

Slip lengths found in these experiments were in the micrometre range, i.e. two orders of magnitude above smooth walls. The slip lengths were also found to be proportional to the distances between nanotubes, similarly to the case of longitudinal bands, discussed above, a result justified by theoretical arguments [74]. In other experiments, impressive slip lengths of up to 400 μm were reported on surfaces patterned with submicrometric pillars or bands [73, 75–77].[10]

[10]These large slip lengths hardly enter the theoretical framework discussed above. See [75, 76].

3.4.7 Micro and nanofluidic Landau Squire jets

The Landau–Squire flow is a jet emerging from a narrow tube into a large reservoir. The flow was first solved by Landau and Squire in two independent works (Landau & Lifshitz [63], Squire [78]). The flow is controlled by the rate of momentum transferred from the tube into the reservoir, i.e. a force, not a flow rate. The solution for the flow produced in the reservoir is given by [63]:

$$v_r = \frac{F}{4\pi\mu}\frac{cos\theta}{r} \; and \; v_\theta = -\frac{F}{8\pi\mu}\frac{sin\theta}{r}$$

in which r, θ are the radial and orthoradial components of a spherical system of coordinate, centered at $r = 0$, F is the force driving the flow at the tube orifice, i.e. the pressure divided by the jet nozzle area, the reference pressure being zero at infinity. The Landau-Squire solution is one of the rare solutions of the nonlinear Navier-Stokes equations.

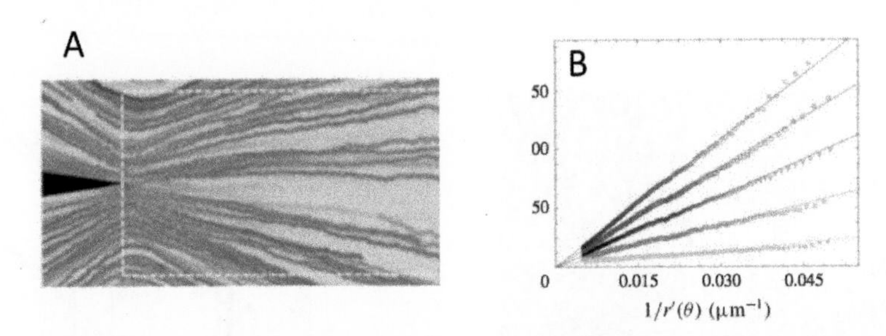

Fig. 3.32: Velocimetry measurements of Landau–Squire jet flows in submicrometric capillaries [79].(A) Tracer trajectories around the nanocapillary, 200 nm in diameter, obtained at ΔP = 140 mbar.(B) Particle velocity v as a function of the parameter $r'(\theta) = 2r\sqrt{1 + 3\cos^2\theta}$ for various ΔP, for a capillary, 320 nm in diameter. From bottom to top, ΔP = 20, 60, 100, 140 and 180 mbar. Landau Squire theory predicts proportionality between v an $1/r'(\theta)$. (Reprinting of Figure from Ref. [79] by permission from J. Fluid. Mech.).

Figure 3.32 shows the flow streamlines, obtained by tracking fluorescent particles, for a nozzle of 200 nm in diameter, pressurized at 140 mbar [79]. As the fluid is ejected from the tube, it forms a jet, which entrains the fluid of the reservoir. This gives rise to the pattern observed in Fig. 3.32 (A). The same figure shows the velocity intensity field, along with measurements of the speed intensity as a function of variable $1/r'(\theta)$ (see caption), showing proportionality, as expected. With additional measurements and theoretical developments it is possible to estimate the flow rate in the nanocapillary or the nanotube [79, 80],[11] and in turn, determine the slip length. This was done in [79]. Measurements indicated hundreds of nanometres for carbon nanotubes and, surprisingly, no slippage for the case of boron nitride nanotubes (BNNTs), despite the

[11]The task needs a separate information, because, in the theory, the jet nozzle has no size and $Q = 0$

fact that the crystallographic structures of the two materials are close. At the time of writing of this book, as underlined in a previous section, the levels of slippage observed in this experiment, and in general in CNTs, is not understood.

3.4.8 Miscible fluids flowing side by side in a Hele-Shaw cell

Here, two miscible fluids are injected in a Hele-Shaw cell. Far from the entry, where the two fluids meet, the flow reaches a steady state, where the fluids flow side by side. The flow structure shown in Fig. 3.33 assumes that mass diffusion is negligible.

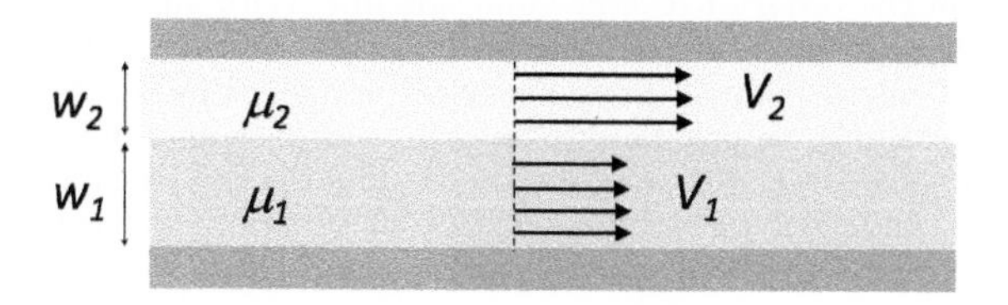

Fig. 3.33: Miscible liquids flowing side-by-side, for which mass diffusion between the two fluids are neglected.

Recall that in a Hele-Shaw cell, the flow is governed by Darcy's law. We thus have:

$$V_1 = -\frac{h^2}{12\mu_1}\frac{\Delta P}{L} V_2 = -\frac{h^2}{12\mu_2}\frac{\Delta P}{L} \tag{3.54}$$

.

which implies that, as long as $\mu_1 \neq \mu_2$, the two speeds V_1 and V_2 are different. The ratio of the two speeds is given by:

$$\frac{V_1}{V_2} = \frac{\mu_2}{\mu_1} \tag{3.55}$$

.

Let us take the particular case where the flow rates of the two fluids are the same. We have $Q = w_1 h V_1 = w_2 h V_2$, where Q is the flow rate of each fluid. Finally, we arrive at the following relation:

$$\frac{w_1}{w_2} = \frac{\mu_1}{\mu_2} \tag{3.56}$$

.

The fact that velocity is discontinuous at the interface does not violate any law. In fact, there is a thin layer, on the order of the channel height, in which the velocity passes gradually from V_1 to V_2.[12]

Eq. (3.56) also applies to non-miscible fluids, as long as instabilities, droplet formation, or wetting phenomena, do not affect the flow. The system shown in Fig. 3.33 has been used in [81, 82] for measuring viscosities, converting a dynamical measurement into a geometrical one, that of the position of the interface in the channel. In Ref. [81], the sample was less 300 μL and the viscosities ranged from 10^{-3} to 70 Pa·s.

Progress in local velocity measurements

Particle imaging velocimetry (PIV) consists in seeding flows with microparticles and infer, from the analysis of image cross-correlations, the speeds of particle patches. The resolution is on the order of several hundreds μm. With such a resolution, it is impossible to investigate flows in microchannels. Improvements were made in 1999 with the advent of micro particle Imaging velocimetry (μPIV) technique [83]. As in PIV, cross-correlations are calculated, but micro or sub-micrometric fluorescent particles are used. Moreover, the flow is placed under a microscope, with a high aperture objective, and thus small field depths, in order to localize particles. An example is shown in Fig. 3.34 A [84].

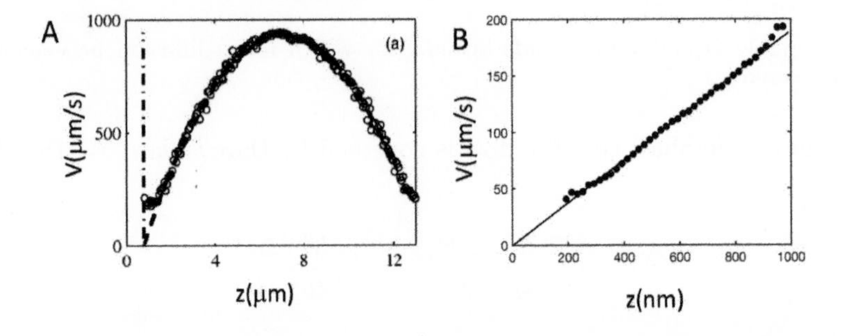

Fig. 3.34: A - MicroPIV measurements of a Poiseuilles flow in a microchannel, with wetting walls [84] ; B - TIRF based measurement of a pressure driven flow profile carried out, in a 18 μm channel high, at less than 1μm from the wall (see [85] for detail). The profile is parabolic, but, close to the wall, it is undistinguishable from a linear profile, i.e. a pure shear flow

Fig. 3.34 (A) shows μPIV measurements of a Poiseuille flow, in a microchannel, using 500 nm particles. With such particles, the wall cannot be approached by less than 1 μm. Decreasing particle sizes to improve the situation is difficult, because of Brownian motion, which tends to cancel correlations and make particle localization inaccurate. Later, investigators coupled smaller particles (typically 100 nm in diameter), and total internal reflection fluorescence (TIRF) [86, 87]. They improved, by one order of magnitude, the spatial resolution of the velocity measurements. Fig. 3.34 B [85] shows that with 100 nm particles, the wall could be approached down to 200 nm. Microfluidics has stimulated progress in velocity measurements.

[12]This is specific to the Hele Shaw geometry. It would not happen in a 2D geometry, i.e. invariant normally to the plane of the figure. In this case, momentum would diffuse and, beyond a certain distance, the velocity would be homogeneous.

3.5 Resistances and capacitances in microfluidics

3.5.1 Flow generators in microfluidics

To drive fluids, we need flow generators. Those mostly used in microfluidics are shown in Fig. 3.35.

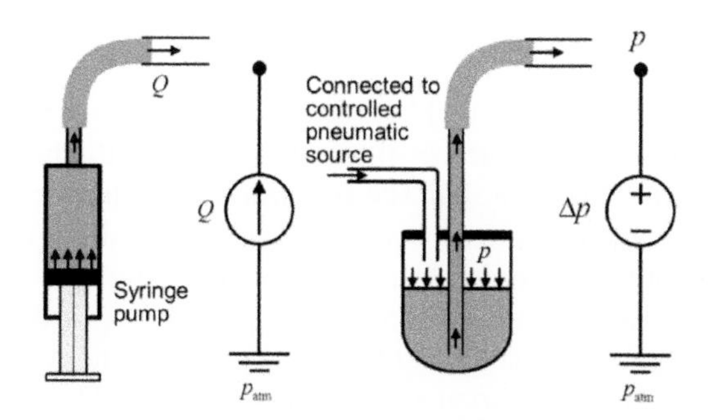

3.35: Microfluidic flow generators. (Left) Syringe pump imposing a fixed flow rate Q;(Right) Pressure source that impose a fixed pressure p upstream.

The syringe pumps include a piston that pushes the fluids inside a cylinder. The flow control can be excellent, and the flow rate oscillations low.[13] The pressure controllers impose a gas pressure in reservoirs partially filled with the working fluid. This system allows liquids in the microfluidic device to be driven by acting on the gas pressure. Pressure controllers impose well-controlled pressures, down to tens of microbars.

Over the years, the field benefited from improvements in the instrumentation, generators, and flowmeters. This progress allowed carrying out of delicate experiments, and undoubtedly, contributed to the development of the field. Fluid columns, imposing a gravitational pressure, were used in the early days of microfluidics. They are much less convenient to operate, but offer a degree of accuracy.

3.5.2 The hydrodynamic resistance

Definition. Microfluidic circuits can have a complex geometry, and, very early, the community sought to harness the concept of hydrodynamic resistance to determine the flow in each of their branches. Contrarily to electrokinetics, where Ohm's law is local, in microfluidics, except in the Hele-Shaw limit,[14] the notion of hydrodynamic resistance is global, i.e. it is attached to a particular geometry. In an earlier section, we saw that, for channels of rectangular cross-section, the pressure drop ΔP is proportional to the flow rate Q (see Eq. (3.38). We thus have the relation:

[13] For low cost syringe pump, a few % is a plausible number. The highest-quality syringe pumps reach 1% fluctuation, roughly

[14] Eq. (3.51) represents a local relation between speed and potential gradient, analogous to local Ohm's law

$$\Delta P = RQ \tag{3.57}$$

where R is called the hydrodynamic resistance. In the case of a shallow channel, where $h << w$, the expression of R is:

$$R = \frac{12\mu L}{h^3 w} \tag{3.58}$$

.

Eq 3.57 points to an analogy with electrokinetics, in which pressure would correspond to electrical voltage, and flow rate to electrical current intensity. At large Reynolds numbers, we can still write a relation of this type, but the resistance is no longer an intrinsic property of the fluids and the geometry, it depends on the flow itself, which reduces the interest of the concept. Going further into the analogy, one can define an hydrodynamic resistivity equal, in the case of shallow channels, to $\frac{12\mu}{h^2}$. This resistivity increases as the inverse of the second power of the height, indicating that there is a 'price' to pay, in terms of pressure drop, for miniaturizing microfluidic devices. The unit of hydrodynamic resistance is $Pasm^{-3}$. This unit does not have a particular name. To provide an order of magnitude for R, it is useful to know that for water ($\mu \approx 10^{-3}$Pas), and a square channel of 100μm side, 1 mm long, we have $R \approx 10^{11} Pasm^{-3}$.

Several geometries, with an without slippage at tha walls, are displayed in Table 4.4.

Shape		R_{hyd}
Shallow	.	$\frac{12\mu L}{h^3 w}$
Rectangle	.	$\approx \frac{12\mu L}{h^3 w}\frac{1}{(1-0.63\frac{h}{w})}$ $(h/w < 1)$
Square	.	$\approx \frac{28\mu L}{h^4}$
Circular	.	$\frac{8\mu L}{\pi r^4}$
Shallow with slip	.	$\frac{12\mu L}{h^3 w(1+6\frac{b}{h})}$
Circular with slip	.	$\frac{8\mu L}{\pi r^4(1+4\frac{b}{r})}$

Table 3.2 Formulas of various resistances for channels of circular and rectangular cross-sections, with and without slip.

Resistances in series and parallel. There are laws for nodes and branches, just as in electrokinetics. For a node, the sum of the flow rates is cancelled out. We thus have the following equation:

$$\sum_{i=1}^{n} Q_i = 0$$

in which Q_i is the flow rate entering or exiting the node. On the other hand, along each closed branch, we have:

$$\sum_{i=1}^{n} \Delta P_i = 0$$

in which ΔP_i is the pressure difference between two points of the closed branch. The two relations form the equivalent of Kirchoff's law in electrokinetics. It follows that hydrodynamic circuits can be treated as electrokinetic circuits. For example, two resistances R_1 and R_2 placed in series are equivalent to a resistance $R = R_1 + R_2$. Such a situation is shown in Fig. 3.36. The series of relations leading to this result is the following:

$$\Delta P_1 = R_1 Q \tag{3.59}$$

$$\Delta P_2 = R_2 Q \tag{3.60}$$

$$\Delta P = \Delta P_1 + \Delta P_2 = (R_1 + R_2) Q = RQ \tag{3.61}$$

In this calculation, we assumed that the transition region between R_1 and R_2, in which the streamlines develop a curved pattern to match the two resistances, has a negligible size. It could be possible to attach a resistance to this region. All this complicates the analysis, however. In practice, when transition regions take a substantial part of the circuit, the notion of resistance is no longer interesting to use. Electrical circuits have the same problem, but, in practice, the transition regions are extremely small and they can always be neglected.

The case of two resistances in parallel are equivalent to a resistance whose expression is:

$$R_{//} = \frac{R_1 R_2}{R_1 + R_2} \tag{3.62}$$

We can formulate the same remarks as previously, concerning the transition regions located upstream and downstream of the two branches. All these analogies are extremely useful when dealing with microfluidic circuits with several well-identified branches, which is most often the case.

3.5.3 Two examples of microfluidic resistance networks

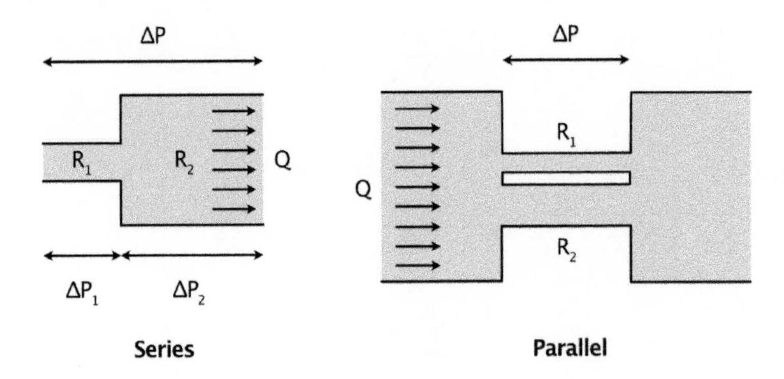

Fig. 3.36: Two hydrodynamic resistances placed in series (left) and in parallel (right).

Example 1: The hydrodynamic focusing geometry. A paper published in 1998 [88] reported a micromixer with a mixing time in the μs range. The work received a large echo, because this mixing time was three orders of magnitude faster than the most rapid mixers used at that time. The experiment is shown in Fig. 3.37. The geometry is the intersection of two microchannels. On the left, an aqueous solution of fluorescein, is injected at pressure P_i, while on the sides, pure water is introduced at pressure αP_i. The idea is that the side flows squeezes the fluorescein stream, down to deca-nanometric dimensions [88]. Now, this type of geometry is frequently used to form droplets, as will be explained in the next chapter. Its name is 'hydrodynamic focusing'.[15]
Fig. 3.37 (right) shows the equivalent electrical circuit.

On resolving the currents and pressure drops in the network, one is led to the following expression for the ratio of the injected current in the central branch I_i to the total current I_0.

$$\frac{I_i}{I_0} = \frac{1 + 2\sigma - 2\alpha\sigma}{1 + 2\alpha\gamma}$$

in which the meanings of the variables are displayed in Fig. 3.37. For $\alpha > \frac{1+2\sigma}{2\sigma}$ (for instance, the main branch highly resistive, and large side pressures), I_i is negative. In such a case, the flow in the main branch is inverted. Before inversion, a weak stream of fluorescein, strongly squeezed by the side-flows, penetrates into the collecting channel, as shown in Fig. 3.37. In the experimental work, the authors succeeded to reduce its width down to less than 100 nm, which led to diffusive times lying in the tens of μs range. These were the conditions for which an extremely fast dilution of fluorescein with water could be obtained.

[15] Sometimes it is called 'X cross-channel'

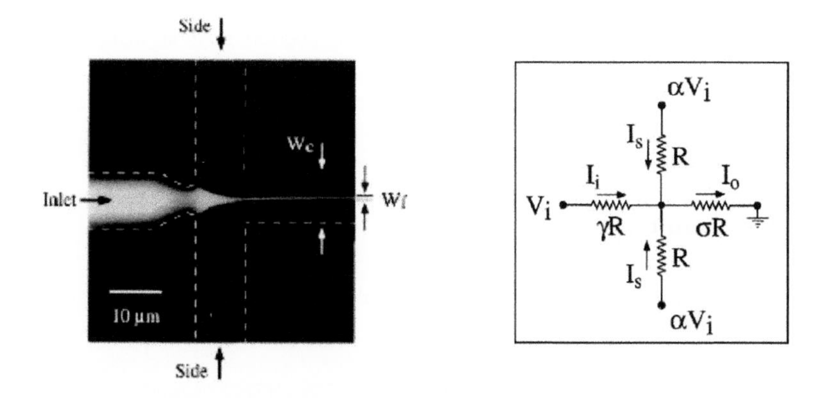

Fig. 3.37: Hydrodynamic focusing is a geometry for which fluids penetrate the device through three entries, and, depending on the flow conditions, are collected or not by a main channel. The left figure was taken from [88]. In the reference, the flow was visualized by using fluorescein. On the right, the equivalent hydrodynamic circuitry, with the original notations, that we keep for the calculation. (Reprinted figure with permission from Ref. [88] Copyright (2022) by the American Physical Society with permission of R. Austin.)

Example 2: The 10-fold log levels solute distributor. G. Whitesides developed a device enabling the distribution of tracers in multiple levels. Years later, Furlani et al. [89] invented a device, in which a solution can be distributed with 10-fold dilutions, spanning two decades in the solute concentration (see Fig. 3.38).

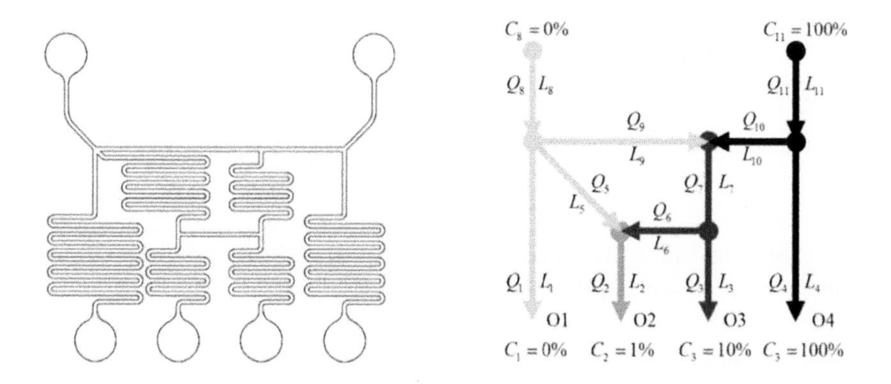

Fig. 3.38: The ten-fold dilution device [89]. (Left) Design of the network, incorporating branches of different lengths, so as to implement the set of resistances calculated by the theory; (right) Detail of the network, with the definitions of the lengths and flow rates in each branch. The calculation leads to: $Q_1 = Q_2 = Q_3 = Q_4 = 1\ \mu$l /min; $Q_5 = 0.9; Q_6 = 0.1; Q_7 = 1.2; Q_8 = 2.9; Q_9 = 0.99; Q_{10} = 0.11; Q_{11} = 1.11\ \mu$l/min; $L_1 = 25.2; L_2 = 10; L_3 = 10.2; L_4 = 21.4; L_5 = 16.8; L_6 = 2; L_7 = 10; L_8 = L_{11} = 2; L_9 = 4; L_{10} = 2$ mm.

Here, one has two entries, with two colors, and four outlets, that redistribute the colour levels in a series of 10-fold log levels. The authors succeeded in designing the circuit, i.e. determining the lengths of each branch, to obtain this remarkable result. The right part of Fig. 3.38 shows the various flow rates circulating in each branch. For detail, see [89].

3.5.4 Hydrodynamic capacitance

In the same way as for resistance, the notion of hydrodynamic capacitance of an element can be introduced with the relation:

$$Q = C\frac{\mathrm{d}P}{\mathrm{d}t}$$

where Q is the volumetric flow rate, C is its capacitance, and P is the pressure drop across the element in question. The idea of hydrodynamic capacitance is interesting when the channels are deformable. An example is the soft plastic tubing that conveys the fluids into the device. In all cases, under the action of the pressure, the volume changes, generating additional flow rates, which are precisely accounted for in the definition of the capacitance. The unit of hydrodynamical capacitance is $m^3 Pa^{-1}$, which, as for the resistance, does not have a specific name.

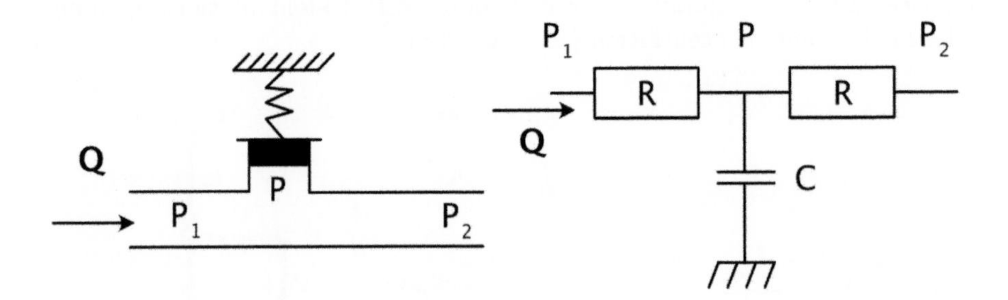

Fig. 3.39: (Left)Flow in a microchannel, with a piston-spring system. (Right) Electrokinetic model representing the flow.

To illustrate the notion of capacitance, consider a case where the flow is driven in an elastic tube. Suppose that, in a lapse of time δt, the global pressure in the tube is suddenly changed by an increment $\delta P > 0$. The fluid volume V will expand by an amount δV, generating an additional flow rate q that fills the new available space. Let us calculate q. One has:

$$q = \frac{\delta V}{\delta t}$$

On the other hand, the tube is made in elastic material, with a Young modulus E, a thickness e and a diameter ϕ. We have:

$$\delta P = \frac{2e}{\phi} E \frac{\delta V}{V}$$

Then, combining the two equations, we end up with the relation:

$$q = C\frac{\delta P}{\delta t}. \text{ with } C = \frac{\phi V}{2eE}$$

This important example, which will be used later, illustrates the concept of capacitance.

3.5.5 Two examples of applications of the concepts of resistance and capacitance

Example 1: The long transients caused by trapped bubbles. In the preceding example, the capacitance was not localized, but distributed along the tube. Take the case, now, where the tube deformation is localized. Figure 3.39 shows a piston, held by a spring of stiffness k, placed on the side of a channel (see Fig. 3.39). The equivalent circuit of this system is shown on the same figure. One has a first resistance R along which the fluid is driven. Then there is a node, at which the flow separates in two branches, one pushing or pulling the piston, the other flowing through another resistance R. Calculations show that the capacitance C is given by the expression:

$$C = \frac{S^2}{k}$$

in which S is the piston area.

By resolving the circuit of Fig. 3.39), one finds the following equation for the current $i(t)$ circulating in the branch of the second resistance:

$$\frac{\Delta P}{R} = i + RC\frac{di}{dt}$$

From this equation, one concludes that the time needed to reach a stationary state is RC. This type of calculation can be repeated to determine the behaviour of a device in which a bubble is trapped. By replacing the piston/spring system by a bubble, one finds, for the expression of the capacitance, $C = \frac{V_0}{P_0}$, where V_0 is the volume bubble. On considering typical microfluidic values, one concludes that, with trapped bubbles, the time constant RC, which controls the time it takes to reach a stationary state, is large: let us take a resistance of 10^{11} $Pa.sm^{-3}$ (see Section 3.57) and a bubble of 100 μm in diameter. One finds a time constant RC of 1000 s. It will thus take 15 mins to reach a steady state after a change in the operating conditions. If there are several bubbles, in the device, which evolve, for instance, dissolve or reform, one may never reach a steady state. It can be concluded that the presence of trapped bubbles in a microfluidic device is extremely detrimental to its functioning. Trapped bubbles are the 'enemies' of microfluidics.[16] .

In practice, bubbles are more likely to appear when there is a leak, and during the filling of the device. In order to avoid bubble trapping, one usually works with hydrophilic materials (if aqueous solutions are the working fluids), and avoid geometrical dead zones, corners, intersections. Evacuating before filling, or pre-wetting with short-chain alcohols, can also be strategies. It is interesting to note, working with a material permeable to gas, such as polydimethylsiloxane (PDMS), facilitates the removal of bubbles, enabling it to work more easily with complex designs.

Example 2: The long transients induced by syringe pumps. There is an important effect, in microfluidics, that I called 'bottleneck effect' in the first edition of this book, and which must be known to avoid long transient in microfluidic devices. The situation is shown in Fig. 3.40.

A syringe pump pushes a liquid in a microfluidic devices, represented as a simple microchannel. The source of flow includes the syringe pump, the tubing, which is deformable and possesses a low resistance. On the microdevice side, the system is rigid and highly resistive. Since the device is highly resistive, the piston must build up a substantial pressure in the chamber and the tubing for pushing the liquid. The equivalent circuit is shown in the left-hand part of Fig. 3.40. The source of flow includes a current source, that delivers a flow rate Q in parallel with a capacity C that takes into account the fact that part of the flow rate produced by the piston is sucked by the volume increase of the tubing. The two contributions, Q and the capacity flow rate add up to enter the device, characterized by R.

Replacing R by its expression for channels of square cross-section (see Table 4.4) and C by its expression for an elastically deformable tube, one obtains, for the time constant $\tau = RC$ of the circuit:

$$\tau = RC = \frac{3.6\pi\mu l\phi^3 L}{Eeh^4}$$

[16]In the blood circulation of the body, micrometric bubbles dissolve rapidly, and thereby do not generate problems. On the other hand, it is accepted that bubbles greater than 7 μ m in diameter may cause embolisms. One may have sorts of embolisms in microfluidic devices, when, for instance, because of a leak, trapped bubbles are so large that they block the main channel.

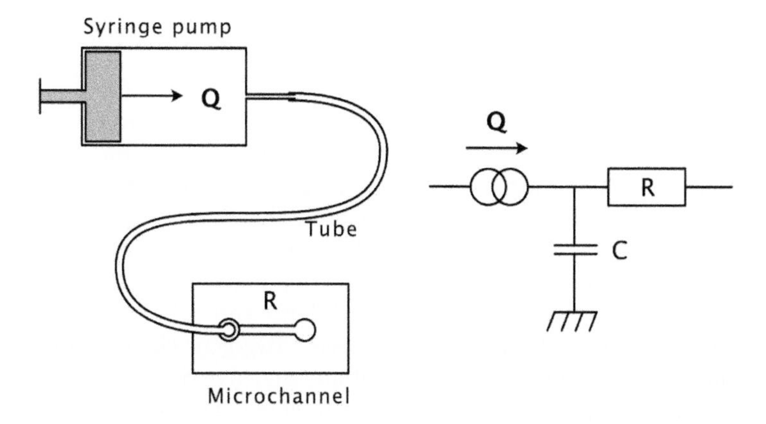

Fig. 3.40: When a syringe pump is used for driving a fluid in a microfluidic device, the deformability of the tubes may generate long transients. This was called bottleneck effect in the first version of the book, because, during the transient regime, no liquid penetrates in the microfluidic channel, even though the piston keeps advancing. Somewhere, a bottleneck needed to be unlocked to feed the device The right figure shows the hydrodynamic circuitry equivalent to the syringe/microchannel system.

in which ϕ and L are the diameters and lengths of the tubing, l and h are the lengths and side of the microchannel, and K is the Young modulus of the tube material. In practice one obtains a time constant on the order of 20 min, for a 5x5 μm square microchannel, 1 cm long, a tube of 1 mm diameter, 0.5 m long, 0.1 mm thick plastic tubing (1 GPa of Young modulus), and water (0.9 10^{-3} Pa.s). This means that when we start operating a microfluidic device of that size, one must wait for 10 min for reaching a steady state, which is obviously a long time. There are ways of improving the situation, by working with larger channels, using more rigid tubes, or with smaller diameters. However the most elegant way to resolve the problem is to replace the syringe pump by a pressure controller, i.e. to work at fixed pressure, not at fixed flow rate. The explanation is provided by Fig. 3.40, where we see that, by replacing the current source by a pressure source, the device is submitted, without any transient regime, to a fixed pressure. Indeed, the source will have to deliver a current in the capacitance to compensate for the inflation of the tubes. But this derivation will not affect, in any respect, the fact that the device R will be submitted immediately by the pressure imposed by the source. In this manner, we completely suppress the transient state.[17] Accurate measurements of the effect are shown in Ref. [91].

[17]This type of idea is developed in other domains. For instance, in hot wire anemometry, the probe is operated at fixed resistance and thus fixed temperature, and not at fixed electrical current, in order to eliminate long transient regimes related to the thermal inertia of the probe [90].

3.6 Inertial microfluidics and millifluidics

3.6.1 Introduction

We have noted that, due to scaling laws, the Reynolds number is inherently small in microsystems, and, as a consequence, flows developing in them are governed by the Stokes equations. However, it is necessary to qualify this line of thought. Many microsystems do not work at small Reynolds numbers. This is the case micro of heat exchangers. We will see, in Chapter 5, that they operate at moderate or high Reynolds numbers to achieve high efficiencies. Moderate Reynolds numbers are also involved in inkjet printers, which were presented at the beginning of this book. In such devices, moderately high Reynolds number are needed to achieve large throughputs without triggering hydrodynamic instabilities. In other microdevices, divergent channels function at high Reynolds numbers to generate flow irreversibility and give rise to pumping. In other microsystems, eddies are produced at moderate Reynolds numbers to favour mixing, without developing turbulence. These examples show that, in a number of cases, moderate or high Reynolds numbers are used in miniaturized systems, and, consequently, they must be considered in this book.

Another topics, which surged in the last decade, associates high or moderate Reynolds numbers to particle or cell transport. This domain has been called 'inertial microfluidics'. In this book, we will not adopt this definition. We will consider that 'inertial microfluidics' covers all situations where, in a micro or millifluidic system, inertia forces are important, i.e. the Reynolds number is well-above unity.

3.6.2 The phenomenon of boundary-layer separation

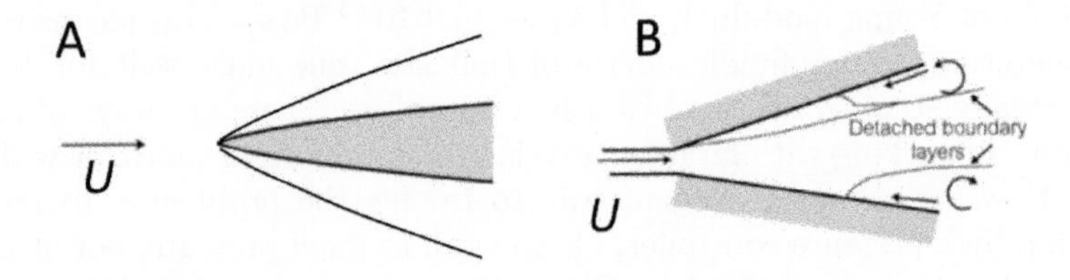

Fig. 3.41: (A) Convergent geometries are favourable, in the sense that they keep the boundary layer 'attached' to the wall, and are associated to small hydrodynamic losses.(B) Divergent flow cause substantial hydrodynamic losses, due to the development of detachment eddies triggering turbulence. This case is called unfavourable.

Laminar boundary layers form a classical topics in fluid mechanics, and one can find detailed descriptions in many textbooks. We consider here a laminar boundary layer developing over a wedge of small angle (see Fig. 3.41). This situation could be, for example, the flow along the walls of a convergent or divergent microchannel. We can distinguish two situations often called 'favourable' and 'unfavourable'. The favourable case is sketched in Fig. 3.41 (A).

Here we have the following two conditions:

$$\frac{\mathrm{d}U}{\mathrm{d}x} > 0 \quad \text{and} \quad \frac{\mathrm{d}P}{\mathrm{d}x} < 0,$$

where $U(x)$ represents the flow external to the boundary layer (which mainly depends on the streamwise direction x because the wedge angle is small), and $P(x)$ is the corresponding pressure. The flow $U(x)$ increases along x because of mass conservation. On the other hand, Bernoulli's theorem, which is valid at high Reynolds numbers (which we consider here) and for stationary flows, stipulates that $P(x) + \frac{1}{2}\rho U(x)^2$ is constant along a fluid trajectory. This implies that $P(x)$ decreases downstream. Pressure $P(x)$, which also holds in the boundary layer, thus tends to accelerate the flow therein. In this configuration, the boundary layer remains 'attached' to the wall. Also, for reasons that we have no space to develop here, the flow is stabilized. This case is considered as 'favourable'.

The unfavourable case is shown in Fig. 3.41(B), corresponding to a divergent channel. In this case, the velocity $U(x)$ decreases (where we recall that x is the streamwise coordinate), implying (due to Bernoulli theorem) that the pressure $P(x)$ increases downstream. Consequently, pressure acts against the flow. There is thus the possibility to trigger the appearance of an internal recirculation, i.e. an eddy, within the boundary layer. The formation of this eddy is called boundary-layer detachment, because, in such a situation the streamlines emitted from the apex seem to detach from the wall. It is also called boundary layer separation, because, as sketched in Fig. 3.41 (B), the boundary layer splits into two regions. In geometries like that of Fig. 3.41 (B), detachment of the boundary layer is unstable, so that separation generally triggers turbulence, which considerably increases the hydrodynamic losses. In the case where the detachment point is anchored at a corner, like in Fig. 3.42 [92], there is no turbulence triggering, but hydrodynamic losses are still substantially more important than in the favourable case. This feature was exploited in the 1990s to fabricate micropumps. Eddies frequently occur in diverging geometries [92].

These eddies have also been used in Ref. [92] to separate particles.

Pumps working at moderate Reynolds number. Separation can intervene in microsystems functioning at moderate Reynolds numbers it the geometry includes edges, divergent parts, or ruptures. An example is shown in Fig. 3.43 [93].

The principle consists of pushing and pulling the fluid with an electrostatically driven membrane, in an asymmetric chamber. When the fluid is sucked by the membrane, separation of the boundary layer occurs at the right because there is an abrupt opening of the geometry. On the left, thanks to the divergent, the speed is decreased before the flow arrives at the rupture, weakening the nonlinear phenomena associated to

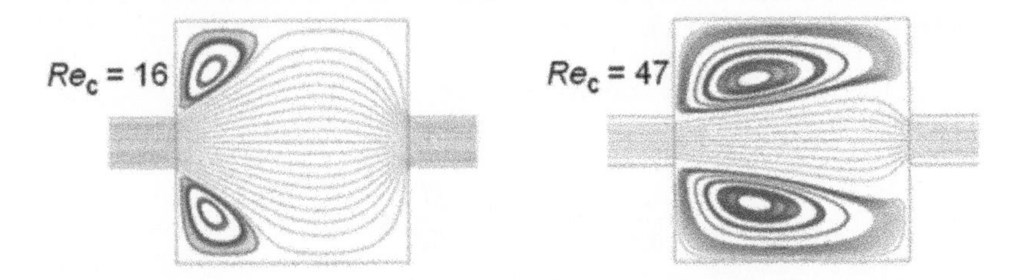

Fig. 3.42: Eddy pair developing in a rectangular cavity crossed by a flow at moderate Reynolds numbers [92]. The pair appears at the entry, i.e. where geometry is diverging, not close to the outlet, where geometry is converging. (Reprinted from Ref. [92], with permission from the Royal Chemical Society. Copyright 2022.)

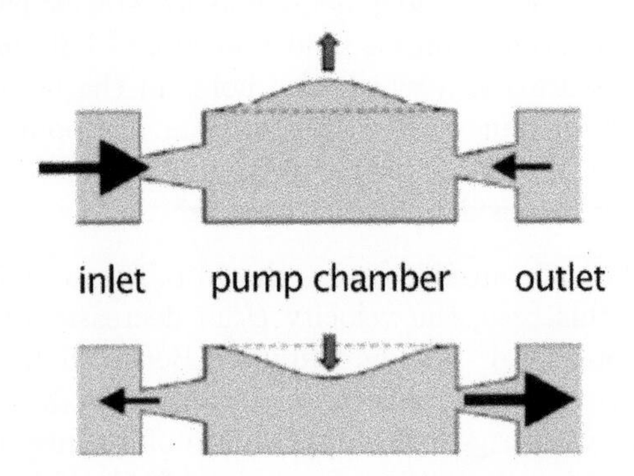

3.43: Micropump working at moderate or high Reynolds numbers, exploiting a geometrical asymmetry between the left and the right, to produce a net flow towards the right (see explanations in the text) [93].

it, so that the rupture of the geometry generates smaller losses. Similar phenomena occur when the movement of the membrane is inverted, i.e. from pulling to pushing, so that eventually, as indicated in Fig. 3.43, a net flux, from left to right, develops. One limitation of such pumps, beyond the difficulties of microfabrication, and their cost, is their low efficiency, meaning that hey are rarely used in practice.

3.6.3 Centrifugal microfluidics

Centrifugal microfluidics has a long history [94–96]). It started in the 1970s and was reexpanded in 1998, fostered by contributions from M. Madou and G. J. Kellogg, along with industrial involvements of Abaxis and Gyros [94]. The field is now considered to be moving towards a mature technology. The idea is to exploit centrifugal forces to move fluids radially outwards on a disk. In practice, microchannels are moulded in plastic cartridges, which look like CD disks [94]. Even though the scaling laws associated with inertial terms are unfavourable in microsystems, the high rotation speeds allow the development of appreciable centrifugal forces, permitting the controlled movement of fluids. The rotation speeds used here are on the order of 3,000 turns per minute. The principle of the functioning of such a system is represented in Fig. 3.44.

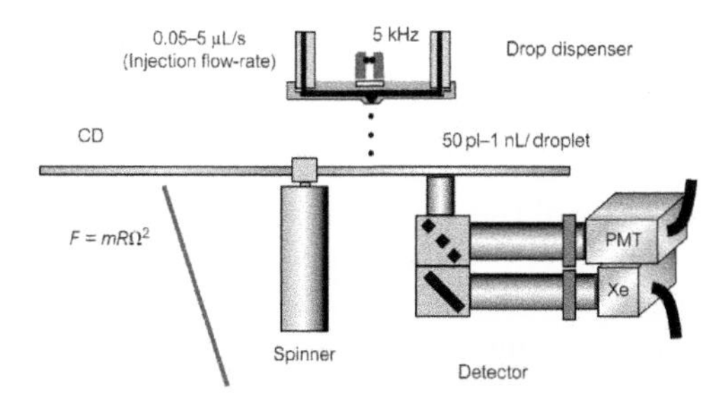

3.44: Microfluidic system using centrifugal forces to drive fluids on a disk. The system shown here was developed by the company Gyros, about twenty years ago.

In the figure, we see a dispenser supplying fluid to a rotating disk. The details of these canals and reservoirs in the disk are shown in Fig. 3.45.

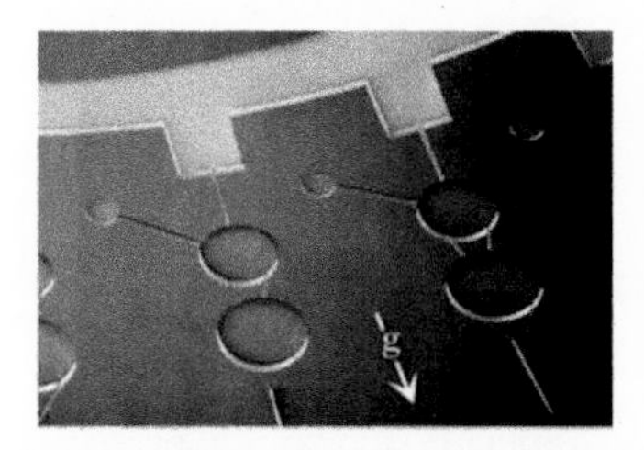

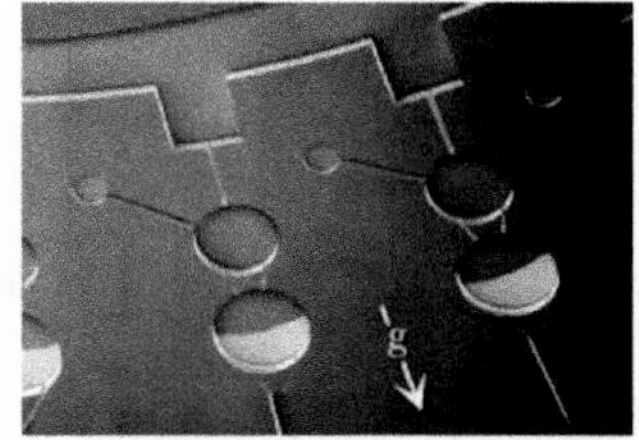

Fig. 3.45: Detail of a rotating disk, showing a valve formed by a hydrophobic restriction, in which a meniscus is held by capillary forces.(Left) At small rotation speeds, the meniscus is immobile, and the white fluid is blocked. (Right) At larger speeds, the meniscus cannot sustain the centrifugal pressure, and the (white) fluid flows radially outwards.

Each reservoir is separated by an hydrophobic restriction, which plays the role of a valve. They are successively filled using an incremental augmentation of the rotation speed of the disk.

The following relation, which provides the expression of the centrifugal forces f driving the fluid, allows to describe, schematically, how a centrifugal valve functions. Let us consider a fluid volume v, enclosed in a chamber, it will be able to flow outwards if the following condition is satisfied:

$$f = \rho v \Omega^2 R > f_{\text{cap}}$$

where Ω is the rotation velocity, R is the radial position of the valve, and f_{cap} is the capillary forces that holds the fluid meniscus. This inequality thus compares capillary forces, which retain the fluid meniscus, with the centrifugal forces, which pushes the fluid towards the periphery of the disk. The valve is open if centrifugal forces are higher than the capillary forces, which is expressed by the inequality above. The capillary force can be controlled by adjusting the dimensions of the canal connecting the two reservoirs, or the properties of the surfaces exposed to the fluid, or both. The smaller the channel, the larger the forces keeping the liquid on the disk must be. By increasing the rotation speed, fluid volumes are drivent outwards to fill new reservoirs.

Thus, centrifugal microfluidics elegantly resolves the delicate problem of pumps and valves, despite unfavourable scaling laws.

Over the last two decades, an important number of functionalities has been developed by the centrifugal microfluidic community (mixing, separation, aliquoting, RNA extraction, amplification, etc.) [94]. The integration of these functionalites on centrifugal devices enabled to reach a certain level of complexity, attracting, in the recent years, the interest of companies, including major ones such as Abbott and Samsung. An example of application of the technology to the diagnostic field is shown in Fig. 3.46.

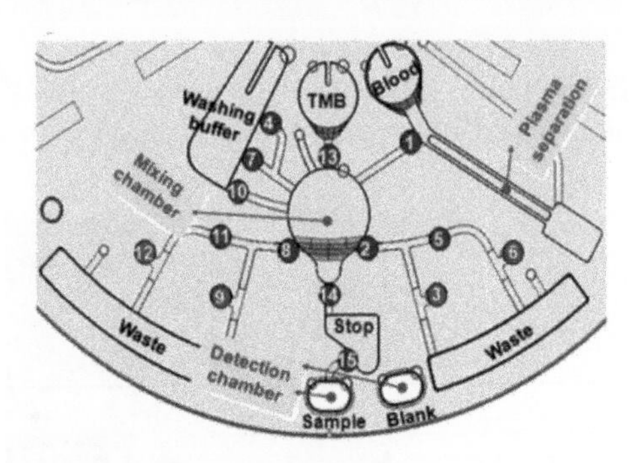

3.46: An integrated ELISA on a disc from Samsung.

The future will tell whether the technology will give rise to major microfluidic applications.

3.6.4 Dean vortices in curved channels

When a flow is driven in a curved rectangular channel, as shown in Fig. 3.47, recirculating flows appear. They superimpose to the mean motion, so that fluid particles spiral out, as they move downstream. This is what J. Eustice [97] discovered in 1911, by injecting ink in water flows driven in curved tubes. The secondary motion was calculated sixteen years later, by W. R. Dean [98], for the case of small Reynolds numbers and circular tubes of small curvature.

A flow driven in a curved channel of rectangular cross-section is subjected to centrifugal forces, whose radial component f_r is given by:

$$f_r = \frac{\rho {U_\theta}^2}{r}$$

in which U_θ is the orthoradial (or circumferential) flow component. The problem is that, because of the presence of horizontal walls (see Fig. 3.47), U_θ varies with the vertical coordinate z. This implies that this force cannot be balanced by a pressure gradient. One way to show that is to calculate the rotational of the centrifugal force, aligned along θ, and equal to $\frac{1}{r}\frac{\partial}{\partial z}U_{\theta,r}(z)^2$ which is not null, precisely because of the presence of these walls. This implies that it cannot be balanced by a pressure gradient (whose rotational is zero), and, therefore a motion within the fluid, develops. Due to the fact that the rotational of the driving force is along θ, a recirculating flow will develops in the (r, z) plane, as sketched in Fig. 3.47. This flow is called Dean flow' after the pioneering paper written by him. Physically, the flow is pushed radially outwards by the centrifugal forces, more strongly in the center than on the top and bottom parts of the channel, forcing the flow to recirculate.

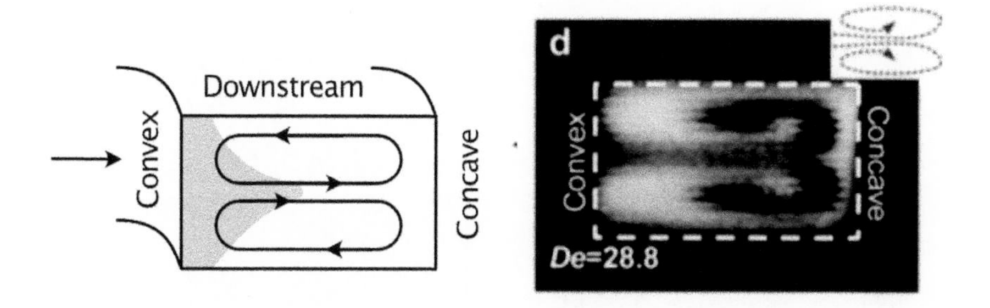

Fig. 3.47: Dean flows are recirculating flows that superimpose to a mean flow driven through a curve tube. At the lowest Dean numbers, they take the form of two symmetric eddies, as sketched in the lefthand part of the figure. The existence of these eddies was confirmed experimentally (right).(Reprinted from [99] under Creative Commons licence.)

The order of magnitude of the Dean flow can be estimated by using the Navier-Stokes equations, that must be written in cylindrical coordinates for this purpose (expressions can be found in [3]). Assuming the secondary flows to be weak, and balancing centrifugal forces and viscous terms, one finds, for the r and z component of the secondary flow:

$$u_r, u_z \sim \frac{\rho {U_\theta}^2 h^2}{\mu r}$$

The Dean number De is constructed as a Reynolds number based on the secondary flow components:

$$De = Re\sqrt{\frac{h}{2r}} \text{ in which } Re = \frac{Uh}{\nu}$$

This number provides an indication on the Dean flow intensity along with the inertia forces associated to it. In the years 1990–2000, it appeared that the primary Dean flow structure, sketched in Fig. 3.47 is replaced, as the Dean numbers was increased, by more complex vortical structures, including larger number of vortices. Describing these structures would lie outside the scope of this book.

In practice, in microfluidics, the secondary flows are exploited mostly for mixing and separation. An example is shown in Fig. 3.48 [100]. Blood cells are separated from circulating tumor cells (CTC) under high throughput conditions. As cancer cell sizes are larger, it is considered that they separate under the action of the Dean vortices and centrifugal forces. As the CTCs are rare (1–10 per ml), important blood volumes need to be screened.

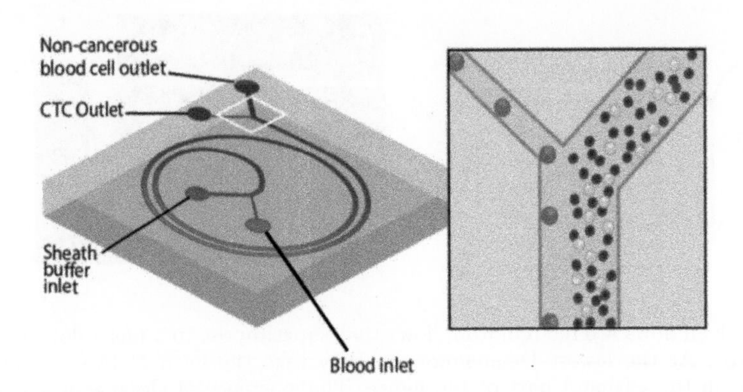

Fig. 3.48: The spiral microfluidic chip, exploiting the action of centrifugal forces and Dean recirculations, enables an efficient, size-based separation and isolation of circulating tumor cells (CTCs) from whole blood [100].

3.6.5 Millifluidics and micro reaction technology (MRT)

Micro reaction technology was mentioned in Chapter 1. The domain develops reactors working at large throughput. Therefore, they need to function at moderate or large Reynolds numbers.[18] Channel dimensions pertain to the millimetre range, and flows developing in these systems can be qualified as 'millifluidic'. An example is given in Fig. 3.49 [101].

The flow direction is signaled by the arrow shapes of the channel structures. The geometry favors the development of eddies at the edges of the (white) end of the arrow, inducing mixing. Typical flow rates are milliliters per minute, and typical speeds are metres per second.

[18] In these systems, it is preferable to work below the turbulence onset. We often hear, in the microfluidic community, that below a Reynolds number of 2,000 or so, there is no turbulence. This statement is not accurate. The onset of turbulence depends on the flow geometry, and for subcritical bifurcations (for instance a pipe), on the initial level of perturbation. The example of pipe flows illustrates this point. In linear instability theory, the flow is stable at all Reynolds numbers. A perturbation of substantial amplitude is needed to trigger turbulence. In practice, turbulence appears at Reynolds numbers of 3,000 - 10,000. For flows behind blunt bodies, the critical Reynolds numbers are typically a few hundreds or less. These examples show that the onset of turbulence strongly depend on the geometry. A critical Reynolds number of 2,000 provides a general estimate, valid within a factor ten or so, of the onset of turbulence.

3.49: Mixer commercialized by Corning, functioning at moderate and high Reynolds numbers (Reprinted from Ref. [101], under Creative Commons licence). The channel widths are millimetric.

References

[1] L. Boltzmann, *Leçons sur la théorie des gaz, Gauthier-Villars (1902-1905)*. Réédition J. Gabay (1987).
[2] D. Burnett, *Proc. Lond. Math. Soc.*, **2**, 382 (1936).
[3] G. Batchelor, *An Introduction to Fluid Dynamics*, Cambridge University Press (2012).
[4] W. Sparreboom, A. van den Berg, J. C. T. Eijkel, *New J. Physics*, **12**, 15004 (2010).
[5] L. Bocquet, E. Charlaix, *Chem. Soc. Rev.*, **39**, 1073 (2010).
[6] H. Navier, *Memoire de l'Académie Royale de l'Institut de France*, 389 (1723).
[7] J. C. Maxwell, *Phil. Trans. Roy. Society*, **170**, 231 (1879).
[8] E. Guyon, J. P. Hulin, L. Petit, *Hydrodynamique Physique, CNRS Edition*, 2nd edn (2001).
[9] G. Karniadakis, A. Beskok, *Micro Flows*, Springer Verlag (2002).
[10] S. Schaaf, P. Chambre, *Flows of Rarefied Gases*, Princeton University (1961).
[11] S. Schaaf, in *Modern Developments in Gas Dynamics*, W. Loh (ed.), Plenum Press, 235 (1969).
[12] M. Gad-El-Hak, *The Fluid Mechanics of Microdevices*, *J. Fluid Eng.*, **121**, 5 (1999).
[13] F. Sharipov, *J. Vac. Sci. Technol.*, **A 17**, 3062 (1999).
[14] D.A. Lockerby, J. M; Reese, *J. Fluid.Mech.*,**604**, 604 (2008).
[15] T Ewart, P Perrier, I.A. Graur, J.G. Méolans, *J. Fluid Mech.*, **584**, 337 (2007).
[16] J. Pitakarnnop, S. Geoffroy, S. Colin, L. Baldas, *International Journal of Heat and Technology*, **26**, 167 (2008).
[17] E. Arkilic, K. Breuer, M. Schmidt, *J. Fluid. Mech*, **437**, 29 (2001).
[18] C. Aubert, S. Colin, *Microscale Therm. Eng.*, **5**, 1, 41 (2001).
[19] J. Maurer, P. Joseph, H. Willaime, P. Tabeling, *Phys. Fluids*, **15**, 2613 (2003).
[20] T. Ohwada et al., *Phys. Fluids A*, **1**, 2041 (1989).
[21] Y. Zhang, R. Qin, DR. Emerson *Phys.Rev.E*, **E 71**, 047702 (2005).
[22] A. Sangmin, S. Corey Stambaugh, K. Gunn Kim, L. Manhee Lee, K. Yonghee Kim, L. Kunyoung, J. Wonho, *Nanoscale.*, **4**, 6493 (2012).
[23] L. Li, L. H. Li, Y. Chen, X. J. Dai, T. Xing, M. Petravic, X. Liu, *Nanoscale Research Letters.*, **7**, 417 (2012).
[24] J. A. Thomas, A.J. McGaughey, *Nano Letters*, **8**, 2788 (2008).
[25] J. A .Thomas, A. J. McGaughey,. *Phys. Rev. Lett*, **102**, 1 (2009).
[26] G. Hummer, J. Rasaiah, J. Noworyta, *Nature*, **414**, 188 (2001).
[27] L. Bocquet, *Ann. Rev. Fluid. Mech.*, **53**, 377 (2020).
[28] J. Israelachvilii, *J. Colloid Interface Sci.*, **110**, 263 (1986).

[29] J. Israelachvili, *Intermolecular and Surfaces Forces*, Academic Press, 2nd edn (1991).
[30] T. D. Li, J. Gao, R. Szoszkiewicz, U. Landman, E. Riedo, *Phys. Rev. B*, **75**, 115415 (2007).
[31] S. Bendid, O. Francais, P. Tabeling, H. Willaime, *Proc.12th Micromechanics Eur. Work. MME2001*, 16-18 September, 23 (2001).
[32] S. Goldstein, *Ann. Rev. Fluid Mechanics*, **1**, 1 (1969).
[33] P-G. De Gennes, *Rev. Modern Physics*, **57**, 827 (1985).
[34] M. Knudsen, *Annal. Phys.*, **28**, 75, 130 (1909).
[35] M. Knudsen, *Kinetic Theory of Gases*, London (1934).
[36] P. Tabeling, *Introduction to Microfluidics*, 1st Edition, Oxford University Press (2005).
[37] C. Cercignani, R. Illner, M. Pulvirenti, *The Mathematical Theory of Dilute Gases*, Springer-Verlag, New-York (1994).
[38] N. G. Hadjiconstantinou, *Microscale Therm. Engin.*, **9**, 137 (2005).
[39] N. G. Hadjiconstantinou, *Phys. Fluids*, **15**, 2352 (2003).
[40] L. Bocquet, J-L. Barrat, *Phys. Rev. Lett.* , **82**, 23 (1999).
[41] F. H. Stillinger, *J. Sol. Chem.* **2**, 141 (1973).
[42] M. Jensen, O. G. Mouritsen, H. G.Peters, *J. Chem. Phys.* **120**, 9729 (2004).
[43] A. P. Willard, D. Chandler, *J. Chem. Phys.*, **141**, 18C519 (2014).
[44] C. Sendner, D. Horinek, L. Bocquet, R. Netz, *Langmuir*, **25**, 2768 (2009).
[45] J-L. Barrat, *Faraday Discussions*, **112**, 119 (1999).
[46] L. Bocquet, J. L. Barrat, *Soft Matter*, **3**, 685 (2007).
[47] E. Lauga, M. P. Brenner, H. A. Stone, *arXiv preprint cond-mat/0501557*, (2005).
[48] C. H. Choi, K. J. A Westin, K.S. Breuer, *Physics of Fluids*, **15**, 2897 (2003).
[49] L. Leger, H. Hervet, R. Pit, *Phys. Rev. Lett.*, **85**, 980 (2000).
[50] Y. Zhu, S. Granick, *Phys. Rev. Lett.*, **87**, 9 (2001).
[51] D. Tretheway, C. Meinhart, *Lett. Phys. Fluids*, **14**, 3, L9 (2002).
[52] C. Cottin-Bizone, B. Cross, A. Steinberger, E. Charlaix, *Phys.Rev.Lett.*, **94**, 056102 (2005).
[53] C. Cottin-Bizone, A. Steinberger, B. Cross, O. Raccurt, E. Charlaix, *Langmuir.*, **24**, 1165 (2008).
[54] C. Cottin-Bizone, J-L. Barrat, L. Bocquet, E. Charlaix, *Nature Mater.*, **2**, 237 (2003).
[55] P. Joseph, P. Tabeling, *Phys. Rev. E*, **67**, 056313 (2005).
[56] Z. Li, L. D'Eramo, C. Lee, F. Monti, M. Yonger, P. Tabeling, B. Chollet, B. Bresson, Y. Tran, *Journal of Fluid Mechanics*, **766**, 147 (2015).
[57] P. Attard, M. P. Moody, J. W. G.Tyrrell, *Physica A*, **314**, 696 (2002).
[58] P-G. De Gennes, *Langmuir*, **18**, 3413 (2002).
[59] L. Bocquet, P. Tabeling, *Lab on a Chip*, **14**, 3143 (2014).
[60] S. K. Kannam, P. J. Daivis, B. D. Todd, *MRS Bulletin*, **42**, 283 (2017).
[61] L. Bocquet, *Nature Mat.*, **19**, 254 (2020).
[62] T. Rodrigues, F. J. Galindo-Rosales, L. Campo-Deano, *Materials*, **12**, 1086 (2019).
[63] L.Landau, E. Lifschitz, *Mécanique Des Fluides*, Mir (1967).
[64] H. K. Moffatt, *Jour. Fluid. Mech.*, **18**, 1 (1964).

[65] H. K. Moffatt, *Cours Ecole Polytechnique, Microhydrodynamics.*, (1995).
[66] E. Lauga, H. A. Stone, *J. Fluid Mech.*, **489**, 55 (2003).
[67] J. R. Philip, *Z. Angew. Math. Phys.*, **23**, 353 (1972).
[68] J. R. Philip *Z. Angew. Math.*, Phys.**23**, 960 (1972).
[69] C. Cottin-Bizonne, J-L. Barrat, L. Bocquet, E. Charlaix, *Nat. Mater.*, **2**, 237 (2003).
[70] J. Ou, J. B. Perot, J. P. Rothstein, *Phys. Fluids*, **16**, 4635 (2004); **17**, 103606 (2005).
[71] P. Tsai, A. M. Peters, C. Pirat, M. Wessling, R. G. H Lammertink, D. Lohse, *Physics of Fluids*, **21**, 112002 (2009).
[72] C. Ybert, C. Barentin, C. Cottin-Bizonne, P. Joseph, L. Bocquet, *Physics of Fluids*, **19**, 123601 (2007).
[73] C. H. Choi, C.J.. Kim *Phys. Rev. Lett.*, **96**, 066001 (2006).
[74] P. Joseph, C. Cottin-Bizonne, J.-M. Benoit, C. Ybert, C. Journet, P. Tabeling, L. Bocquet, *Phys. Rev. Lett.*, **97**, 156104 (2006).
[75] L. Bocquet, P. Tabeling, S. Manneville, *Phys. Rev. Lett.*, **97**, 109601 (2006).
[76] C. H. Choi, C.J. Kim, *Phys. Rev. Lett.*, **97**, 109602 (2006).
[77] C. H. Choi, U. Ulmanella, J. Kim, C. M. Ho, C. J. Kim *Phys. Fluids*, **18**, 087105 (2006).
[78] H. B. Squire, *Quart. J. Mech. Appl. Math.*, **4**, 321 (1951).
[79] E. Secch, S. Marbach, A. Niguès, A. Siria, L. Bocquet *J. Fluid Mech.*, **826**, R3, (2017).
[80] N. Laohakunakorn, B. Gollnick, F. Moreno-Herrero, D. G. A. L. Aarts, R. P. A. Dullens, S. Ghosal, U. F. Keyser, *Nano Lett.*, **13**, 13 (2013).
[81] P. Guillot, P. Panizza, J-B. Salmon, M. Joanicot, A. Colin, C-H. Bruneau, T. Colin, *Langmuir*, **22**, 14 (2006).
[82] P. Nghe, P. Tabeling, A. Ajdari *Journal of Non-Newtonian Fluid Mechanics*, **165**, 313 (2010).
[83] C. D. Meinhart, S. T. Wereley, J. G. Santiago, *Experiments in Fluids*, **27**, 414 (1999).
[84] P. Joseph, P. Tabeling, *Physical Review E*, **71**, 035303 (2005).
[85] A. Vilquin, V. Bertin, P. Soulard, G. Guyard, E. Raphaël, F. Restagno, T. Salez, J. D. McGraw, *Phys. Rev. Fluids*, **6**, 064201 (2021).
[86] C.-H. Choi, K. J. A. Westin, K. S. Breuer, *Phys. Fluids*, **15**, 2897 (2003).
[87] Z. Li, L. D'eramo, C. Lee, F. Monti, M. Yonger, P. Tabeling, B. Chollet, B. Bresson, Y. Tran, *Journal of Fluid Mechanics*, **766**, 147 (2015).
[88] J. B. Knight, A. Vishwanath, J. P. Brody, R. H Austin, *Phys. Rev. Lett.*, **80**, 3863 (1998).
[89] K. W. Oh, K. Lee, B. Ahn, E. P. Furlani, *Lab on a Chip*, **12**, 515 (2012).
[90] C. G. Lomas, *Hot wire anemometry*, Cambridge University Press (1986).
[91] Fluigent, https://www.fluigent.com/.
[92] J. S. Park, S. H. Song H-I. I. Jung, *Lab Chip*, **9**, 939 (2009).
[93] E. Stemme, G. Stemme, *Sens. Act. A. Phys.*, **39**, 167 (1993).
[94] O. Strohmeier, M. Keler, F. Schwemmer, S. Zehnle, D. Mark, F. von Stetten, R. Zengerle, N. Paust *Chem. Soc. Rev.*, **44**, 6187 (2015).

[95] R. Gorkin, J. Park, J. Siegrist, M. Amasia, B. S. Lee, J. M. Park, J. Kim, H. Kim, M. Madou, Y. K. Cho , *Lab on a Chip.*, **14**, 1758 (2010).
[96] M. Madou, J. Zoval, G. Jia, H. Kido, J. Kim, N. Kim, *Ann. Rev. Biomed. Eng.*, **8**, 601 (2006).
[97] J. Eustice, *Proc. Roy. Soc. Lond.*, **A 85**, 119 (1911).
[98] W. R. Dean *Phil. Mag*, **4**, 208 (1927).
[99] N. Nivedita, P. Ligrani, P. I. Papautsky, *Sci. Rep.*, **7**, 44072 (2017).
[100] M. E. Warkiani, B. L. Khoo, L. Wu, A. K. Ping Tay, A. A. S Bhagat, J. Han, C. T. Lim, *Nature Protocols*, **11**, 134 (2016).
[101] K. J. Wu,V. Nappo, S. Kuhn, *Ind. Eng. Chem. Res.*, **54**, 7554 (2015).

4

Hydrodynamics of microfluidics 2: droplets

4.1 Liquid–vapour interfaces

4.1.1 Microscopic phenomenology

Interfacial phenomena often defy intuition, and it has proved useful, along the years, to use molecular dynamic (MD) simulations to get a better sense of the physical mechanisms at work. Figure 4.1 shows a molecular simulation of water in equilibrium with its vapour [1] A. At the interface, molecules leave the liquid by evapouration and vapour renews the population through condensation. A permanent layer takes place, whose properties are different from the bulk because the density is neither that of the gas nor the liquid, and the molecules pertaining to the layer develop less cohesive bonds than in the bulk.

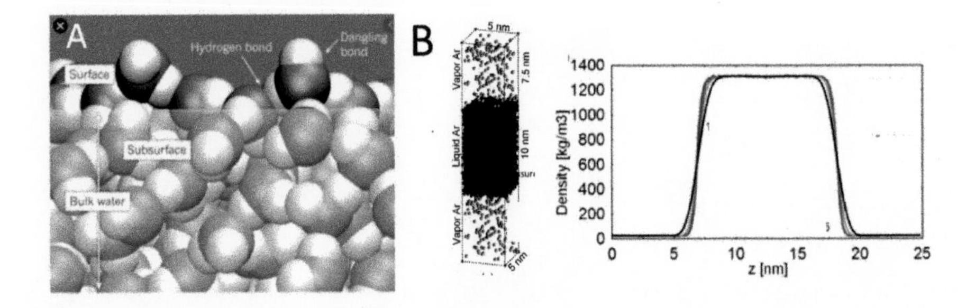

Fig. 4.1: Simulations of water vapour interfaces.(A) The interface layer is underlined with darker molecules. Molecules in this layer develop less cohesive bonds than their colleagues in the bulk.(Reprinted from [1], with permission from Springer Nature. Copyright 2022.) (B) A 10 nm thick argon film suspended with 7.5 nm thick vapour on both sides along the z direction. The different curves correspond to different scales characterizing the pairwise interaction energy. (Reprinted from [2], with permission from Elsevier. Copyright 2022.)

Fig. 4.1 (B) represents another MD simulation, modelling a 10 nm thick liquid argon film, in the presence of its vapour [2]. The curve shown in the figure reveals that the interfacial layer is extremely thin, a fraction of nanometre or so, i.e. a few intermolecular distances. How can we describe these systems? Here, we will limit ourselves to energetical and mechanical approches.[1]

4.1.2 Liquid vapour interfaces: energetical approach

In Chapter 2, we learned that Van der Waals liquids and solids are materials composed of molecules interacting through Lennard–Jones (LJ) potential $V(r)$, whose expression is:

$$V(r) = 4\epsilon\left(\left(\frac{\sigma}{r}\right)^{12} - \left(\frac{\sigma}{r}\right)^{6}\right) \tag{4.1}$$

where ϵ is a cohesion energy and σ is the molecule size. We recall that the attractive part is associated to the negative term and the repulsive part, which ensures the non interpenetrability of the molecules, to the positive one. For the example of argon, which we considered in Chapter 2, $\sigma = 3.045$ Å and $\epsilon = 1.67\ 10^{-21}$ J.

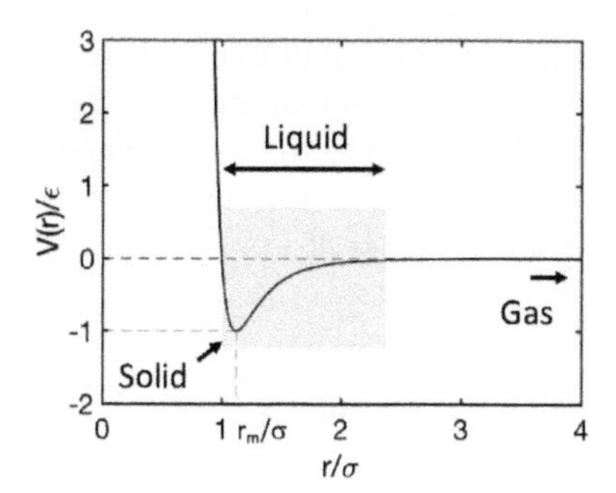

4.2: Lennard Jones potential, indicating the range of scales explored by the molecules, at ambient temperature, in different states of matter: solid, liquid, and gas. In solids, apart from small fluctuations, molecules remain at the minimum. In liquids, positions fluctuates (see the grey zone). In gases, the mean intermolecular distance, indicated by the horizontal arrow, and not shown here, is typically ten σ.

Pairs of molecules will tend to place themselves at the minimum of the LJ potential, i.e. at the equilibrium point given by $r_m = 2^{1/6}\sigma$, and $V(r) = -\epsilon$. In solids and liquids, molecules have several neighbours and energies must be added up to characterize the equilibrium state. For example, in a simple cubic structure, the energy associated to the equilibrium state will not be $-\epsilon$, as in Fig. 4.2, but -8.38ϵ. This is the energy that we must provide to extract one molecule from the material. With this reasoning, one sees that molecules located at the surface are energetically at a higher level than in the bulk. This leads to the metaphor of 'happy' molecules (in the bulk) and 'unhappy' molecules (at the surface), often taken in Ref. [4], and which we will also use in this book. Taking the example of the simple cubic structure, as just said, a molecule in the bulk 'feels' attractive forces corresponding to an energy of -8.38ϵ, while, at the

[1] The question was discussed in depth by J. S. Rowlinson and B. Widom [3]. Their book, which includes an interesting account of the controversies held on the subject, in the nineteen century, analyses the question from different prospectives: mechanical, thermodynamical, and statistical.

interface, the corresponding energy is smaller (-5.88ϵ). There is thus an excess of energy of 3.5ϵ for molecules located at the surface. The corresponding energy, per unit of area, is called γ. For Van der Waals materials, γ is on the order of a few tens of mN/m.

4.3: Herge's drawing showing that, in the absence of gravity, liquids spontaneously adopt a spherical shape. A question that could be addressed is that of the initial conditions, i.e. filling the cup in the absence of gravity. Another question concerns the material. Should it be glass, whisky, which wets it, would perhaps prefer to stay inside. (Reprinted with permission from Moulinsart SA.)

Interfaces will adopt a shape that minimizes the number of 'unhappy' molecules. This leads to spectacular phenomena in liquids. An illustration is given by Captain Haddock, in 'On a marché sur la lune' [8]. In the microgravity environment of the rocket the captain was flying in, the whisky poured in the cup spontaneously reshapes into a sphere floating in the air. Again, we recover the fact that fluids seek to reduce their exposed surface area, the sphere representing the minimum exposed area for a given volume. On earth, gravity inhibits the process, and we can swim in lakes, for instance, without fearing that the water level suddenly jumps to form a huge drop. In microfluidic systems, gravity forces, as we saw in Chapter 2, are negligible against the interfacial forces we discuss here. We will thus be exposed to phenomena similar to those observed by Captain Haddock.[2]

4.1.3 The liquid–vapour interface: mechanical approach

To describe the interfacial region from a mechanical viewpoint,[3] we use the illuminating paper published by M. Berry in 1971 [5].[4]

In gases, molecules are far apart and pressure is only kinetic. In liquids, because molecules are agitated and remain extremely close to each other, the pressure has two components: one is 'kinetic', as in gases and the other one, due to Van der Waals cohesive forces, is 'static'. These two contributions must be taken into account for understanding the mechanics of the interfacial layer. So, let us consider the system of Fig. 4.4), and call 'T' the coordinate parallel to the interface and 'N' the one normal

[2]The search for a minimum of the exposed area often represents a mathematical challenge. Ref. [4] provides a nice presentation of the subject

[3]In fact, the mechanical description of liquid vapour interfaces is subtle, and in almost all cases, on the web, and in textbooks, it is incorrectly presented. In fluid mechanical textbooks, the problem is eluded by taking a mathematical membrane model [39], which is abstract.

[4]The work was well–explained by B. Andreotti [6, 7]. I am indebted to him for his critical remarks on a first version of this Section.

to it. The components of the forces (per unit of area) exerted on the faces of the cube shown in the figure are p^T (along T) and p^N(along N) (see Fig. 4.4).

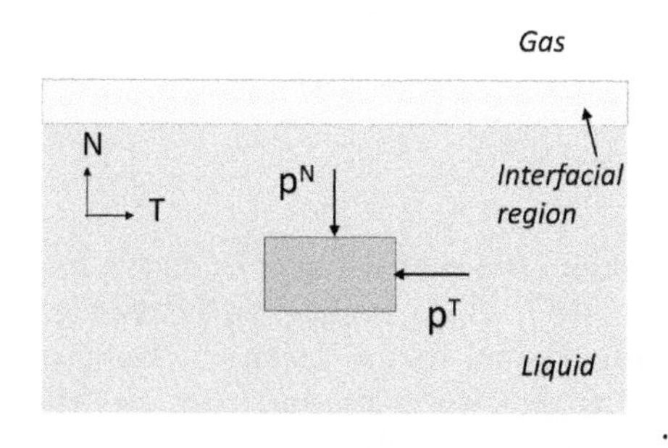

4.4: Representation of the components p^T and p^N, in which coordinates are indexed by T (parallel to the interface) and N (normal to it).

Let us first consider the bulk, and note the pressure components by p_L^N and p_L^T. These quantities decompose into kinetic and static components as follows:

$$p_L^N = p_k^N + p_s^N = n_L kT + p_s^N \text{ and } p_L^T = p_k^T + p_s^T = n_L kT + p_s^T \tag{4.2}$$

in which the indices k and s mean 'kinetic' and 'static', and n_L is the number density of molecules (i.e.the number of molecules par unit of volume). In the above equations, we have considered that the kinetic terms obey Mariotte's law, even though we are in a liquid. Since the system is at equilibrium, p_L^T and p_L^N are equal to P_0, the atmospheric pressure, that we take equal to 1 bar. Let us look at the orders of magnitude: the kinetic term is, roughly, 1000 bars,[5] while the static components is –1000 bars. Remarkably, the two terms almost compensate each other, so that their sum is equal to P_0. If the compensation was not possible, the liquid/gas system would not be stable. In such a case, the liquid would solidify or evaporate.

In gases the situation is much simpler. There is no static terms. Pressure is only kinetics. It is given by the relation:

$$p_G = n_G kT$$

where n_G is the number density of molecules in the gas. This is again Mariotte's law. Gases being one thousand times less dense than in liquids, the kinetic term will be one thousand times smaller, which leads again to 1 bar, i.e. P_0.

What happens at the liquid–vapour interface? The response, given by M. Berry [5], is schematized in Fig 4.5.

In the interfacial region, normal pressure component p^N keeps being equal to P_0. Indeed, p_k^N and p_s^N change, because the fluid density decreases, and there are less

[5] $n_L kT$, with $n_L \sim 3\ 10^{28}$ molecules/m^3 (for water) and $kT \sim 4\ 10^{-21}$ J (at ambiant temperature) is on the order of 1000 bars

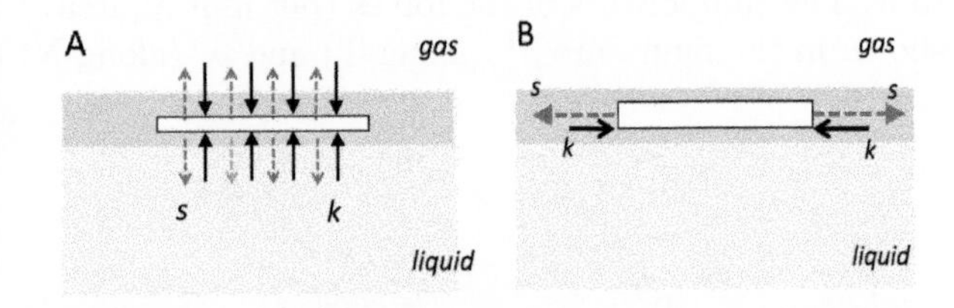

4.5: (A) Normal pressure components in the interfacial layer (black arrows: kinetic; dashed grey arrows: static). (B) Tangential pressure components in the interfacial region, with the same symbols.

molecules to interact with, but the sum must remain the same, otherwise mechanical equilibrium would not be achieved (see Fig. 4.5 (A)). This is possible because the kinetic term p_k^N and the static term p_s^N are proportional to the density.[6] Therefore, as the density decreases, they can compensate, as they do in the bulk. In the other direction, i.e. along the tangential direction T, things are different. Concerning the static part of the tangential pressure component p_s^T , we note that the edges of the rectangle of Fig. 4.5 are pulled by the liquid nearby. If we assimilate this situation to a sphere-plane system, the Van der Waals forces that pull these edges vary as the fluid density to the power 2/3.[7] Concerning the kinetic term p_k^T, the pressure it exerts onto the rectangle edges varies as the fluid density. Therefore, as the fluid density decreases, the static term decreases less rapidly than the kinetic term. Consequently, the sum $p^T = p_k^T + p_s^T$ becomes negative in the interfacial region (see Fig. 4.5 (B)). As a result, a tension appears, which is the surface tension we introduced before. This conclusion was formulated by M. Berry [5] without calculation, noting that static forces are long range and kinetic ones short range and taking the geometry of the problem into account. Complete discussion is given in [5, 7]. Surface tension thus results from a subtile mechanism. Note that, in the interfacial layer, surface tension is dominated by Van der Waals forces. It is thus legitimate to neglect the kinetic contribution. This is done in the literature, without justification, as we did in the previous section.

Numerical simulations confirm Berry's analysis (see Fig.4.6) [7].

The evolution of the density across the interface is shown in Fig. 4.6 (a) and (b) [7]. As in Fig. 4.1, the interfacial region is two or three molecular thick. Much information is conveyed by the generalized' pressure $\Pi(z)$, defined by:

$$\Pi(z) = p_L^N(z) - p_L^T(z)$$

where z is the normal to the interface. $\Pi(z)$ is also (minus) the extra stress developed in the interfacial region. It allows to calculate the surface tension γ through the relation

$$\gamma = \int_{-\infty}^{+\infty} (p_L^N - p_L^T) dz$$

.

[6] In fluids, the Van der Waals force between two planar surfaces varies as their separation to the power −3, i.e. as the fluid density.

[7] In fluids, the Van der Waals force between a sphere (representing the edge), and a planar surface varies as the separation to the power −2, i.e. as the fluid density to the power 2/3.

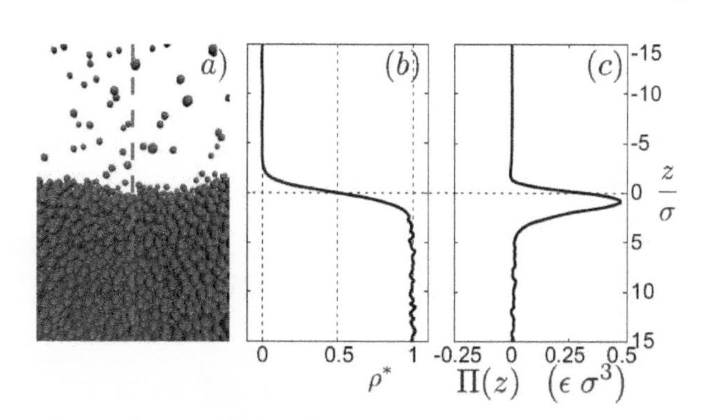

Fig. 4.6: (a) Numerical simulation of the water vapour interface [7]; Evolutions of the density (b) and generalized pressure $\Pi(z)$ (c) across the interface, in dimensionless units (Reproduced from Ref. [7], with permission of American Association of Physics Teachers. Copyright 2022 AIP Publishing.)

$\Pi(z)$ is represented in Fig 4.6(c). In the gas, $\Pi(z)$ is zero, In the liquid bulk, p_L^N and p_L^T are equal to the vapour pressure P_0 and therefore $\Pi(z)$ vanishes too. In the interfacial region, $\Pi(z)$ is found positive (see Fig. 4.6 (c)) and its magnitude is a fraction of ϵ/σ^3, i.e. hundreds of bars. This implies that p_L^T is negative and larger than p_L^N. We may thus conclude that the fluid layer is under tension. It is like a stretched skin, an image often taken in the literature. Should we sample out an element, it would immediately retract. According to Formula 4.1.3, the surface tension would be on the order of 0.25 ϵ/σ^3, which, for ordinary fluids, leads to a range of 10–30 mN/m for γ. We see that the numerical results are in excellent agreement with Berry's description.

A misleading presentation of the molecular origin of surface tension

A presentation of the molecular origin of surface tension, often given on the web, in textbooks and, sometimes, in scientific reviews, is shown in Fig. 4.7. In these presentations, pressure is only static (there is no kinetics). Accordingly, a force appears at the interface, oriented towards the interior of the liquid. The problem is that there is no way to balance this force. With such an unbalanced force, either the fluid is pressurized above the atmospheric pressure or the interface moves; This would contradict the hypothesis of equilibrium. The error here is the omission of the kinetic contribution to the pressure.

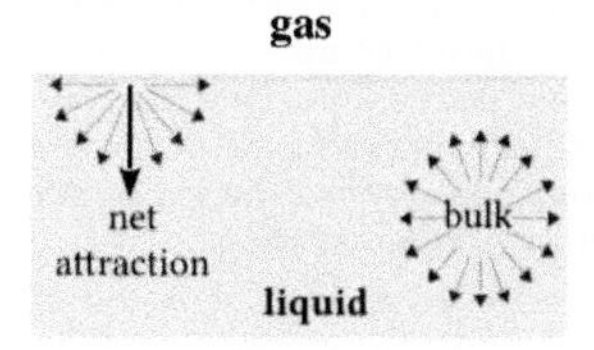

Fig. 4.7: Misleading presentation of the molecular origin of surface tension.

4.1.4 Experiments

Liquid film retracting in a rectangular frame. We now have tools for understanding simple, or apparently simple, experiments. The one shown in Fig. 4.8 is subtle, and offers an elegant illustration of the notions previously discussed.[8] It is borrowed from Ref. [4].

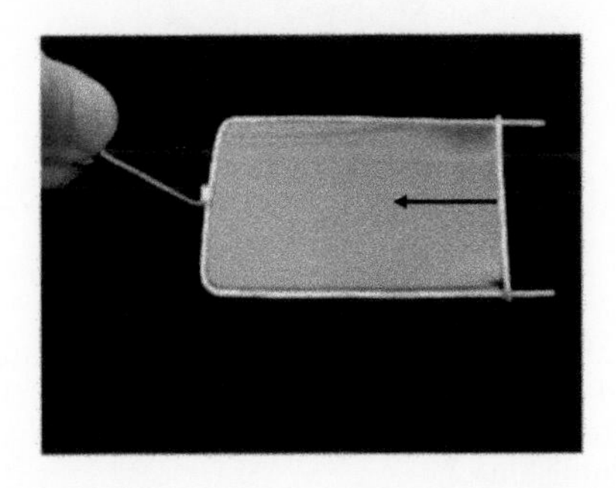

4.8: A soap film is enclosed in a rectangular metallic wire, with a mobile side. The mobile bar spontaneously moves in the direction of the soap film [4,9].

A soap film is held in a metallic rectangular frame. There is a mobile bar, forming one side of the rectangle, also metallic (see Fig. 4.8). One observes that the bar spontaneously moves inwards, reducing the film area.[9] The experiment brings evidence that, in this configuration, a pulling force exists, directed normally to the bar, and parallel to the plane of the film, which works at reducing the soap film area. This is the surface tension we previously discussed.

One can play with this observation and establish an equivalence between surface energy and surface tension (in liquids). The work dW done during an infinitesimal displacement of the bar is dx is $2\gamma L dx$ (with L the bar length), the factor 2 being due to the fact that the soap film is composed of two surfaces. This work is equal to the force $2F$ exerted by each interface onto the bar. We thus have:

$$dW = 2\gamma L dx = 2F dx \tag{4.3}$$

In these expressions, γ can be viewed as the energy needed to increase the area by one unit, or the tension per unit of length needed to perform the same task. Since we have

[8] An infinitely thin membrane is often taken, in mechanical textbooks, to expose the concept of surface tension. But this is quite abstract. How to suspend a monomolecular layer in the air? Where would the tensile force apply ? On the molecules forming the edges? Soap films can be used to illustrate the concept, but the experiment is more complex than it looks. Soap films are composed of two surfactant monolayers sandwiching a water film. How can the concept be connected to this complex structure? In practice, the soap film experiment is never presented or discussed. The consequence is that students have no idea about the physical origin of surface tension, and are left with elegant equations describing a strange physical phenomenon.

[9] As the film contracts, where do the soap and the liquid of the film, escape? The answer is inside the menisci attached to the frame and the moving bar. These regions contain most of the liquid volume, and their exposed area is small compared to the planar part. They can be filled and emptied without consuming volumetric (if we neglect viscous losses) nor surface energy. In this manner the film can freely reduce its surface area and therefore decrease its energy without cost.

$F/L = \gamma$, the system can be used as a tensiometer for bilayers. In practice, however, it is delicate to use.

Thread spontaneously forming a circle in a soap film. Another spectacular manifestation of surface tension is shown in Fig. 4.9. Here again, a soap film is held in a rectangular frame. This time, all the sides are fixed and a thread of arbitrary shape is introduced in the film (see Fig. 4.9 (A). The thread does not move because, whatever its shape, the total surface area of the soap film, and then the surface energy, remains the same. In other words, no gain of energy is obtained by changing the shape. Now, the film inside the thread is pierced by a needle (see Fig. 4.9(B)). It bursts and disappears, leaving a hole. Suddenly, the thread adopts a circular shape. The simplest explanation is based on forces: as shown above, each thread element is subjected to the same force. This force applies in the plane of the film, normally to the thread, towards the liquid. These forces work at reducing the soap film area. The system reaches an equilibrium state by adopting a circular shape, where all forces balance each other. If we look at this system from an energetic standpoint, we may suggest that in 2D, the biggest area enclosed in a loop of fixed length is obtained with the circle. Alternatively, the shortest loop enclosing a fixed area forms a circle.

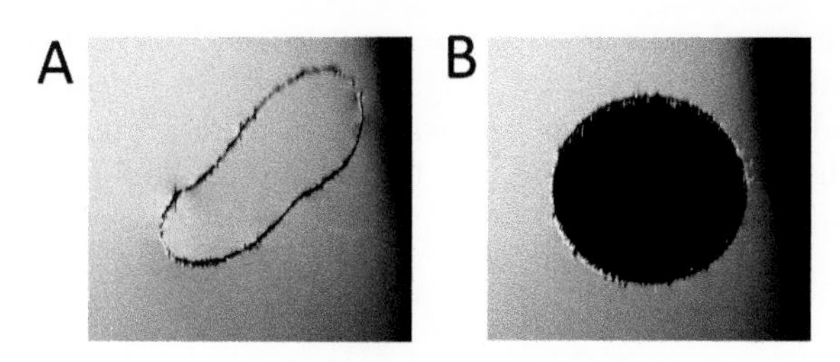

4.9: A: A closed thread is inserted in the soap film. B: With a metallic piece, the film inside the thread is burst. In a fraction of second, the thread adopts a circular shape (vertical scale adjusted). (Courtesy of Leo Boulot (Amaco). Reprinted from [4] under Creative Commons licence.

4.2 Laplace's law

4.2.1 Establishment of Laplace's law

Laplace's law stipulates that, in order to maintain a curved fluid interface in mechanical equilibrium, it is necessary to exert a higher pressure on the convex side, as for a balloon we would blow inside to keep it spherical. Two methods exist for establishing Laplace's law, one based on surface tension, the other on energy.

Method based on surface tension for a bubble. Let us consider a bubble pressurized at pressure ΔP, surrounded by a liquid. Our hypothesis is that, in order to maintain the bubble in mechanical equilibrium, one must pressurize it. Let us define a subsystem, formed by half of the bubble. According to the Euler–Cauchy principle, one must replace the missing part by the forces it exerted on the subsystem, before the separation. Let us consider the equator: it is pulled by a surface tension equal to γ. The corresponding pulling force is equal to $2\pi R\gamma$. On the other hand, the equatorial plane is pushed by pressure ΔP. The corresponding force is $\pi R^2 \Delta P$. The subsystem being in equilibrium, one must have:

$$\Delta P = \frac{2\gamma}{R}$$

This is Laplace's law for a bubble.

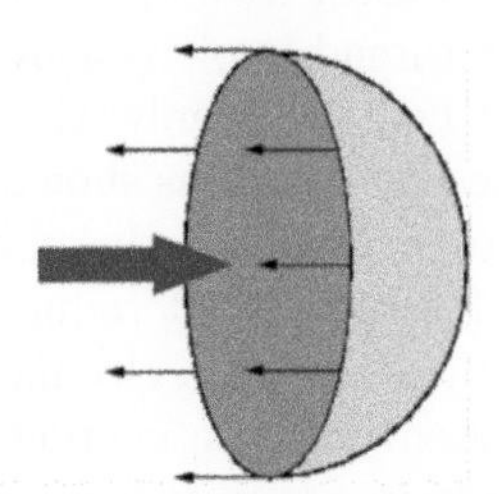

4.10: In a spherical droplet at rest, half of the structure is subjected to tensile forces exerted by the other part.

Method based on energy for a sphere. Let us now use the energetic method, and, for this purpose, reuse our bubble, still pressurized at ΔP (Fig. 4.11). Here, we do not cut it in two parts, but increase its radius by δR, for instance with $\delta R > 0$. In this process, we assume that the volume of the continuous phase (i.e. the liquid surrounding the bubble), along with its pressure, does not change.

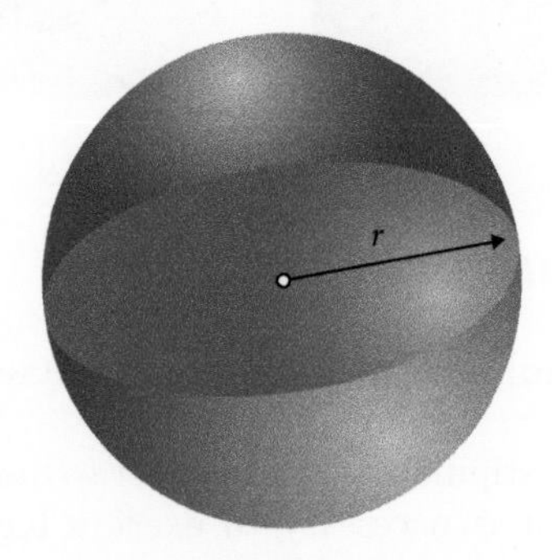

4.11: Geometry used for the derivation of Laplace's Law, for a sphere. Pressure ΔP is larger inside the bubble, in order to maintain mechanical equilibrium. The corresponding pressure jump, across the interface, is called Laplace pressure, or capillary jump.

By swelling, the bubble creates a new interfacial area δS. This costs energy. This energy is, by definition of γ:

$$\delta E_{\mathrm{S}} = \gamma \delta S = 8\gamma\pi R \delta R$$

By increasing the volume, we develop a negative work, inside the bubble equal to $\delta E_{\mathrm{V}} = -\Delta P \delta V$, whose expression is:

$$\delta E_{\mathrm{V}} = -\Delta P \delta V = -4\pi R^2 \Delta P \delta R$$

When the bubble swells, it displaces the liquid around it. This displacement, nonetheless, does not cost any energy because, as observed above, liquid pressure is fixed it is performed at fixed volume of the continuous phase. The system being at equilibrium, the variation of the total energy $\delta E_S + \delta E_V$ should be zero. This implies:

$$\Delta P = \frac{2\gamma}{R}$$

The result is identical to the one obtained with the forces. The law, established for bubbles, applies for any pair of immiscible fluids, i.e. for droplets.

General expression of Laplace's law in 3D. It is a matter of mathematics to generalize the results obtained above, to the case of an arbitrarily curved interfacial element in 3D (see for instance Landau's textbook on fluid mechanics [137] or Refs. [12, 24]). For this purpose, one must introduce the two principal radii of curvature, $R_1 = \kappa_1^{-1}$ and $R_2 = \kappa_2^{-1}$, where κ_1 and κ_2 are the principal curvatures, i.e. the maximum and minimum values of the curvature at point P of a surface (see Fig. 4.12).

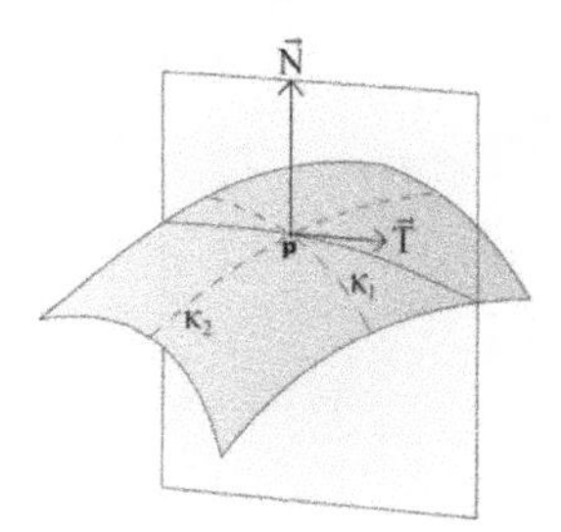

4.12: At point P, the two principal curvatures κ_1 and κ_2 are the maximum and minimum values of the curvature. On the figure, a plane, normal to the surface, with its normal and tangential vectors, is shown.

At point P, in each principal plane, surface tension develops a force directed inwards, equal to $\kappa\gamma\delta S$, where δS is a surface element and κ the curvature. To reach equilibrium, a pressure ΔP must build up in he fluid. Since we have two principal planes of curvature, we must have:

$$\Delta P = \gamma \left(\frac{1}{R_1} + \frac{1}{R_2} \right).$$

This is Laplace law. For 2D systems, Laplace's law reduces to $\Delta P = \frac{\gamma}{R}$, in which R is the local curvature radius.

In terms of order of magnitude, the Laplace jump for a bubble of 1 μm in radius, in water, is 70 kPa. In microfluidics, and in small systems in general, Laplace pressure gives rise to important and sometimes spectacular phenomena.

4.2.2 Is Laplace's law valid at the nanoscale?

Considering that the interface has a finite thickness (of nanometric size, as observed in numerical simulations), it has been argued that surface tension is not a constant, but some function of the geometric characteristics of this layer, for instance its mean curvature, or its thickness. The argument indeed impacts Laplace law. In 1949, J. Tolman [15] proposed the following formula:

$$\gamma = \gamma_\infty (1 - \delta/R) \tag{4.4}$$

in which R is the mean curvature radius of the interface, γ is the true' surface tension, γ_∞ is the value of γ for R infinite and δ is the Tolman length. The latter can be positive or negative. How large is δ? The question has been discussed for decades. Recent simulations [16], based on molecular dynamics (MD), have shown that the Laplace relation is valid down to 1.3 nm; therefore, δ is smaller than this limit. The geometry is shown in Fig. 4.13.

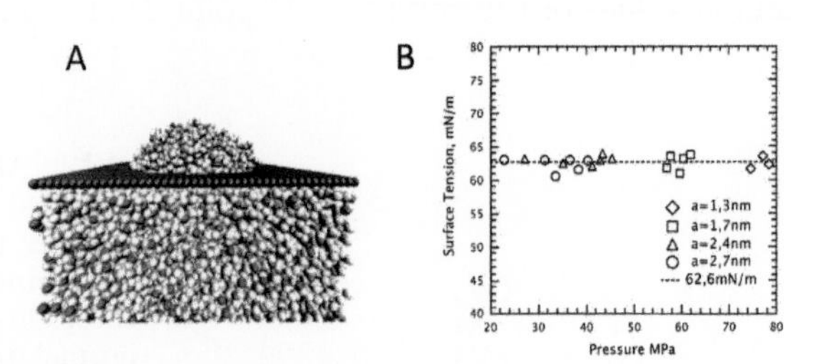

4.13: (A) Molecular simulation of a liquid held through a hole by capillary forces.(B) Measured surface tension (Reprinted from [16], under Creative Commons Attribution 4.0 International License.)

Thousands of molecules, interacting through a Lennard Jones potential, are placed below a plate pierced by a nanohole. The fluid develops a semi–spherical dome. A subtle feature is that the angle formed at the intersection between the semi–sphere and the horizontal plane is constrained by the dome curvature and not by the wetting properties of liquid/solid system. This is due to the extreme thinness of the plate. In the molecular dynamics simulation, pressure P is monitored, and the dome radius R is measured. It was observed that the Laplace relation, i.e. $P = \frac{\gamma}{R}$, holds in a range of hole radii down to 1.3 nm, i.e. several intermolecular scales, with the same γ (see Fig.4.13 (B)).

The work thus shows that Laplace law is valid down to 1.3 nm. Series of numerical simulations, performed in recent years, suggested δ on the order of 0.1 nm. Recent experimental investigations, studying nucleation in water [21], or measuring curvatures of nanometric menisci [20], provide a similar order of magnitude.[10]

[10]This does not imply that Tolman length should always be neglected. For instance, it plays a considerable role in bubble nucleation [21].

4.2.3 Manifestations of Laplace's law

Capillary adhesion. A first example is capillary adhesion (see Fig. 4.14).

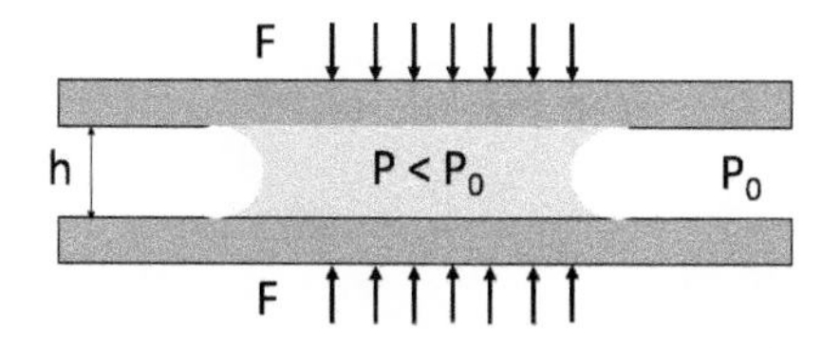

4.14: Capillary adhesion, in which a water droplet, held between two glass plates, develops an attracting force which tends to squeeze them.

In this case, a small volume of water is trapped between two plates (for instance clean glass). Because water wets clean glass (we shall discuss later the general question of wetting), the meniscus is oriented in the direction shown in Fig. 4.14. On calculating the Laplace pressure jump across the menisci, and therefore the pressure P inside the water film, one finds the following expression:

$$P = P_0 - \frac{\gamma}{R}$$

where P_0 is the atmospheric pressure. As P is below P_0, a force of attraction develops between the two plates. It is equal to $\frac{2\gamma}{h}S$, h being the separation distance between the plates and S being the wetted area by the liquid. A film of 1 μm, with a wetted area of 10 cm^2, can hold a weight of 7 kg. The effect is thus important.

Capillary collapse. Capillary adhesion raises issues for the fabrication of suspended microstructures, such as (micro) bridges or (micro) cantilevers. It can lead to capillary collapse. Illustrations are shown in Fig. 4.15 [17]. Droplets or films are often present on the wafers, during microfabrication, such as after rinsing. These droplets or films may establish a liquid bridge between the cantilever beam and the substrate, as sketched in Fig. 4.15 (left). Owing to the small sizes, Laplace adhesion is substantial and can trigger the collapse of the structure. Fig. 4.15 (right), shows a situation where, due to the phenomenon, the teeth of a micro–comb stick together, which was not the objective.[11]

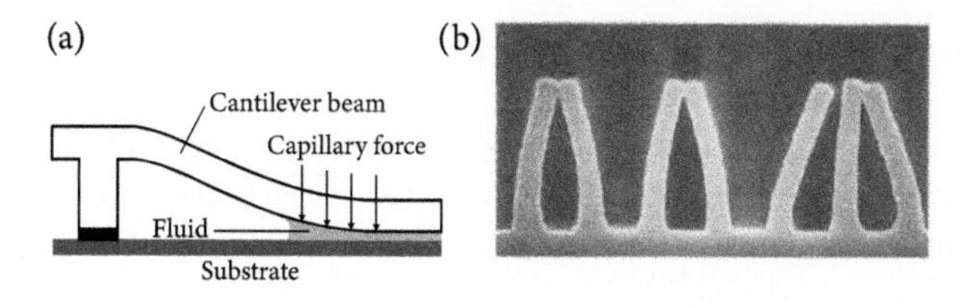

4.15: (A) Capillary collapse of a microcantilever beam (Reprinted from Ref. [17] with permission from IOP Publishing.).(B) Capillary collapse of an array of micropillars [17].

[11] To realize such a structure, one needs to perform several lithographic steps, the final step involving Deep Reactive Ion Etching (DRIE), a long process, which usually takes one night to be performed. These teeth therefore take much labour and time to fabricate.

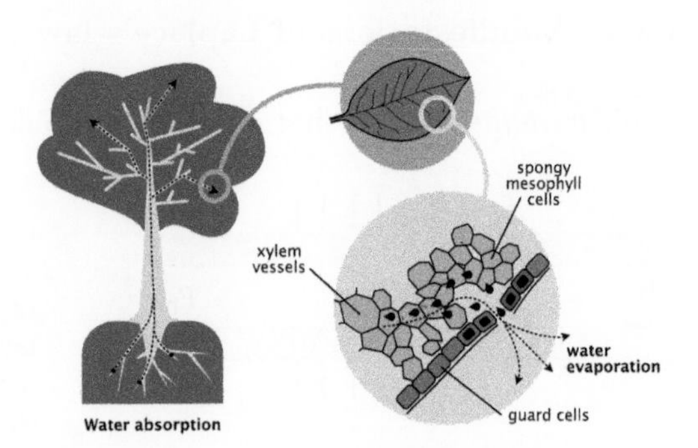

4.16: Leaf and the structure through which water evapourates.

Capillary pumping in trees. Giant trees, like sequoias, possess the capacity of lifting sap at heights above 100 m. This fact indicates that somewhere, there is a pump delivering a pressure drop of, at least –10 bars between the top of the tree and its roots. This is precisely what Laplace's law provides. Sap is driven along microchannels (called xylems), in the trunk and the branches (see Fig. 4.16). At the end of the journey, which takes a week, sap penetrates in the leaves. The journey ends with the imbibition of a porous region, formed by the packing of spongy mesophyll cells, within which sap evapourates (the phenomenon is called transpiration). In this process, the spongy medium encloses a large number of menisci. Since the pores are 10 –50 nm in diameter, the Laplace pressure in the sap, just behind the menisci, will be on the order of –20 to –100 bars. With such pressure differences, between the roots and the top, there is no difficulty for the tree to lift sap up to 100 m, which corresponds to the height of the highest trees on earth.[12]

Bird drinking water. How do birds drink? How do they manage to drag water drops along their beak, against gravity, without tongues?[13] The answer is provided by Fig. 5.79. In the beak, the droplet possesses two different curvatures. Close to the head, the meniscus is more curved than at larger distances. Consequently, because of Laplace's law, the pressure, inside the water drop, is smaller close to the head. The droplet will thus move towards the throat and the bird will be able to drink. We will see later that as wetting is partial, drops can get pinned at the beak surface. The bird solves the problem by developing more complex movements of its beak [13, 14].

4.2.4 Methods of measurement of interfacial tension

Static surface tension measurements in liquids. Measuring surface tension of liquid/gas and liquid/liquids interface is performed using a variety of methods that are well–described in textbooks (see [4] and references therein). We will thus succinctly

[12]One difficulty, however, is to operate liquids at negative pressures. Liquids at negative pressures are unstable. Bubbles (positive pressure) grow and provoke embolism. Trees have thus developed networks of mechanical valves, which isolate bubbles as soon as they appear

[13]Some birds, like the hummingbird, possess a tongue. In this case, the drinking process must be explained differently [13]. Others, like phalaropes, studied in Ref. [111], do not have tongues

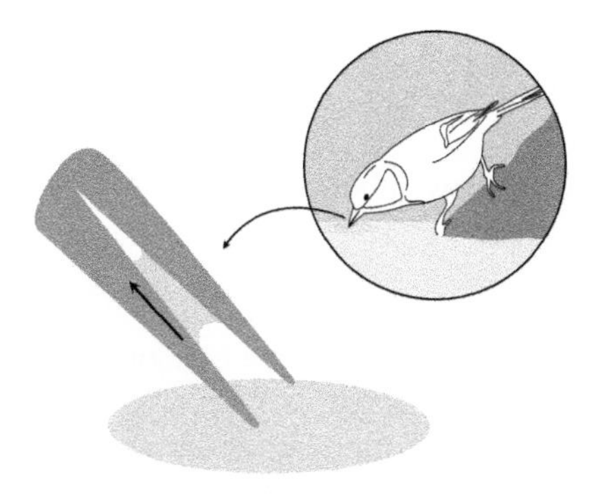

4.17: Bird drinking water. It exploits the wedge geometry of its beak to swallow the droplet.

summarize the topics. Figure 4.18 shows the most current techniques used in the laboratories.

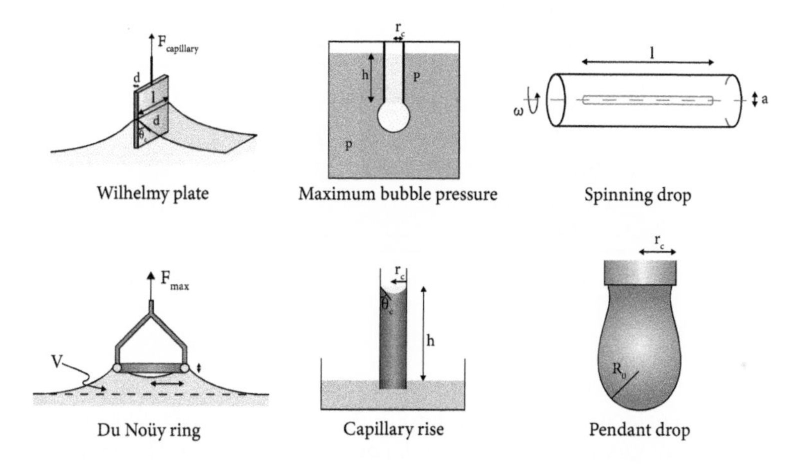

Fig. 4.18: Current methods of measurements of the surface tension

– Whilelmy plate technique consists of immersing a thin plate into the liquid, and measuring both the contact angle θ formed with the interface and the interfacial force exerted by the fluid onto it. In the wetting case, the interfacial force has an expression similar to that involved in the rectangular frame experiment (see above), i.e. γl, in which l is the plate length. By measuring this force, γ can be determined.
– The ring method is based on the same principle, but in a circular geometry.
– The maximum bubble pressure technique consists of blowing a bubble up to the point where the bubble radius becomes equal to the capillary radius. In such circumstances, the internal pressure reaches a maximum $P_{max} = 2\gamma/r$ in which r is the capillary radius. From this expression, γ can be determined.
– The pinning drop exploits the Rayleigh–Plateau instability. This will be described later.
– Capillary rise represents the first experiment ever made on capillarity, with the ex-

perimentalist being Leonard de Vinci (see [4]). We will demonstrate later the following expression: $h = \frac{2\gamma cos\theta}{\rho g r}$, in which h is the column height, ρ is the fluid density, θ is the contact angle, and r is the capillary radius. In the expression, all quantities are known, except γ. This parameter can thus be determined.

– Finally, the pendant drop method, used in many laboratories, is based on the analysis of the droplet shape, which depends on γ.

A number of these techniques are commercially available, offering a catalogue of measurement methods accessible to laboratories. Most of them can be used with two fluids, allowing the measurement of surface tension of liquid–liquid interfaces. Table 4.1 shows a selection measurements of γ reported in the literature, often relevant to microfluidics.

Name	Surface energy (mN/m)
Acetone	23
Ethanol	22
Glycerol	92
Water	72
Toluene	28
Silicon oil	22
Hexadecane	28
Mercury	474

Table 4.1 Surface tensions of a selection of liquids.

Dynamical surface tension measurements in microchannels. What about dynamical measurements of surface tension? In microfluidic devices, surface tension may suddenly change, induced by changes in surfactant concentrations (see the next section). The time scale of change is below 1 s, and the methods described previously are too slow to follow the dynamics. The system of Fig. 4.19 takes advantage of miniaturisation for measuring sub–second dynamical changes of the surface tension.

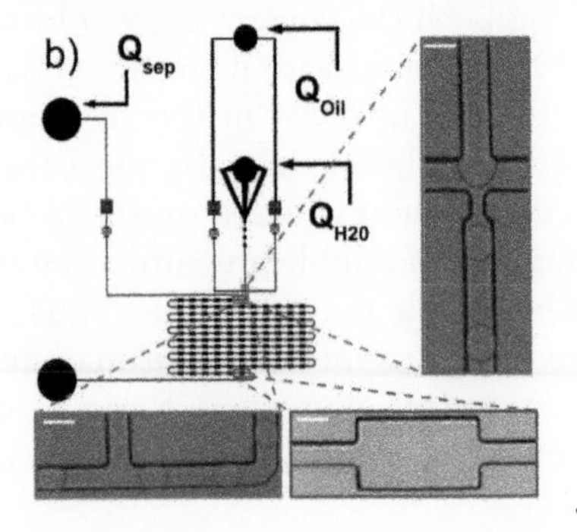

4.19: Droplets are produced by focusing a stream of water (Q_w)in a stream of fluorinated oil–surfactant solution Q_{oil} and spaced by a separation stream Q_{sep}. Droplets flow in a series of expansions where the droplet shape is recorded and interfacial properties are measured. The white scale bar has a length of 100 mm. (Reprinted from [22] under Creative Commons Attribution 3.0 International License.)

Droplets produced in a flow–focusing junction flow in a series of expansions/contractions where the droplets, driven at constant speeds, and subjected to a known pressure gradient, undergo sequences of deformations, during which the surfactant concentration, and therefore the surface tension, changes. From the analysis of the droplet deformation, the dynamical evolution of the surface tension was measured [22, 23]. The temporal resolution of the method is on the order of fractions of a second.

Micropipette interfacial area–expansion method. The method for measuring dynamical interfacial tension reported in Ref. [24] uses systems whose sizes pertain to the microfluidic range of scales. The device, based on micropipettes, is shown in Fig 4.20.

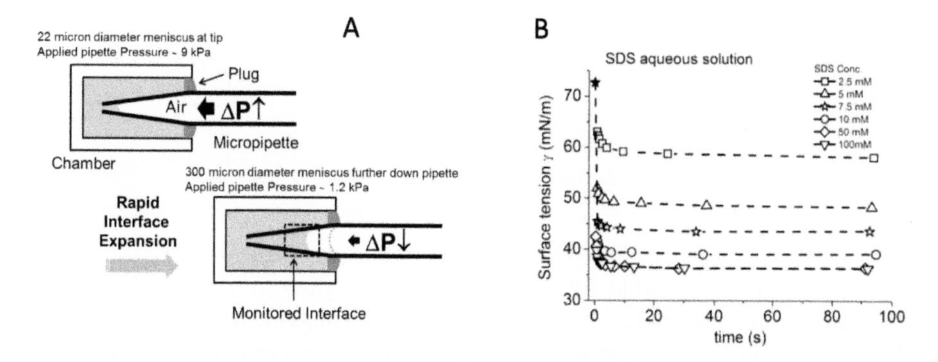

Fig. 4.20: (A) Dynamical surface tension (DST) measurement using the micropipette interfacial–area expansion method. (1) The micropipette is inserted into the solution under high positive applied pipette pressure. The interfacial area was quickly expanded 200–fold by decreasing the applied pipette pressure.(B) Measurements of the temporal evolution of the surface tension, at different surfactant concentrations. (Reprinted with permission from [24]. Copyright 2022, Elsevier.)

The micropipette is inserted into the solution by applying pressure. The interfacial area, inside the pipette, is quickly expanded 200–fold by rapidly (0.1– 0.3 s) decreasing the pressure. The measurements shown in Fig. 4.20 involves time constant larger than the temporal resolution of the method. We learn from the experiment that for interfaces whose size and curvature radius are in the 10 μm range, the time required to re–establish the equilibrium interfacial tension lies in the second range.

Surface tension measurements in solids. The measurement of surface energy in the cases of solid/liquid or solid/gas interfaces is much less straightforward, because (except with soft solids) surface–tension–induced deformations are not measurable. Many of the methods used for solids are thus indirect. They are often based on contact angle measurements made with series of fluids, using the Zisman plot [18], or on detailed models of intermolecular forces. Other techniques are based on working at high temperatures, or analysing heat sublimation. Direct methods nonetheless exist, based on measuring the force needed to open a crack in the material. The case of soft solids deserves particular attention because the measurement could be made more directly, with excellent accuracy, by analysing the deformation of polydimethylsiloxane (PDMS) interfaces placed in contact [19]. Describing the techniques would lie beyond the scope of this book. Table 4.2 shows a selection of surface energies obtained with one or another approach, for materials currently used in microfluidics.

Abbreviation	Name	Surface energy (mN/m)
PC	Polycarbonate	46
PMMA	Polymethylmethacrylate	41
PVA	Polyvinyl alcohol	39
PS	Polystyrene	34
PDMS	Polydimethylsiloxane	23
Co	Copper	2450
Au	Gold	1500
C	Carbon (graphite)	68000
Glass	(Clean) glass	4400

Table 4.2 Surface energies of a selection of solids.

Consistently with the preceding discussion on the order of magnitude of surface tension of solid/air interfaces, Table 4.2 shows that plastics, close to Van der Waals materials, have low surface energies, on the order of 20– 30 mN/m, much lower than metals. Clean glass also has a large surface energy, which decreases by a factor of 2 or so, when exposed to an ambient atmosphere. This is due to the deposition of wetting films (water, for instance) or dust, which work at decreasing surface energies. Dust, made of low–energy organic materials, floating in the air and being deposited on high energy surfaces, is happy to stay on these.

4.2.5 Surface tension and gravity

Capillary length. Dimensional analysis allows an interesting scaling law for capillary phenomena to be established, which has important implications in microfluidics. The question is: what is the typical length l_c for which gravity and capillarity balance? Such a length would be determined by the following condition:

$$l_c = f(\gamma, \rho, g) \tag{4.5}$$

where f is a function. Here, we have four physical quantities, with three units (masse, time and length). Using the Π theorem, we conclude that the above expression reduces to:

$$l_c = \sqrt{\frac{\gamma}{\rho g}} \tag{4.6}$$

l_c is called the capillary length. Another manner to obtain the same result is to note that gravitational forces are on the order of $\rho g l^3$, and capillary forces on the order of γl. Systems of size l_c achieve the balance between the two forces. As l_c is on the order of a few millimetres (2.6 mm for water), we can also infer, from this order of magnitude reasoning, that in microfluidics, gravity is negligible against surface tension.

Shape of the meniscus. In contact with a smooth vertical plane, partially wetting liquids form a meniscus that matches the plane with a contact angle θ, assumed here to be acute (see Fig. 4.21).

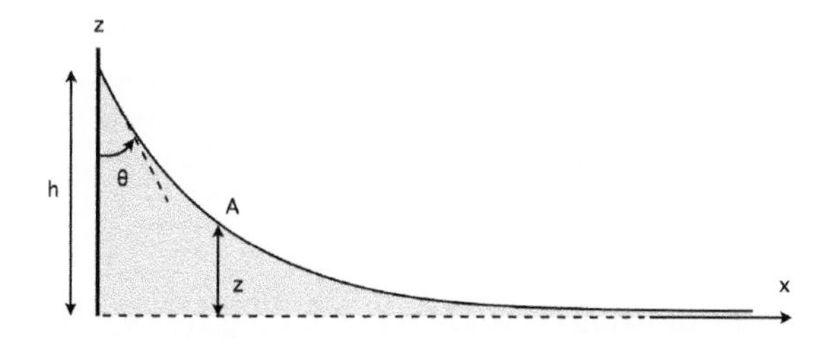

4.21: Profile $z(x)$ of a meniscus taking place on a vertical plate, with an angle θ.

Since the fluid is at rest, capillary pressure (which elevates the fluid), must balance the hydrostatic pressure $\rho g z(x)$, where z is the liquid elevation and x is the horizontal coordinate. We thus have:

$$\frac{\gamma}{R} = \rho g z(x) \tag{4.7}$$

Using a known mathematical relation between R and $z'(x), z''(x)$, the relation reads:

$$\frac{\rho g}{\gamma} z(x) = \frac{z''(x)}{(1 + z'(x)^2)^{3/2}} \tag{4.8}$$

There is no general analytical solution to this equation. Let us thus simplify the problem, by considering the case where θ is close to $\pi/2$. In this case, we can assume that the slope of the water/gas surface is small everywhere. Under such an assumption, the differential equation becomes:

$$z = l_c^2 z'' \tag{4.9}$$

With the boundary conditions $z \to \infty$, $z = 0$, and $z = 0$, $dz/dx = -\cot\theta$, the solution reads:

$$z(x) = l_c \cot\theta \exp(-\frac{x}{l_c}) \tag{4.10}$$

The formula shows that the height of the meniscus, i.e. the elevation $z(x)$ at $x = 0$, equals $l_c \cot\theta$. This result underlines the role of l_c, which naturally emerges as *the* characteristic length of the problem. The fact that for $\theta = 0$, $z(0)$ is infinite signals that at small contact angles, the surface is strongly curved; consequently, the approximation we used is no longer acceptable. The rigorous calculation provides, for the maximum elevation h of the meniscus at $z = 0$, the following expression:

$$h = \sqrt(2) l_c (1 - \sin\theta)^{1/2} \tag{4.11}$$

There is no more divergence at $\theta = 0$. In this situation,i.e.complete wetting, $h = \sqrt(2) l_c$ For clean water, this leads to $h \approx 3.7$ mm. This small elevation requires careful visual observation to be observed, but it is much larger than the size of a microchannel. As noted in the precedent subsection, the fact that l_c (and thus, h) is much larger than a microchannel, implies that in microfluidics, gravity can be neglected against capillarity.

Jurin's law. This law is the oldest law of capillarity. (The fascinating history of the law is told in [4].) For acute contact angles, a circular capillary of internal radius r placed in contact with a liquid bath spontaneously forms a column in the tube. Its height h is given by the relation:

$$h = \frac{2\gamma \cos\theta}{\rho g r} = 2\cos\theta \frac{l_c^2}{r_0} \tag{4.12}$$

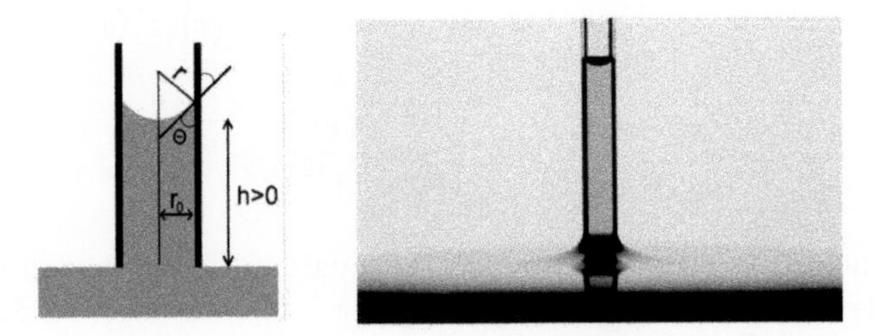

Fig. 4.22: On the left, the situation for which Jurin's law applies, with the definitions of the geometrical parameters. On the right, an experimental realization (courtesy by D.Quere).

This law results from the balance between the hydrostatic pressure $\rho g h$ and the capillary pressure $\frac{2\gamma\cos\theta}{r_0}$, the principal radius of curvature of the meniscus r being $r_0/2\cos\theta$.[14]). In most situations, liquids in contact with high energy surfaces (such as

[14]The geometrical calculation is done by determining the radius of the sphere for which, in the vertical plane, the meniscus forms an arc

glass) develop acute contact angles and therefore rise. Mercury on glass is a counter–example. The contact angle is 135°, and the meniscus moves downwards.

4.3 Surfactants

4.3.1 What is a surfactant?

Surfactants[15] are molecules made of two parts: a hydrophilic part which 'likes' water (energetically speaking), and a hydrophobic/lipophilic part which 'dislikes' water and likes' oil. These molecules are called amphiphilic, because they like simultaneously being in one phase and out of that same phase. Their names come from the Greek 'amphi', meaning double. Chemistry has synthesized and is still synthesizing many surfactants [25,26]. The domain is considerable. There are numerous applications, such as in cosmetics, the food industry, biology, the oil industry, etc. with a multi–billion dollar market. Research started more than a century ago, and is still extremely active. Obviously, without surfactants, the domain of droplet–based microfluidics would not exist.

Surfactants are classified in a number of ways; one of these, established according to the charges they convey, is shown in Fig. 4.23.

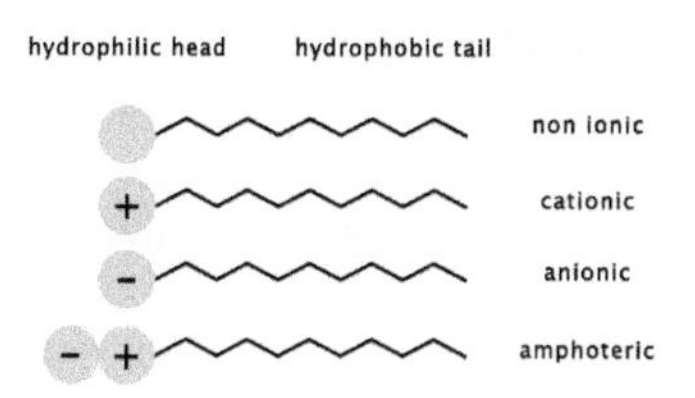

4.23: Classification of the surfactants, according to the charges they convey: non ionic, anionic, cationic and amphoteric

Often, the hydrophilic part of the surfactant is an ion forming a polar or charged head, and the hydrophobic part is one aliphatic chain (or several) $CH_3(CH_2)_n$ forming a hydrophobic (lipophilic) tail. The hydrophobicity of the aliphatic chains originates from the fact that they force water molecules to reorganize around them in an entropically unfavourable manner. In ionic and amphoteric surfactants, the hydrophilicity of the head is linked to its polarization, or charge, which, for water, leads to the creation of hydrogen bonds and therefore strong attractive forces. For non–ionic surfactants, the hydrophilic part is a short chain of neutral units soluble in water. Examples of surfactants, frequently used in microfluidics, are: Span 80, Tween 20 and Krytox (non ionic), SDS (anionic), and CTAB (cationic). Table 4.3 shows the hydrophilic and hydrophobic groups of several surfactants.

Because the molecules are amphiphilic, i.e. their hydrophobic part likes oil, and the hydrophilic part likes water, it is argued that the best energetical solution for them is to stay at the interface (we will return to this point later). We have experimental

[15]Surfactant is a blend of two terms: surface and active agent.

Abbreviation	Hydrophobic part	Hydrophilic part
SDS	$CH_3(CH_2)_{11}$	SO_4^{--}
SPAN80	$CH_3(C_{18}H_{38})C_5H_3O_6$	$C_6H_{11}O_5$
KRYTOX	$C_2F_5(C_3F_6)O)_nOC_2F_4$	COOH
CTAB	$CH_3(CH_2)_{15}$	$N^+(CH_3)$

Table 4.3 Chemical formulas of surfactants currently used in microfluidics, along with hydrophobic and hydrophilic groups.

evidence for this, for instance by observing, in a bathtub, the propagation of soap at the water's surface. The same experiment can be carried out in a dish, using pepper to visualize the surface flow, as will be discussed later. The numerical simulation of Ref. [27] confirms this feature. Figure 4.24 represents a water/dodecane interface, with sodium dodecyl sulfate (SDS) as a surfactant.

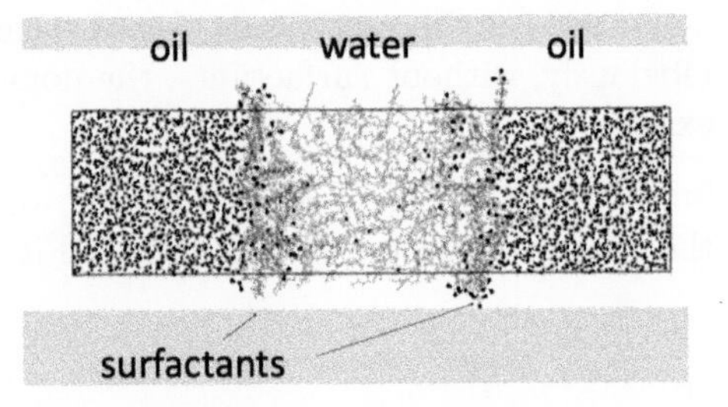

4.24: Dodecane– water interface with SDS simulation. Water and dodecane molecules are drawn as lines, and surfactant molecules are drawn as 'balls–and–sticks'. (Reprinted from [27] with permission from John Wiley and Sons. Copyright 2022.)

The simulation shows that surfactant molecules form a layer, approximately 20 nm thick. By populating the interfaces in this manner, they change their surface properties, in particular their surface tension.

4.3.2 Surfactant and solubility

Fig. 4.25 shows a container filled with water and oil, in which surfactant molecules, at low concentrations, have been introduced. What happens?

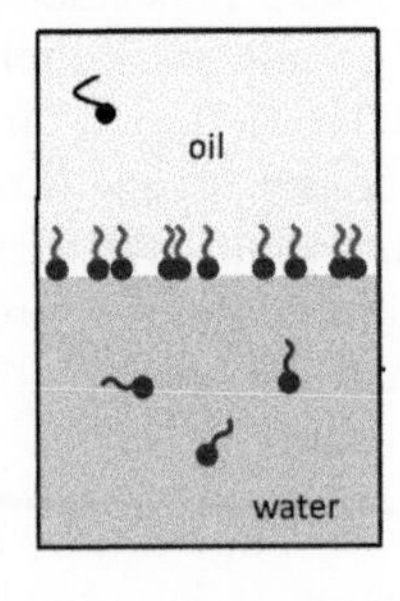

4.25: Surfactant introduced at low concentration in a water/oil system, in a container.

As we previously saw, most of the surfactants adsorb at the interface. However, some are dispersed in the bulk, in oil and in water, in proportions that depend on their solubility. This is partitioning. An equilibrium takes place, characterized by a distribution factor K, which reads:

$$K = \frac{C_w}{C_o} \tag{4.13}$$

in which C_w and C_o are the bulk surfactant concentrations in water and oil, respectively. Instead of using K, it is usual, in the domain, to use Hydrophilic Lipophilic Balance (HLB). The definition of HLB is:

$$HLB = 7 + \Sigma(\text{hydrophilic groups number}) - \Sigma(\text{hydrophobics group number}) \tag{4.14}$$

in which the group (for example, hydroxyl or methyl groups) numbers involved in the definition are determined empirically. More detail can be found in [4]. The idea behind this definition is to establish a link between solubility and molecular structure. Long hydrophobic chains are associated to small HLB. In this case, the surfactant will be poorly soluble in water. In the opposite case (large HLB), the chain is short and solubility in water will be substantial. The knowledge of HLB also allows us to determine whether an emulsion will be direct (oil droplets in water) or inverse (water droplets in oil). This is Bancroft rule, which states: 'The phase in which an emulsifier is more soluble constitutes the continuous phase' [28, 29]. Other properties, such as foaming, spreading or solubilization are also correlated with HLB. Table 4.4 below lists HLB values, and some properties to which these values correlate.

HLB	Properties	Examples
<10	Lipid–soluble (water–insoluble)	Span 80
>10	Water–soluble (lipid–insoluble)	Tween, SDS
3– 6	W/O (water in oil) emulsifier	
13– 16	Detergent	Span 80
8– 16	Oil in water	

Table 4.4 HLBs of a selection of surfactants, with their current usages.

In the field of microfluidics, HLB is not frequently used because channel walls largely control the system behaviour. The occurrence of water–in–oil or oil–in water droplets depends on the wettability of the walls with respect to the working fluids, not on the HLB. For example, in microfluidic devices, one can produce water–in–oil emulsions with HLB values indicating that oil in water emulsions should form. An example is shown in Ref. [30]. In these experiments, SPAN 80 was used. Instead of obtaining oil droplets, the authors observed stable water droplets. The reason was that, in this case, microchannel walls being lipophilic, oil wetted the walls, and therefore was selected as the external phase, against Bancroft rule. We will return later to this type of phenomenon. In general, in droplet based microfluidics, Bancroft rule is irrelevant.

4.3.3 Surfactants and surface tension

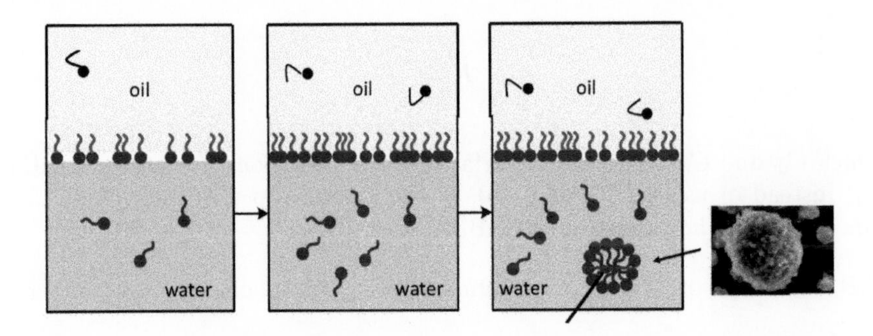

Fig. 4.26: As more surfactant is introduced in the container (centre), the surface saturates and, on further increase (right), above a threshold called critical micellar concentation (CMC), micelles appear. A typical image of a spherical micelle, several tens nanometre in size, is shown on the right

The case of liquids. Let us take again Fig. 4.25 and add surfactants (see Fig. 4.26). Indeed, as more surfactant is added, the interface tends to saturate. At $C = C_S$, saturation is achieved. As the amount of surfactant is increased further, the concentration, in the bulk, increases. But we have to reach a new threshold, called critical micellar concentration (CMC), to see micelles appearing in the solution.

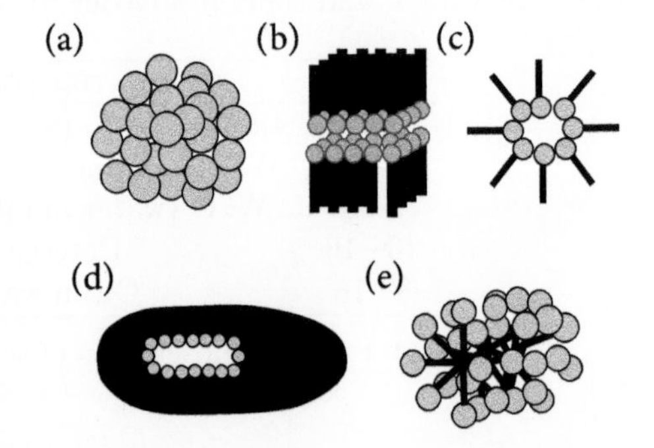

4.27: Five types of micelles, as interpreted from experimental data: (A) spherical; (B) lamellar; (C) inverted (or reversed); (C) disk; (E) cylindrical or rodlike. (From [26].)

Micelles are self–assembled nanometric structures that protect the hydrophobic tails from the aqueous environment. They are energetically advantageous but they need to jump over a nucleation barrier to form. Often, micelles are represented as spherical objects, but as shown in Fig. 4.27, a variety of morphologies exist[16] [26].

So, how does the system respond energetically to the addition of surfactants and the formation of micelles ? Fig. 4.28 show the evolution of the water–oil interfacial tension γ as a function of the surfactant concentration C_S. γ decreases as the surfactant concentration C_S increases [30]. Why is this the case ? The argument is that, when

[16]Each being associated to specific CMCs. The same system can thus have several CMCs.

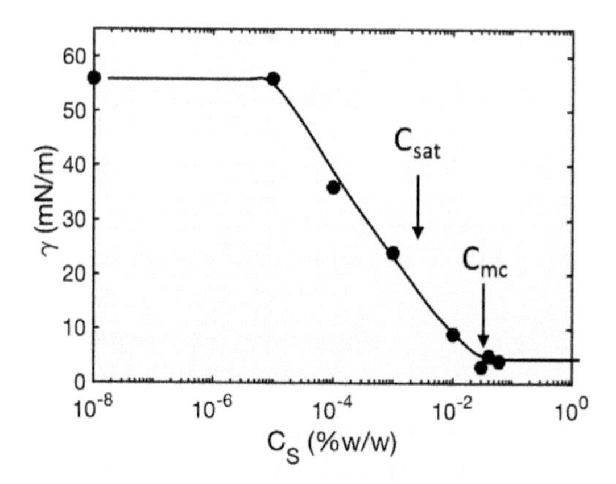

Fig. 4.28: Evolution of the water/hexadecane interfacial tension as a function of the surfactant concentration C_S, the surfactant being SPAN 80. The concentration axis has units of %w/w (w for weight) of solute in solvent. To convert in mMole/litre, a unit often used in the literature, one must multiply by $10^3/M$ where M is the molar mass of SPAN 80 (429 g/mole). In this case, we have a CMC of 3 10^{-4} w/w, then $0.6mM/l$. (Taken from [30].)

surfactants absorb at the surface, they replace the air/water interface by an organic phase/air interface, whose cohesive (between themselves) and adhesive (with water) energies are weaker than for water/water interfaces. The interface is thus less energetic. Therefore, surface tension decreases. Measurements show that above a concentration called C_{sat} (see Fig. 4.28), for which the interface is saturated in surfactant, γ decreases linearly with the logarithmic of C_S, consistently with Gibbs theory [25]. As C_S reaches the CMC, γ levels off. [17] The plateau is the energetic signature of micelle formation (see Fig. 4.26).

The explanation of this behaviour is subtle. The approach that is usually taken is macroscopic and is based on the concept of chemical potential, that we do not develop in the book. We recommend Ref. [4] to readers for a pedagogical presentation. We may nonetheless envisage that micelle creation does change much the energy of the system. Micelles do not perturb the bulk properties, because, in the range of C_S we address, their concentration is low. Neither do they perturb the interface, which is saturated. Therefore, we may understand, at least intuitively, that in Fig. 4.28, a plateau forms above CMC.

In practice, the knowledge of CMC is fundamental. Depending on the surfactant, in particular the length of the chain (the longer the chain, the higher the free energy gain, the smaller the CMC), CMC can vary over several decades, [26]. For SDS, SPAN 80, and CTAB, in water/alkane systems (tetradecane, hexadecane, etc.), CMC is on the order of one or several mM/L, or 10^{-4} w/w (surfactant mass over liquid mass in the same volume). It is approximately 4 μML, i.e. three orders of magnitude smaller, for

[17] Often, C_{sat} and CMC are considered as identical, which is incorrect

Surfactant	Liquid 1	Liquid 2	CMC (M/l)
SDS	Water	Air	$8\ 10^{-3}$
SPAN 80	Water	Hexadecane	6.10^{-4}
Krytox	Water	PFC	4.10^{-6}
CTAB	Water	air	9.10^{-4}
Tween20	Water	Air	10^{-5}

Table 4.5 CMC at 20°C for a surfactants currently used in microfluidics.

Krytox in water/PFC systems (i.e. perfluorinated oil), frequently used in microfluidics, for their compatibility with PDMS (no swelling). Table 4.5 lists a few values.

In practice, working at several CMC guarantees that we stand on the plateau of Fig 4.28 and therefore γ is constant. Around CMC, the surfactant concentration is low and viscosity is essentially that of the solvent. At much larger C_S, this ceases to be true, and non–Newtonian behaviour takes place. We will not address this type of situation, for which, interestingly, microfluidics provided accurate data and proposed novel concepts [31, 32].

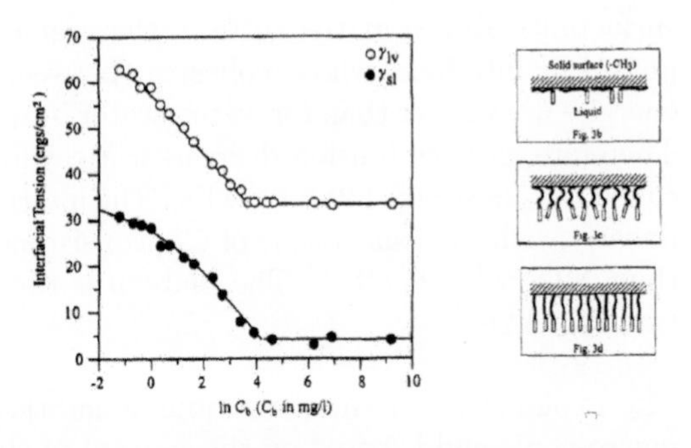

4.29: Evolution of the interfacial tensions of PDMS in water (closed circles), and water in air (open circles), with surfactant (heptaethylene glycol mono–n–dodecyl ether) concentration C_s. The units here are expressed in the old system CGS (erg/cm^2). One has 1 erg/cm^2=1 mJ/m^2. (Reprinted with permission from [33]. Copyright 2022 American Chemical Society.)

The case of solids. For reasons similar to those presented above, surfactants adsorb at the walls. They form mono layers, or multilayers, that affect the surface tension γ_{LS}. The effect was demonstrated by Chaudhury [33], who exploited the deformability of PDMS to measure surface tensions as a function of bulk surfactant concentrations. Fig. 4.29 show the existence of a critical micellar concentration, above which, similarly as for liquid/gas interfaces, surface tension reaches a plateau. Interesting is to note that the CMC of the liquid/solid interface is different, but of the same order of magnitude as that of the liquid–vapour interface (see Fig. 4.29). On the right–hand side of the figure, attempts of the authors are made to figure out the molecular structures adsorbed at the interface [33].

4.3.4 Marangoni flow

When one deposits a small drop of liquid soap (for dish cleaning; for instance) on a water surface, almost immediately, the surface gets covered with a soap film (see Fig.

4.30). In the experiment shown in Fig. 4.30 pepper powder was used to visualize the phenomenon [34]. One observes that the speed of the water/soap front is on the order of 1 m/s. So, what happened?

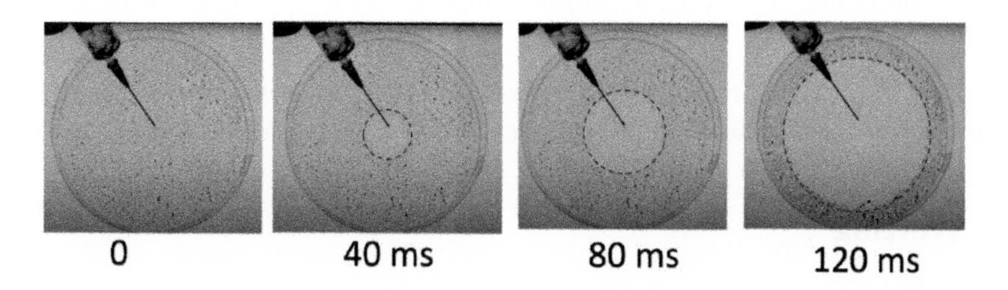

Fig. 4.30: Pepper experiment: water is covered by pepper (for the sake of visualization). A drop of soap, introduced in the system, spreads at fast speed, pushing the pepper [34]. (Credits: Chirag Kalelkar, Abhijeet Kant Sinha.)

The theory of Carlo Marangoni, built in the nineteenth century, allows us to explain the phenomenon [35]. In the presence of a gradient of surface tension (here, the gradient is produced by the soap patch initially deposited on a fresh water surface), similarly as the soap film confined in a rectangular frame, the system searches to decrease its energy. A flow starts over, in order to spread the surfactant over the surface, since, as explained above, soap reduces surface tension. The phenomenon can be modelled by noting that the continuity of the tangential stress at the interface imposes the following relation:

$$\mu \frac{\partial U}{\partial z} = \frac{\partial \gamma}{\partial r}$$

in which μ is the fluid viscosity, U is the flow speed at the surface $z = h$, with z the vertical dimension, γ is the surface tension and r is the radial coordinate. For uniform γ we have the usual boundary condition for a free surface, i.e. $\mu \frac{\partial U}{\partial z} = 0$. However, in the pepper experiment, with the addition of the soap, γ is no longer homogeneous. The problem is a century old, but it was analyzed comprehensively only recently [36]. The stress, driven by $\nabla \gamma$, generates a flow in the bulk, which pushes the pepper radially outwards. A hole (i.e. a surface without pepper)thus grows at speed U. As the patch radius r grows, the momentum diffuses in the depth of the water bath, down to a distance $\delta \sim \sqrt{\frac{U}{\nu r}}$ from the free surface, in which ν is the kinematic viscosity. By equating the viscous stress $\mu \frac{U}{\delta}$, where $\Delta\gamma = \gamma_w - \gamma_s$ (here γ_w is the water surface tension and γ_s the surface tension with soap) with the surface tension gradient $\frac{\partial \gamma}{\partial r} \sim \frac{\Delta \gamma}{r}$, one obtains the following expression for U:

$$U \sim \frac{\Delta\gamma}{\mu} Oh^{1/3}, \text{with } Oh = \frac{\mu^2}{\Delta\gamma\rho r} \text{the Ohnesorge number}$$

For soap spreading on water, the expression leads to $U \sim 1m/s$, which is consistent with the observation.

We will see that the Marangoni effect plays in important role in microfluidics, for the questions of droplet stability and coalescence.

4.4 Wetting

4.4.1 The three fundamental cases of wetting

There exist three basic cases of wetting: complete (or full) wetting, partial wetting, and dewetting [37, 48]. The first two cases, complete and partial wetting are shown in Fig. 4.31.

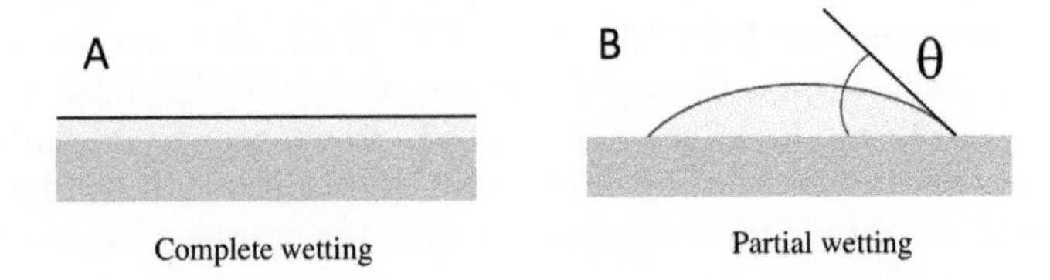

Fig. 4.31: Two fundamental cases of wetting: (A) complete and (B) partial.

In the case of complete wetting (Fig. 4.31 (A)), a droplet deposited on the surface spontaneously spreads and forms a film. Using a vocabulary employed before, this behaviour suggests that the liquid likes' the surface.

The second case is partial wetting (Fig. 4.31 (B)). Here, the droplet does not spread and a contact angle appears.

In the literature, one distinguishes between two sub–cases: contact angle smaller than 90° or larger than 90° (see Fig. 4.32). In the former case, the fluid is declared 'wetting' (a name which must not be confused with 'complete wetting'), and, in the latter, 'non–wetting'.

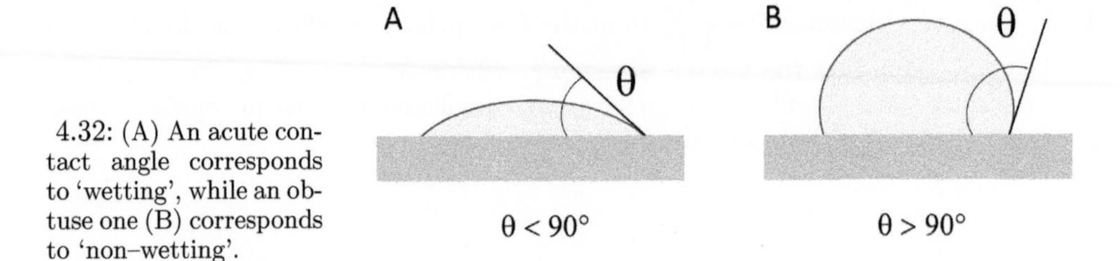

4.32: (A) An acute contact angle corresponds to 'wetting', while an obtuse one (B) corresponds to 'non–wetting'.

We will see that, from an energetic perspective, there is no reason to single out the 90° angle. However, from a dynamical perspective, the two situations are fundamentally different. A wetting fluid introduced in the entry of a capillary tube, or a microchannel, will develop a convex meniscus (i.e. whose curvature is oriented against the fluid) and consequently, as was seen earlier, the Laplace pressure will work to generate a flow that pushes the meniscus further into the device. In such circumstances, filling is spontaneous. In the other case (non–wetting), the meniscus will recede backwards, against filling. We will return to this behaviour later.

An important case is when the contact angle is 180° (Fig. 4.33). In this case, called dewetting, the droplet minimizes its contact with the surface. This suggests that in this case, the liquid hates' the surface. In air, it never happens. In liquid–liquid systems, dewetting can be achieved by using surfactants. This is detergence. We will describe this situation later.

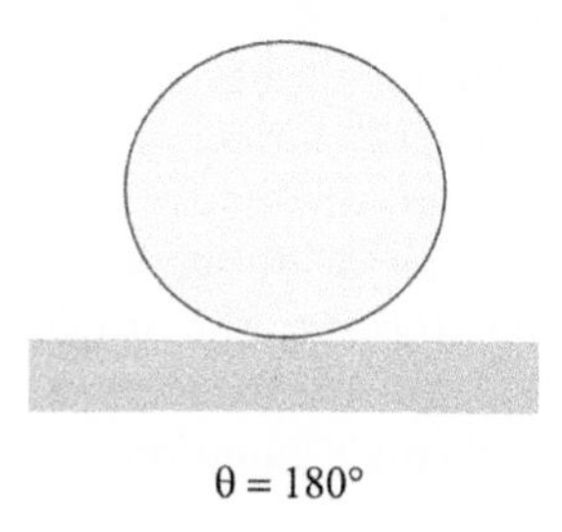

4.33: Dewetting: in this case, a droplet sitting on a surface develops a contact angle equal to 180°.

4.4.2 The spreading coefficient

How can we describe the two situations shown in Fig. 4.31? We again use reasoning based on energy [37]. We have here three interfaces: solid/gas, liquid/solid, and liquid/air. An interfacial energy corresponds to each of these interfaces. To describe the system, we will make use of a fundamental parameter: the spreading coefficient S, which is defined by:

$$S = \gamma_{\mathrm{SG}} - \gamma_{\mathrm{SL}} - \gamma$$

in which γ_{SG}, γ_{SL} are the energies of the solid–gas and solid–liquid interfaces respectively, and γ, as usual, is the energy of the liquid–vapour interface. There is a subtlety here: γ_{SG} assumes the exposed surface to be in equilibrium with the vapour, which in turns implies that, owing to volatility, a few molecules adsorb to the surface. The surface is thus moist. The difference between a dry and a moist surface can be large. Ref. [37] mentioned the case of metallic oxides: in such a case, the moist surface is on the order of 60 mN/m, while the dry one is ten times larger. In the case of organic materials, liquids and solids, or non–volatile liquids, the difference is smaller.

The spreading parameter S allows us to quantify the two situations outlined in the preceding section. We will assume that the fluid is not volatile, and consequently S_{LG} represents the surface energy of a dry solid.

- $S > 0$: complete wetting. We may view S as representing the difference between the surface energy of the dry surface (γ_{SG}), and the energy of the surface covered by a film ($\gamma_{LG} + \gamma_{SL}$). $S > 0$ means that a surface entirely covered by a liquid film is energetically more favourable than a dry surface. Solid gas interfaces 'cost' too much. To reduce the price to pay, liquid blobs placed on dry surfaces spread spontaneously.
- $S < 0$: partial wetting or dewetting. In partial wetting, film spreading is not energetically favourable, and a droplet, characterized by an equilibrium contact angle θ, forms on the surface. When S is strongly negative, for instance when γ_{SL} is large, a gas layer, sandwiched between the liquid and the solid, tends to form, in order to avoid contact between the two phases. This configuration could physically occur in microgravity. On earth, liquids do not spontaneously levitate and dewetting corresponds to a situation where the contact angle is equal to 180°. This discussion shows that complete wetting and dewetting are the two faces of the same coin. If the dispersed phase (the droplet) is dewetting with respect to a surface, then the continuous phase (the fluid exterior to the droplet) is completely wetting with respect to the same surface. We will study these situations in more detail below. For that, we need to establish Young's equation.

4.4.3 Young's equation

Let us concentrate on partial wetting, again with non volatile liquids, so that γ_{SG} corresponds to a dry surface exposed to gas. We take here the demonstration made in Ref. [37]. To express the equilibrium, we again use the method of virtual displacements, considering that, in the process, liquid and gas volumes do not change. This implies that the only energies that matter are surface energies. Let us assume that the fluid interface is flat down to the wall, an hypothesis we will comment later, and calculate the variation in interfacial energy produced by a displacement δx of the contact line on the solid surface (Fig. 4.34(A)). During this displacement, in the case shown in the figure (displacement on the right), the solid liquid interface looses a length δx, while the solid gas interface gains δx. On the side of the liquid–vapour interface, there is a gain $\delta x \cos\theta$, in which θ is the contact angle. Why is it the case? One may think that, far from the wall, the liquid interface is held at a fixed point. Then, a geometrical calculation shows that the length of the liquid–vapour interface must increase by a quantity equal to $\delta x \cos\theta$ in order to displace the contact line by a quantity equal to δx.

The work δW associated to the contact line displacement is:

$$\delta W = (\gamma_{\mathrm{SG}} - \gamma_{\mathrm{SL}})\delta x - \gamma \cos\theta \delta x \tag{4.15}$$

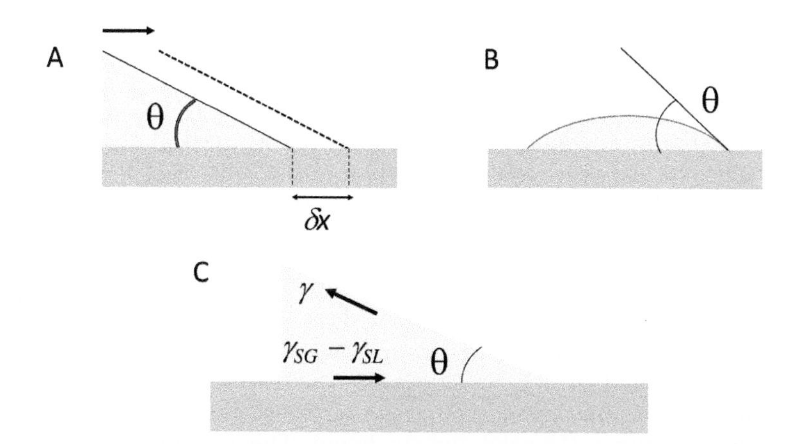

Fig. 4.34: Solids are in dark grey, liquid in light grey, and gas in white.(A) To demonstrate Young's equation, the contact line is displaced by δx [39].(B) Young's equation can also be established by considering a droplet placed on a solid and minimizing the surface energy of the system [38].(C) Mechanical demonstration of Young's equation. The tension $\gamma_{\mathrm{SG}} - \gamma_{\mathrm{SL}}$ applies to the liquid/solid interface, close to the contact line [5,7].

At equilibrium, this incremental work must be equal to zero. On expressing this condition, we obtain Young's equation [39]:

$$\gamma_{\mathrm{SG}} - \gamma_{\mathrm{SL}} = \gamma \cos\theta \tag{4.16}$$

or:

$$\cos\theta = \frac{S}{\gamma} + 1 \tag{4.17}$$

In this demonstration, the liquid–vapour interface can curve in as we approach the contact line, but this does not change the result as long as the curved region occupies a negligible fraction of space.

Young's equation can also be demonstrated by considering a droplet, deposited on a flat surface, and minimizing, at fixed volume, the surface energy of the system [38] (see Fig.4.34 (B)). The mathematics are not complicated, but it takes one page to perform the calculation. It allows us to avoid handling infinite volumes of fluid, infinite surface energies, seen from a point located at infinity. The approach is thus less abstract than in Ref. [37], but less elegant.

Young's equation can also be established by reasoning with line tensions, but the demonstration is more delicate.[18] According to Refs. [5,7], the force (per unit of length) exerted by the liquid/solid interface on the contact line is found equal to $\gamma_{SG} - \gamma_{SL}$

[18]Reasoning with energy allows to avoid asking questions, such as: where do forces apply? In which directions do they point? How to incorporate kinetic pressure? The reason is that energy is a global quantity, which integrates all these questions.

(see Fig. 4.34 (C), for the case $\gamma_{SG} - \gamma_{SL} > 0$). On the other hand, the liquid pulls the contact line with a tension γ. The system being at equilibrium, these forces must equilibrate. By projecting the equilibrium condition onto the horizontal axis, one obtains Young equation (4.16).[19] The projection on the vertical axis yields a pulling force normal to the wall, oriented upwards, which is balanced by the elasticity of the material. This gives rise to a slight deformation of the substrate. This deformation is extremely small in hard solids and, consequently, is difficult to measure. The task is easier in soft solids, such as rubber, PDMS, or fresh paint [4].

Whether obtained energetically or mechanically, Eq. (4.17) stipulates that if the spreading coefficient S is positive or less than -2γ, there is no solution, and, consequently, there is no contact line. For positive S, the system forms a film completely covering the solid. This is the case of oil on clean glass (see the discussion below). For $S \leq -2\gamma$, the liquid adopts a shape that minimizes contact with the solid, for instance a gas layer sandwiched between the solid and the liquid. This is dewetting. As said above, between these two extremes, the contact angle is prescribed by Eq. (4.17).

The link between θ and the material characteristics. This is also a long story and we will content ourselves to summarize an important conclusion [37]. There is a link between θ and polarizability, which, as we saw in Chapter 2, controls the intensity of the Van der Waals interactions. The more polarizable the solid, the more wetting the surface. This statement, justified in [37], rests on a number of simplifications and hypotheses that restrict its domain of validity. Nonetheless, from this, it can be considered that poorly polarizable liquids, such as oil, wet highly polarizable materials, such as glass, and metals. The intuitive notion that fluids with low surface tensions wet high energy surfaces, is a consequence of this result.

4.4.4 Experimental validation of Young equation

In the spirit of this book, which emphasizes real measurements, and due to its importance, it is interesting to mention a direct experimental validation of Young's equation, established for the particular case of soft solids. This was done by Chaudhury and Whitesides in the 1990s [19]. Data is shown in Fig.4.35. Verification of Young's equation achieved with mixtures of water and methanol on PDMS. The graph shows the evolution of $\gamma \cos\theta$ with γ_{SL}, for advancing and receding angles (two different symbols), which are barely distinguishable, within experimental error. All quantities are measured independently, using standard techniques for the liquid/vapour interface, and, at that time, a novel method, already mentioned in this chapter, based on the analysis of the deformation of PDMS, for the γ_{SL} measurement. The data shows that $\gamma \cos\theta$ varies with γ_{SL} linearly, with a slope -1, in excellent agreement with Young's equation.

[19] Fig 4.34 (C) differs from the figures shown in textbooks. In the traditional approach, each interface is associated to a tension that pulls the contact line. In Berry's theory, only liquid/vapour and liquid/solid interfaces pull the contact line.

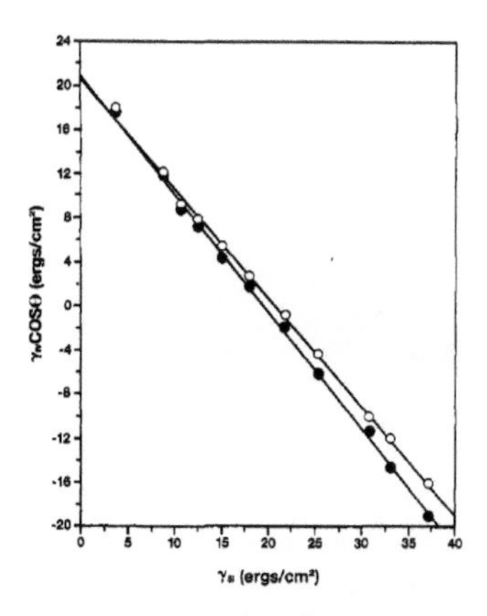

4.35: Verification of Young's equation, made with mixtures of water and methanol on PDMS. The graph shows the evolution of $\gamma \cos\theta$ with γ_{SL}, for advancing and receding angles (two different symbols), which are barely distinguishable, within experimental error. All quantities are measured independently. The data shows that $\gamma \cos\theta$ varies with γ_{SL} linearly, with a slope -1, in excellent agreement with Young's relation (Reprinted (adapted) with permission from [19]. Copyright 2022 American Chemical Society).

4.4.5 Is Young's equation valid at the nanoscale?

We saw that, in the case of partial wetting, the region where gas, solid and liquid meet defines a line, the contact line. Physically, the contact line has some thickness, and a legitimate question is raised as whether the corresponding region has some special properties, different from the bulk and the interface, which should be incorporated, in some way, in Young's equation. Gibbs suggested long ago [40], that this line is subjected to a tension τ, which must be directly introduced in Young's equation as a force applied onto the contact line, that prevents, or favour its displacement.These considerations led to the proposal of a modified Young's equation, which reads:

$$\gamma_{\mathrm{SG}} - \gamma_{\mathrm{SL}} - \frac{\tau}{\gamma R} = \gamma \cos\theta$$

in which τ is the tension of the contact line (which can be stretched or compressed, depending on the sign of τ), and R is its curvature radius of the same line. If the drop is cylindrical, the contact line is straight and the correction vanishes out. Only curved contact lines are concerned by the correction. When τ is positive, the 'true' contact angle is larger than the Young angle. One century of discussion has been conducted on the value of τ, but, today, there is a consensus, essentially based on numerical simulations, that the correction is extremely small. It was found that the length scale $l=\frac{\tau}{\gamma}$ does not exceed a fraction of nanometres, so that the ratio l/R is negligible in all cases of practical interest.

4.4.6 Dewetting induced by surfactants – detergence

We learned that surfactants decrease the surface tension of liquid/air interfaces. We will see here that this feature generates spectacular effects. An example is detergence, a phenomenon that laundry companies exploit for cleaning clothes and that turns out to be extremely important for microfluidics.

Without surfactant

With SPAN80

Fig. 4.36: Dewetting in a water–tetradecane system, in which a water droplet, 100 μm in size, sitting on a smooth glass surface, is created. On increasing the SPAN 80 concentration (up to several CMCs), the system shifts from left to right, i.e. from partial wetting to dewetting.

The photo on the left of Fig. 4.36 represents a droplet of water in tetradecane, without surfactant. We have here a situation of partial wetting. On the right, we introduced SPAN 80 at a concentration three times above CMC. Data on this surfactant were shown in Fig 4.28. The effect of Span 80 is spectacular: it brings the system from a state of partial wetting to dewetting.

Why this is the case? In fact, the phenomenon is a spectacular consequence of Young's equation, which we recall here (in the case of a smooth surface):

$$\gamma_{SW} - \gamma_{SO} = \gamma_{WO} \cos\theta$$

where the symbols S, W, and O signify solid, water, and oil, respectively, and θ is the contact angle formed between the solid and water, defined, by convention, *inside* the droplet. Without surfactant, this angle is about 110 degres. Adding a surfactant above CMC, as we have seen (see Fig. 4.28), reduces the interfacial tension γ by a factor of 20. Indeed, surfactant molecules will adsorb at the solid interfaces, and change interfacial tensions (see Fig. 4.29). However, Span 80 is almost insoluble in water. Therefore, we have γ_{SO} decreasing, and γ_{SW} approximately constant. Consequently, with γ decreasing, Young's relation imposes $\cos\theta$, initially negative, to become more negative. As it reaches -1, θ adopts a value of 180°and dewetting occurs. At this stage, the contact line vanishes and Young's equation no longer holds. The system is locked in this state. This is detergence.

Laundry products harness this effect: the droplet is an oil stain initially adsorbed onto a piece of clothing. The surfactant transforms the stain to a drop, then the hydrodynamic currents of the washing machine remove these droplets from the clothing. In a microfluidic device where flows develop, forced by syringe pumps or pressure controllers, as explained in Chapter 3, droplets are kept away from the walls in a similar manner. In permanent regimes, a lubricating film covers the walls, preventing the droplets to establish direct contact with them.

4.4.7 The effect of surface roughness on the contact angle

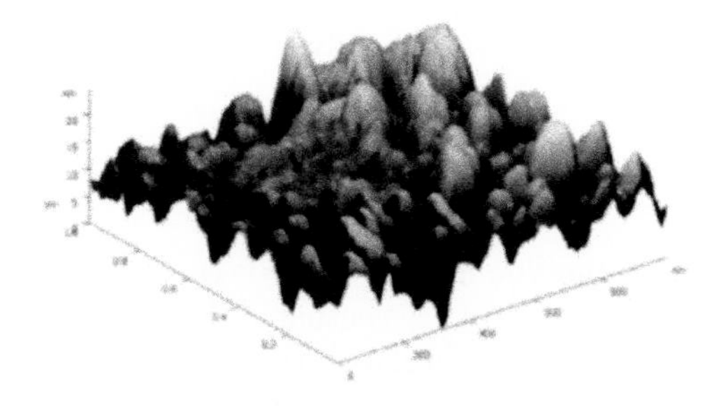

4.37: Typical image of a glass etched surface, imaged by Atomic Force Microscopy (AFM). Vertical axis unit is nanometre and horizontal units are micrometres(Reprinted from [41] by permission of Springer Nature.)

Thus far, we have assumed smooth surfaces. In reality, surfaces are rough. Silicon surfaces are atomically rough and glass or polymers (such as PDMS), more frequently used than silicon, have surfaces with roughnesses in the nano– or deca–nanometre range. When the surface is coated, with a hydrophobic layer for instance, the roughness increases up so several tens of nanometres. Figure 4.37 shows an AFM image of a glass surface that has been etched, without particular effort deployed for minimizing roughness [41]. One can see a roughness in the order of 10–20 nm, with a wavelength of 100 nm. The parameter r, also called roughness', defined as the surface area over the projected area, will be on the order of 1.1. Although small, this roughness must be seriously taken into account.

Discussing the effect of roughness on wetting is a complex task. Decades of discussion have been conducted on the subject. Over the years, progress in theory, simulation and surface observation and control, allowed a consensus of knowledge to be built up. In the domain, one usually distinguishes between two types of roughness: geometrical roughness and chemical heterogeneities (see Fig. 4.38). Each type needs a specific model.

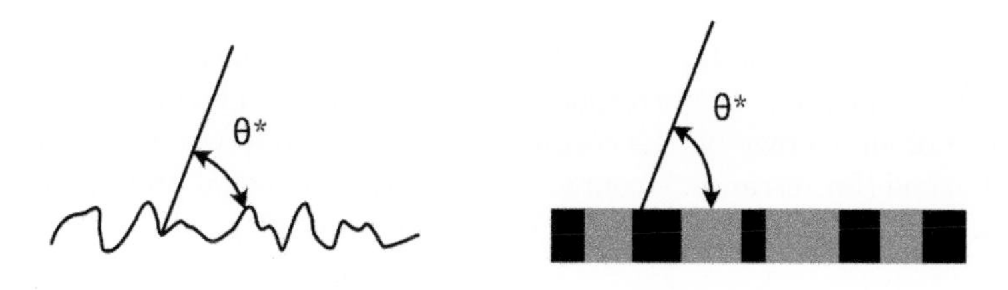

Fig. 4.38: Two types of roughness: geometrical (left) and induced by chemical heterogeneities (right).

Geometrical roughness: Wentzel model. The Wentzel model is well reviewed in the literature, and we content ourselves with discussing the most important outcome. R. N. Wentzel [43,44] showed that the apparent contact angle $\theta*$ (see Fig. 4.38) is related to the equilibrium contact angle θ_E and the roughness r, as given by the following relation:

$$\theta* = r\cos\theta_E \tag{4.18}$$

in which $\theta*$ is the apparent contact angle, θ_E is the Young angle and r is the roughness. As shown in Fig. 4.39, since $r > 1$, the roughness always accentuates the wettability or non wettability of the material. therefore, for wetting surfaces, the base of drop, at the molecular scale, develops an averaged contact angle *smaller* than the macroscopic equilibrium contact angle. For small roughnesses, $\theta*$ will differ from θ_E by a quantity $\approx -(r-1)\frac{\tan\theta_E}{r}$. For $r \sim 1.1$ and a contact angle of 60°, the deviation is on the order of 10°.

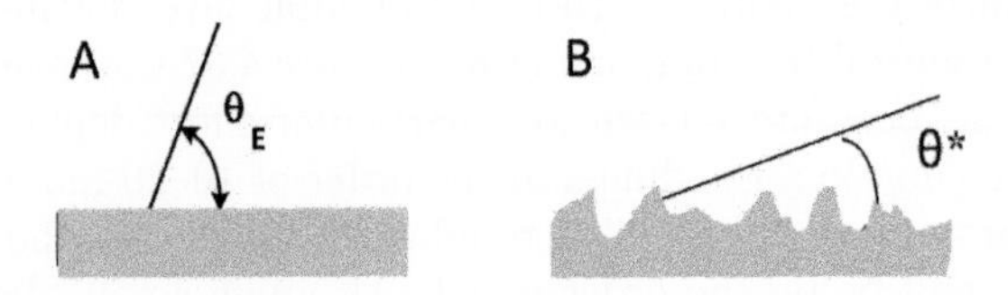

Fig. 4.39: (A)Smooth wetting surface, characterized by a contact angle θ_E.(B) Same material, but rough. The averaged contact angle $\theta*$ is smaller than θ_E.

Chemical heterogeneity: Cassie–Baxter model. In the Cassie–Baxter model [45], we have the following relation:

$$cos\theta* = f_1\cos\theta_1 + f_2\cos\theta_2 \tag{4.19}$$

in which f_1 and f_2 are the fractions of the surfaces of chemical type 1 and 2, respectively, and θ_1 and θ_2 are the corresponding Young angles. The apparent angle is thus the outcome of an average on the cosinus. If we mix a fully wetting material with an hydrophobic one (for instance, a contact angle of 120°, in equal proportions, the result will be a wetting surface with an angle of 75° (and not 60°).

Pinning. We saw before that microfluidic surfaces are rough. We recall that their typical roughness factors are on the order of 1.1 and their maximal amplitudes are 10 – 30 nm. Here, we shall see that, even though r looks small, and amplitudes lie in the nanometre range, roughness considerably affects the dynamics of the droplet.

Fig. 4.40 shows the mechanism of pinning. In order to move forward, the contact line must travel in a landscape composed of hills and valleys. During the journey, the

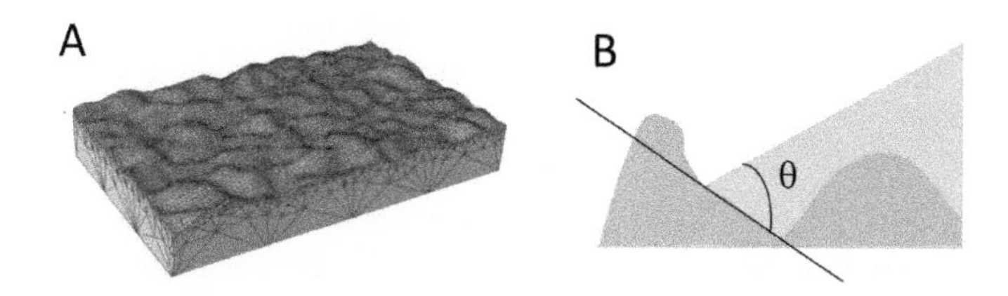

Fig. 4.40: (A) An example of a rough surface [42].(B) Zoom of the contact line region over a rough surface.

contact angle departs from its equilibrium value θ shown in the figure, by an amount $\Delta\theta \sim (r-1)$, for $r-1$ small and $\tan(\theta) \sim 1$. The force working at bringing the meniscus back to equilibrium is $\Delta\theta\sigma h$, where h is the side length of the square cross–section and σ is a representative of the surface forces. On the other hand, the fluid is subjected to an overpressure ΔP, which pushes the interface downstream with a force $\Delta P h^2$. For the interface to advance, one must have:

$$\Delta P h^2 > \Delta\theta\sigma h \tag{4.20}$$

Let us estimate, naively, the various terms of this inequality. Overpressures developed in microfluidic flows are on the order $\mu UL/h^2$ in which U is a typical speed, L is the channel length, and μ is the fluid viscosity. On the other hand, for Van der Waals materials, $\sigma \sim \gamma$, γ being a typical surface tension for such systems (we estimated on the order of 20 mN/m). As a consequence, in order for the meniscus to move forward, one must have:

$$\Delta\theta < \frac{\mu UL}{\gamma h} \tag{4.21}$$

We shall see later that the group $\mu U/\gamma$ plays an important role in droplet–based microfluidics. It is called capillary number and its value is on the order of 10^{-5}. With L $\sim$ 1 cm and $h \sim 100\mu m$, one gets $\Delta\theta < 10^{-3}$, which in practice is impossible to achieve with current materials, such as polymers or glass.

To summarize, the naive model developed here indicates that even though roughness is small, the meniscus remains trapped in valleys and the droplet stops. In order to avoid this problematic situation, one needs to work in completely wetting conditions (or dewetting conditions for the dispersed phase).

Hysteresis. Let us make another remark on Fig. 4.40. As the contact line advances, it floods a dry landscape, while when it recedes, the landscape is already wetted by itself. Liquid is adsorbed on the surface, or trapped in valleys, so that surface energy is different from the dry state. This asymmetry is at the origin of the hysteresis of the contact angle. The hysteresis must be taken into account carefully in the measurement of contact angles. Hysteresis of several degrees, is frequently encountered on unprepared surfaces. Smaller hysteresis, of a fraction or degree or so, is currently achieved in wetting studies.

Superhydrophobicity. In reality, surfaces must be viewed, in general, as including both chemical heterogeneities and geometrical roughness. For hydrophobic rough surfaces, air bubbles can be trapped in the valleys, giving rise to superhydrophobicity. The subject of superhydrophobicity has been studied for decades (see, for instance, Ref. [46]). It has given rise to many interesting applications, such as in textile industry, with repelling tissues. We saw in Chapter 3 that, with these surfaces, slip lengths could be enhanced by one or several orders of magnitude. Superhydrophobic surfaces have first been created by using microfabrication techniques but, over the years, much simpler techniques have been invented [47]. Even simplified, these techniques represent serious microfabrication constraints, which must be balanced with the gain obtained from working with superhydrophobic walls. Today, in the specific context of microfluidics, superhydrophobicity is still awaiting applications.

4.5 Droplets advancing on a surface

4.5.1 The physics of displacement

What happens when a drop is in motion over a plate? It took decades to answer this apparently simple question [37,48,49]. Electro wetting on dielectrics (EWOD) technology [50], which now is an important component of microfluidics, added a technological facet to the subject. A droplet moving over a surface seems a simple phenomenon. Actually, it uncovers considerable complexity.

A now well–accepted representation of the structure of the region around the contact line (the 'foot') of an advancing droplet is shown in Fig. 4.41.

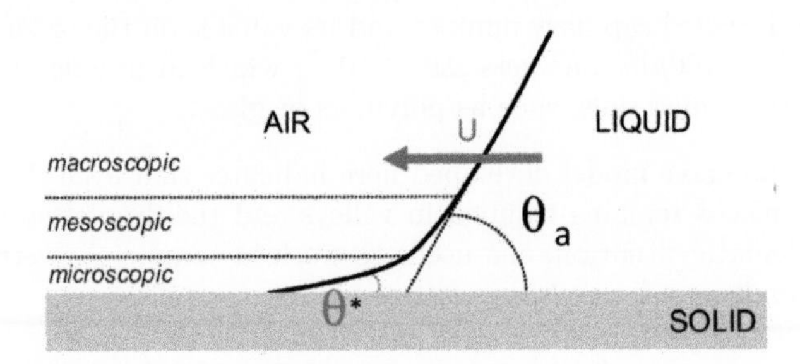

Fig. 4.41: The three regions appearing when a droplet advances over a smooth flat plate: microscopic, mesoscopic, and far region. In the microscopic region, roughness is represented by an apparent acute contact angle $\theta*$ smaller than the equilibrium angle θ_E (see above).

Close to the contact line, the interface for smooth wetting surfaces can be divided into three regions:

Microscopic region : concerns scales comparable to the microscopic, or nanometric roughness. As seen above, the averaged contact angle is smaller than the equilibrium angle.

Mesoscopic region : intermediate region, of size $Ca \sim R/\theta_E$ (where R is the system size (for instance, the droplet radius) and θ_E the Young angle), between the microscopic and the far region.

Far region : region whose size is comparable to the system size.

With droplets of centimetric sizes, the three regions embrace *seven* decades of scale. At the smallest (nanometric) level, the surface is rough and an apparent contact angle, $\theta*$, given by the Wentzel model, develops. This angle is different from the equilibrium contact angle θ_E, not shown on the figure. The far region is characterized by an apparent angle θ_a which depends on the droplet speed. This angle concerns submillimetric, millimetric, or centimetric dimensions, and is thus accessible to the measurement. The relation between the two angles is given by the Cox–Voinoi law [51, 52]. The demonstration uses matched asymptotic expansions, balancing viscous and capillary forces in each of the three regions, and matching them. It is well described in several places [4, 37, 48, 49], and lies beyond the scope of the book:

$$\theta_a^3 = \theta *^3 + 9Ca \; log\frac{\alpha l_0}{l_i} \tag{4.22}$$

in which $Ca = \frac{U\mu}{\gamma}$ (in which U is the droplet speed, and μ is the fluid viscosity), α is a geometry dependent factor, l_0 a length characterizing the droplet size (on the order of R, its radius), and l_i is a microscopic scale. The formula tells us that the faster the droplet, the more obtuse the apparent contact angle. In terms of order of magnitude, for a water droplet 100 μm of radius, moving at 100 μm/s, which is quite fast, the difference between the static and the apparent dynamical contact angles do not exceed a few degrees. The size of the mesoscopic region, roughly estimated to be $\sim Ca\ R/\theta_E$, is, in practice, on the order of tens of nanometres.

The structure was challenging to establish, because of considerable experimental difficulties (the regions to analyse are extremely small and roughness is difficult to quantify and control) along with the limitation of numerical simulations, which cannot fully address the multi–scale nature of the problem. Now, the structure is well–accepted. One conceptual difficulty that the community had to face, and which is reflected by the existence of a phenomenological microscopic length l_i in the Cox–Voinoi formula, is the slippage of the contact line. The droplet advances as a caterpillar [53], thus without slippage, but the contact line must slip. Because of the logarithmic dependence in Eq. 4.22, different phenomenological estimates of slip lengths work equally well, as far as the global movement of the droplet is concerned [4, 48, 49]. We will not discuss this point any further as it stands well beyond the scope of this book.

4.5.2 Displacing droplets by patterning surfaces

One way to move a liquid drop placed on a plate is to pattern the wettability of the surface. Figure 4.42 shows an example where, on the left–hand side, the surface is hydrophilic, and, on the right–hand side, it is hydrophobic. This pattern can be obtained on plastic or PDMS surfaces, for instance, by exposing them to deep UV through a quartz mask, before closing the device. We will discuss the delicate question of surface patterning later, in Chapter 7.

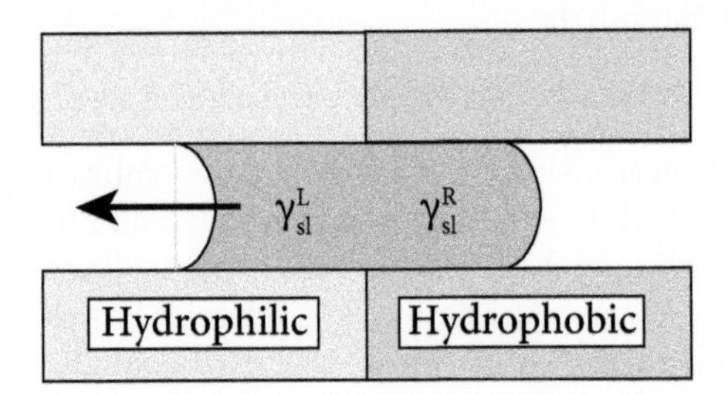

Fig. 4.42: Difference between the contact angles of two sides of a drop produced by a gradient in the surface chemistry. Due to Laplace pressure, the drop will move to the side where the contact angle is smaller.

As shown in Fig. 4.42, Young's equation imposes different contact angles at the left and the right of the drop, creating a pressure difference inside the drop, from the right to the left. The drop will advance towards the hydrophilic region, which it 'likes' most. At some point in its journey, the contact line, initially located on the hydrophobic part, touches the hydrophilic zone. Then, the meniscus inverts its curvature, the pressure difference disappears, and the drop stops. To obtain a continuous movement, one must renew permanently the surface energies of the substrate. This cannot be done mechanically, nor chemically, but thanks to the phenomenon of electrowetting, coupled to electro wetting on dielectrics (EWOD) technology, that we will discuss in Chap. 6, the task is feasible.

4.5.3 The important role of pinning and dewetting in droplet microfluidics

A microfluidic experiment carried out in 2003 illustrates two phenomena discussed above: pinning of the contact lines and dewetting [30]. Experiments involved injecting tetradecane and water into a hydro–focusing junction made in glass (see Chapter 3), with tetradecane in the centre and water on the sides. Two cases are considered: with surfactant (SPAN 80 above CMC) and without surfactant. With surfactant, water dewets the wall (see Fig. 4.36, right). In this case, well–defined morphologies are obtained (see Fig.. 4.43 (left)): isolated drops, drop pairs, necklaces. Note that the system produces water droplets, not oil droplets (as would be expected from Bancroft's rule). This is because channel wall wettability controls the type of droplet that is obtained, through the detergence effect.

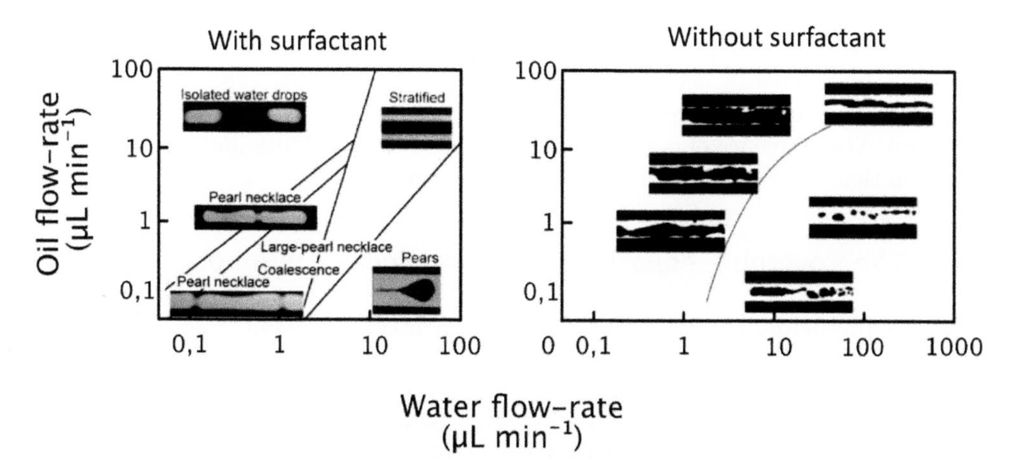

Fig. 4.43: Different regimes observed in a microcanal 20 μm deep and 200 μm wide, for tetradecane–water systems, with a surfactant (Span 80) whose concentration is over the CMC [30].

In the absence of surfactant, wetting is partial (see Fig.4.36). For the same flowrates as in the preceding case, different morphologies and dynamical behaviours are obtained (Fig. 4.43). Featureless water blobs, and sticking intermittently to the walls, are observed. The flow produced is therefore meandering, and its structure is erratic and uncontrolled. The reason is that in this case, as mentioned above, contact lines develop and, from time to time, are pinned by the roughness of the glass walls, however small it is (on the order of 10 nm). Varying flowrates by four orders of magnitude does not modify the situation in any respect.

These observations can be illustrated in Fig. 4.44. Fig 4.44 (A) shows a stable situation: a water droplet circulates in a microchannel filled with oil, in which walls are hydrophobic. Since oil completely wets the walls, the oil film, which separate the droplet from the walls is stable and the droplet can steadily move downstream.

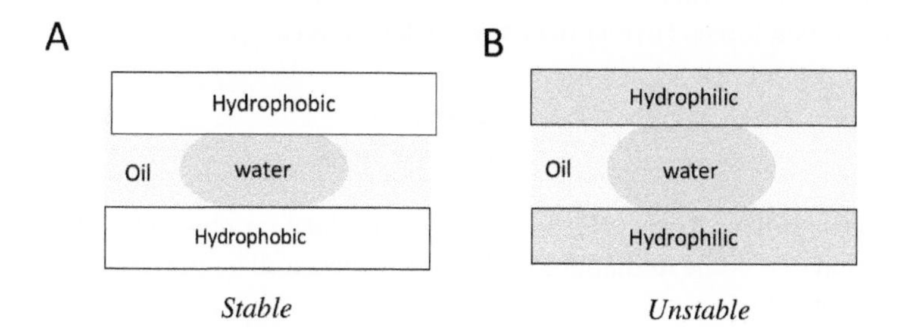

Fig. 4.44: (A) Water droplet in a channel, with hydrophobic walls, i.e. fully wetting with respect to oil. The system is stable.(B) The same system, but with hydrophilic walls, i.e. fully wetting with respect to water. The system is unstable

Fig 4.44 (B) shows the opposite situation, in which we assume water droplets in oil, but the microchannel walls are hydrophilic. In this case, the system is unstable. If the droplet touches the wall, water will spread over it and destroy its structure. After some time, phase inversion will take place, in which oil droplets, instead of water droplets, will form. This sequence of events illustrates that wall wettability imposes the type of emulsion that can be produced in microfluidic channels.

4.6 The governing equations and the capillary number

4.6.1 Dimensional reasoning and physical interpretation

In Chapter 3, we showed that, at small Reynolds numbers, the flow pattern only depends on the geometry. We could arbitrarily change the flow rate without changing the flow pattern. Now, what happens when capillarity comes into play?

Let us consider a flow in a system including two immiscible fluids 1 and 2, characterized by their viscosities μ_1 and μ_2, and the interfacial tension γ. The geometry is characterized by a single scale l and it is driven by a speed U. In this problem, inertia is negligible (because the Reynolds number is small), and therefore the fluid density ρ does not play any role. On applying the Π theorem (with local velocity $\mathbf{u}$ and position $\mathbf{x}$, the two viscosities, the characteristic length and the surface tension, we have six physical variables and three units. Therefore, three dimensionless numbers characterize the flow. Consequently, local velocity $\mathbf{u}$ can be written as follows:

$$\mathbf{u}(\mathbf{x}) = Uf(\frac{\mathbf{x}}{l}, \lambda, Ca) \tag{4.23}$$

in which $\lambda = \frac{\mu_1}{\mu_2}$ is the viscosity ratio and Ca is the capillary number, defined by:

$$Ca = \frac{\mu_1 U}{\gamma} \tag{4.24}$$

When the system is characterized by a single scale, the physical meaning of Ca is the ratio of the viscous forces ($\mu_1 Ul$) over the surface forces (γl).

$$Ca \sim \frac{\text{Viscous forces}}{\text{Capillary forces}} \tag{4.25}$$

The larger the capillary forces, the smaller the capillary number. Despite this disturbing construction, Ca is the undisputed reference dimensionless parameter for describing capillary phenomena.

The equations that govern capillary flows are the Stokes equations for each phase, supplemented by boundary conditions. One is kinematical, prescribing where the interface goes. The others include velocity and tangential stress continuity, along with pressure discontinuity, caused by the Laplace jump. This system becomes nonlinear, and, as a consequence, all the virtues' of Stokes equations, established with rigid boundaries, vanish. The stability of the solutions, their unicity, the flow reversibility are gone. Instabilities are now allowed, along with irreversibility. The presence of free interfaces changes profoundly the dynamics of the system.

4.6.2 Application to droplet break–up

The preceding discussion is well–illustrated by the problem of droplet break–up in a shear flow at low Reynolds numbers [54]. Figure 4.45 shows the situation.

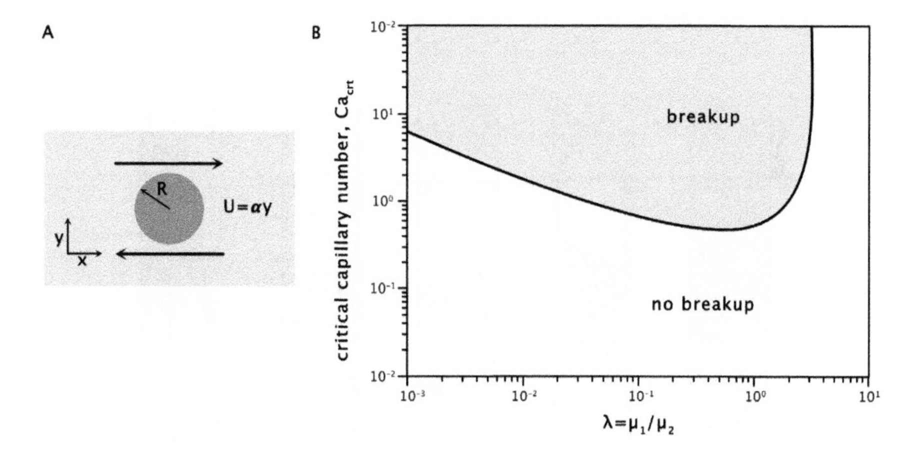

Fig. 4.45: Grace curve: Ca vs λ diagram, in which $Ca = \frac{\mu_1 \alpha R}{\gamma}$, where R is the initial droplet radius and $\lambda = \frac{\mu_1}{\mu_2}$ the viscosity ratio, with μ_1 that of the dispersed phase (e.g. gas if there are bubbles), and μ_2 that of the continuous phase (for instance air for rain).

Under the action of the shear, the droplet rotates and elongates. As noted above, the process depends only on Ca and λ, with λ the ratio of the viscosity of the dispersed phase over the continuous phase. Fig. 4.45 is a Ca vs λ diagram, in which $Ca = \frac{\mu_1 \alpha R}{\gamma}$, where R is the initial droplet radius. Below a certain curve called the Grace curve [56], the system reaches a steady state, for which the droplet undergoes steady elongation. Above the Grace curve, the droplet increasingly elongates and eventually breaks up. Physically, in this regime, capillary forces can no longer preserve the integrity of the droplet, against the viscous forces which work at tearing it off. It is easier to break–up a droplet when the two fluids have comparable viscosities. Above a critical $\lambda \approx 3.5$, no break–up is possible. This is the case of rain drops. Their λ is on the order of 10^3. However large their speed may be, rain drop do not break up.

The phenomena are many and it is remarkable that two numbers suffice to describe them, throughout a considerable range of scale, from R (let us say 1 cm) to the nanometric scale (filament thicknesses preceding break–up[20]). We thus have here an illustration of the capacity of the capillary number to characterize the wealth of phenomena. Droplet break–up moreover illustrates three properties of the governing equations: nonlinear behaviour (droplet deformation varies nonlinearly with the shear), presence of instabilities (break–up) and irreversibility (coalescence and break–up are not symmetric).

4.6.3 What happens at large capillary numbers ?

In microfluidics, as we saw, capillary numbers are small. Typical values range in the domain 10^{-6} to 10^{-4}. This result was anticipated in Chapter 2, when we pointed out the prevalence of capillary forces in small systems.

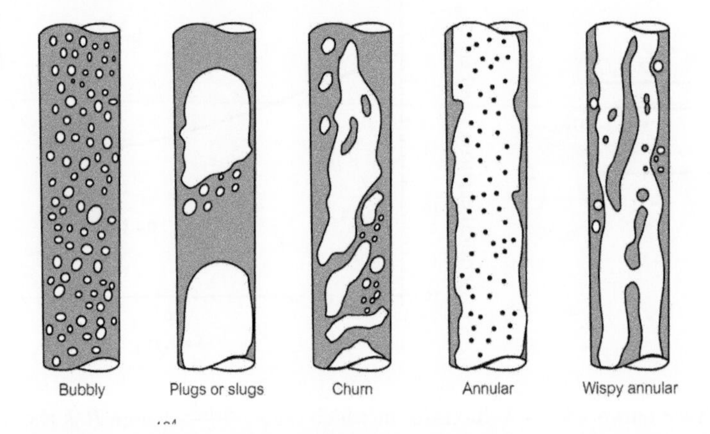

Fig. 4.46: Different flow regimes. The vertical pipe diameter is 0.20 m and allows speed of several metres per seconds. The main regimes are, from left to right: bubbly, plugs or slugs, churn, annular and wispy annular. Many subregimes exist.

Imposing small Reynolds numbers and, in the meantime, large capillary numbers is difficult. For the capillary number to be large, U must be larger than a speed $U_c \sim \gamma/\mu$. With typical surface tensions (say, 10 mN/m), and ordinary fluids($\mu \approx 10^{--3}$ Pa.s), one has $U_c \sim 10$ m/s. At such speeds, in most situations of practical interest, the Reynolds number is large, and the flow is turbulent. Systems with extremely small surface tensions exist, but they are used in very specific contexts.

Here, we consider gas–water flows at high capillary and large Reynolds numbers. Such a situation is often encountered in the oil and nuclear industries [58]. We use it to

[20]The statement is reinforced by an additional complexity of the phenomenon, revealed by studes carried out in the 1990s. To achieve break–up, one needs to break Van der Waals bonds inside the fluid. How can this be done The response is through a cascade of instabilities, taking place in the filaments that join, in the last moments preceding break–up, the daughter droplets together [57].

illustrate the idea that at large capillary numbers, interfaces are strongly deformed by the flow. Fig. 4.46 thus shows different regimes for water/air flows circulating at high Reynolds and capillary numbers. In this case, Reynolds numbers and capillary numbers are on the order of 10^5 and 1–10, respectively. One can see that the dynamics of the system is extremely rich and the interface shapes between water and gas adopt a considerable variety of time dependent shapes, undergoing turbulent behaviour. In microfluidics, these numbers are small, there is no hydrodynamic turbulence, and the droplets that form, owing to the prevalence of capillarity, are spherical or (due to the confinement) oblong, unless some specific process forces them to depart from these two basic morphologies.

4.7 The Landau–Levich and Bretherton films

4.7.1 The Landau–Levich film

When a plate (e.g. glass) is pulled upwards, from a bath filled with a completely wetting fluid (e.g. oil), the fluid is entrained by the solid and gives rise to a film of constant thickness that covers its surface. In 1942, L. Landau and B. Levich provided a rigorous explanation of the phenomenon [59], giving their names to the film.[21]The situation is sketched in Fig. 4.47.

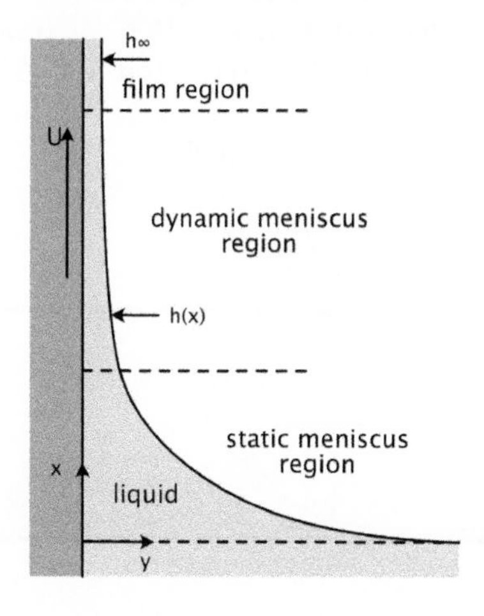

4.47: Landau levich film, with the definition of the four regions that must be introduced to solve the flow equations

The flow can be divided into four zones:

- the film of constant thickness h_∞, where the interface is flat
- the dynamical meniscus, in which the interface starts to curves in
- the static meniscus, whose shape is identical to the one in the absence of flow
- the horizontal free surface

[21]The name of Derjaguin [60], who made the calculation in 1943, is also often linked to the discovery.

Of particular interest is the film thickness h_∞ developing over the plate. Its expression is:

$$h_\infty \approx 0.95\, l_c\, Ca^{2/3} \tag{4.26}$$

where $Ca = \frac{\mu U}{\gamma}$ and l_c is the capillary length, introduced in this Chapter. Typical thicknesses of the deposited layer is a few μm. To get a sense of the orders of magnitude, suppose you swim in a lake. If you decide to leave the lake leisurely and walk, you will have to support a water film of 100 μm thick over your body, increasing your weight by 100 g, until it dries out. If you rush, the weight will be larger.

Formula 4.26 thus shows that the film thickness is controlled by the speed and the fluid properties. We may infer that an entity of micrometric size (the film) can be obtained, in a simple manner, by utilizing large objects (plate, container), while achieving excellent control. An important utilization of this capacity, which we will describe in Chapter 7, is spin coating.

4.7.2 The Bretherton film

We will not analyse the way in the which Landau–Levich film equations are resolved. The reader can refer to Refs. [3, 59, 60], for detailed analysis (Ref. [3] provides an excellent tutorial). Instead, we pay more attention to the Bretherton film [62], which is more relevant to microfluidics. Bretherton extended the Landau–Levich analysis to the case of a bubble moving in a tube at a constant speed[22] [62].

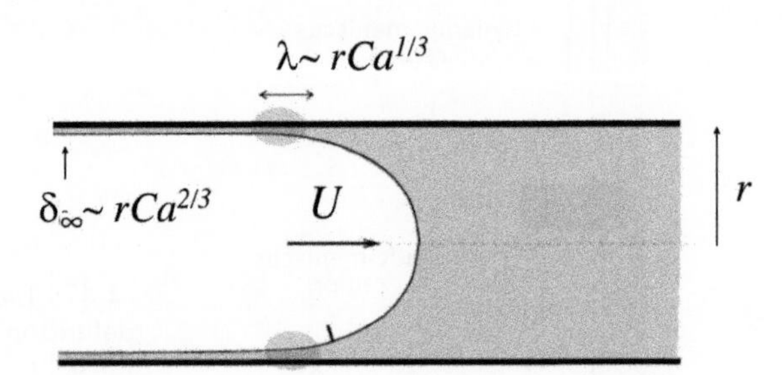

4.48: Bretherton film deposite by a bubble moving at constant speed in a tube

The constant thickness δ_∞ of the film, far from the finger nose (i.e. the interfacial region, approximately circular, around the symmetry axis of the channel), can be found by orders of magnitude arguments [63, 64]. Similarly as for the Landau Levich film, several regions can be outlined: the nose, a transition region, of length λ and thickness δ, and the film, far from the nose, of constant thickness δ_∞ (see Fig. 4.48). First, the curvature of the transition region must match that of the nose, implying:

[22] In fact, Bretherton neglected the curvature of the tube, and thus carried out the calculation as if the film developed on a plane.

$$\frac{\delta}{\lambda^2} \sim \frac{1}{r}$$

and therefore $\lambda \sim \sqrt{r\delta}$. On the other hand, as the bubble moves downstream, a flow develops in the transition region to keep the film supplied with fluid. In this region, on employing the Stokes equation, we have, in terms of order of magnitude:

$$\frac{\mu U}{\delta^2} \sim \frac{1}{\lambda}\frac{\gamma}{r}$$

By combining the two equations, we end up with:

$$\delta \sim rCa^{2/3} \ ; \ \lambda \sim rCa^{1/3}$$

A detailed analysis of the governing equations led Bretherton to the following result:

$$\delta_\infty \approx 0.66 \ r \ Ca^{2/3}$$

We may conclude that the micrometric film thickness can be controlled accurately. An important point is the asymmetry between the front and the rear of the bubble, which represents another example of irreversibility of the flow equations. This asymmetry was also analysed by Bretherton [62] (see Fig. 4.49). The physical origin is that, at the front, the capillary flow in the transition region develops against the flow, reducing the curvature, while the situation is opposite at the rear. We thus have a larger local curvature in the transition region in the rear than in the front. This effect translates into a larger nose radius at the rear than at the front. The dynamical consequence of this asymmetry is the existence of a pressure drop ΔP along the channel, which will play an important role later. The amplitude of ΔP is not straightforward to calculate. The result found by Bretherton, in terms of order of magnitude, is:

$$\Delta P \sim \frac{\gamma}{r} Ca^{2/3}.$$

4.7.3 An application of the Landau–Levich film: dip coating technology

An application of Landau–Levich film is dip coating technology.

An object is immersed in a formulated bath, then lifted at constant speed, and eventually dried out (Fig. 4.50). This technology is used in the industry, for coating glass

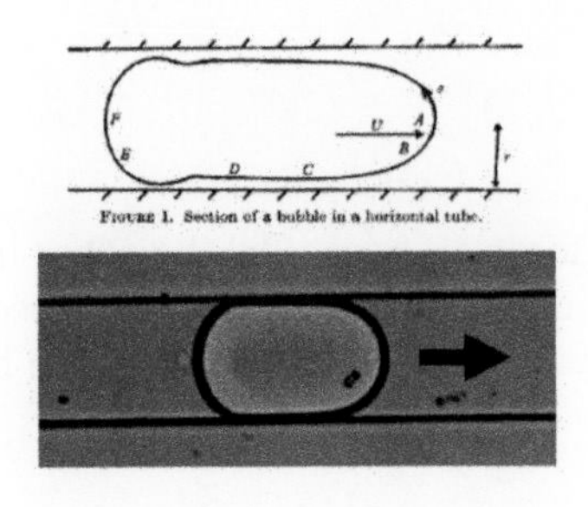

4.49: Bretherton bubble asymmetry: because of the film, the nose of the bubble is different from the rear. This difference, calculated by Bretherton in a tube, also exists in a microchannel, as shown in the bottom picture, in which a water droplet circulates in a 100 μm square microchannel at 1 mm/s. (Courtesy of O.Caen.)

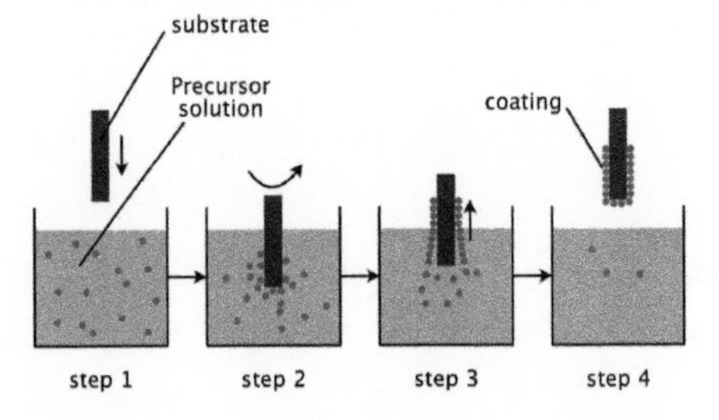

4.50: Dip coating technology. An object is immersed in a solution containing nanoparticles, and pulled upwards at constant speed, forming a Landau–Levich film. As the film dries, the particle agglomerates and form a solid coating. Different formulations can be used (monomers, polymers with cross–linkers, etc.)

fibres for instance, but it also belongs to the catalogue of microfluidic technologies. Dip coating is helpful, for instance, for depositing micrometric gel structures onto surfaces.

4.8 The Rayleigh–Plateau instability

4.8.1 Physical mechanism

A column of liquid ejected from a nozzle spontaneously creates droplets. This happens when we turn on a tap: the jet spontaneously breaks up into droplets (see Fig. 4.51).

The same break–up instability develops in microfluidics. An example is shown in Fig. 6.50 [68].

In this case, a droplet is transported at a T–junction. Along the side and the main channels, the liquids are pumped. The droplet splits in two parts, one advancing in a first channel and the other in the second. At some point, the droplet builds up a bridge which thins out and eventually breaks up, giving birth to two droplets.

All this is due to an ubiquitous instability: the Rayleigh–Plateau instability [65, 66]. Its physical origin can be explained with a reasoning based, as usual, on interfacial

4.51: Tap producing drops. (Courtesy of L. Bocquet, T. Maimbourg.)

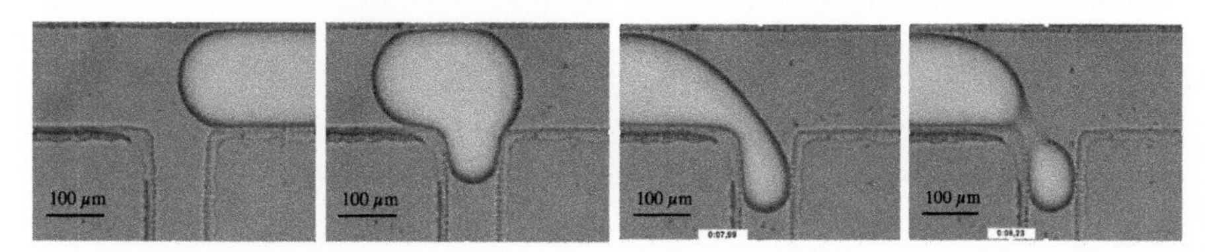

Fig. 4.52: break–up of a droplet at a T–junction. The droplet give rise to a daughter droplet, which will further move downstream in the side channel [68].(Courtesy of L. Menetrier-Deremble.)

energy. Let us consider the interfacial excess energy E of a column of fluid of volume V. One has:

$$E_C = 2\pi R L \gamma = \frac{2\pi V \gamma}{R}$$

We now consider the excess energy associated with a collection of N drops of radius R':

$$E = 4\pi R'^2 N \gamma = \frac{3\gamma V}{R'}$$

We thus see that the droplets with radii:

$$R' \geq \frac{3}{2\pi} R$$

are less energetic than the column. The column thus 'prefers' to form a string of drops.

We now give more detail on this instability, in 3D and 2D.

Rayleigh–Plateau instability in 3D. The calculation of the instability is traditionally presented in the limit where the fluid is perfect, i.e. with zero viscosity. Although microfluidics is dominated by viscosity, the inviscid calculation is interesting because it allows us to figure out the mechanism of instability. Details can be found in the original papers and in reviews such as Refs. [54, 67, 115]. Calculation shows that the column is destabilized by infinitesimal perturbations of wavelength λ, subjected to the following condition (see Fig. 4.53): :

$$\lambda \geq 2\pi R$$

Why is it the case? In the presence of a perturbation (see Fig. 4.53), the radii of curvature of the column, in the radial and longitudinal directions, R_1 and R_2, respectively, vary, in opposite manners. As R_1 increases, R_2 decreases, and vice versa. For wavelengths above $\geq 2\pi R$, the Laplace pressure field induced by the disturbance generates a flow that amplifies the perturbation. Crests and dips grow. This is the mechanism. It provides a dynamical pathway for decreasing the interfacial energy. Now, among all the unstable wavelength, which grows the fastest? The most unstable wavelength is $\lambda_c = 2\pi R\sqrt{2}$. The radius R' of the droplets produced in this manner is given by the relation:

$$\frac{R'}{R} = \left(\frac{3\pi}{\sqrt{2}}\right)^{1/3} \quad \text{implying} \quad R' \approx 1.85R$$

The same calculation provides the length L of the column above which drops will appear:

$$L = U\left(\frac{8\rho R^3}{\gamma}\right)^{1/2}$$

where U is the jet speed. This length can also be found by the Π theorem, considering that it is only function of ρ, U, and R. For inkjet printers, speeds are large, and the distance of formation of droplets may be problematic. In such situations, we typically have $R \approx 25\mu$m (for the jet radius), $U = 10$m/s, so that droplets are about 50 μm in radius. The length of formation of the droplets is ≈ 3 mm. The formula indicates that, increasing the printing speed may lead the ink be splashed too early onto the paper, before a droplet can be formed, which is not desirable.

Now, what happens when viscosity comes into play ? Rayleigh [65] also resolved the viscous case in his seminal paper. As expected, viscosity slows down the instability in a manner that depends on a dimensionless number, the Ohnesorge number, introduced before. In microfluidics, viscosity and confinement come into play and modify the conditions of development of the disturbance. It may be suggested, although in a very naive manner, that in such a constrained geometry, the time of droplet formation τ_D depends on μ, the viscosity, the channel transverse dimension l, and γ. Using the Π theorem, one obtains:

$$\tau_D \sim \frac{\mu R}{\gamma}$$

Guillot et al [69] solved the problem analytically, in the case of the co–flow, and in a cylindrical geometry. We will come back to this calculation later. It led, for the so–called dripping regime, to an estimate for τ_D consistent with the preceding reasoning. In reality, the situation is more complicated. The flow may entrain the perturbation and inhibit the instability; both of these actions increase the length of appearance of the droplet, giving rise to the 'jetting' regime, or, taking the words currently used in hydrodynamics, convective instability'. We will return to these phenomena later on.

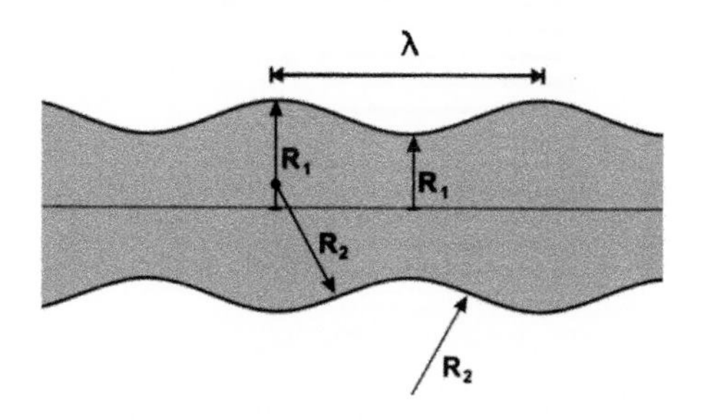

4.53: Perturbed interface giving rise, in 3D, i.e. for a fluid column, to the Rayleigh–Plateau instability, the source of droplets. In 2D, the same perturbation does not amplify and consequently does not lead to the formation of a droplet.

No Rayleigh–Plateau instability in shallow microchannels. In 2D, the same calculation shows that a straight band (the equivalents of a jet) is stable. The result can be understood from Fig. 4.53. Surface tension will always tend to kill small perturbations, because they increase the interfacial area. Consequently, the interface will return to rest. This does not imply that the band is the least energetic structure. For instance, a disk (the equivalent of a drop) is less energetic than a band. The problem is that, if we restrict ourselves to small perturbations, there is no dynamical pathway that would reshape a band into a disk. The band is in a metastable state. This result has important implications for microfluidics. It implies that, in shallow channels, which achieve conditions close to two–dimensionality, immiscible fluids flow side by side in a stable manner [55].

4.8.2 The break–up of nanojets

What happens when the jet becomes nanometric? Numerical simulations based on molecular dynamics [70], discussed by [71] have addressed this question (see Fig. 4.54).

The simulations show that at the nanometre scale, instead of capillarity, thermal noise is responsible for the breaking up of the jet. To quantify this, a fundamental scale of the problem, the so–called thermal–capillary' scale l_{CT}, for which the orders of magnitude of thermal energy (given by kT), and capillary energy (given by γl_{TC}^2) are comparable, can be introduced. The scale is defined by the following relation:

$$l_{\mathrm{TC}} = \sqrt{\frac{kT}{\gamma}}$$

For ordinary fluids (oil, water, etc.) l_{TC} is a few nanometres. Therefore, for jet diameters much larger than this length, Rayleigh instability dominates, while for smaller jets, on the order of tens of nanometres or less, thermal noise prevails. Incidentally, it

4.54: Nanojet emitted by a nozzle of 6 nm. (Reprinted from Ref. [70] with permission from American Association for the Advancement of Science. Copyright 2022.)

is interesting to note that, from a flow control perspective, microfluidics operates in an optimal range of scales: small enough to suppress turbulence, and large enough to smooth out the effect of thermal noise.

4.9 Washburn law and paper microfluidics

4.9.1 Washburn law

A meniscus is pushed in a capillary, as depicted in Fig. 4.55.

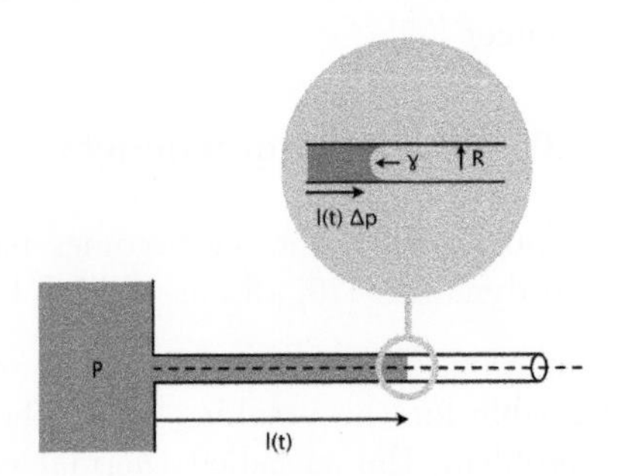

4.55: Washburn law. A fluid is pushed from a reservoir, and fills a capillary. The zoomed view shows a meniscus with a contact angle equal to zero.

The reservoir, on the left, is pressurized at pressure P above the atmospheric pressure P_0. The overpressure $\Delta P = P - P_0$, which drives the fluid in the capillary tube at a mean speed U, is equal to (see Chapter 3):

$$U = \frac{r^2}{8\mu}\frac{\Delta P}{l} \tag{4.27}$$

in which r is the tube radius, μ the fluid viscosity and $l(t)$ the position of the meniscus in the capillary. Noting that $U = \frac{dl}{dt}$, and assuming full wetting of the fluid with respect to the capillary (for instance, oil in glass), one obtains, after integration, the position of the meniscus in the capillary, at time t.

$$l(t) = \sqrt{\frac{\gamma r t}{2\mu}} \tag{4.28}$$

This is Washburn law [72][23] The square–root law comes from the fact that, as the meniscus advances, the hydrodynamic resistance of the system increases. In practice, with liquids like water, and capillary 20 μm in radius, it takes 10 s to push the fluid 8 cm.

The calculation can be generalized to the case where the contact angle between the fluid and the tube walls, is arbitrary, say θ. In this case, assimilating the meniscus to a spherical cap, the tube radius r must be replaced by $r/cos\theta$. With this generalization, we conclude that, for $\theta > 90°$, a situation we previously called non wetting', the meniscus recedes. Then filling a small tube is possible only if one applies a pressure, at the tube entry, working against capillarity. For tubes or in a straight microchannels, this is possible. However, in practice, in geometries with a minimum of complexity, working with non wetting fluids unavoidably leads to bubble trapping. As we saw in Chapter 3, these bubbles perturb the functioning of the system considerably. This is why it is highly preferable to work with wetting fluids, which spontaneously fill or tend to fill cavities and branched channels, owing to the action of capillarity.

4.9.2 Paper microfluidics

The domain of paper microfluidics was initiated by G. Whitesides [73, 74]. The first paper–based microfluidic device, developed by his group in 2007, simultaneously detected protein and glucose via colour–change reactions. In paper microfluidics, the substrate is paper, the motor is capillarity, and the underlying law that controls the flows is Washburn law. Because paper is inexpensive and widely available, it is possible to create inexpensive diagnostic devices, reducing health–related cost in general, along with providing testing accessibility to low–resources countries, which cannot afford glass or PDMS. The philosophy of paper microfluidics is similar to low–cost immunoassays, also made in paper. However, with paper microfluidics, it becomes possible to increase the complexity of the test (for instance, by integrating different elements on the same device) [75].

Fig. 4.56 shows a scanning electron micrograph (SEM) image of an enlarged part of a paper sheet.

[23] E. W. Washburn established his law date in 1921, but several researchers established it before: L. Lucas in 1918 and J. M .Bell, F. K. Cameron in 1906. Nonetheless, the name of the law was given to Washburn.

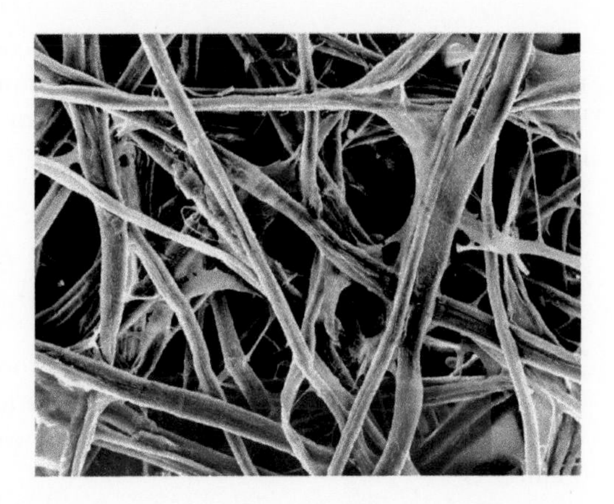

4.56: SEM of Indian print paper surface with cellulose fibres. Cellulose or cellulosic fibres are fibres structured from plant cellulose, a starch–like carbohydrate. They are created by dissolving natural materials such as cellulose or wood pulp, which are then regenerated by extrusion and precipitation. Magnification: x130 when shortest axis printed at 25 millimetres. (Science Photo Library.)

This type of paper comprises pores of 20–50 μm transverse dimensions, which is a typical range of values for paper, although filtration paper may have much smaller pores. It is possible to confine liquids inside boundaries by using wax technology [74]. The technology consists in depositing, with a wax printer, wax droplets on one side of the paper, heating up the sheet at 60°C or so, and melting the wax to force it to cross the paper thickness. In this manner, the wax, which is hydrophobic, forms a barrier allowing, in a way similarly as in microfluidic channels, aqueous solutions to be confined in a restricted space, within the paper. Fig. 4.57 shows a fluorescein solution confined in the tree, without leaks, so that the other parts of the sheet remain dry.

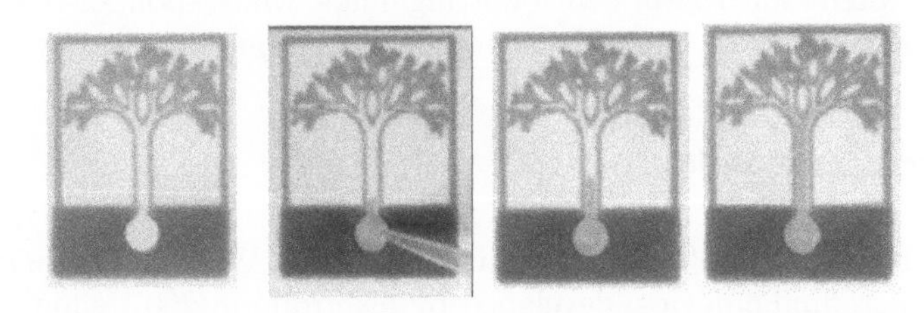

Fig. 4.57: Paper tree, patterned with green ink (grey on the figure), imbibed by a small water drop, coloured with fluorescein. (Courtesy of L. Magro (MMN, 2010)).

The experiment shows that fluoresceinspeed decreases in time, consistently with Washburn law.

A remarkable feature is that, despite the complexity of the paper matrix, which could have given birth to a wealth of different regimes[24], Washburn law applies over a broad range of time and conditions, semi–quantitatively. This was shown by P. Yager et al. (see Fig.4.58) [76].

In the experiments of Ref. [76], paper sheets are cut in channels of different geometries. Sheets are dipped in a coloured water bath. Imbibition starts over, and one observes

[24]This wealth of regimes appears in the drainage case, i.e. when the fluid is non–wetting with respect to the paper. We will not discuss them in this book.

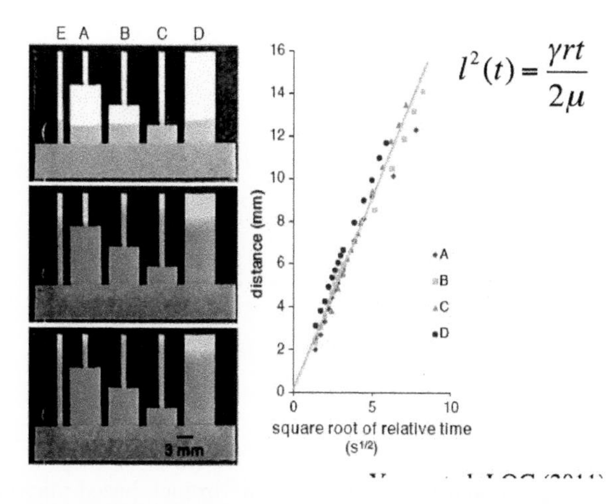

4.58: Series of experiments made in the same paper, thus with the same pore size r, cut in different forms, showing that in all cases, Washburn law applies with the same pre-factor, which turns out to be close to the theory (see Ref. [76]) for details on the relative time s definition. (Reprinted with permission from [76]. Copyright 2022. Springer Nature.)

that all water fronts penetrate the channels at the same speed, independently of their geometries. The observation confirms Washburn's law, which stipulates that $l(t)$ only depends on the pore size r, and not on the channel width. Figure 4.58 (right) also shows that the front dynamics agrees with Washburn law, at a semi–quantitative level, despite the fact that the paper internal structure, including pores of different sizes, geometries and spatial organizations, is incomparably more complex than a cylindrical capillary.

4.10 Production of microfluidic droplets and bubbles

4.10.1 Early experiments

The domain of droplet–based microfluidics took off at the beginning of the twenty–first century, after precursor experiments (See, for instance, [77, 78]) suggested the potential of the coupling between droplets and microfluidic technology. In particular its capacity to produce large numbers of droplets with unprecedented size control. Among the few pioneering contributions made at that time, two experiments turned out to be particularly inspiring (see Fig. 4.59).

In S. Quake group's experiment [79] oil and water meet at a T–junction. Wetting was perhaps not optimal (this could explain the strange shape of the interface at the junction). Nonetheless, the experiment showed that large quantities of monodisperse droplets could be produced under excellent control, either well separated from each other, or forming regular patterns.[25] Another series of experiments, performed in H. Stone's lab [81], revealed the same type of phenomena.

[25] A curious situation occurred here. The paper was titled 'Dynamic pattern formation in a vesicle–generating microfluidic device'. If we consider the usual definition of vesicles (i.e. a lipid layer forming a topologically closed structure in the same medium), there was not a single vesicle in the experiment. The author notes that another definition of the word vesicle', given in *Oxford English dictionary*, may be more compatible with the title given to the article: 'A minute bubble or spherule of liquid or vapour, esp. one of those composing a cloud or fog' [80].

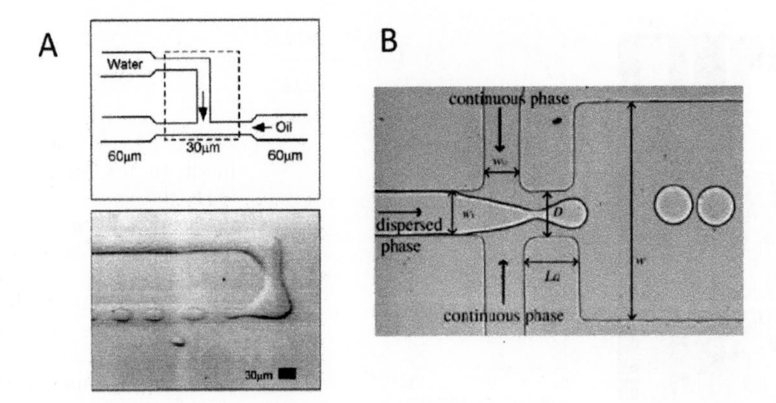

Fig. 4.59: Two early experiments that prompted the development of droplet–based microfluidics.(A) Production of droplets in a T-junction. (Reprinted figure with permission from [79]. Copyright (2022) by the American Physical Society.).(B) Production of droplets in a hydrofocussing geometry. (Reprinted figure with permission from [81]. Copyright (2022) by the AIP Publishing.)

The take–off of the domain came soon after, when it was realized that these droplets could be used as microreactors and material templates, opening new avenues for biology, chemistry, and material science.

4.10.2 General behaviours

Although emitter geometries are diverse, it is possible to figure out common behaviours, imposed by the hydrodynamical and geometrical constraints all systems are subjected to. We use Ref. [82] to classify the different cases, adopting the same vocabulary.

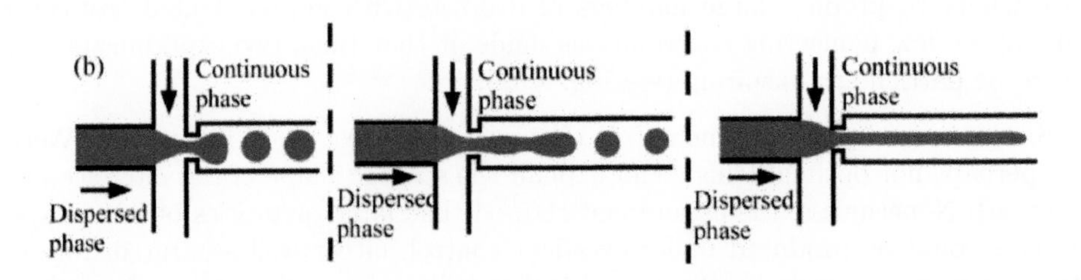

Fig. 4.60: The three basic regimes, omnipresent in microfluidic droplet emitters, as flow rates are raised up: from left to right, dripping, jetting, and stable co–flow. (Reprinted from Ref. [82] with permission from IOP Publishing.)

The three basic flow regimes are sketched in Fig. 4.60. They are shown for the case of flow focusing, but they hold for all emitters. As the flow rates of the phases are increased, we successively have:

1. *Dripping regimes.* The fluids form droplets where they meet. We shall stick to this definition, which is slightly different from Ref. [82]. It helps simplifying the descriptions of the various regimes holding in microfluidic emitters. In such conditions, an excellent control of the droplet characteristics can be reached.

2. *Jetting regimes.* As speeds increase, the disturbance giving rise to the droplet is transported downstream. The fluids flow side by side, up to a certain distance, largely uncontrolled, where a droplet forms. Although interesting from the viewpoint of the throughput, this regime is not used for producing droplets, because of the poor control of the droplet characteristics.

3. *Stable co–flow regimes.* At higher speeds, fluids flow side by side and no droplet forms in the device.

As discussed in Ref. [82], these behaviours are general because they all result from the competition between capillary and shear forces, in contexts where, because of the confinement, a single characteristic scale emerges. Crudely, at small speeds, capillary prevails, imposing a formation of the droplet close to the entry, giving rise to dripping regimes, while at large speeds, transport phenomena come into play, the instability is transported downstream, which leads to jetting regimes. This naive reasoning may help us to understand, intuitively, why the sequence of regimes presented above is so generic. In Ref. [82], general estimates for the cross–overs, expressed as capillary numbers, between different regimes, are suggested. In practice, geometry must be taken into account to estimate these numbers.

There can exist sub–regimes, within dripping or jetting regimes, or regimes related to them, giving rise to rich phase diagrams. However, these subregimes organize themselves around the general trends we presented. It is interesting to note that the notions of dripping and jetting regimes resonate with those of absolute and convective instabilities, which are well–documented in fluid dynamics [83], and for which, depending on the flow conditions, the perturbation giving rise to the instability (here, the droplets) grows locally (dripping) or is transported downstream (jetting).

4.10.3 The T–junction

The T–junction represents the most documented geometry. The principle is shown in Fig. 4.61.

Let us start with the dripping regime (also called shearing' or 'squeezing' regime, names we do not use herein). A fluid (the continuous phase) is driven along the main channel (horizontal on the figure), at a flow rate Q_c. The other, non miscible, fluid (the dispersed phase) is driven along the side–channel, at a flow rate Q_d. The two fluids meet at the junction and form droplets. The dripping regime has not received a first principle–based theoretical analysis, owing to the complexity of the hydrodynamics. Garstecki et al. [84] proposed a physical model. Schematically, the idea is to divide the process of formation of the droplet into three steps:

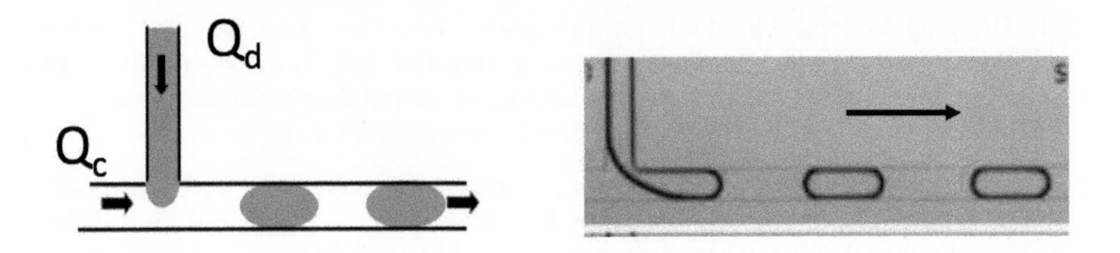

Fig. 4.61: Formation of droplets in T-junctions

1. penetration of a tongue into the main channel until obstruction;
2. transportation of the rear interface towards the right–hand corner of the junction; and
3. break–up at the corner, droplet formation, and return to step 1;

These steps are schematized in Fig 4.62 [84].

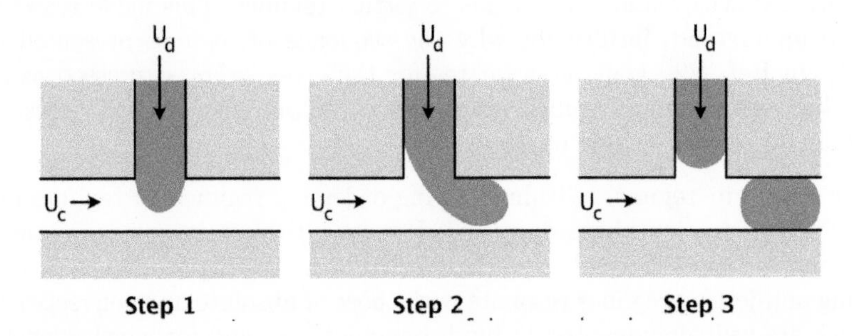

Fig. 4.62: The formation of the droplets can be divided into three steps: penetration, advection, and break–up [84].

Step 1 takes a time τ_1 equal to w/U_d to be completed, in which w is the side–channel width, and $U_d = \frac{Q_d}{w^2}$ is the mean speed of the dispersed phase, called, in the jargon of two–phase flow, superficial speed'. Step 2 takes a time τ_2 equal to w/U_c to be completed, where $U_c = \frac{Q_c}{w^2}$ is the mean speed of the continuous phase. The volume of fluid collected by the droplet, during the two steps is:

$$V = Q_d(\tau_1 + \tau_2) = w^3(1 + \frac{Q_d}{Q_c}) \tag{4.29}$$

This is Garstecki's relation. As usual, corrections are needed to adapt the simple model to the experiments. The measurements of the droplet lengths as a function of the flow rates, carried out for channels of square cross–sections [84], could be accurately fitted by the expression: $\frac{L_d}{w} \approx 1 + 2\frac{Q_d}{Q_c}$ which is exactly the structure of the model. The law holds for liquid and gases as well. Since the theory relies on transport only, there is no evolution of L_d with the capillary number. This is one of the most remarkable features. With rectangular cross–sections, the structure of the relation remains valid, but the constants must be changed.

As stated in the preceding section, at larger flow rates, the jetting regime takes place.[26] This is shown in Fig. 4.63 (see [82]).

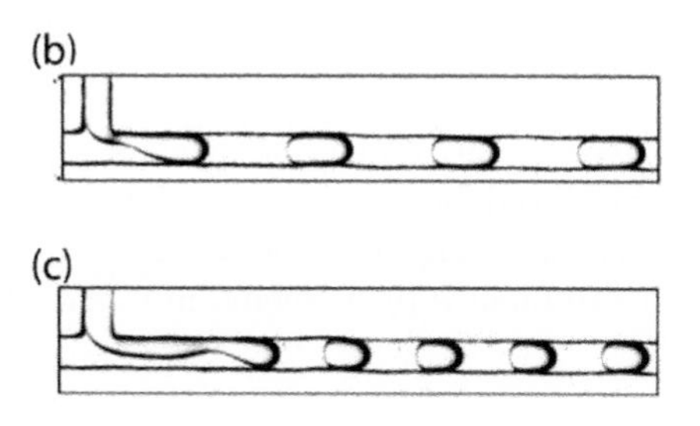

4.63: Jetting regime in T junction: liquids flow side by side and, at some distance from the junction, they form a droplet (Reprinted from Ref. [82] with permission from IOP Publishing.)

The disturbance giving rise to the droplet is transported downstream. At a certain distance, largely uncontrolled, a droplet forms. As explained above, this regime is not used for producing droplet, because of the poor control of the droplet formation mechanism. As speeds are increased further, the droplet forms at larger and larger distances, and we enter a stable co–flow regime.

Fig. 4.64: Formation of droplets in shallow T–junctions [89].

T–junction in shallow channels. It is tempting to consider shallow geometries, for which, as explained in Chapter 3, the equations take a much simpler form. This was done very recently [89]. As compared to square channels, a new regime, called 'tearing regime', dominates the system: droplets move along the lower channel side, being, in the meantime, torn off by the shear developed at the junction (see Fig.4.64)). Further downstream, the droplet is only weakly distorted. The length L_d of these tears follows a power law, in the form $\frac{L_d}{w} \approx 0.3 \pm 0.1 Ca^{--1/3}$, in which w is the channel width, and Ca is the capillary number, based on the droplet speed and the continuous phase viscosity. Theoretical analysis [89] provided justifications for this law.

[26] In the T–junction, our jetting regime is called 'dripping' in Ref. [82]. In the jetting regime, the droplet forms far from the entry. Therefore, it is preferable not to call it 'dripping'. Jetting is the appropriate word because the disturbance is transported downstream before forming a droplet. All this is somewhat confusing indeed.

Phase locking of V–junction emitters. V–junctions are variants to T–junctions. On Fig 4.65, two V–junctions are placed in parallel, producing alternatively red (shown in dark grey) and droplets with no dye (shown in white).

4.65: Two V–junctions placed in parallel, producing alternated droplets. (Courtesy of MMN team (ESPCI)).

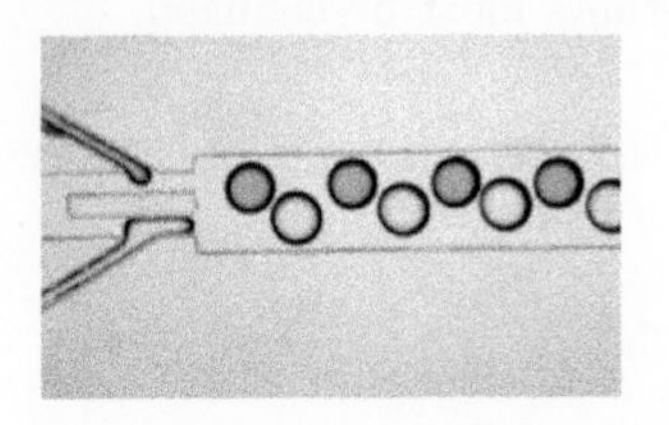

The emitters are phase–locked in an out–of–phase regime. When one droplet is produced in the upper V–junction, no droplet is produced below, and vice versa. The synchronization of the emitter is stable. It results from the hydrodynamic interactions building up between the two injectors, leading to resonance phenomena, in a manner shown, in other contexts, by Refs. [85, 86].

4.10.4 Flow focusing

Flow focusing geometry was discussed in Chapter 3 and its capacity to focalize fluid streams was underlined. Here, the same geometry is considered, but, this time, for producing droplets. Now, flow focusing represents the most currently used droplet emitter. The way in which it functions is shown in Fig. 4.66.

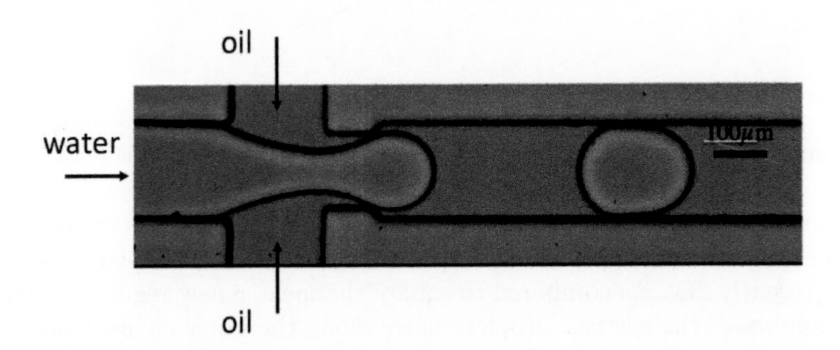

Fig. 4.66: Flow focusing geometry. Droplets of aqueous solutions are produced at 10 kHz in PFC oil, using Krytox surfactant for working in dewetting conditions. The device material is PDMS. (Courtesy of O.Caen.)

Oil is injected at the sides, and water at the centre. The organic phase focusses the aqueous stream in the restriction (see Fig.4.66). The elongated shape of the stream favours break–up. In such circumstances, droplets form. In the experiment of Fig. 4.66, the continuous phase is fluorinated oil (PFC). A surfactant (Krytox) is added to the external phase to force water droplets to dewet the channel walls and avoid pinning phenomena, as explained earlier.

As mentioned in the previous subsection, there exist several regimes. They are shown in Fig 4.67 [82]. In the dripping regime, break–up of the water stream occurs at the restriction. At higher flow rates, break–up occurs downstream, marking the advent of the jetting regime. At higher flow rates, water and oil flow side by side.

4.67: Regimes in the flow focusing geometry. From bottom to top: dripping, jetting, and stable co–flow (Reprinted figure with permission from [87]. Copyright (2022) by AIP Publishing.)

Despite its apparent simplicity, there is no general model expressing, for instance, the droplet sizes as a function of the flow rates, partly because of the many aspect ratios involved in the problem. Empirically, it is observed that flow focusing gives rise to substantial, even large, throughputs. Typical frequencies lie in the range 1– 50 kHz, i.e. more than one order of magnitude higher than the T–junction. This range is typical for droplets, and, for bubbles, it can reach several MHz (an exemple at 300 kHz, 5 μm bubbles, is shown in Fig. 4.68 [88]). Typical bubble sizes are 10– 200 μm and the volumes are typically a few picolitres.

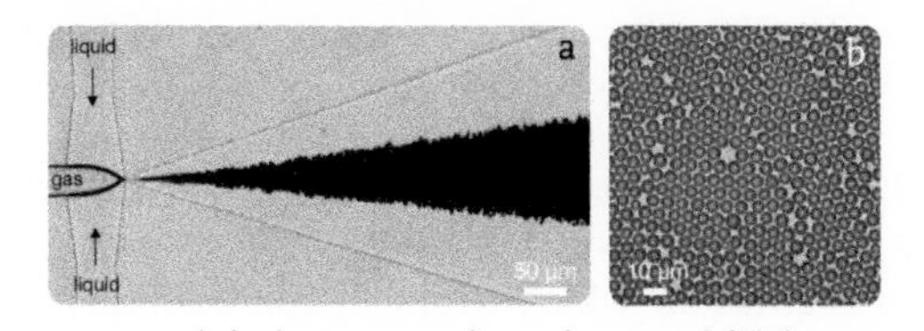

4.68: 5 μm in diameter bubbles produced at 300 kHz in a flow focusing device. (Left) Snapshot of the production.(Right) Harvested bubbles, showing excellent monodispersity. [88]. (Courtesy of T. Segers.)

4.10.5 Step emulsification

Using a geometrical rupture to produce droplets in microdevices was demonstrated by the turn of the century [90] and developed later on by a number of contributors [91,92]. The principle of this technique, called 'step emulsification', is shown in Fig. 4.69.

Two immiscible fluids are injected at the entries of a shallow flow focusing geometry and penetrate in a deeper channel after having passed an abrupt step. As mentioned previously, in shallow channels, immiscible fluids flow side by side without developing Rayleigh–Plateau instability. The pair of immiscible fluids thus arrives at the step as a stable co–flow. At the step, the interface is destabilized, and droplets form.

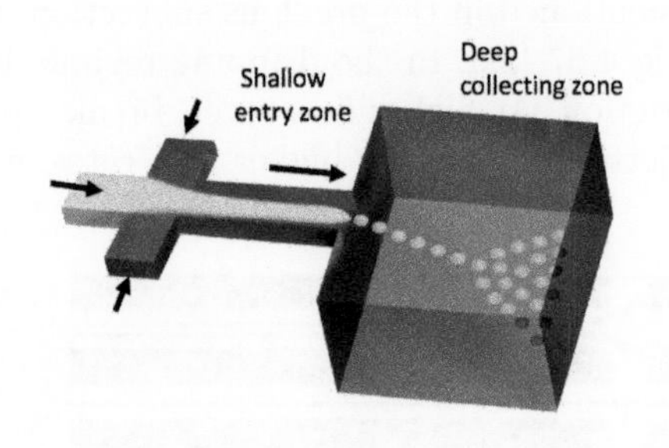

4.69: Geometry of step mulsification. Two immiscible fluids are injected in a shallow flow focusing geometry and penetrate in a deeper channel after having passed the step.

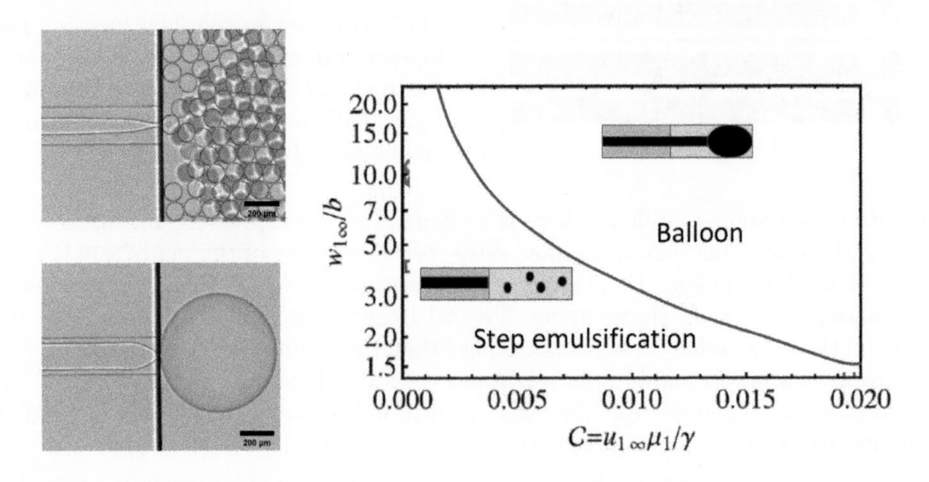

Fig. 4.70: (Left) Step emulsification (top) and balloon (bottom) regimes. (From X. Ge and Z. Li [94]) (Right) Phase diagram showing the domains of existence of the two regimes. [93]

The mechanism of droplet formation was figured out by Leshansky et al. [93]. Its description lies beyond the scope of this book. The analysis showed that two regimes exist (Fig. 4.70): the step emulsification regime, in which droplets (or bubbles) are produced at the step, and the balloon regime, where the incoming fluids fill the deeper channel, as if one would inflate a balloon in a cavity. The two regimes represent the equivalents of the dripping and jetting regimes, respectively. The following formula for the droplet diameter d, in the step emulsification regime, was established:

$$\frac{d}{h} \approx 3\left[1 + \frac{(1+k)}{w/h}\right]^{-1} \tag{4.30}$$

in which h is the inlet channel height, w its width, and $k = \mu_1 Q_1/\mu_1 Q_1$, where μ and Q stand for fluid viscosity and flow rates, and 1 and 2 for the dispersed and continuous phases, respectively.

In practice, d/h lies in the range 3 – 5. We will see, in Chapter 7, that there is not much difficulty microfabricating channel heights ranging from 500 nm to 200 μm.

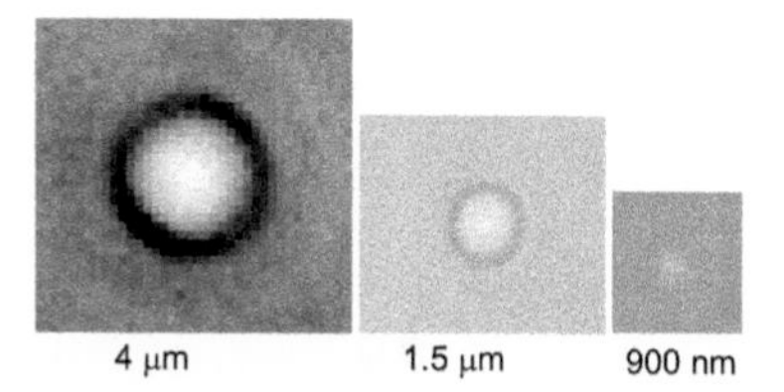

4.71: Droplets of 33, 2, and 0.4 femtolitres (i.e. 4, 1.5, and 0.9 μm in diameter respectively), obtained with the step emulsification emitter. The smallest size obtained in the experiment is 900 nm in diameter [91]. The world record is 400 nm [95].

As demonstrated in Fig 4.71 [91], forming submicrometric droplets, i.e. femtoliter droplets, is therefore feasible with the step emulsion emitter. In this category, the world record is, at the time of writing, 400 nm in diameter [95]. Step emulsification is moreover favourable for the obtention of large throughputs, for dynamical reasons (the large depth of the collecting channel, just after the step, associated to a small hydrodynamic resistance, facilitates the application of large flow rates). Frequencies in the range 10–100 kHz are common for this type of system. Double emulsions have also been obtained, with three different fluids [91], and, very recently, with extra–thin shells [96].

4.10.6 Producing droplets in glass capillaries

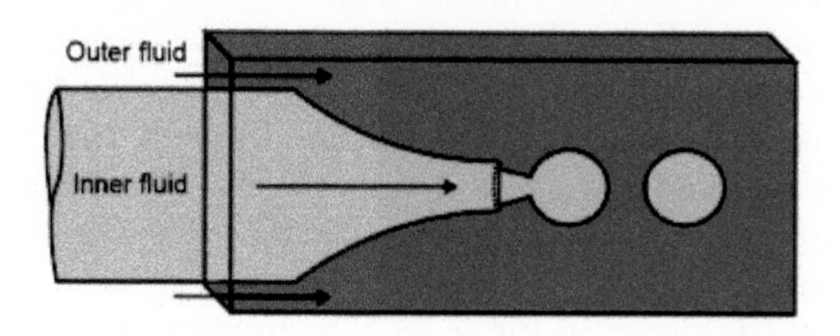

Fig. 4.72: Schematic of a co–flow microcapillary device for making droplets. Arrows indicate the flow direction of fluids and droplets.

An unconventional approach emerged in 2005 [96]. The device was microfluidic, in the sense that microfluidics manipulates fluids in small volumes, but, unlike the preceding devices, it did not use any microfabrication technique. The geometry is shown in Fig. 4.72. The internal cross–section of the large capillary tube is square. The design thus centers the smaller tube in the larger one, with a precision of 50 – 100 μm or so. Fluids are injected in the two capillaries, and meet at the end of the smaller one, forming a co–flow that destabilizes into a droplet stream. The technology is suitable for producing submillimetric droplets (i.e. above 50 μm), at moderate throughputs (10 –100 Hz). Even though manual skills are needed to operate it, glass capillary technology is attractive, because it does not require any clean room equipment. Important to note, since each tube has its own wettability, complex multiple emulsions can be produced much more easily than in microfabricated channels.

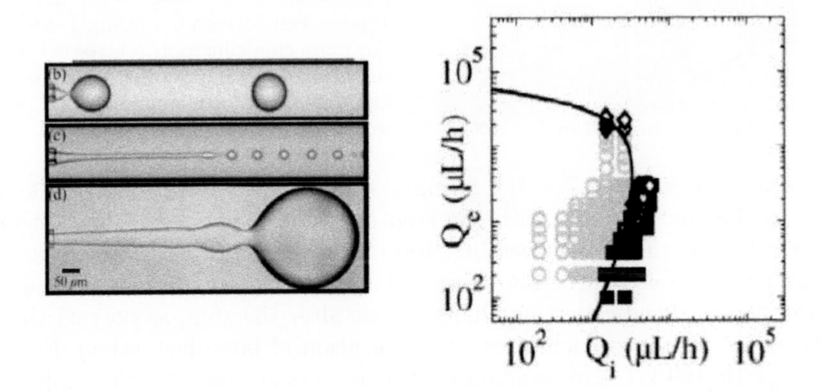

Fig. 4.73: A – Different flow regimes. From top to bottom: dripping, jettting and 'balloon' regimes (not described here); B – Comparison between theory [69] and experiment: grey points: dripping (experiment); black points: jetting (experiment); the full line is the theory (reprinted from [69] with permission of American Physical Society. Copyright 2022.

Different flow regimes exist, as shown in Fig. 4.73(A). From top to bottom, the dripping regime, jetting regime, and, at larger flow rates, a balloon regime. An analytical theory, based on a number of approximations, succeeded to describe the onset of droplet formation, along with the transition between the dripping and jetting regimes [69]. This is the only case where microfluidic droplet formation is described analytically. The formulas for the growth rate of the perturbations are complicate and not particularly inspiring, and it is out of the scope of this book to present them. A comparison between theory and experiment is shown in Fig. 4.73. The frontier, between the dripping and jetting regimes, takes the form of an inverted L, indicating that flow rate increases of any of the two liquids brings us outside the dripping regime. Agreement between theory and experiment is excellent. From a practical perspective, it is noteworthy that the most interesting regimes for droplet production (dripping regime) occupies a broad space, extending over roughly two orders of magnitudes in flow rates.

4.10.7 Other methods of droplet production

Since 2000, several methods of droplet production have been invented and there is not enough space here to review them. We mention here a technique, that exploits a gradient of confinement along the flow. We already discussed the effect of a gradient of confinement when discussing bird drinking. This gradient induces a pressure drop, which drives the droplet into the throat of the animal. In the droplet emitter we discuss here, the same physics is at work. The system is shown in Fig. 4.74 [99].

The dispersed phase is pushed through the inlet channel into a reservoir, whose top wall is inclined, with respect to the floor. A capillary pressure drop, generated by

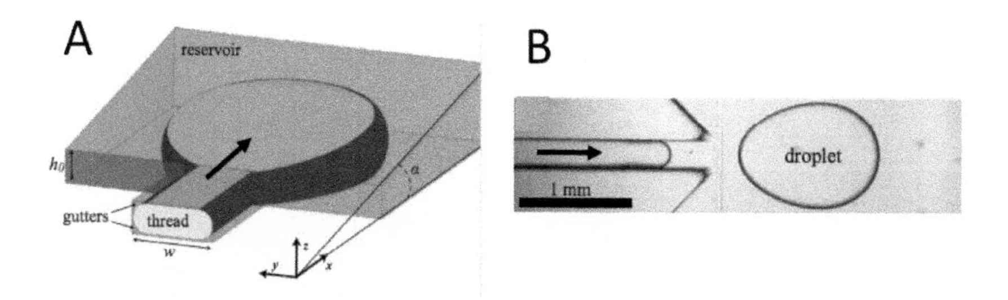

Fig. 4.74: (A) Sketch of the droplet emitter. The dispersed phase is pushed through the inlet channel into a reservoir, whose top wall is inclined with respect to the floor. (B) The capillary pressure drop, generated by the inclination, drives the flow and produces break–up. (Reprinted from [99] with permission of PNAS. Copyright 2022.)

the inclination, drives the dispersed phase into the continuous phase, and, for subtle reasons explained in Ref. [99], produces break–up, giving rise to a droplet. With this mechanism, no pressure source is needed to produce droplets.

4.11 Characteristics of microfluidic droplets and bubbles

4.11.1 Obstructing and non–obstructing droplets

The structures that can be produced, in typical microfluidic conditions, are droplets or bubbles in a liquid phase. We recall here a question of vocabulary: **droplets are liquid** and **bubbles are gaseous**, notions that are often mixed up by beginners. Bubbles and droplets are most often treated in the same manner. Nonetheless, in a limited number of situations, a distinction is made between droplets and bubbles that takes into account gas compressibility (negligible in droplets), or internal pressure drops (negligible in bubbles).

In typical microfluidic conditions, it is impossible to envisage droplets as the dispersed phase, and gas as the continuous one.The reason is related to wetting. A droplet confined in a microfluidic channel, moving in a gas flux, would be unstable. Should the droplet touch the wall, it would immediately stick to it. As explained earlier, in solid–liquid–gas systems, the droplet never dewets the wall. In such circumstances, after the droplet has touched the surface the contact line sticks to it. Entrained by the flow, the structure moves downstream, in an erratic and uncontrolled manner, in a way similar to that shown in Fig. 4.43. No stable droplet could be formed in such conditions. Indeed, at very large speeds, the droplet may detach and return to the flux. However, this situation is not representative of microfluidic conditions.

This being said, Fig. 4.75 shows the two types of droplets or bubbles that can be produced, in a stable manner, in microfluidic channels: they are either 'non–obstructing' or 'obstructing'. As just noted, they can be droplets or bubbles, but in all cases, they are immersed in a liquid. In the non–obstructing case, the droplet (or bubble) size is smaller than the channel transverse dimensions. In this case, the droplet is spherical; it travels downstream, at an uncontrolled distance from the walls and a poorly controlled speed, because its velocity depends on its distance from the walls.

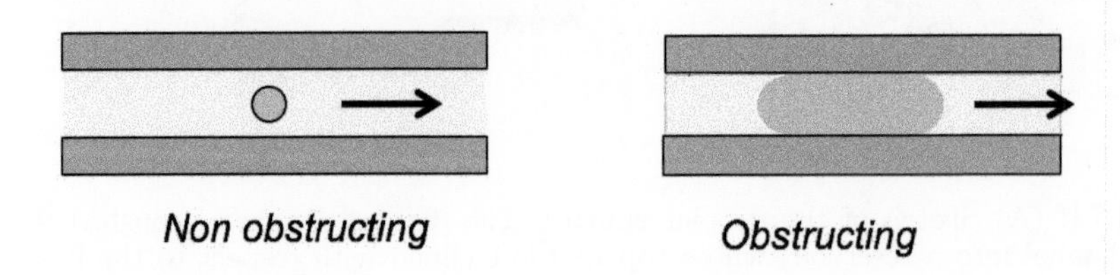

Fig. 4.75: Non obstructing and obstructing microfluidic droplets

The other case (obstructing) is more interesting, from the viewpoint of control: the droplet position with respect to the walls is fully controlled, along with its speed, fixed by the flow rate. The three–dimensional structure of such a droplet (or bubble) was established in the 1990s [100]. It is shown in Fig. 4.76.

In Ref. [100], the bubble is at rest. The front and the rear form a nose which is almost spherical. At the channel corners, gutters form, leaving a small – but often dynamically important – space filled with the continuous phase. The bubble is separated from the walls by a thin film, whose shape and thickness evolves slowly over time. This film is composed of a dome at the centre and a convex film closer to the gutters, both evolving extremely slowly over time, due to the action of capillarity.

How can we understand the structure of this bubble? Again, by invoking Laplace's law. The bubble is pressurized, because of the curvature at the nose and at the corners. At the nose, assuming a spherical shape, the internal overpressure is $\approx \frac{4\gamma}{h}$, in which γ is the surface or interfacial tension and h is the height of the square cross–section of the channel. To ensure mechanical equilibrium, the pressure in the bubble must be homogeneous. This implies that the curvature radius r of the interface at the corners

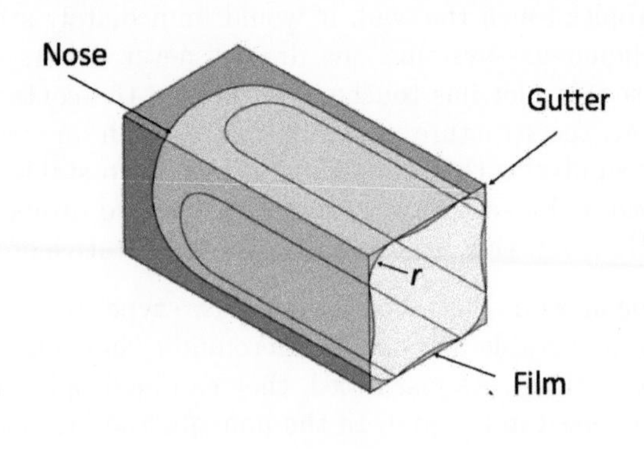

4.76: Most frequently used type of droplet : the obstructing droplet

should be $\approx \frac{h}{4}$. With such a curvature, and in the case where the liquid fully wets the solid, the meniscus matches the wall, or the film at a distance $\approx h/2$ from the channel mid–plane. The corresponding bubble shape is shown in Fig. 4.76.[27] The corner region, outside the bubble is called gutter'. It sometimes plays a role in the bubble dynamics. The arguments given here also apply for droplets.

4.11.2 The excellent monodispersity of microfluidic droplets

A striking feature of microfluidic droplets is their high monodispersity. Millions of droplets per hour, departing only slightly from their mean sizes, are currently produced in microdevices. An example is shown in Fig. 4.77 [102].

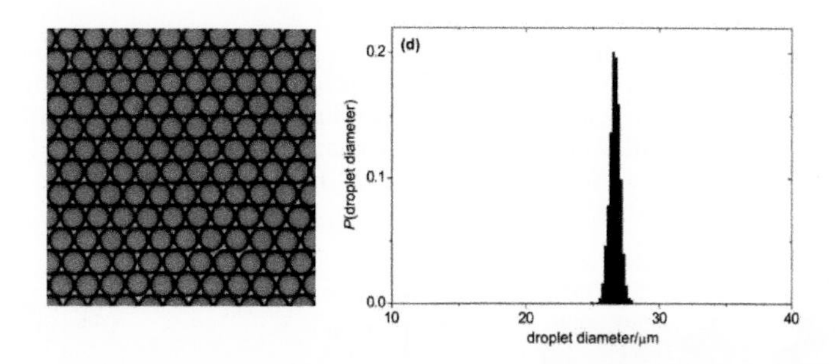

Fig. 4.77: monodispersity of microfluidic droplets. (Left) Typical oil in water droplet population obtained in microfluidic devices. (Right) Typical size distribution of microfluidic droplets (dodecane in water). In this case, the mean value is 26.6 μm and the standard deviation is around 1.5% [102].

Typical standard deviations of droplet sizes are less than 5%. This small number reflects the excellent control of the hydrodynamic flows in microfluidic systems. The break–up process, like all instabilities, is triggered by noise, which can be a source of polydispersity. However, this noise concerns only the ultimate times preceding break–up, when the structure of the droplet is already established. Consequently, it does not impact the characteristics or the droplet that is emitted.

Monodispersity and high throughput are critical for applications. For example, millions of identical reactors, where reactions take place under the same conditions, can be created, enabling massive screening. Another example concerns the production of solid particles with uniform properties, as shown in Fig. 4.78.

Most often, industrial emulsions are highly polydisperse. An example is shown in Fig. 4.79.

Broad size distributions, 100 % or more, are common. When comparing microfluidic to standard emulsions, the question that is most often raised concerns the throughput.

[27] If the liquid partially wets the solid, the meniscus must find a shape that satisfies Young Equation and pressure homogeneity condition

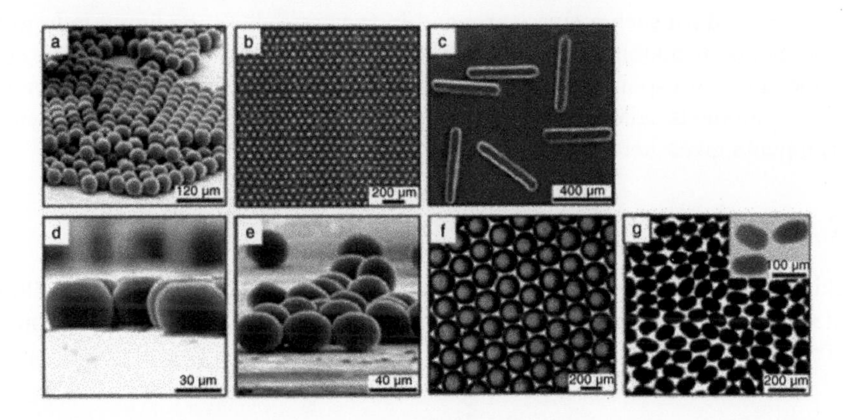

Fig. 4.78: Optical microscopy images of polyTPGDA particles: (A) microspheres, (B) crystal of microspheres, (C) rods, (D) disks, and (E) ellipsoids. Optical microscopy images of (F) agarose disks and (G) bismuth alloy ellipsoids produced using thermal solidification; inset: micrograph of the bismuth alloy ellipsoids at higher magnification. (Reprinted from [101] with permission of John Wiley and Sons. Copyright 2022.)

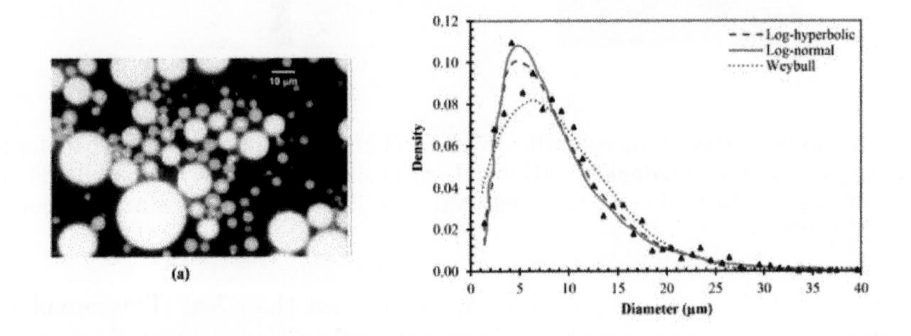

Fig. 4.79: A typical emulsion produced in the industry, showing a large polydispersity, with typical standard type deviations on the order of 100%. (Reprinted from [103] with permission of John Wiley and Sons. Copyright 2022.)

Producing thousands of droplets per seconds, hundreds of micrometres in diameter, leads to a mass production of, roughly, 30 kg per year, while industry currently produces tons in the same period of time. The gap seems large, but one must emphasize that progress is made on parallelization. Sometimes, the fabrication of complex, fragile, objects needs microfluidics or millifluidics to be produced. This is the case of the lipid nano particles (LPN) involved in the COVID vaccine, which must be produced in huge quantities. Today, use is made of millifluidic technology. We may envision that micro or millifluidic technology will progressively compete with traditional emulsion techniques, even in domains currently inaccessible to them.

4.11.3 How fast obstructing droplets move ?

The behaviour of obstructing droplets is surprising: in cylindrical tubes and in 2D, they move faster than the mean flow, while in square channels, it is the opposite. How can we understand this behaviour?

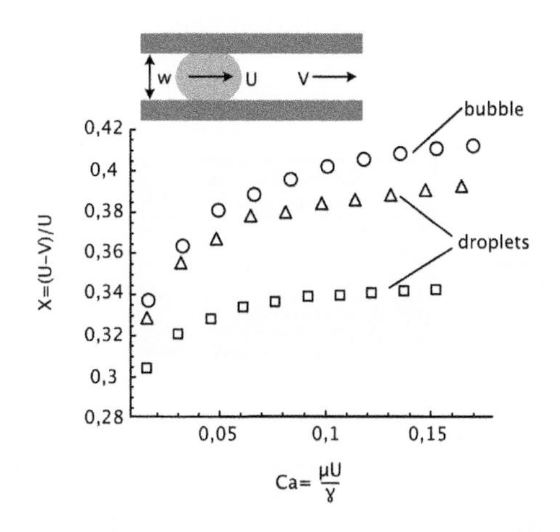

4.80: 2D calculation, showing the evolution of the relative speed excess $X = (U - -V)/U$, where U the droplet speed and V the mean flow speed, versus the capillary number, for different viscosity ratios .(Reprinted from [104] with permission from the American Physical Society. Copyright 2022.)

Let us start with Fig. 4.80, which shows numerical results obtained in a 2D geometry [104], representative, for the problem we address, of the cylindrical geometry, i.e. the geometry of the tube. In this case, the droplet speed U is larger than the mean flow V and its expression is given by:

$$U = V(1 + \beta Ca^{1/3}) \tag{4.31}$$

in which $Ca = \frac{\mu U}{\gamma}$ is the capillary number, and β is a coefficient depending on the viscosity ratio between the disperse and the continuous phase. Gas bubbles slide at faster speed than viscous droplets. The excess speed, $U - -V$, is substantially above the mean velocity. The phenomenon can be explained by invoking mass conservation: as the fluids are incompressible, the volumetric flow rate is conserved along the channel. Far from thee droplet, and in 2D, it is Vh (with h the channel height) while at the droplet level, it is $U(h - -2\delta)$ considering that the drained film, of thickness δ, is practically at rest. This quantity can be calculated, using the law we established earlier for the Landau–Levich film. We have $\delta \sim Ca^{1/3}$. By applying flow rate conservation, and proceeding to a development at small Ca, one obtains the observed relation between U and V. The origin of the excess speed is thus related to the presence of a Bretherton film at rest.

In square microchannels, the presence of gutters, at the corners, not taken into account in the previous reasoning, inverts the conclusion. These gutters act as resistances placed in parallel with the bubble. This resistance is much lower than the Bretherton film. The consequences are important: Wong et al. [107] showed that at small capillary numbers, and for moderate lengths, bubble speeds are on the order of $VCa^{1/3}$, i.e. significantly below the mean flow speed V.

4.11.4 Internal mixing of microfluidic droplets

As droplets move downstream, they generate internal recirculations [108, 109]. This is shown in Fig. 4.81:

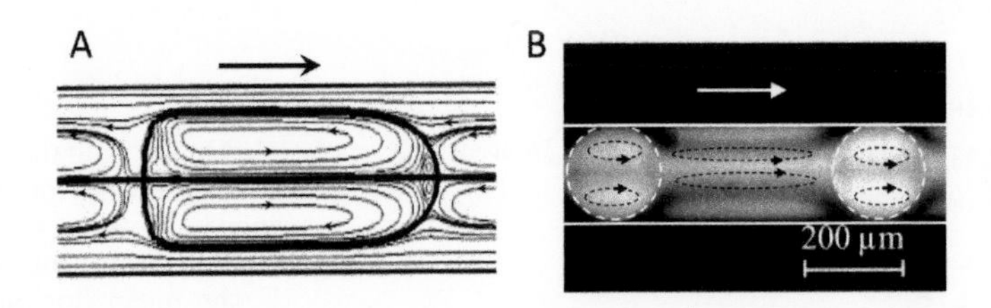

Fig. 4.81: (A) Numerical simulation of the recirculations developing inside and outside the droplet, in its own frame of reference, as it moves rightwards in a square microchannel. (Reprinted from [109] with permission of John Wiley and Sons. Copyright 2022.) (B) Visualization, with rhodamine initially introduced in droplets, of the recirculations. External phase is octanol and dispersed phase (droplets) is water [110].

The origin of the recirculations is purely kinematical. In the frame of reference of the droplet, walls move in the opposite direction, all around. These walls entrain the fluids and give rise to the internal recirculations shown in the figure. Depending on the aspect ratio of the channel cross–section, the speed, the droplet can include different numbers of eddies. In Fig. 4.81, four eddies are shown.

Those recirculations are extremely important for the applications. The eddies mix the fluids contained in the droplets. Being purely kinematical, they do not depend on thee physico–chemistry of the system, and therefore are robust. They are vigorous. Their turn–over time is typically h/U, in which h is the channel height and U the droplet speed. A droplet moving at U = 1 mm/s, in a channel of 100μm high, will produce a recirculation time of 0.1 s, which is comparable to the best conventional (i.e. non microfluidic) mixers. If two reactants are present in the droplet, a homogeneous reaction can take place, in a fast time. This property, coupled to others (monodispersity, throughput) opened, for microfluidics, avenues in the fields of chemistry and biology.

4.11.5 Large pressure drops generated in channels by microfluidic droplets or bubbles

As droplets or bubbles circulate in microchannels, they develop a pressure field. A counter–intuitive result is that in typical microfluidic conditions (i.e. small Reynolds numbers, small capillary numbers), even though droplets occupy a small fraction of the channel, however small the viscosity can be (e.g. gas for bubbles), the pressure

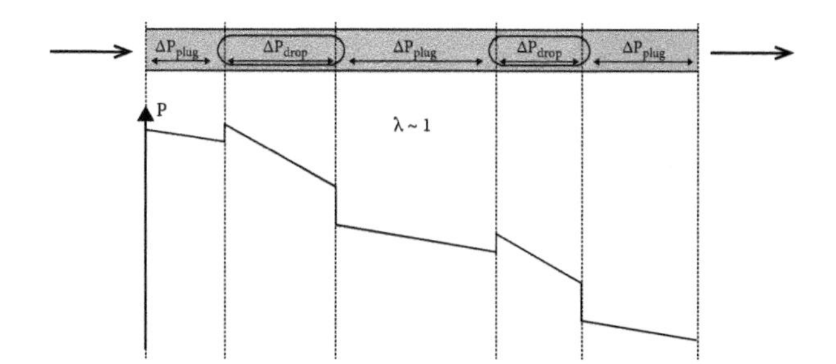

4.82: Pressure drops induced by droplets along a microchannel (from [65]).

drop induced by them dominates the other contributions. In other words, the cost, in terms of pressure, of driving droplets or bubbles in microchannels is high. This feature was analysed in [106] and reviewed in [65]. It is represented in Fig. 4.82.

The pressure decreases downstream, as in any channel, but in a complicated manner. In between the two droplets, the pressure ΔP_c (c for continuous) decreases linearly, at a rate controlled by the hydrodynamic resistance of the channel. The same remark holds for ΔP_d, the pressure drop inside the droplets (d for the interior of the dispersed phase). Droplets move over a Bretherton film, which is at rest. Figure 4.82 shows that between the film and the droplet nose regions, there is a sudden pressure change. This feature was discussed earler. We saw that the pressure drop ΔP_i (i for interface) is on the order of $\frac{\gamma}{h}Ca^{2/3}$. It is positive in the rear of the droplet and negative at the front. Should the droplet be perfectly symmetric, these pressure jumps would add up and cancel each other. But they are not. The interface at the front is distorted differently than in the rear, for subtle reasons discussed in Ref. [62]. Eventually, the imbalance between the front and the rear cause a net pressure drop.

All this can be quantified in terms of order of magnitude. Let us take the case of a channel of length L including one bubble of length L_d, so that the remaining channel length is $L_c = L - -L_i$ (see Fig. 4.82). The expression of the pressure drop along the channel reads:

$$\Delta P = \Delta P_c + \Delta P_i \tag{4.32}$$

in which we put $\Delta P_d = 0$, because here, we consider a bubble. The two terms can be developed:

$$\Delta P = R_c Q + \frac{\gamma}{h} Ca^{2/3} \tag{4.33}$$

where R_c is the resistance of the continuous phase and $Ca = \frac{\mu U}{\gamma}$ is the capillary number. In Eq. 4.34, for the sake of simplicity, we assume that the coefficient in front of the interfacial contribution is unity. R_c was established in Chapter 3. Its expression reads, without considering the pre-factor, $R_c \sim \frac{\mu L_c}{h^4}$. Eq. 4.34 can be rewritten in the following manner:

$$\Delta P \sim \frac{\mu U}{h}(\frac{L_c}{h} + Ca^{--1/3}) \tag{4.34}$$

With capillary numbers on the order of 10^{--6}, one sees that, for channels where $\frac{L_c}{h}$ is a few units, the interfacial contribution dominates the pressure drop. In other words, we need a channel length one hundred times its height to overcome the bubble contribution. The conclusion that we can draw from this order of magnitude analysis is that, in general, pressure drops are dominated by the droplets contributions.[28]

Boolean algebra with droplet microfluidics. An amazing application of droplet–based microfluidics is Boolean algebra [111]. The principle is shown in Fig. 4.83.

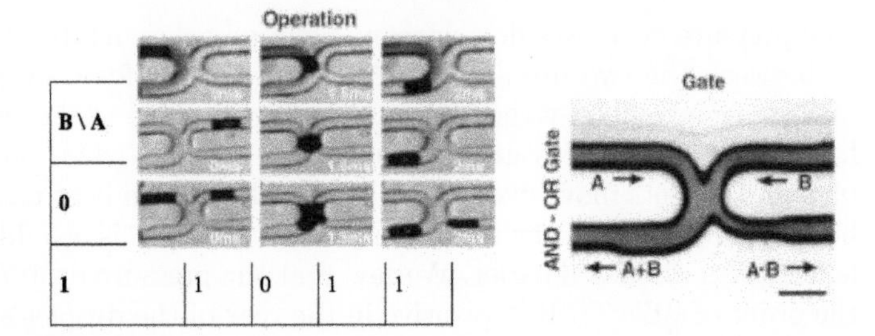

4.83: Boolean algebra with droplet microfluidics (see text)

Let us examine Fig. 4.83 (right). Assume a bubble comes from the left, entry A, i.e. $A = 1$ and $B = 0$. The bubble will join outlet A+B, because the channel is wider, and therefore its resistance is smaller. We thus confirm that this outlet is $A+B$ $(0+1 = 1)$ and the other is $A * B$, because, in this case is $0 * 1 = 0$. The same reasoning for the case where a bubble is introduced at entry B, i.e. $A = 0$ and $B = 1$. Now, assume that $A = 1$ and $B = 1$; In this case, the two bubbles are just behind each other. The first will join the outlet $A + B$. However, by doing this, it increases the resistance of the channel, as we saw previously. Consequently, the second droplet will joint the outlet $A * B$. We show, then, that the device works as a Boolean operator, as summarized in Fig 4.83 (left).

4.11.6 How do droplets break up?

In the first years of the development of droplet microfluidics, i.e. at the beginning of this century, investigators desired to build up a toolbox that would include all the operations needed to manipulate droplets. One operation that came to their attention was droplet break–up. But how can this be done? At first sight, this seems impossible. Small pearls at the dew are very difficult to break up. So, how could it be possible to break up even smaller droplets? How to reproduce the Grace break–up conditions,

[28]The analysis made here neglects the gutters, which, as we saw earlier, may play a substantial role. For short bubbles, the gutters tend to reduce the bubble speed, and,consequently the pressure drop induced by them. However, this effect tends to become negligible in long bubbles.

which require capillary numbers on the order of unity, impossible to achieve in microfluidic devices? In fact, in microfluidics, droplets break up in many circumstances. So, how to explain this feature?

The simplest geometry in which the problem can be addressed is the T–bifurcation (see Fig.4.84) [112]:

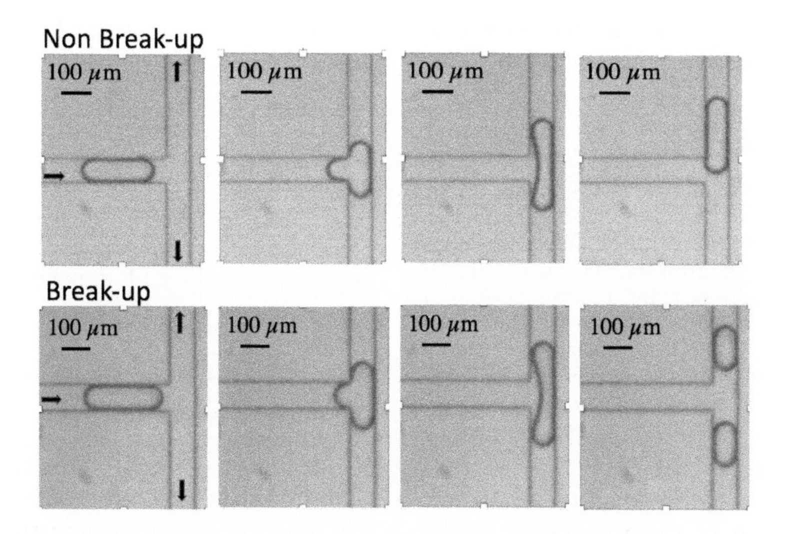

Fig. 4.84: Droplet break–up in a T-bifurcation. (Top) Typical situation at small flow rates: no break–up.(Bottom) Typical situation at large flow rates: breakup [112].

A water droplet is driven along a channel and arrives at a T–bifurcation. At small flow rates (see Fig.4.84 (top)), the droplet elongates and takes the shape of a dumbbell. This dumbbell is symmetric. The dumbbell stays there for some time, a few seconds or so, and, suddenly, it looses its symmetry, moves on the left branch, adopts an oblong shape, and is evacuated by the flow. At larger flow rates (see Fig.4.84 (bottom)), the sequence of events is different: as previously, a dumbbell forms in the junction, but, instead of adopting a stationary shape, it keeps elongating and eventually breaks up, forming two daughters that further on, are advected downstream.

Why is it so? The answer was given by A. Leshansky and L. Pismen [113].

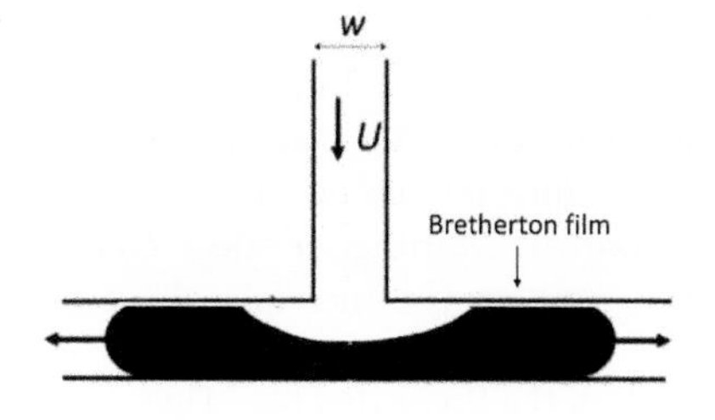

4.85: Sketch of a droplet arrived at the T–junction. A thin Bretherton film confined between the droplet bulges and the walls controls the dynamics of the system

The analysis relied on the presence of Bretherton films, confined between the bulges and the lateral walls (see Fig.4.85).[29] In this geometry, when the speed U is imposed, the flow is forced to pass through small gaps. Large viscous stresses develop, which, above a critical speed, entrain the bulge downstream. In such circumstances, the droplet elongates and eventually breaks up. This explains why droplet break–up occurs in microfluidic devices even though the capillary number is small. The paradox is thus solved. In Ref. [113], it was shown that, when $l/w >\sim Ca^{0.2}$ (where l is the droplet length, w is the channel width, and Ca is calculated by using U, w and the interfacial tension), break-up occurs.

The T–bifurcation can be asymmetric, as in Fig. 4.86 (see Ref. [117], from which the image is taken).

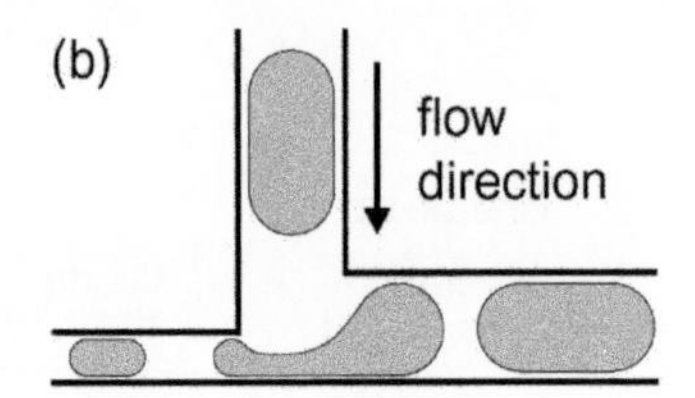

4.86: Non–symmetric T–bifurcation, generating small droplets. The size reduction can be increased by placing several T–bifurcations in cascade [114]. The process is limited by the lithographic resolution. In practice, the smallest droplet size obtainable with such a process is a few micrometres

By placing several T–bifurcations in cascade, smaller and smaller droplets can be produced [114]. However, the smallest size is limited by the smallest channel that can be made, i.e. by the lithographic resolution – in practice, several micrometres –.

4.11.7 Coalescence of microfluidic droplets

Physics of coalescence. Droplet merging or coalescence has been studied for decades [67,115]. Today, although considerable progress has been made, questions remain. It is difficult to describe the various regimes occurring at the early times of the coalescence process. Even more difficult is the incorporation, in the analysis, of a surfactant. It is challenging to carry out experimental studies, owing to the fast times and small scales involved, along with numerical simulations, due to the multiscale nature of the problem. We shall thus restrict ourselves to basic ideas.

We learned that surfactants control wettability. We will see here that they also stabilize the droplets. The stabilization mechanism is shown in Fig. 4.87.

Surfactants prevent coalescence by raising electrostatic and steric barriers. Therefore, to merge two droplets, holes must be made in the surfactant layers. This is precisely what happens when two droplets approach. As sketched in Fig. 4.87, the displacement generates a recirculating flow that tends to deplete the surfactant concentration locally. The surfactant film is drained by the flow, fresh surface is created, and again, due to energy minimization, coalescence may occur. However, Marangoni effect counteracts the process by swiftly rehomogenizing the surfactant at the interface, in a manner

[29] The idea was new: previous studies overlooked the role of the Bretherton film [114].

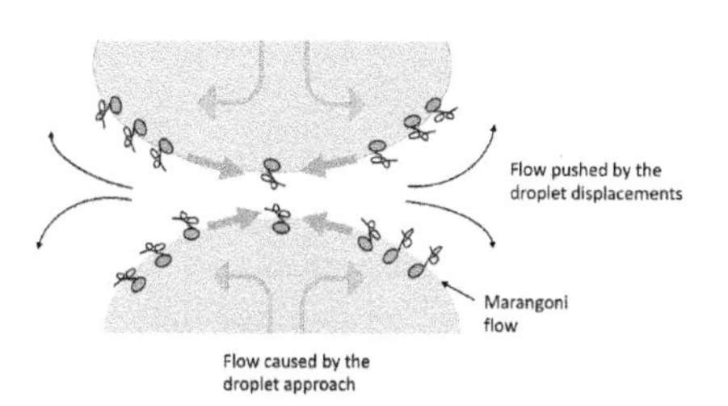

4.87: In an infinite medium, when two droplets approach, they generate a flow that tends to deplete the surfactant in the contact region, thus exposing fresh surfaces. However, Marangoni stresses counteracts this effect, generating flows that prevents coalescence (adapted from [23]). In microfluidic systems, these flows are damped by the walls. Therefore, conditions for obtaining coalescence are easier to achieve.

similar to the pepper experiment. This is, in short, the mechanism invoked in reviews and books to explain that emulsions are stabilized by surfactants (see, for instance, Ref. [23]). So, how droplets can be coalesced in microfluidic systems, in which, most often, surfactants are present? This looks difficult.

The answer is that in microfluidic channels, confinement inhibits the Marangoni restoring mechanism. No large flows can develop, and we are left with diffusion for rehomogenizing the surfactant at the droplet interface. A reasoning, presented in Ref. [116], allows to characterize the dynamics of the process: to fill a fresh interface, just depleted in surfactant, one must recruit the molecules located around it. Their bulk concentration is c_s while, at the interface, at saturation, the area per molecule is a_s. The size l of the recruitment volume is then $l \sim (a_s\, c_s)^{-1}$. In typical conditions, well above CMC, this length is tens of micrometres. The diffusive time $\tau = \frac{l^2}{D}$, in which D is the diffusion coefficient of the surfactant, will thus be $\tau \approx 1$ s. Therefore, if the droplet approach process is faster than this time, coalescence has a chance to occur. With such times, one may conclude that coalescence is possible. One must nonetheless add that flow transport, due to the presence of local pressure differences, induced by shape change, may also play a role. The conclusion is that we have, at the moment, a qualitative picture of the coalescence process, which helps in the design of devices dedicated to this task, but no comprehensive description.

Microfluidic strategies for coalescing droplets. In practice, the strategies used for coalescing droplets consist of forcing them to collide in constrictions, openings, or combs. The various solutions proposed in the literature were reviewed by Seeman et al. [117], from which Fig. 4.88 and its caption are sourced.

A practical remark should be made here: the strategies shown in Fig. 4.88 are sometimes difficult to integrate in a device because they perturb the pressure field, the flow conditions and, sometimes, need fine tuning to operate efficiently. Therefore, in a number of cases, in order to circumvent these difficulties, electric fields are used for achieving merging (an example is Ref. [119]). We will return to this subject in Chapter 6.

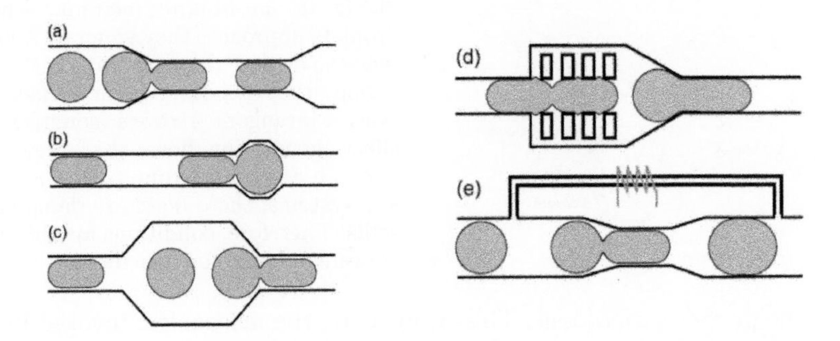

Fig. 4.88: Schematic of different geometries to coalesce droplets by (A) stopping a droplet at a narrowed channel, (B) stopping a droplet in a widened channel, (C) slowing down the drop movement in a widening channel, (D) slowing down a droplet in a comb geometry, and (E) slowing down a droplet by a controlled bypass. (From [117].)

4.11.8 Microfluidic multiple emulsions

Double emulsions (one droplet engulfed in another one, often called a globule'), are a century–old subject. The domain is involved in many industrial applications, and the reader may refer to [120, 121] for general presentations. Here, we define two types of double emulsions: (i) with three different liquids, and (ii) with two different liquids (oil and water).

(i) Double emulsions with three different liquids. When three different immiscible liquids are put together, they spontaneously form structures. The structures can be Janus, separated droplets or double droplest (see Fig.4.89).

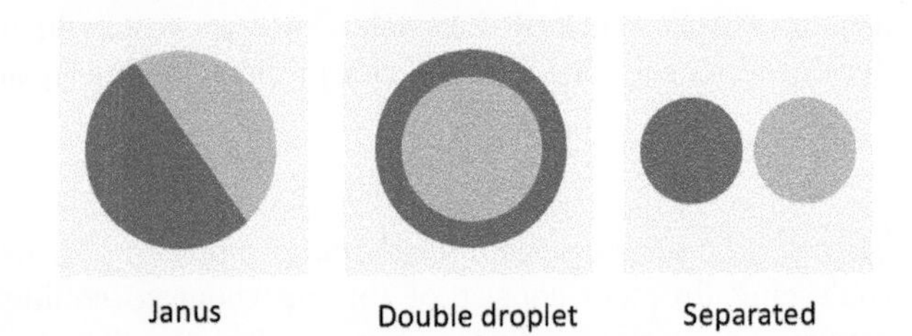

4.89: Janus, separated droplets, or double droplet.

These structures are controlled by spreading coefficients [120–122]. In the domain of microfluidics, the first double emulsions appeared soon after droplet–based microfluidics took off. One example is shown in Fig. 4.90 (A).

In the experiment of Ref. [122], TPGDA, tetradecane and water/SDS are used. The channels are hydrophilic. The system was stable because TPGDA wets the walls more than tetradecane and, further downstream, water/SDS wets the walls more than TPDGA. Moreover, for the tetradecane/TPGDA/water system, double droplets represent an energetically favourable state.

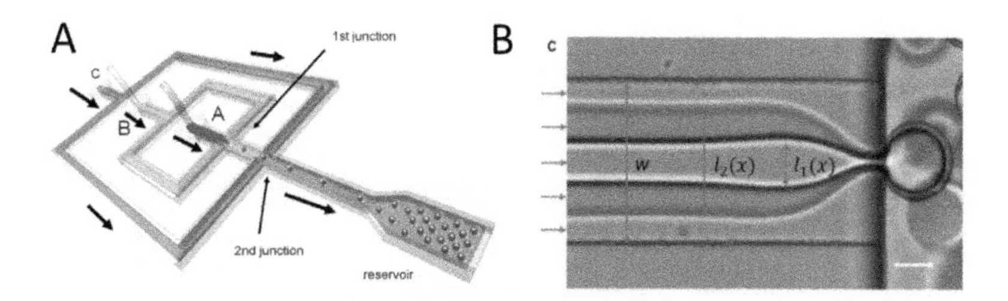

Fig. 4.90: (A) Tetradecane droplets in TPGDA are produced at the first junction. At the second junction, water/SDS engulfs the droplets. The structure further moves into a reservoir [122].(B) Water/oil/water (W/O/W) double emulsion produced in the step emulsion geometry [96]

(ii) O/W/O and W/O/W emulsions. The second type of double emulsion uses only two fluids: oil, or, more generally, a lipophilic phase, and water. An example is an oil droplet engulfed in a water droplet, in an oil medium (O/W/O emulsions) or the opposite, i.e. water encapsulated in oil, in water (W/O/W emulsions). These double droplets are metastable (the lowest energy state would be an oil or a water droplet). They survive because they are stabilized by surfactants, as previously noted.

Let us now revisit the strategy of Fig. 4.90. To fabricate a W/O/W emulsion (water in oil in water), a water droplet in oil must first be created. Further downstream, the droplet must be encapsulated in an oil droplet, immersed in water. The problem is that, in the upstream part, water droplets must circulate in *hydrophobic* channels, otherwise they are unstable, while, downstream, the oil droplet engulfing the water droplet must circulate in *hydrophilic* channels, for exactly the same reason. The conclusion is that wetttability must be patterned inside the device to operate in stable conditions. Today, wettability patterning remains challenging despite a number of efforts (see, for instance, Ref. [124]). This is one of the bottlenecks that today, microfluidic technology is facing. These difficulties could be circumvented in Ref. [96] by transporting the three fluids in a co–flow regime at high speeds, in order to reduce the detrimental effect of inappropriate wetting conditions (see Fig.4.90 B).

With glass capillary technology, there is no difficulty to solve the wettability problem, by using bare glass in hydrophilic regions and silanized glass in hydrophobic zones. Moreover, droplets are formed in the center of the capillaries and they do not touch the walls. This prevents complications from arising close to the nozzles. These features considerably facilitate the creation of multiple emulsions. D. Weitz's group succeeded, in this manner, by using combinations of tubes of different wettabilities, to produce impressive multiple emulsions [123]. An example is shown in Fig. 4.91.

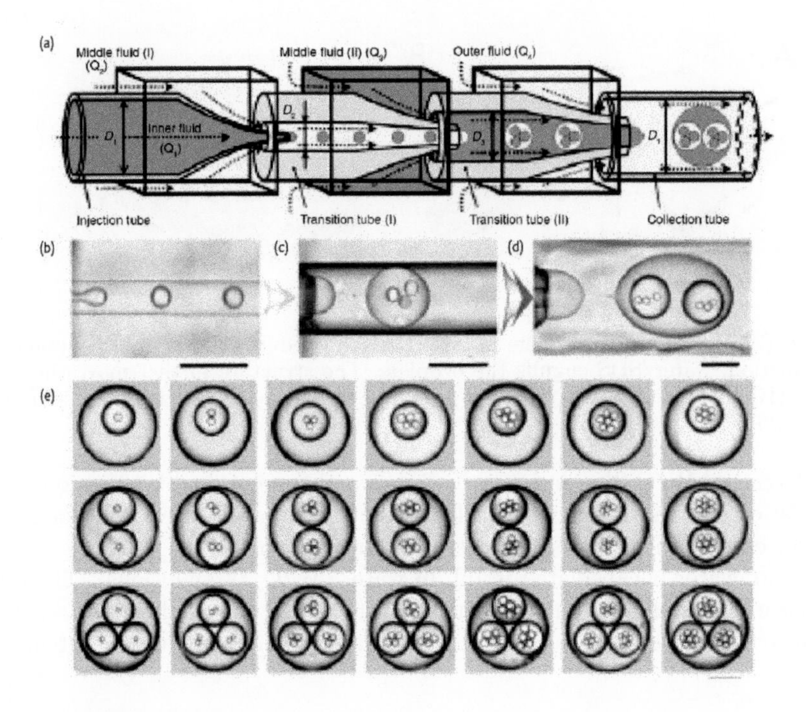

Fig. 4.91: Fabrication of triple emulsions in a capillary device. (A) Schematic of the device. (B)–(D) Optical micrographs taken with a high–speed camera displaying the (B) first, (C) second, and (D) third emulsification stages. (E) Optical micrographs of triple emulsions that contain a controlled number of inner and middle droplets [123].(Reproduced with permission of Wiley and Sons (2007).)

4.11.9 The microfluidic droplet toolbox

We saw that droplets can be used as microreactors, and these reactors can be manipulated in a number of ways: for instance, droplets can be split or merged, or put at rest to serve as micro–incubators or storage units, awaiting further processing. One may imagine a tool box, which gathers all droplet functionalities, and provides bricks for building miniaturized chemical factories, capable of running complex chemical or biochemical processes in a continuous and automatized manner. An example of such a toolbox, imagined by Ref. [125], is shown in Fig. 4.92.

In the tool box, we have: coalescence, generation, mixing, storage, detection, sorting, re–injection, splitting, and off–chip incubation. To this list, one may add droplet content modification, in which a reagent is added in the droplet (this will be explained in Chapter 6), diluting, heating (e.g. for polymerization), or thermal cycling (for amplifying DNA, using PCR). The list is long; therefore, in principle, a high level of complexity can be achieved by assembling these functionalities in the manner of Lego, as suggested by Fig. 4.92.

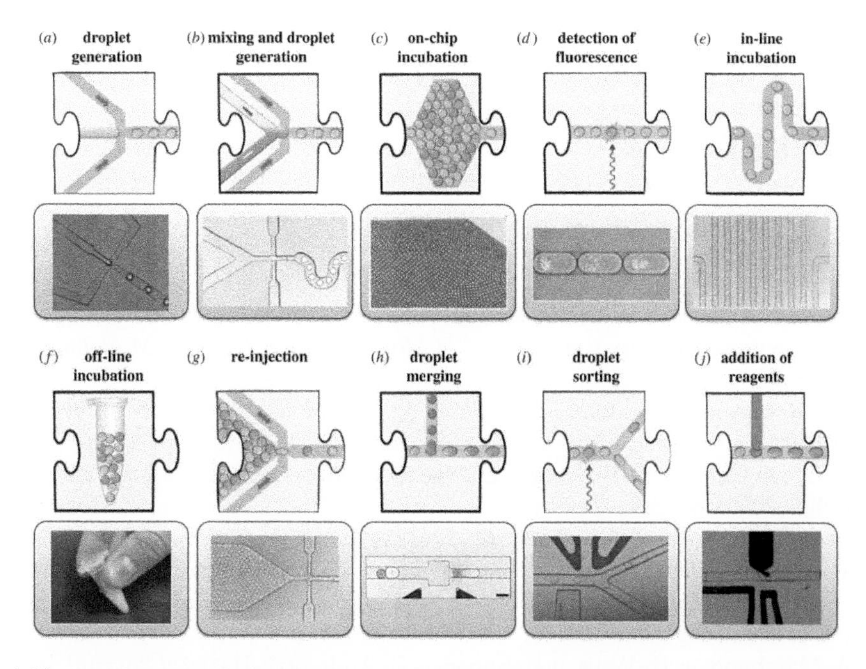

Fig. 4.92: Droplet microfluidic toolbox: coalescence, generation, mixing, storage, detection, sorting, re–injection, splitting, and off–chip incubation. (Reprinted from Ref. [125], under Creative Commons licence 4.0.)

In practice, so far, the number of functionalities that have been successfully integrated on the same device has remained limited. This is because, as the device becomes more complex, it is increasingly difficult to control its behaviour. Unlike with electronics, all elements are coupled to the others, via the pressure field, which propagates hydrodynamic perturbations across the system at the speed of sound. Local pressure changes, for instance, due to the formation of a new droplet at a T-junction, are felt everywhere in the device. This fact, inherent to hydrodynamics, represents a bottleneck tor integrating a large number functionalities. Although the number of fluidic operations that can be integrated on a single device is limited, the capacity of performing complex operations in microfluidic droplets remains considerable, so that, over the years, droplet–based microfluidics has become a major component of the field, as shown in Chapter 1.

References

[1] P. Jurgins, *Nature*, **474**, 168 (2011).
[2] Y. D. Sumith, S.C. Maroo, *Journal of Molecular Graphics and Modelling*, **79**, 230 (2018).
[3] J.S. Robinson, B. Widom, *Molecular theory of capillarity*, Oxford University Press (1989).
[4] P. G. De Gennes, F. Brochard, D. Quéré, *Gouttes, Bulles, Perles Ondes*, Edition Belin (2002).
[5] M. Berry, *Physics Educations* , **6**, 79 (1971).
[6] J. C. Berg, https://fr.scribd.com/document/337031200/Berg–Chap02 or *An Introduction to Interfaces and Colloids: The Bridge to Nanoscience*, World Scientific Publishers, Singapore (2010).
[7] A. Marchand, J. H. Weijs, J.H. Snoeijer, B. Andreotti *Americal Journal of Physics*, **79**, 999 (2011).
[8] Herge, 'On a marche sur la lune', Casterman.
[9] http://www.funsci.com.
[10] L. Landau, B. Levich, *Mécanique des Fluids*, Mir (1967).
[11] H. Bruus, *Theoretical Microfluidics*, Oxford University Press (2008).
[12] V. Shkolnikov, *Principles of Microfluidics* (2018).
[13] K. Wonjung, J. W. M. Bush, *J. Fluid Mechanics*, **705**, 7 (2012).
[14] M Prakash, D. Quéré, J. W. M. Bush, *Science*, **320**, 931 (2008).
[15] J. Tolman, *Chem. Phys.*, **17**, 333 (1949).
[16] H. Liu, G. Cao, *Scientific Reports*, **6**, 23936 (2016).
[17] D. W Bassett, *ECS Trans.*, **92**, 95 (2019).
[18] W. Zisman in *Contact Angle Wettability and Adhesion* in *Chemical Series*, **43** Washington (1964).
[19] M. K. Chaudury, G. M. Whitesides, *Langmuir*, **7**, 1013 (1991).
[20] S. Kim, D. Kim, J. Kim, S. An, W. Jhe, *Phys.RevX*, **8**, 41046 (2018).
[21] N. Bruot, F. Caupin, *Phys.Rev.Lett.*, **116**, 56102 (2016).
[22] Q. Brosseau, J. Vrignon, JC. Baret, *Soft Matter*, **10**, 3066 (2014).
[23] J–C. Baret, *Lab on a Chip*, **12**, 422 (2012).
[24] K. Kinoshita, E. Parra, D. Needham, *Journal of Colloid and Interface Science*, **504**, 765 (2017).
[25] D. Myers, *Surfaces, Interfaces, and Colloids*. 2nd ed. New York, Wiley (1999).
[26] D. Myers, *Surfactant Science and Technology*, 3nd ed. New York, Wiley (2006).
[27] B. Li, L. Zhang, S. Liu, M. Fan, *J Surfact Deterg.* , **22**, 85 (2019).
[28] W. D. Bancroft, *J. Phys. Chem.* **17**, 501 (1913).
[29] W. D. Bancroft, *J. Phys. Chem.* **19**, 275 (1915).

[30] R. Dreyfus, P. Tabeling, H. Willaime, *Phys. Rev. Lett.*, **90**, 14 (2003).
[31] P. Nghe, G. Degré, P. Tabeling, A . Ajdari, *Applied Physics Letters* , **93**, 204102 (2008).
[32] J Goyon, A. Colin, G. Ovarlez, A. Ajdari, L. Bocquet, *Nature*, **454**, 84 (2008).
[33] H. Haidara, M. K. Chaudhury, M. J. Owen, *J.Phys.Chem.*, **99**, 8681 (1995).
[34] Several pepper experiments of this type can be found on the web. The address of the one shown in Fig. 4.30 is https://www.youtube.com/watch?v=4dVHf7IVloI; credits: C. Kalelkar, A. K. Sinha.
[35] V. G. Levich, V. S. Krylov, *Annu. Rev. Fluid Mech.*, **1**, 293 (1969).
[36] M. Roché, Z. Z. Li, I. M. Griffiths, S. Le Roux, I. Cantat, A. Saint–Jalmes, H. W. Stone, *Phys. Rev. Lett.*, **112**, 208302 (2014).
[37] P. G. De Gennes, *Rev. Mod. Phys.*, **57**, 827 (1985).
[38] K. Seo, M. Kim, D. H. Kim, http://dx.doi.org/10.5772/61066.
[39] T. Young *Phil. Trans. Roy. Soc. (London)*, **95**, 55 (1805).
[40] J. W. Gibbs, *The Collected Papers of J. Willard Gibbs*, **Vol. I**, London: Yale University Press, (1957).
[41] H. W Lee, D. C. S. Bien, S. A. M. Badaruddin, A.S. Teh, *Microsyst. Technol.* **19**, 253 (2013).
[42] K. Kartini, E. Saputra , R. Ismail , J. Jamari , A.P. Bayuseno, *Matec*, **58**, 4007 (2006).
[43] R. N. Wentzel, *Ind. Eng. Chem.*, **28**, 988 (1936).
[44] R. N. Wentzel, *J. Phys. Coll. Chem.*, **53**, 1466 (1949).
[45] A. B. D. Cassie, S. Baxter, *Trans. Far. Soc.*, **40**, 546 (1944).
[46] A. Lafuma, D. Quéré, *Nature Mat.*, **2**, 457 (2003).
[47] F. Chen, D. Zhang, Q. Yang, J. Yong, G. Du, J. Si, F. Yun, H. Xun, *ACS Appl. Mater. Interfaces*, **5**, 6777 (2013).
[48] D. Bonn, J. Eggers, J. Indekeu, J. Meunier, E. Rolley, *Rev.Mod.Phys.*, **81**, 739 (2009).
[49] J. H. Snoejes, B. Andreotti, *Ann. Rev. Fluid. Mech.*, **45**, 269 (2013).
[50] J. Lee, H. Moon, J. Fowler, T. Schoellhammer, C.–J. Kim, *Sens. Actuators*, **A95**, 259 (2002).
[51] R. G. Cox, *Jour. Fluid. Mech.*, **168**, 169 (1986).
[52] O. V. Voinov, *Fliud.Dyn.*, **11**, 714 (1976).
[53] E. Dussan, *Annual Rev. Fluid. Mech.*, **11**, 371 (1979).
[54] H. A. Stone, *Annu. Rev. Fluid Mech*, **26**, 65 (1994).
[55] K. J. Humphry, A. Ajdari, A. Fernandez–Nieves, H. A. Stone, D. A. Weitz, *Phys. Rev*, **E 79**, 056310 (2009).
[56] H. P. Grace, *Chem. Eng. Commun*, **14**, 225 (1982).
[57] X. D. Shi, M. P. Brenner, S. R. Nagel, *Science*, **265**, 219 (1994).
[58] G. Hetsroni, *Handbook on Multiphase Flows*, Clarendon Press (1980).
[59] L. Landau, B. Levich, *Acta Physicochim.* **17**, 42 (1942).
[60] B. V. Derjaguin, S. M. Levi, *Acta. Physicochim. USSR*, **20**, 349 (1943).
[61] https://www.utwente.nl/en/tnw/pcf/education.
[62] F. P. Bretherton, *J. Fluid Mech.*, **10**, 166 (1961).
[63] R. Krechetnikov, G. M. Homsy, *J. Fluid Mech.* **559**, 429 (2006).

[64] P. Aussillous, D. Quéré, *Physics of fluids*, **12**, 2367 (2000).
[65] F. Rayleigh, *Proc. Lond. Math. Soc*, **s1–10**, 4 (1878).
[66] J. Plateau, J., *Experimental and Theoretical Steady State of Liquids Subjected to Nothing but Molecular Forces*, Gauthiers–Villars, Paris (1873).
[67] J. Eggers, *Reviews of Modern Physics.*, **69**, 865 (1997).
[68] L Ménétrier–Deremble, P. Tabeling, *Phys. Rev.*, **E74**, 035303 (2006).
[69] P. Guillot, A. Colin, A. S. Utada, A. Adjari, *Phys. Rev. Lett.* **99**, 10452 (2007).
[70] M. Moseler, U. Landman, *Science*, **289**, 1165 (2000).
[71] J. Eggers *Phys. Rev. Lett.*, **89**, 84502 (2002).
[72] E. W. Washburn, *Phys. Rev.*, **17**, 3, 273 (1921).
[73] A. W. Martinez, S. T. Phillips, J. Manish, G. M. Whitesides *Ang. Chemie (Int. Ed.)*, **46**, 8, 1318 (2007).
[74] E. Carrilho, A. W. Martinez, G. M. Whitesides, *Anal. Chem.*, **81**, 16, 7091 (2009).
[75] L. Magro, C. Escadafal, P. Garneret, B. Jacquelin, A. Kwasiborski, J-C Manuguerra, F. Monti, A. Sakuntabhai, J. Vanhomwegen, P. Lafayee, P. Tabeling, *Lab on a Chip.*, **17**, 2347 (2017).
[76] E. Fu, S. A Ramsey, P. Kauffman, B. Lutz, P. Yager, *Microfluidics and nanofluidics*, **10**, 29 (2011).
[77] A. M. Gañán–Calvo, *Phys. Rev. Lett.*, **80**, 285 (1998).
[78] I. Kobayashi, M. Nakajima, J. Tong, T. Kawakatsu, H. Nabetani, Y. Kikuchi, A. Shohno, K. Satoh, *Food Sci Techn Research*, **5**, 350 (1999).
[79] T. Thorsen, R. Roberts, F. Arnold, S. Quake, *Phys. Rev. Lett.*, **86**, 4163 (2001).
[80] S. Quake, private communication (2021).
[81] S. Anna, N. Bontoux, H. Stone, *Appl. Phys. Lett.*, **82**, 4163 (2003).
[82] J. K. Nunes, S. S. H. Tsai, J. Wan, H. A. Stone, *J. Phys. D: Appl. Phys.*, **46**, 114002 (2013).
[83] P. Huerre, P. A. Monkewitz, *Annual Review of Fluid Mechanics*, **22**, 473 (1990).
[84] P. Garstecki, M. J. Fuerstman, H. A. Stone, G. M. Whitesides, *Lab on a Chip*, **6**, 437 (2006).
[85] V. Barbier, H. Willaime, P. Tabeling, F. Jousse, *Physical Review*, **E 74**, 046306 (2006).
[86] H. Willaime, V. Barbier, L. Kloul, S. Maine, P. Tabeling, *Phys. Rev. Lett.*, **96**, 054501 (2006).
[87] T. Cubaud, T. G. Mason, *Phys. Fluids*, **20**, 53302 (2008).
[88] T. Segers, dissertation thesis (2015).
[89] I. Chakraborty, J. Ricouvier, P. Yazhgur, P. Tabeling, A. M. Leshansky, *Phys. Fluids*, **31**, 22010 (2019).
[90] I. Kobayashi, K. Uemura and M. Nakajima, *Langmuir*, **22**, 10893 (2006).
[91] F. Malloggi, N. Pannacci, R. Attia, F. Monti, P. Mary, H. Willaime, P. Tabeling, B. Cabane, P. Poncet, *Langmuir*, **26**, 2369 (2010).
[92] C. Priest, S. Herminghaus, R. Seemann, *Appl. Phys. Lett.*, **88**, 024106 (2006).
[93] Z. Li, A. M. Leshansky, M. Pismen, P. Tabeling, *Lab Chip*, **15**, 1023 (2015).
[94] X. Ge, Z. Li, unpublished (2021).
[95] D. L. Shui, A. van den Berg, J. C. T. Eijkel, *Microfluid. Nanofluid.*, **11**, 87 (2011).

[96] X. Ge, B. Y. Rubinstein, Y. He, F. N. Bruce, L. Li, A. M. Leshansky, Z. Li, *Lab on a Chip*, **21**, 1613 (2021).
[97] A. S. Utada, E. Lorenceau, D. R. Link, P. D. Kaplan, H. A . Stone, D. A. Weitz, *Science*, **308**, 537 (2005).
[98] L.Y. Chu et al., *Angew. Chem. Intl. Ed.*, **46**, 8970 (2007).
[99] R. Dangla, S. C. Kayi, C. N. Baroud, *PNAS*, **110**, 853 (2013).
[100] H. Wong, C. J. Radke, S. Morris, *J. Fluid. Mech*, **292**, 71 (1995).
[101] S. Xu, Z. Nie, M. Seo, P. Lewis, E. Kumacheva, et al., *Angew. Chem.*, **117**, 734 (2005).
[102] W. A. C. Bauer, J. Kotar, P. Cicuta, R. T. Woodward, J. V. M. Weaverd, W T. S. Huck, *Soft Matter*, **7**, 4214 (2011).
[103] F. Goodarzi, Z. Sohrab, *The Canadian Journal of Chemical Engineering*, **97**, 281 (2019).
[104] M. De Menech, *Phys. Rev. E*, **73**, 031505 (2006).
[105] C. N. Baroud, F. Gallaire, R. Dangla, *Lab on a Chip* , **10**, 2032 (2010).
[106] M. J. Fuerstman, A.L. Meghan, E. Thurlow, S. S. Shevkoplyas, H. A. Stone, G. M. Whitesides, *Lab on a Chip*, **.7** , 1479 (2007).
[107] H. Wong, C.J.Radke, S. Morris, *J. Fluid. Mech*, **292** 95 (1995).
[108] R. Lindken, L. Gui, W. Merzkirch, *Chem. Eng. Technol.*, **22**, 202 (1999).
[109] F. Sarrazin, L. Prat, C. Gourdon, K. Loubiere, T. Bonometti, J. Magnaudet, *AichE J.*, **52**, 4061 (2006).
[110] P. Mary, V. Studer, P. Tabeling, *Anal. Chem*, **80** 2680 (2008).
[111] M. Prakash, N. Gershenfeld, *Science*, **315**, 822 (2007).
[112] M. C. Jullien, M. J. M. M Ching, C. Cohen L. Menetrier, P. Tabeling, *Phys. Fluids*, bf 21, 072001 (2009).
[113] A. M. Leshansky, L. M. Pismen, *Phys. Fluids*, **21**, 023303 (2009).
[114] D. R. Link, S. L. Anna, D. A. Weitz, H. A. Stone, *Phys. Rev. Lett.*, **92**, 054503 (2004).
[115] J. Eggers, J. R. Lister, H. A. Stone, *Journal of Fluid Mechanics*, **401**, 293 (1999).
[116] N. Bremond, A. R. Thiam, J. Bibette, *Phys. Rev. Lett.*, **100**, 24501 (2008).
[117] R. Seeman, M. Brickmann, T. Pfohl, S. Herminghaus, *Rep. Prog. Phys.*, **75**, 016601 (2012).
[118] C. Holtze, A. C. Rowat, J. J. Agresti, J. B. Hutchison, F. E. Angile, C. H. J. Schmitz, S. Koster, H. Duan, K. J. Humphry, R. A. Scanga, J. S. Johnson, D. Pisignano, D. A. Weitz, *Lab on a Chip*, **8**, 1632 (2008).
[119] A. R. Abate, T. Hung, P. Mary, J. J. Agresti, D. A. Weitz, *PNAS*, **107**, 19163 (2008).
[120] F. Leal–Calderon, V. Schmitt, J. Bibette, *Springer Science & Business Media* (2007).
[121] J. Bibette, F. Leal–Calderon, P. Poulin, *Rep. Prog. Phys.*, **62**, 696 (1999).
[122] N. Pannacci, H. Bruus, D. Bartolo, I. Etchart, T. Lockhart, Y. Hennequin, H. Willaime, P. Tabeling *Phys. Rev. Lett.*, **101**, 164502 (2008).
[123] S. H. Kim, D. A Weitz, *Angewandte Chemie*, **123** 8890 (2011).
[124] V. Barbier, M. Tatoulian, H. Li, F. Arefi-Khonsari, A. Ajdari, P. Tabeling, *Langmuir*, **22**, 5230 (2006).

[125] L. D. van Vliet, P. Y. Colin, F. Hollfelder, *Interface Focus*, **5**, 20150035 (2015).

5

Transport in microfluidics

5.1 The microscopic origin of diffusion

5.1.1 Brownian motion

The idea that diffusion of small objects in liquids results from numerous collisions with molecules' was expressed by Robert Brown in 1827[1] [1]. He noticed that pollen grains do not stay immobile on the surface of water, but undergo spontaneous erratic movements, as in Fig. 5.1. After having thought that this could be the expression of life, Brown hypothesized that the origin of the phenomenon was due to collisions between pollen grains and extremely small particles that he called 'molecules'. This was an important step, because his work allowed, for the first time, a bridge to be established between the molecular structure of matter and a macroscopic phenomenon. Brown's experiment generated passionate discussions for more than a century.

The theory of molecular diffusion, anticipated by Brown, was established 180 years later, by Einstein [4], Langevin [5], and Smoluchowski [6].[2] Soon after, in 1909, Perrin [3] confirmed the theory, also reporting the first estimate of the Avogadro number (Fig. 5.1).

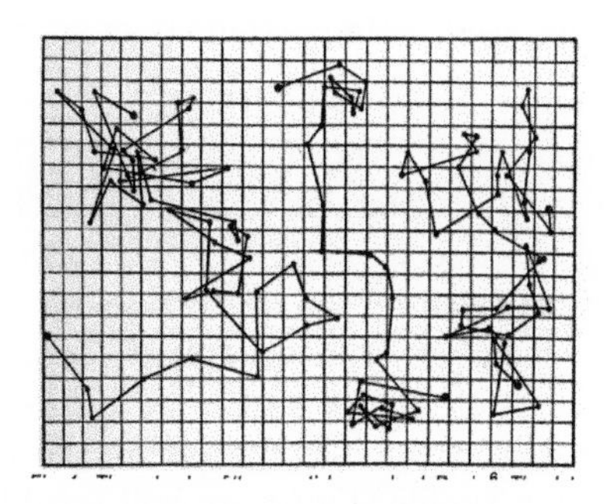

5.1: Trajectories of particles obtained in Perrin's experiment (1909) [3].

[1]Credit for the discovery of Brownian motion should probably be given to dutch doctor Jan Ingenhousz: in 1785, he noticed the erratic movement of charcoal powder on the surface of an alcoholic solution. Concerning the notion of molecules, it traces back to Lucrece (see a discussion by Y. Pomeau and J. Piasecki [2]).

[2]The work of Sutherland [7], performed at the same time, is commented in [2].

To develop an intuitive understanding of the phenomenon, let us consider a walker on a line, taking steps of length $\pm l$ in a random manner, like a drunken sailor who decides to pay a visit to the bars on his street, starting from home $X = 0$ at $t = 0$, where t is time (see Fig. 5.2). Assume the bars are distributed homogeneously along the street and the drunkard has no memory of what he did before. We also assume that he walks at a constant speed V and the time for which he stops in the bars is negligible, being invited to leave the establishments promptly (FIG. 5.2).

5.2: Brownian walker performing random steps of lengths l, along a line.

A Brownian walker like this occupies a position $X(N)$ that is equal to the sum of the N steps that it has made:

$$X(N) = \sum_{1}^{N} l_i$$

where l_i is a random variable that can take two values: $+l$ or $-l$. Over time, the drunkard seems barely to depart from the home position $X = 0$, because he walks, with equal probabilities, on the right and on the left. Still, over time, he explores larger and larger distances, given by the variance of X, i.e. the sum of the squared displacements. We have :

$$X^2(N) = \sum_{1}^{N} l_i^2 = Nl^2$$

Since $N \approx Vt/l$, we have:

$$X^2(N) \approx 2Dt$$

where $D = \frac{1}{2}Vl$ is the diffusion coefficient . The result is important: it shows that the drunkard is governed by a diffusive law.

Now, suppose that the drunkard has many colleagues, who decide to visit the same bars, but in an uncorrelated manner, i.e. without being influenced by each other. How can we describe their distribution along the line ? Said differently, how can we calculate $p(X,t)$, the distribution of distances travelled by the walkers ? This is where a fundamental statistical theorem comes into play: the central limit theorem, which stipulates that at late times, the spatial distribution of the walkers is Gaussian. For a spot of dye initially injected in a fluid at rest, this signifies that at long times, the concentration field of the dye adopts a Gaussian profile. Later, we will establish the same result in a different manner.

With such populations of walkers, Fick's law, i.e. proportionality between matter flux and concentration gradients, can be demonstrated. Take N walkers located on the same site, and assume that the next site, located on their right, is empty. Half of the population will move to the empty site, so that the flux J, defined as the number of walkers that jump to the empty site per unit of time, will be:

$$J = \frac{1}{2}NV$$

On the other hand the gradient of walker number G is $-N/l$. With $D = \frac{1}{2}Vl$, we have:

$$J = -DG$$

which is Fick's law[3].

5.1.2 Normal and anomalous diffusion

The notion of anomalous diffusion

The limits of the preceding theory, called 'normal diffusion', were discussed by Levy in the 1930s [8,9] (see the review by JP. Bouchaud and A. Georges [10]). Normal diffusion assumes the absence of correlations between the walker jumps, and narrow distributions of their characteristics (for instance, step lengths, or waiting times between two successive steps). However, in a number of cases, such assumptions do not hold. Take a traveller who walks to the taxi station, reaches the airport by car, takes a flight from Paris to Chicago, then takes another taxi and, eventually, reaches the hotel by walking to the lobby. The total travelled distance, which is the sum of all the steps he made, is dominated by one single event: the flight. We could neglect the others, this will not change the result significantly. The example illustrates the case where the distribution of the walker's steps is broad. In such cases, averaged quantities are dominated by a few large events. In the example of the preceding section, the step distribution was narrow and averages result from the contributions of many events. The existence of broad distributions may completely change the statistical behaviour of the system, leading to 'anomalous' diffusion. In anomalous diffusions, the central limit theorem no longer applies and walkers do not disperse as $\sqrt{t}$, but in a different manner, no longer universal, but dependent on the particular process at work. This is not just abstraction. Many physical situations, relevant to microfluidics, involve anomalous diffusion.

Hyperdiffusion in the presence of a shear

An important case of anomalous diffusion is hyperdiffusion. A model is shown in Fig. 5.3.

The walker moves in a 2D space, in the presence of a shear flow $U(y) = \gamma y$, where U is the flow component along x. Let us assume that it makes leaps in the y direction, and, in the meantime, is transported in the x direction by the shear flow . We have here a coupling between a diffusive and a ballistic process, and there is no reason that

[3]Note that the law can also be obained by using Π theorem: writing $J = f(D, G)$, we have a number of units equal to 2 and a number of variables equal to 3. With that, only 3-2=1 expression is possible, $J \sim DG$

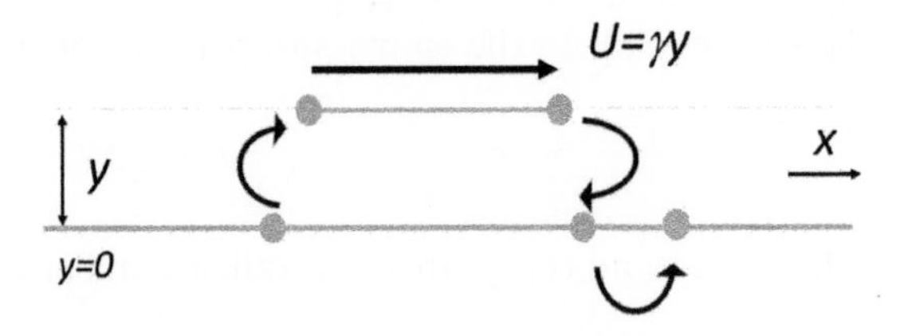

5.3: An example of hyperdiffusion: walker wandering in the plane in the presence of a shear flow.

the walk would be diffusive. The walker trajectory $(x(t), y(t))$ can be estimated with the following equations:

$$x(t) \sim \gamma y t; y \sim \sqrt{Dt}$$

in which D is the diffusion coefficient. From this equation, we find:

$$x(t) \sim \gamma D^{1/2} t^{3/2}$$

An assembly of such walkers thus spreads, along the flow streamlines, as time to the power of 3/2. This is 'hyperdiffusion'. It is interesting to note that x and y are linked by the relation $y \sim (\frac{D}{\gamma})^{1/3}\, x^{1/3}$. With this relation, one may infer that the cloud of particles adopts an S shape. If the shear was zero, the pattern formed by the particles would be a disk.

The hyperdiffusive behaviour of the walkers, in the presence of the shear, originates in the existence of correlations: the movement of the walker in the x direction is coupled to the leaps in the y direction. This correlation breaks the assumption of independency of the events, under which the central limit theorem can be established. Later, we will encounter several cases of anomalous diffusion.[4]

5.1.3 The Stokes-Einstein law

Langevin equation. Einstein established the Stokes-Einstein law in 1905 [4]. To obtain this law, he applied a concept introduced in 1883 by J.H. Van't Hoff [12] – the osmotic pressure – to particle suspensions.[5] A simpler demonstration was given later, in 1908, by P. Langevin [2, 5, 11]. In his demonstration, P. Langevin introduced, for the first time, the notion of stochastic equations, which later had a deep influence on contemporary physics, as underlined in Ref. [2]. We shall take his approach.

In Langevin's theory, a spherical particle of mass m, radius r, immersed in a fluid of viscosity μ, is subjected to a random force $f(t)$ that drives it in an erratic manner. With $u = \frac{dx}{dt}$ its speed, we have the relation:

[4] In the context of microfluidics, due to confinement, after some time, another regime takes place, which is, most of the time, diffusive.

[5] In fact, in the period 1905-1910, Einstein produced three demonstrations of his law [11]

$$< m(\frac{dx}{dt})^2 >= kT \tag{5.1}$$

in which the brackets mean 'statistical averaging', i.e. averaged over many particles. This is the equipartition theorem. At equilibrium, the kinetic energy of the particles must be equal to the thermal energy of the liquid around, i.e. 1/2 kT per degree of freedom.

Applying the fundamental law of mechanics, we have:

$$m\frac{du}{dt} - \beta u = f(t) \tag{5.2}$$

for which we will assume, for simplicity, the following initial conditions: at $t = 0$, $x = 0$ and $v = 0$.

In Eq. 5.2, $\beta = 6\pi\mu r$ (where r is the particle radius and μ the fluid viscosity) is the Stokes friction coefficient (determined in Chapter 3). $f(t)$ is a stochastic force. In the traditional presentation of Langevin equations [2], $f(t)$ represents the momentum transferred by fluid molecules hitting the particles, and pushing them in an erratic manner. The time of variation of f is molecular. To provide an idea, we show, in Fig. 5.4 an approximate mathematical model of $f(t)$. The graph represents a function which adopts, at each increment of time, a random value comprised between -1/2 and 1/2.

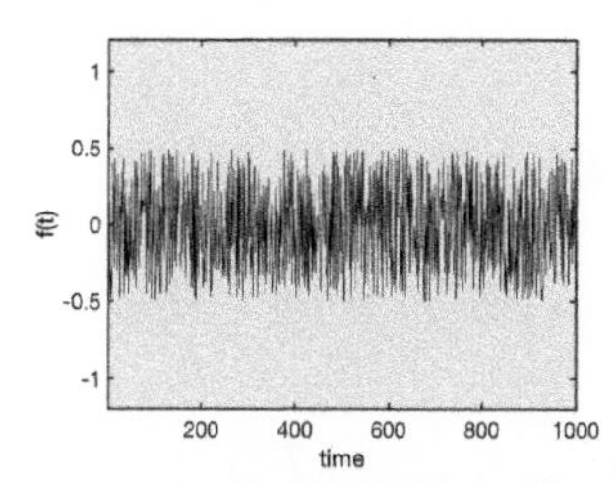

5.4: An example of f(t).

In the simplest situation, the random forces exerted against the particles are considered instantaneous, of zero statistical average, and without memory. These properties are expressed by the following relations:

$$< f(t) > 0 \text{ and } < f(t)f(t+\tau) >= F\delta(\tau)$$

in which F is a force which we will determine later. In the above equation, the Dirac function is a generalized function whose properties are: $\delta(\tau) = 0$ for $\tau \neq 0$ and $\int_{-\infty}^{+\infty} \delta(\tau)d\tau = 1$. The mathematical function shown in Fig. 5.4, divided by its standard type deviation, and applied to a population of particles, possesses such properties in the limit where the time increment is zero.

One might think that, since $f(t)$ is impossible to express analytically, it will be difficult, by working out Eq. (5.2), to gain valuable information on $v(t)$. P. Langevin [5] solved it by postulating, without much explanation, zero correlation between the particle position x and the fluctuation force f. Here, we take the modern' approach, which requires a more elaborate mathematical formalism, but does not need this assumption (see comments in [2,11]). To perform the calculation, we follow the presentation given in 'colloidal dispersion', by W. B. Russel et al. [13].

By integrating Eq. 5.2 with the boundary conditions, we find:

$$v = \frac{1}{m}\int_{-\infty}^{t} f(t')exp(-\beta(t-t')/m)dt'$$

By multiplying the equation by $v(t+\tau)$ and taking the statistical average, one obtains the relation[6]:

$$< R(\tau) >=< v(t)v(t+\tau) >= \frac{F}{2\beta m}\exp(-\frac{\beta}{m}\tau) \tag{5.3}$$

In liquids, the viscous time $\tau_v = m/\beta = \frac{m}{6\pi\mu r} \approx \frac{2\rho r^2}{9\mu}$, with ρ the fluid density, is for water, and for particles 1 μm in diameter, approximately 200 ns. After a few τ_v, the self correlation of v vanishes out, which means that, after this time, particles forget the speed at which they moved. Further, on comparing, for $\tau = 0$, Eq. (5.5) to Eq. (5.1), we find:

$$F = 2\beta kT$$

Bu using Eq. (5.3), it is not difficult to calculate the product $\frac{1}{2} < x(t)x(t) >$. We find:

$$\frac{1}{2}\frac{d}{dt} < x(t)x(t) >=< x(t)v(t) >= \int_{0}^{t} R(\tau)d\tau = \frac{kT}{\beta}(1-\exp(-\frac{\beta}{m}\tau))$$

Therefore, by restricting ourselves to times much larger than τ_v, which defines the so-called overdamped approximation', we obtain, after integration, the following relation:

$$< x^2 > (t) = \frac{2kT}{\beta}t \tag{5.4}$$

which is the Stokes–Einstein law. Particles, animated by uncorrelated impulsions, undergo a diffusion process, characterized by a diffusion constant equal to:

[6] To perform the calculation, we use the identity $\int_{-\infty}^{+\infty}\delta(t')g(t')dt' = g(0)$.

$$D = \frac{kT}{6\pi r \mu}.$$

The law was used by Perrin in 1914 to estimate the Avogadro number [3]. The observation he made is shown in Fig. 5.1. This shows three trajectories of particles placed at the surface of a water tank. Perrin could provide an estimate of D by analyzing them. From the knowledge of D, kT and therefore the Avogadro number $N_A = R/kT$ (where R is the ideal gas constant) could be estimated. Perrin found 6.8 10^{23}, which is close to the modern value. It was the first time that a quantitative link could be established between molecules and a macroscopic phenomenon.

Calculating dispersion phenomena in the presence of a flow on a computer. To calculate trajectories of Brownian particles, in the presence of a flow, one should, in principle, use Langevin equations and add terms representing the flow. In this approach, it is tempting to neglect the acceleration term, which, *in fine* is neglected in the last part of the resolution of the problem, the justification being that we are not interested in times smaller than τ_v, i.e. extremely short times. However, if we set $m = 0$ at the beginning of the problem, owing to the equipartition theorem, there is no particle agitation. So, how to proceed?

What is done in practice is to suppress the inertial term and, in the meantime, adjust the forcing so as to keep molecules agitated. In this spirit, by assuming that particle and flow speed are equal (an hypothesis we will return to later), the equations governing particle speed $\mathbf{v}$ reads:

$$\mathbf{v} = \mathbf{g(t)} + \mathbf{U(x,t)} \tag{5.5}$$

in which $\mathbf{g}(t)$ is an effective random forcing and $\mathbf{U}(\mathbf{x}, t)$ is the local flow speed. In the equations, $\mathbf{g(t)}$ is such that for a fluid at rest, particles are subjected to the Stokes–Einstein law. This approach raises a number of subtle questions that we do not address here.

Let us now go to the computer and decompose the trajectories in n steps. In the computation, $\mathbf{g_n}$ is a random, uncorrelated function whose variance is $< g_n^2 >= 2D\tau$, in which τ is the temporal increment used in the computation. Defined in this manner, $\mathbf{g_n}$ guarantees that in the absence of flow, particles are subjected to Stokes Einstein law.

To take an example, let us consider an ensemble of particles immersed, in a two-dimensional space, in a shear flow. Defining x_n and z_n as the particle position at step n and setting $\mathbf{U} = (\gamma z_n, 0)$ (with γ the shear rate), we have:

$$x_{n+1} = x_n + \gamma z_n \tau + g_{xn} \tag{5.6}$$

$$y_{n+1} = y_n + g_{yn} \tag{5.7}$$

From Eq. (5.27), we infer that $z_{n+1} = \sum_1^n g_{y_n}$ and thus $< y_{n+1}^2 >= \sum_1^n < g_{y_n}^2 >= 2Dn\tau = 2Dt$, where $t = n\tau$. Thereby, the movement of the particles, in the direction normal to the shear, i.e. along y, is brownian. In the absence of the shear, i.e. with $\gamma = 0$, the same property holds for the movement along the shear, i.e. along x. These results confirm the choice of g we made. The left plot of Fig. 5.5 shows a superposition of 100 trajectories, obtained without shear, all with the same initial conditions, i.e. $x = y = 0$ for $t = 0$, and after 1,000 steps. They are calculated with $D = \frac{1}{2} < g^2 >= 0.0424$, One sees that the trajectories develop isotropically. The cloud' width can be estimated by $\sqrt{2Dt} \approx 9$.

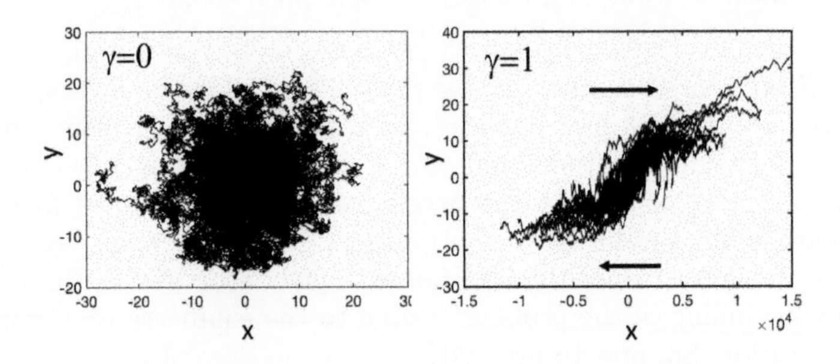

Fig. 5.5: Two simulations, conducted with the same initial conditions. In one case, the fluid is at rest ($\gamma = 0$) and in the other case, a shear flow is present ($\gamma = 1$).

When the flow is present (see Fig. 5.5, γ=1), the movement of the particles along y remains the same, as we saw, but that along x changes. Trajectories spread orders of magnitude faster along the flow, than normally to it, developing an S shape. The movement is no longer isotropic.

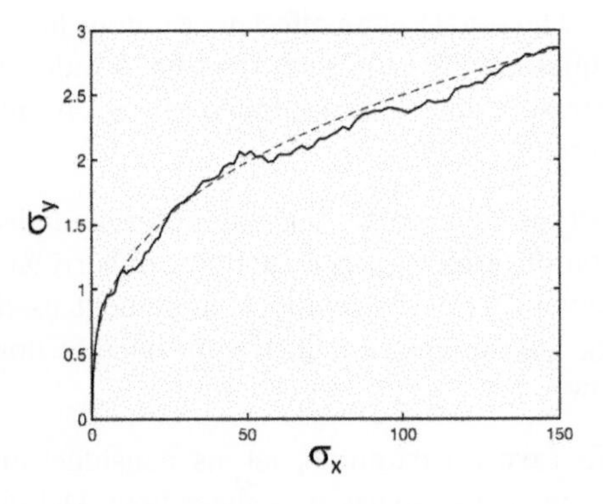

5.6: Width σ_x, along x, of the particle cloud, formed by particles injected at the origin $x = z = 0$, as a function of the cloud width σ_y, along y. Averaging is performed over 100 trajectories. The dashed line represents a fit given by $\sigma_y = 0.54\sigma_x^{1/3}$

The simulation leads, in the case γ=1, to the following relations:

$$< x^2 >\approx 0.025\ t^3 \text{ and } < y^2 >\approx 0.085\ t$$

The fact that $< x^2 >$ varies more rapidly than t indicates the hyperdiffusive character of the regime. Fig. 5.6 represents the standard type deviations of y, i.e. $\sigma_y = \sqrt{< y^2 >}$, as a function of $\sigma_x = \sqrt{< x^2 >}$. These standard deviations provide estimates for the widths of the 'clouds' shown in Fig. 5.5. The empirical law we found is $\sigma_y \approx 0.54\sigma_x^{1/3}$. As said above, the tracer spreads rapidly along the flow, and slowly normally to the flow, adopting a S shape. The exponent we find is consistent with the walker model we developed earlier, which led to $y \sim (\frac{D}{\gamma})^{1/3}\, x^{1/3} \approx 0.36x^{1/3}$.

The concept of mobility

A force present in the system, for instance a Coulomb force produced by an electric field on charged particles, can affect the particle movement. How to introduce such a force theoretically? This is achieved by using the notion of mobility, i.e. a parameter μ_m defined by:

$$\mathbf{F} = \mu_m \mathbf{v}$$

An example of mobility is the Stokes factor β which we used earlier. With this notion, forces, translated into speed (with the relation $\mathbf{v} = \mathbf{F}/\mu_m$), can be directly introduced in the equations of the problem.

5.2 Advection-diffusion equation and its properties

5.2.1 Fick law

Here, we are no longer interested in determining the trajectories of each tracer particle. We work in a continuum framework, characterize the particle population by its mass concentration $C(\mathbf{x}, t)$, and attempt to determine the evolution of C. The massic flux $\mathbf{J}(\mathbf{x}, t)$ is, by definition, a vector oriented along the mass flux of particles, i.e. along their speed, and whose intensity is defined by the ratio δm crossing, per unit of area and time, a surface δS, placed normally to $\mathbf{J}$, during a time δt. It is thus defined by the relation:

$$\mathbf{J}.\mathbf{n} = \frac{\delta m}{\delta t\ \delta S}$$

in which $\mathbf{n}$ is the unit vector normal to the surface element δS. Fick's law relates $\mathbf{J}$ to the local concentration gradient $D\nabla C$. Its expression is:

$$\mathbf{J} = -D\nabla C$$

where D is the coefficient of diffusion. For reasons similar to those given in Chapter 3, the domain of validity of this law concerns, for gases, systems larger than the mean free path, and for liquids, systems roughly larger than 1 nm. In a number of situations, the system may be anisotropic, i.e. the diffusion constant may be different in different directions. This happens close to a wall. In this case, D is a diagonal tensor with two components, one normal and the other parallel to the wall. We will describe this situation later. For the moment, unless specified, we will consider only isotropic situations, where the diffusion process is characterized by a single parameter, the diffusion constant (or coefficient) D.

The units of the diffusion coefficient are m^2/s. Table 5.1 shows a selection of values.

Diffuser	Medium	Diffusion constant (m^2/s)
Fluorescein	Water	$4\ 10^{-10}$
Sucrose	Water	$5\ 10^{-10}$
Albumin	Water	$6\ 10^{-11}$
20 bp DNA strand	Water	$5\ 10^{-12}$
6 Kbp DNA strand	Water	$8\ 10^{-13}$

Table 5.1 Diffusion constants of a selection of molecules in water.

The orders of magnitude of the diffusion coefficients, for small molecules in simple liquids (e.g. alcanes, alcohol), are on the order of 10^{-10} m^2/s. The diffusion coefficients found for albumin and DNA strands in water are smaller because of the larger sizes of the molecules, which is consistent with the Stokes–Einstein law. More generally, the values of Table 5.1 agree, within 30% or so, with the Stokes-Einstein equation, which we rewrite here:

$$D = \frac{kT}{6\pi R\mu}$$

5.2.2 Advection-diffusion equation

The advection-diffusion equation expresses mass conservation. In a fixed frame of reference, the local flux of matter includes two components: one, $\mathbf{u}C$, where $\mathbf{u}$ is the local velocity, is the advective part – sometimes called 'convective' –, and the other, $-D\nabla C$, is the diffusive one. By using the divergence formula, one finds that the mass change δm, in a volume δV, caused by these fluxes, is equal to $\mathrm{div}(-D\nabla C + \mathbf{u}C)\delta V$. With this, by noting that mass change of the control volume is also $\frac{\partial C}{\partial t}\delta V$, and that mass is conserved, one obtains, for incompressible fluids, the advection-diffusion equation:

$$\frac{DC}{Dt} = \frac{\partial C}{\partial t} + \mathbf{u}\nabla C = D\Delta C \tag{5.8}$$

The phenomenon described by this equation is called dispersion'. In the case where a source or a sink is present in the system, additional terms must be incorporated in the right–hand side of the equation. This possibility, however, is difficult to realize experimentally. Thus, for the sake of simplicity, we will not consider these cases. The boundary conditions are:

$$C = 0 \text{ for adsorbing walls and } \nabla C = 0 \text{ for impermeable walls}$$

In a closed system, with impermeable walls, Eq. (5.8) possesses a trivial solution, in which concentration C is uniform and stationary. The system always tends to this solution. This intuitive statement is justified mathematically by considering the variance of the concentration field, defined by:

$$\Theta = \int C^2 d\mathbf{x}$$

where $d\mathbf{x}$ is the elementary volume and the integral is taken over the total space occupied by the fluid. By multiplying each term of Eq. (5.8) by C, with the boundary conditions indicated above, and integrating, one obtains the folllowing equation for Θ :

$$\frac{\partial \Theta}{\partial t} = -D \int (\nabla C)^2 d\mathbf{x}$$

The equation states that the variance always decreases with time. In the ultimate state, $\int (\nabla C)^2 d\mathbf{x} = \mathbf{0}$, which implies that C is uniform. The tracer is said to be 'completely mixed' with the fluid. In closed systems, dyes and tracers, will thus always tend to get completely mixed with the fluid. The question is: how much time do we need to reach this state? This important question will be the focus of the next sections.

When $D = 0$, we have:

$$\frac{DC}{Dt} = 0 \, .$$

The equation states that the local concentration $C(\mathbf{x}, \mathbf{t})$ is conserved along the fluid trajectories. In practice, this property holds for short times, i.e. during an interval of time well below the diffusion time l^2/D, in which l characterizes the system size. As discussed earlier, we need to allow this time for diffusion to come into play.

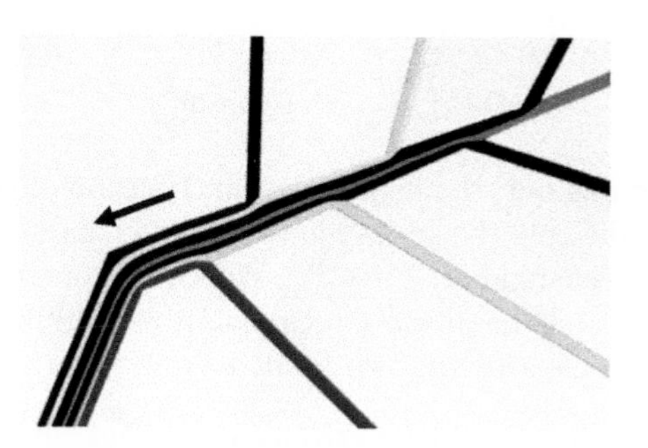

Fig. 5.7: Microfluidic experiment in which seven aqueous solutions flow side by side in a microchannel with several entries. The experiment shows that, over short times, dye particles follow, in practice, the flow trajectories, or equivalently, dye concentration is conserved along the flow streamlines. (Reprinted from [14], with permission from American Association Advanced Science. Copyright 2022.)

Figure 5.7 illustrates such a situation: seven aqueous solutions are driven, from right to left, in a system including six bifurcations [14]. Dyes remain in the device only a fraction of second, so diffusion did not have time to act. In such conditions, the aqueous solutions flow side to side, keeping sharp interfaces between them. The dye particles follow the flow trajectories, or equivalently, dye concentration is conserved along the flow streamlines. Should diffusion come into play, the interfaces between the streams would blur substantially.

The opposite limit, $\mathbf{u} = 0$ is pure diffusion. In this case, Eq. (5.8) reduces to:

$$\frac{\partial C}{\partial t} = D\Delta C \tag{5.9}$$

and the system behaves as if the fluid were at rest.

5.2.3 The Peclet number

One fundamental dimensionless number is the Peclet number. Its definition is:

$$Pe = \frac{Ul}{D}$$

where U is a characteristic flow velocity, l is a characteristic scale of the system (e.g. the channel height), and, as we saw above, D is the diffusion coefficient. The order of magnitude of the advection terms is $U\delta C/l$ (where δC is a typical variation of the concentration over a distance l), and the order of magnitude of the diffusion terms is $D\delta C/l^2$. The Peclet number represents the ratio of the first term with the second. It thus estimates the relative importance of advection with respect to diffusion. Using minimal wording, we could write:

$$Pe \sim \frac{Advection}{Diffusion}$$

The higher the Peclet number, the higher the contribution of the flow to the transport process. If we applied directly the scaling lows given in Chapter 2, we would get following relation for the Peclet number: $Pe \sim l^2$. We would conclude that in microfluidics, only diffusion matters, and the advection terms are negligible. This would imply that in microsystems, it is useless to stir two fluids to mix them.

However, these conclusions would be erroneous: in fact, diffusion coefficients D are typically on the order of 10^{-10} m^2/s for molecular tracers in water for instance (see Table 5.1). Thus, with velocities on the order of 1 mm/s, in a canal 100 μm high, the Peclet number is on the order of $1,000$. We could envision a situation where the Peclet number is small. However, this would concern channels much smaller, for instance, a micrometre in size, which is not current in microfluidic. In nanofluidic devices, the situation is different. The Peclet number is small and diffusion prevails.

5.3 Analysis of diffusion phenomena

5.3.1 Diffusion of a tracer in an infinite medium

Let us place ourselves in a one-dimensional space, and inject a tracer, a solute, or a dye spot in a liquid tank (i.e. a line), 'filled' with a liquid at rest. In such conditions, Eq. (5.8) reduces to:

$$\frac{\partial C}{\partial t} = D\frac{\partial^2 C}{\partial x^2} \tag{5.10}$$

In order to simplify the mathematics, we will consider that, initially, the concentration profile $C(x,t)$ is represented by the following function:

$$t = 0,\ C = C_0\delta(x) \tag{5.11}$$

where $\delta(x)$ is the Dirac function and C_0 is the total mass of tracer enclosed in the spot. We already used the Dirac function. It is a generalized function that is zero everywhere except at the origin, where it is infinite. Its integral over space is unity. It is important to note that the unity of $C(x,t)$ is a mass per unit of length. In our one-dimensional world, it represents a mass concentration.

The fact that, at $t = 0$, the spot is of zero size and has infinite concentration may look unphysical. In fact, it is possible to argue that taking this initial condition is acceptable at late times, i.e. when the dye has spread so much that its size is much larger than the initial spot. With the initial condition given by Eq. (5.11), the solution to Eq. (5.10) is a gaussian function, whose expression is:

$$C(x,t) = \frac{C_0}{\sqrt{4\pi Dt}}\exp\left(-\frac{x^2}{4Dt}\right) = \frac{C_0}{\sqrt{2\pi\sigma^2}}\exp\left(-\frac{x^2}{2\sigma^2}\right) \tag{5.12}$$

in which σ is the standard type deviation, defined by:

$$\sigma^2 = \frac{1}{C_0}\int_{-\infty}^{\infty} x^2 C(x,t)dx = 2Dt$$

For completeness, let us add two identities

$$C_0 = \int_{-\infty}^{\infty} C(x,t)dx, \quad \int_{-\infty}^{\infty} xC(x,t)dx = 0$$

The solution $\frac{C(x,t)}{C(0,t)}$, plotted as a function of x/σ, is shown in Fig. 5.8. With the dimensionless coordinates we used, the plot is universal, i.e. the same at all times, and independent of C_0 and D.

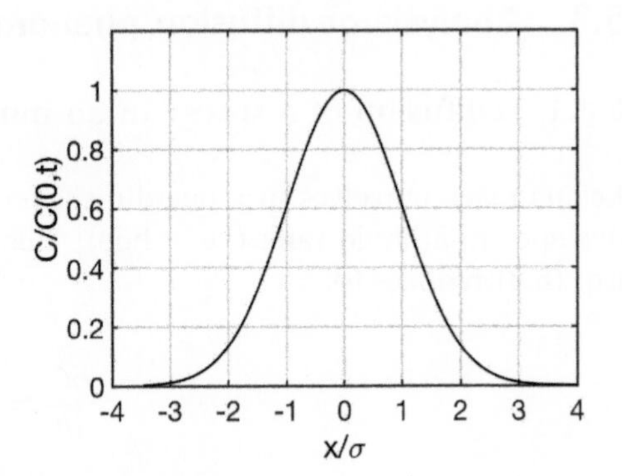

5.8: Universal representation of the gaussian function. For $x = \sigma$, one has $C(x,t)/C(0,t) \approx 0.607$. Mass conservation is satisfied because $\int_{-\infty}^{\infty} C(x,t)dx = C_0$ at all time.

Figure 5.8 can also be viewed as the distribution of concentration levels, at fixed time. This is the so-called Gaussian or normal distribution. The way in which the solution to Eq. (5.12) is obtained is not straightforward. In short, the approach consists of guessing a form of the solution compatible with the boundary conditions. For this task, theorem Π is extremely useful. Owing to linearity of the equation, one must have: $C(x,t) = C_0 F(x,t)$ with F is an unknown function, whose dimension is the inverse of a length. Theorem Π tells us that two forms are possible and, among them only one is compatible with the initial condition. This leads to write F in the form: $F(x,t) = \frac{1}{\sqrt{Dt}} G(\xi)$, in which $\xi = x^2/2Dt$.. This being written, the rest is (tedious) calculation for determining F.

It is usual to take, for simplicity, the standard deviation σ as an estimate of the spot size. A one-dimensional spot of uniform concentration $C(0,t)$ and size σ would includes 80% of the total mass of the tracer. It may therefore represent a model for schematizing the spot. Let us thus write, for the spot size l, in terms of order of magnitude, the following relation: $l \sim \sqrt{Dt}$. This estimate is indeed consistent with our drunkard model. The same relation can also be obtained by using theorem Π: l depending only on D and t, theorem Π imposes $l \sim \sqrt{Dt}$.

Similar Gaussian solutions hold in two and three dimensions, in Cartesian, cylindrical and spherical geometries. Analytical expressions can be found in textbooks and online.

5.3.2 Diffusion of a tracer in a box

Gaussian solution concerns infinite spaces. When a tracer diffuses in a bounded space, for instance, a one-dimensional box, gaussian solution no longer holds. In this case, a solution can still be found, but in the form of developments in Fourier series, as seen in Chapter 3, for the case of flows developing in channels of rectangular cross-sections.

Let us consider a tracer, initially injected, as in the previous subsection, in an infinitely small spot, but now inside a box bounded by two impermeable walls, located at $x = \pm L$. The boundary conditions are thus:

$$t = 0,\ C = 2LC_0\delta(x)\ ;\ x = \pm L,\ \frac{\partial C}{\partial x} = 0 \tag{5.13}$$

in which C_0 is the mass of tracer localized in the spot, divided by $2L$. The units of C_0 and C are mass per unit of length, i.e. in this one-dimensional geometry, mass concentrations. The solution to Eq. (5.10) reads:

$$C(x,t) = C_0 \left(1 + 2\sum_{1}^{\infty} \exp(-Dn^2\pi^2 t/L^2)\cos(n\pi x/L)\right) \tag{5.14}$$

It is represented in Fig. 5.9.

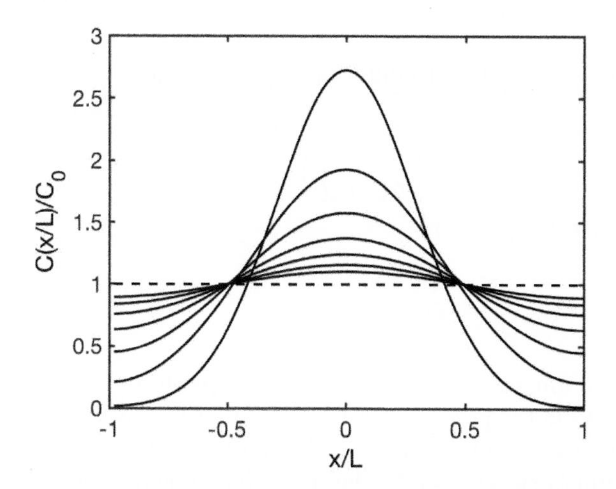

5.9: Concentration profiles calculated in a box, at different times. The curved arrow indicates increasing times. The different times are, in units of $\tau_D = L^2/2D$: 0.043, 0.086, 0.13, 0.17, 0.21, 0.26, and 0.3.

According to Eq. (5.9), t does not act individually, but through the product Dt/L^2. It is thus natural to introduce a characteristic time $\tau_D = L^2/2D$ and describe the behaviour of the solution in terms of the ratio t/τ_D. At short times (e.g. for t=0.03 τ_D), we recover the Gaussian solution, because, at those times, the concentration field has spread over lengths much smaller than the box size. The spot, still localized, does not 'see' the walls. At longer times, the concentration field broadens, and, eventually, tends form a plateau located $C = 1$, which is the mean concentration. We need to wait a time on the order of τ_D to reach a state close to homogeneity. In the process, the total mass of the tracer is conserved, because walls are impermeable. Should we have taken, as the initial condition, $C = 0$, a flux across the wall would develop and, along time, C would decrease to zero.

5.3.3 Interface broadening

Here we return to infinite media, and again, limit ourselves to one dimension. Let us consider that, initially, a front separates two regions: one with a uniform dye concentration C_0 and the other without dye (see Fig. 5.10).

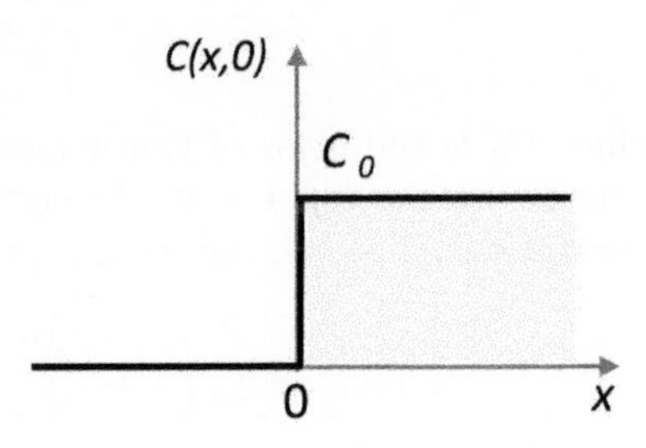

5.10: Initial conditions: the tracer is located on the right, i.e. $x > 0$.

We thus have:

$$t = 0, \;\; C = 0 \text{ for } x < 0, \text{ and } C = C_0 \text{ for } x > 0 \tag{5.15}$$

With such initial conditions, the exact solution to diffusion Eq. (5.9) reads:

$$C(x,t) = \frac{1}{2} C_0 \left(1 + \operatorname{erf}\left(\frac{x}{2\sqrt{Dt}} \right) \right) \tag{5.16}$$

where the error function, called erf, is defined by:

$$\operatorname{erf}(z) = \frac{2}{\sqrt{\pi}} \int_0^z \exp(-v^2) dv \,.$$

The solution, represented by using dimensionless coordinates x/σ (with $\sigma = \sqrt{2Dt}$), and C/C_0, is shown in Fig. 5.11. As in the preceding section, the plot is universal, i.e. it is valid at all times, and is independent of C_0 and D. The thickness of the front is given by the width of the error function, i.e. $\sqrt{2Dt}$. The front thus broadens proportionally with the square root of time.

For a long time, the phenomenon was challenging to investigate experimentally. This was due to the presence of uncontrolled flows, linked to thermal convection or buoyancy, or both, which considerably perturbed the diffusive process. With the advent of microfluidics, it became possible to eliminate these perturbations. The experiments of Refs. [15–17], performed in shallow T-shaped microchannels, called T-sensors', exploited this possibility. The system is shown in Fig. 5.12.

In the experiment of Ref. [15], two miscible fluids are injected in a T-junction and circulate side by side in the main channel. One is a buffer; the other is an albumin

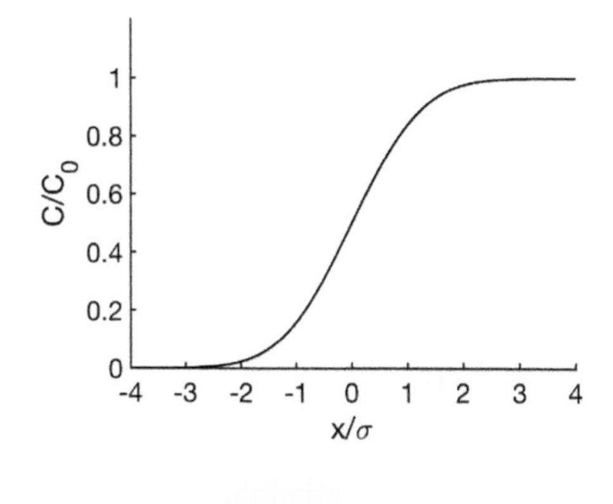

5.11: Universal concentration profile given by Eq (5.16).

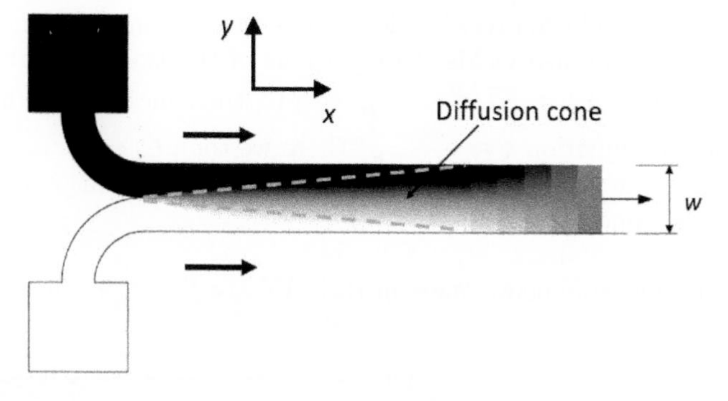

Fig. 5.12: Sketch of the device (T sensor) used in Ref. [15] for carrying out diffusion studies. Two miscible fluids, injected in the device, circulate side by side towards the right. The upper fluid transports a (black) solute, of concentration C_0, and the other is a buffer. Downstream, the system builds up a conical zone, within which the black solute diffuses.

blue solution (AB580). Downstream of the junction, the system builds up a conical zone, within which albumin blue diffuses. How can we model the phenomenon ?

The concentration gradients are mainly along y, i.e. transverse to the flow, in the horizontal plane. There is indeed a variation of the concentration field along x, evidenced by the presence, mentioned above, of a cone of concentration; however, this is weak, and we will neglect it. In the theory, we consider that the flow is 2D. In the experiment, the flow is driven in a shallow channel and develops a parabolic profile in the vertical direction (i.e. normally to the plane of Fig. 5.12), and is uniform horizontally; its mean speed equal to U. As the channel is shallow, we assume that the concentration is homogeneous vertically. We will discuss these approximations later. In these conditions, the stationary diffusion-advection Eq. (5.8) reduces to:

$$U\frac{\partial C}{\partial x} \approx D\frac{\partial C}{\partial y^2}$$

with the initial conditions:

$$\mathrm{x} = 0, \quad C(0, y) = 0 \quad \text{for} \quad y \leq 0, \quad \text{and} \quad C = C_0 \quad \text{for} \quad y \geq 0 \,.$$

For $y << w$, the problem is identical to the one discussed above, with t replaced by x/U. Time is translated into a spatial variable x/U. In such conditions, the solution reads:

$$C(x, y) = \frac{1}{2} C_0 \left(1 + \mathrm{erf}(\frac{y}{2} \sqrt{\frac{U}{Dx}}) \right) \tag{5.17}$$

Remembering that $x = t/U$, Eq. (5.17) tells us that the temporal evolution of C, along the flow streamlines, can be analysed by looking at the stationary dye field in the main channel. Equation (5.17) also yields the equations of the iso-concentration lines. We find $x = k \frac{U}{D} y^2$, in which $k \approx \frac{\sqrt{\pi}}{2} (\frac{2C_l}{C_0} - 1)$, with C_l the concentration level. The cone is thus a parabola of equation $x \approx \frac{\sqrt{\pi}}{4} \frac{U}{D} y^2$ (here we took $C_l = \frac{C_0}{2}$ to define the cone boundaries). The greater the speed, the narrower the width of this parabola, and thus the more acute the cone.

The concentration measurements made in Ref. [15] are shown in Fig. 5.13.

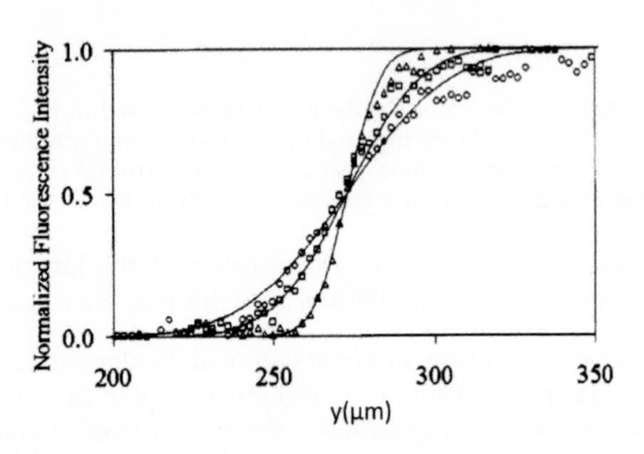

5.13: Fluorescence intensity profiles for the distribution of AB580 across the y dimension at a distance of 5000 μm downstream. Flow rates are 83.3 nl/s (circle), 416 nl/s (square), and 833 nL/s (triangle). Solid lines are analytical fits using D=4.55 10^{-10} m^2/s (Reprinted with permission from [15]. Copyright 2022 American Chemical Society.)

Theory agrees well with experiment. In Ref. [15], the system was used for measuring the diffusion constant of Albumin blue.

The butterfly effect. We just mentioned the good agreement between theory and experiment, but did not discuss the three–dimensional aspects of the problem. In the theory, as noted, we assumed that the concentration is homogeneous in the vertical' direction z, i.e. normal to the flow plane, and that the speed can be replaced by its average along z. In reality, along the vertical axis, the flow is parabolic and the concentration field develops a 'butterfly' structure. This feature was revealed by [15] and analysed by Ismagilov et al. [18] (see Fig. 5.14).

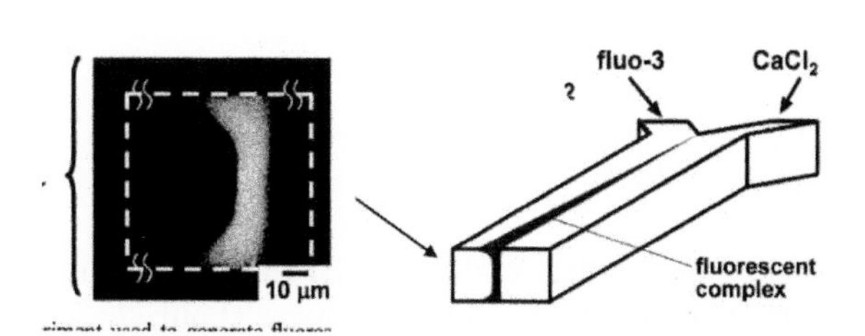

5.14: Butterfly effect. The tracer develops 'wings' close to the floor and the ceiling of the channel, in regions where speeds are low. The phenomenon, revealed in Ref. [15] was analysed by Ref. [18], from which the images are taken. (Reprinted with permission from [18]. Copyright 2022 AIP Publishing.)

Figure 5.14 (left) shows that, in the central part of the channel, the concentration of the fluo-3 complex, at the origin of fluorescence emission, is homogeneous along z. Nonetheless, at the bottom and the top, i.e. close to the horizontal walls, two 'wings' develop. This is the 'butterfly' effect. The phenomenon was explained in Ref. [18]. In short, low speeds close to the walls favour the broadening of the concentration field, compared to the central region, where speeds are greater. This leads to wing formation. Fig. 5.14 obtained a spectacular image of the phenomenon.

5.4 Analysis of dispersion phenomena

5.4.1 Plug flows, pure shear flows, and straining flows

Here, we present three important examples in which tracers are dispersed by flows.

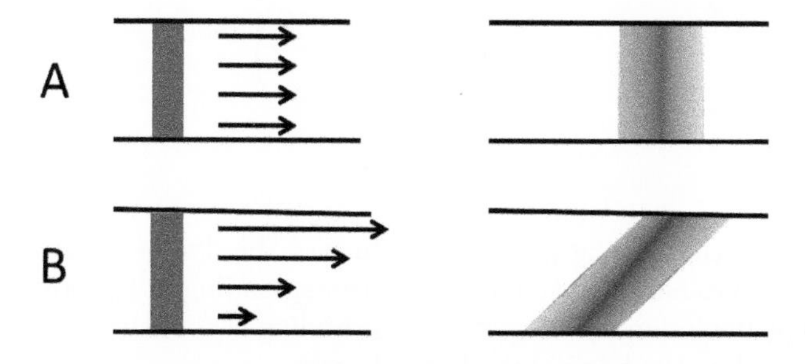

Fig. 5.15: (A) Band of tracer transported by a plug flow at different times (top: initial state; bottom: later times).(B) Band of tracer transported by a shear flow at different times (top: initial state; bottom: later times)

Example A: Dispersion in plug flows. A band of dye is transported by a uniform speed U (see Fig. 5.15 (A)). To solve the governing equations (5.8), one must place ourselves in the frame of reference of the flow. By posing $x' = x - Ut$, in which x is the streamwise coordinate in the laboratory frame, and x' in the moving frame, one obtains:

$$\frac{\partial C}{\partial t} = D\frac{\partial^2 C}{\partial x'^2} \tag{5.18}$$

In the limit where the band width is initially small, the solution is:

$$C(x,t) = \frac{C_0}{\sqrt{4\pi Dt}} \exp\left(-\frac{(x-Ut)^2}{4Dt}\right).$$

In the frame of reference of the flow, the band width δ (the standard type deviation of the gaussian function) varies as $\sqrt{2Dt}$. The flow transports the dye without changing the dynamics of spreading.

Example B: Dispersion in pure shear flows. A band of dye is transported by a shear flow $U(y) = U_0 + \frac{1}{2}\alpha y$, in which y is the coordinate transverse to the mean flow (see Fig. 5.15 (B)). The equation of diffusion advection reads:

$$\frac{\partial C}{\partial t} + (U_0 + \frac{1}{2}\alpha y)\frac{\partial C}{\partial x} = D\left(\frac{\partial C}{\partial x^2} + \frac{\partial C}{\partial y^2}\right)$$

The solution, in the limit where, initially, the band width is infinitely small, can be found in [20].[7] It reads:

$$C(x,y,t) = \frac{C_0}{2\pi\sigma_x\sigma_y} \exp\left(\frac{(x-U_0t-\frac{1}{2}\alpha yt)^2}{2\sigma_x^2} - \frac{y^2}{2\sigma_y^2}\right)$$

in which:

$$\sigma_x^2 = 2Dt\left(1+\frac{1}{12}\alpha^2 t^2\right)$$

and:

$$\sigma_y^2 = 2Dt$$

The solution shows that, along y, the spreading of the dye is diffusive. However, in the flow direction, i.e. along x, at long time, the spreading is hyperdiffusive because the standard type deviation varies as $t^{3/2}$. This finding echoes the anomalous diffusion model we presented at the beginning of the chapter.

Two remarks can be made:

1 After a time $\tau_y = \frac{w^2}{D}$, where w is the channel width, the tracer becomes homogeneous in the y direction. Then, a new regime takes place; this is the Taylor regime, that we will discuss later.

2 In practice, pure shear flows are difficult to realize experimentally, and, so far, no serious attempt has been made to confront theory with experiment. This would necessitate a moving wall, which is challenging to realize in microfluidic devices.

[7] The problem was independently solved, in about the same years, by J. Ottino [21]

Example C: Dispersion in a straining field. Here, we analyse the filamentation and reorientation processes that concentration fields are subjected to in straining flows. For the calculation, as before, we use the notations of Rhines and Young [20]. Let us consider a pure straining velocity field, defined by:

$$U = (\alpha x, -\alpha y) \quad \alpha > 0 .$$

Initially, we have:

$$t = 0, \quad C(x) = A_0 \sin \mathbf{M} \cdot \mathbf{x}$$

where C is the concentration field of a tracer, C_0 is a constant, and $\mathbf{M}$ is a vector with components (M_1, M_2) (see Fig. 5.16).

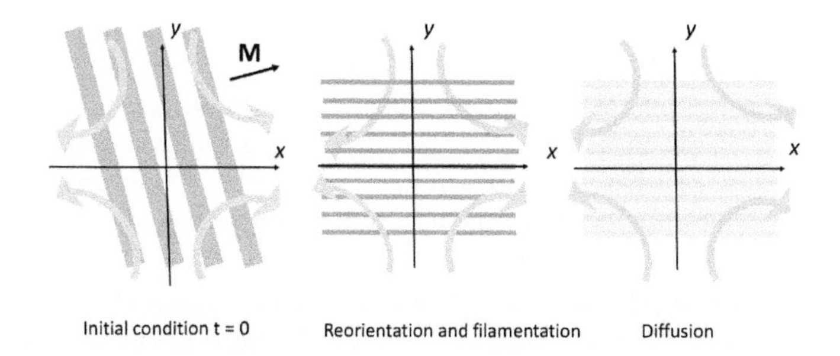

Fig. 5.16: Periodic concentration field immersed in a straining field, undergoing reorientation, filamentation and diffusion.

This initial condition represents a series of bands of dye oriented perpendicularly to $\mathbf{M}$, with 'colours' labelled with positive and negative levels. In order to discuss the different regimes, it is interesting to introduce a Peclet number. The band width is on the order of $\lambda = 1/M$ and a typical speed of the flow field is α/M. We thus define the Peclet number as:

$$Pe = \frac{\alpha}{DM^2}$$

Large Peclet numbers are associated to large straining rates, large bands, or small diffusion constants D. We postulate the following solution:

$$C(x, t) = A(t) \sin(\mathbf{m(t)} \cdot \mathbf{x})$$

where $\mathbf{m}(t) = (m_1(t), m_2(t))$ is a time dependent vector. By inserting it into the advection diffusion equation, we find:

$$\frac{dm_1}{dt} = -\alpha m_1(t); \frac{dm_2}{dt} = \alpha m_2(t) \text{ and } \frac{dA}{dt} = -Dm^2(t)A$$

The solution to these equations reads:

$$m_1(t) = M_1 exp(-\alpha t) \; , \; m_2(t) = M_2 exp(\alpha t) \text{ and } A(t) = A_0 exp(-D \int_0^t m^2 dt)$$

Without flow, the bands keep their orientation, positions, and wave-numbers, while their concentration levels decays, with time, as $exp(-DM^2 t)$. The time constant of this decay is $\tau_D = \frac{1}{DM^2}$. At times well above τ_D, A(t) becomes vanishingly small, and the concentration is homogeneous. This is complete mixing. Taking a width $\lambda = 1/M = 100\mu$m, and $D = 10^{-11} m^2/$s, one obtains $\tau_D = 10^3$ s. This is a long time. This indicates that in microfluidic systems, diffusion can be slow.

Let us now analyse the solution when the flow is present. Several conclusions can be drawn out:

- With $m_1(t)$ decreasing, the vector **m** tends to align along y, the principal axis of the strain. As sketched in Fig. 5.16, the bands rotate and align along y. This is the reorientation process. At large strains, the process is vigorous, because $m_1(t)$ decreases exponentially, with a time constant equal to α^{-1}.

- In the meantime, $m_2(t)$ increases. Thereby, bands become thinner and thinner, while their density, i.e. their number per unit of length in the **m** direction, being proportional to $m_2(t)$, also increases. This is the filamentation process, sometimes called striation. At large strain rates, it is vigorous, because $m_2(t)$ increases exponentially in time, with a time constant equal to α^{-1}. After several α^{-1}, the bands takes the form of closely packed filaments, as sketched in Fig. 5.16.

- With m^2 increasing, the amplitude of the concentration levels decrease. Bands blur. In order to analyse the process, we note that, after several α^{-1}, $m_1(t)$ is vanishingly small, and, therefore, A(t) can be approximated by:

$$A(t) \approx A_0 \exp(-\frac{DM_2^2}{2\alpha} \exp(2\alpha t))$$

  We find a surprising result: the concentration levels decrease at an exponential of exponential rate. This is fast. With α=1 s^{-1}, we find, after 3 s, a decay of $A(t)$ by a factor of 2,400. And after 10 s, the factor becomes 10^{10}. Instead of taking 1000 s, without flow, the mixed state – associated to small A – is reached in a few seconds. It is remarkable that the process is controlled by the strain rate α, independently of the diffusion constant; D influences the process only through a logarithmic correction (this does not mean that D can be replaced by zero).

To conclude, the model shows that, at large strains, mixing is achieved at a rate much faster than diffusion and, in practice, independent of it. The entire process is controlled by the strain, diffusion just 'completing' the homogenization process in a negligible time. These properties will be used to design efficient micromixers.

5.4.2 Taylor dispersion in microchannels

Taylor dispersion, also called Taylor-Aris dispersion [22,23], provides another example for which diffusion is accelerated by the shear. Let us consider a tracer injected in a Poiseuille flow driven between two plates separated by h, at small Reynolds numbers. Fig. 5.17 schematizes the situation. At large Peclet numbers, as discussed earlier, two ranges of time can be defined: at short times, the tracers follow the streamlines, forming an arch, advected downstream. At times greater than h^2/D (where D the diffusion constant of the tracer in the liquid), diffusion will come into play and C will tend to homogenize. As schematized in Fig. 5.17, the concentration field will take the form of a patch of dye, approximately rectangular. How can we describe such a structure ? For this, we again take the diffusion advection equations:

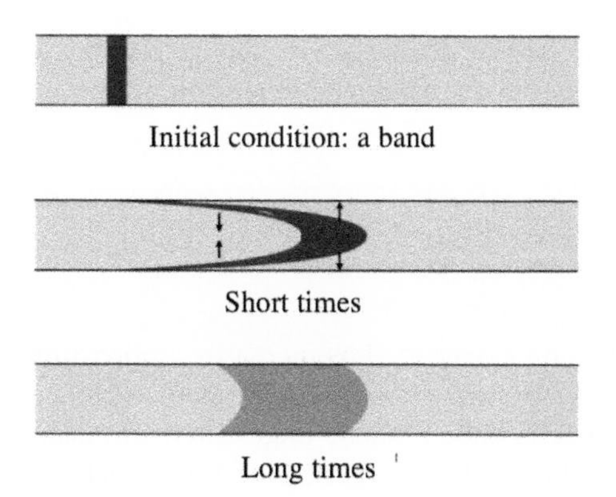

5.17: A dye is injected in a Poiseuille flow, at high Peclet number. First, the dye particles follow the fluid trajectory. Later, the dye spot tends to adopt a rectangular shape, whose spreading rate, in the streamwise direction, is given by the Taylor dispersion equation.

$$\frac{DC}{Dt} = \frac{\partial C}{\partial t} + U(z)\frac{\partial C}{\partial x} = D\Delta C \tag{5.19}$$

where $C(x,z,t)$ it the concentration field, x, z the coordinates, and $U(z)$ the parabolic Poiseuille profile across the channel. We consider here the calculation made by Aris [23], based on a decomposition of the concentration field into two components:[8]

$$C(x,z,t) = \bar{C}(x,t) + c'(x,z,t)$$

in which $\bar{C}(x,t)$ is the concentration averaged across h and $c' = C - \bar{C}$. This decomposition, introduced in Eq. 5.19, generates several terms. There is no analytical solution to it, but over long times, when C is nearly homogeneous in the cross-section, an approximate solution can be found. In such circumstances, $c'(x,z,t)$ is much smaller than the averaged concentration and can be expressed in function of $\bar{C}(x,t)$ and its

[8]The phenomenon is often called 'Taylor-Aris dispersion'.

gradients. The calculation is quite long. It can be found in textbooks (see for example, [24] or the original work of R. Aris [23].). The calculation reveals that the terms at the origin of the enhancement of diffusion is the coupling between $U(z)$ and $\frac{\partial c'}{\partial z}$, and more specifically the (spatial) average of the product $\overline{U(z)\frac{\partial c'}{\partial z}}$. The calculation leads to the following equation:

$$\frac{\partial \bar{C}}{\partial t} + \bar{U}\frac{\partial \bar{C}}{\partial x} \approx D_{eff}\frac{\partial^2 \bar{C}}{\partial x^2}.$$

with

$$D_{\text{eff}} = D\left(1 + \beta\left(\frac{Uh}{D}\right)^2\right)$$

where $\beta = \frac{1}{210}$. For a capillary tube with circular cross-section, the calculation leads to $\beta = \frac{1}{48}$. The above expression can be written by using the Peclet number:

$$D_{\text{eff}} = D\left(1 + \beta Pe^2\right)$$

where $Pe = \frac{Uh}{D}$ is the Peclet number.

At high Peclet numbers, the effective diffusion constant thus increases, approximately, as the square of this parameter. In this range, the spreading of tracers is thus much faster than with diffusion alone. The effect is called 'shear augmented diffusion'. In microfluidics, the effect is considerable. Taking a Peclet number of 100, which is typical, the shear augmented term reaches three orders of magnitude.

We conclude this subsection with four remarks:

- The expression of the effective diffusion constant leads to a counter-intuitive result: the smaller D is, the faster the diffusion is. Thus, without diffusion, the tracer diffuses infinitely fast. On the other hand, in this limit, it takes an infinite time to reach the Taylor regime. To summarize, it takes an infinite time to be infinitely fast.

- Relevant to microfluidics, the calculation of Taylor–Aris has been generalized to arbitrary shapes (triangular, elliptic, etc.), using lubrication approximation [25].

- In practice, Taylor dispersion is difficult to observe. We need to wait a time $\tau \sim h^2/D$ to reach the Taylor regime. With a 100 μm channel, and a diffusion constant of 10^{-11} m^2/s, the waiting time is 1000 s. In the meantime, if the flow speed is 1 mm/s, the liquid will have travelled a distance of 1 m. Thus, in practice, Taylor dispersion rarely holds. This remark is often overlooked by researchers, who often assume that the Taylor dispersion regime is reached, while the system is still in a transient regime. The phenomenon is nonetheless valuable to know, because it treats, in a neat mathematical manner, an example where diffusion is strongly augmented by the shear. It thus provides an important reference.

- Because of Taylor dispersion, and, more importantly, the action of the regimes which precede, it is impossible to transport solutes in a microfluidic device without diluting them. We will see later that the solutions to this problem are droplets and electroosmosis.

5.5 Brief introduction to chaos and chaotic mixing

5.5.1 Chaos

Chaos is an important concept, which deserves a thorough treatment, but this would take us well beyond the scope of this book. We can mention [26,27,29] for descriptions of the subject. We will thus content ourselves with highlighting a few notions, that we consider useful for understanding phenomena important for microfluidics, such as chaotic mixing.

To start, let us introduce the notion of dynamical systems, not in general, but in the context of fluid dynamics. Take the fluid velocity $\mathbf{u}(\mathbf{x}, t)$, with the position $\mathbf{x}(t)$ of a fluid particle. By definition, we have:

$$\frac{d\mathbf{x}}{dt} = \mathbf{u}(\mathbf{x}, t) \tag{5.20}$$

These equations define a dynamical system. One way to describe the behaviour of the solutions is to visualize the particle trajectories $\mathbf{x}(t)$ in the phase space, i.e. in our case, the real space.

Before discussing the nature of the solution to Eq. (5.20), let us comment on flow streamlines and fluid particle trajectories. In stationary 2D flows, streamlines and trajectories are the same. In an (x,y) plane, streamlines are governed by a stream function ψ, defined by:

$$u_x = -\frac{\partial \psi}{\partial y}, u_y = \frac{\partial \psi}{\partial x}$$

By noting that $\frac{D\psi}{Dt} = 0$, one infers that ψ is constant along the trajectory, and, thereby, ψ describe both streamlines and trajectories. One consequence is that, in steady 2D flows, trajectories never cross, because, this would mean that at the same point, one could define two different speeds, which is mathematically not possible. There exist singular points, called hyperbolic or saddle points, where streamlines cross. However, these points are inaccessible to fluid particles because all velocity components are zero. Therefore, even at these singular points, particle trajectories do not cross.

If the flow is time-dependent, things change. Streamlines and trajectories are generally different. An example is shown in Fig. 5.18. Here, we have a time varying field, defined by $u_x = \sin t; u_y = \cos t$. The calculation shows that streamlines are lines, and fluid trajectories are circles. In stationary 3D flows too, streamlines and trajectories can be different. This depends on the flow structure. For some flow structures, however counter-intuitive it looks, streamlines can be steady and fluid trajectories chaotic. We will return to this point later.

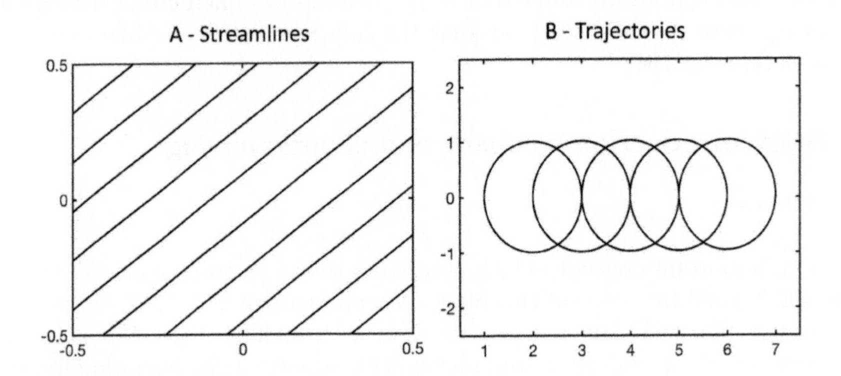

Fig. 5.18: (A) Ten streamlines and (B) five trajectories (circles), associated to the time dependent field defined by: $u_x = \sin t; u_y = \cos t$.

This being said, a solution $\mathbf{x}(t)$ is said to be 'chaotic' if it possesses the property of sensitivity to initial conditions: more precisely, if two nearby trajectories are perturbed, the perturbation grows at an exponential rate. This situation is represented schematically in Fig. 5.19.

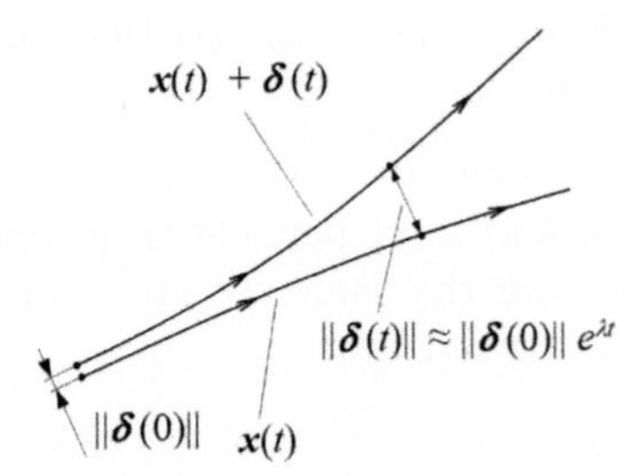

5.19: Divergence of trajectories.

Then, let us consider a point $\mathbf{x}$ placed on a trajectory at time t_0 that we choose as the origin of time. We designate by $\boldsymbol{x} + \boldsymbol{\delta x}$ the vector located at a point near $\mathbf{x}$, at (now) $t = 0$. Over the course of time, the two points evolve on neighbouring trajectories. They can move away from one another, or not. In a chaotic system, a direction must exist for which the two points separate exponentially. The word 'exponentially' is important. Should the divergence be algebraic (for instance, in pure shear flows, separations increase linearly with time), the system would be non–chaotic. Often, in chaotic systems, two directions are diverging, and one is converging. This occurs in the Lorenz model, presented below.

Thus, with this definition, we can write:

$$\delta x_i(t) \approx \mid \delta x_{0i} \mid e^{\sigma_i t}$$

where δx_{0i} the initial perturbation in direction i, δx_i is the modulus of the separation vector at time t, associated to δx_{0i}, and σ_i is the exponential rate. From the preceding formula, one has:

$$\sigma_i = \frac{1}{t} \log \left(\frac{\delta x_i(t)}{\mid \delta x_{0i} \mid} \right) .$$

We can repeat this analysis along the trajectory $\mathbf{x(t)}$ and come up with a collection of parameters σ_i, which depend on the position $\mathbf{x(t)}$ on the trajectory. If there exists a direction i along which the exponent is, on average (i.e. by averaging along the trajectory) positive, the system is chaotic. The corresponding value is known as the Lyapunov number along the direction i, or, if there is only one direction of divergence, the Lyapunov number. If there exist several positive σ_i, the Lyapunov number corresponds with the direction along which σ_i is maximal, i.e. along which the divergence is the most rapid. We can say, to simplify, that a system is chaotic if, on average, trajectories separate from one another at an exponential rate. An example of a chaotic trajectory, obtained in the Lorenz model, is shown in Fig. 5.20.

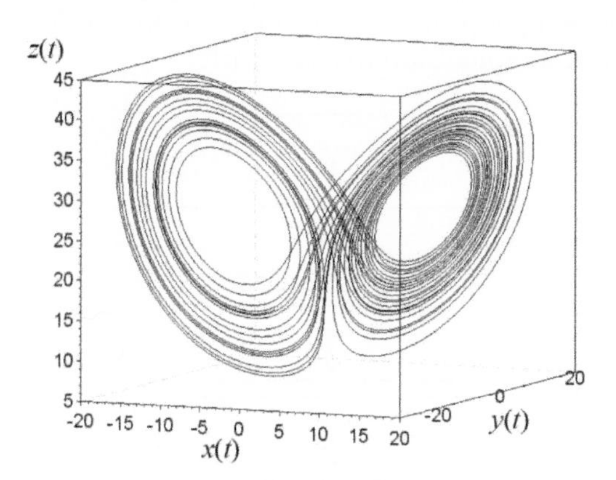

5.20: Lorenz chaotic attractor. The attractor has a fractal dimension of 2.04([32]), so that trajectories evolve on surfaces similar to that of the figure, forming a layer of volume equal to zero. Figure taken from Wikimedias

The Lorenz system was devised to model convection in the atmosphere [30]; at that time, it came as a surprise that erratic trajectories could emerge from apparently simple differential equations (that we call here dynamical system).[9] This numerical discovery put on the scene mathematical ideas developed at the end of the nineteen centuries, in particular by H. Poincaré [31]. It also gave birth to the field of chaos. In Fig. 5.20, a chaotic trajectory of a fluid particle, obtained with the 'classical' values of the parameters of the model, is shown. The Lyapunov number is positive along two directions of space, and negative along the third one. Particles are thus trapped in surfaces, forming layers, characterized by a fractal dimension of 2.04 [32]. This is the Lorenz attractor, which was later called 'strange' by Ruelle [33].

[9]The hypothesis that computers (the work was performed in 1963) were provoking the phenomenon was seriously considered. After some time, it could be ruled out

In practice, the perturbations giving rise to the divergence of the trajectories are, in the experiments, the thermal noise and, in the computations, the numerical noise. We may infer that chaotic systems amplify noise. This property is at the origin of the so-called butterfly effect', in which the flapping of a butterfly's wing in Brazil, being amplified by the chaotic dynamics of the atmosphere, can set off a tornado in Texas [34]. It also leads to the counter-intuitive notion that, although chaotic systems are governed by deterministic equations, their behaviour is erratic.

5.5.2 Horshoe tranform, baker's transform

There is an equivalent way to define chaos. In chaotic systems, sets of particles (for instance blobs of dye), undergo an infinite succession of foldings and stretchings. In chaos theory, this process associated to the presence of a horseshoe transform [35], which repeats indefinitely over time, transforming the blob of particles in a dense, striated, manifold. One example of an horseshoe transform is the baker's transform, which the artisan uses when making mille-feuilles. Initially, we have an object (a segment of some width, as shown in Fig. 5.21). We stretch and fold it, giving rise to a new object, twice as long and half as wide. By repeating the process, we obtain thinner and thinner segments, folded and packed in a multilayered sandwich.

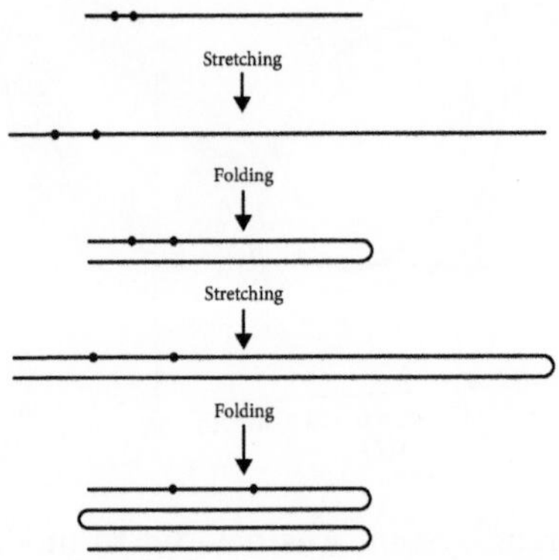

Stretching

Folding

Stretching

Folding

5.21: An example of the horshoe transform: the baker transform, in which a rectangular patch of fluid particles is stretched and folded. At each step the distance separating two neighbour points is multiplied by 2.

The two definitions of chaos, exponential separation, and horseshoe transform, are equivalent. Taking two neighbouring points placed in the initial segment in Fig. 5.21, their distance along the stretching direction is multiplied by two at each step of the process. There is thus an exponential growth of the separations, along one direction, indicating an equivalence between the two definitions.

With the horseshoe transform, material lines extend exponentially and, in the meantime, they fold. This naturally leads to the formation of motifs such as those in Fig. 5.22. One consequence of the process is the exponential growth of the surfaces of exchange and, in the meantime, the exponential increase of the concentration gradients. The two effects, which are linked by a conservation law, intensify the diffusive exchanges. Chaotic mixing can thus be effective for mixing fluids or reagents.

Fig. 5.22: Series of foldings and stretchings produced by the oscillatory movement of one of the container walls [27].

5.5.3 Impossibility of generating chaos in 2D steady flows

Is it possible to know *a priori* if a system is chaotic or not? On this point, several powerful theorems exist [26, 27, 29]. Here, we will restrict ourselves to the case of bounded flows.

An important theorem, called 'Poincaré-Bendixson theorem', stipulates that two dimensional autonomous dynamical systems cannot give rise to chaos. In other words, a steady two dimensional flow cannot generate chaotic trajectories. To obtain chaotic behaviour, one must escape the plane. This can be achieved in two ways:

- Work with stationary *three*-dimensional flows. In this case, chaotic trajectories can develop. This statement is subtle and it contradicts current thoughts on flow reversibility at small Reynolds numbers. These thoughts often result from a misinterpretation of a spectacular experiment, carried out by G. I. Taylor, in the 1950s [37], and frequently presented to students in fluid dynamic courses. The experiment demonstrated that in a fluid driven between coaxial cylinders, whose inner cylinder is rotated several turns, one observes that, as when the cylinder is rotated backwards, tracer particles return to their initial positions. This was Taylor's reversibility experiment [37]. As strange as it seems, in 3D, particles *may not* return to their initial locations when the flow is inverted [38]. This depends, for instance, on whether or not the flow contains hyperbolic points. In such a case, speed is reversible, with respect to flow inversion, but not trajectories. In Taylor's experiment, the flow was regular', i.e. it did not contain hyperbolic points. In such circumstances, the fluid trajectories were reversible.
- Work in two dimensions, but with velocity $\mathbf{u}$ depending on time t. For instance, a flow generated by a wall moving periodically in time. Let us take x and y as the plane coordinates and, just to show, formally, how a third dimension can be introduced, let us define a variable z by the following equation:

$$\frac{dz}{dt} = 1$$

> By replacing t with z in the original 2D dynamical system, we create an autonomous three–dimensional system. In this way, we escape the Poincaré-Bendixson theorem.

Working in three dimensions, in one way or another, i.e. using a 3D geometry or a time dependent excitation, is a necessary but not sufficient condition for developing chaotic trajectories. The flow must possess certain features. We present two of them in the next section.

5.5.4 Trajectories crossing at different times lead to chaos

A necessary condition for obtaining chaotic trajectories in unsteady 2D flows was proposed in Refs. [39, 40]. The demonstration is technical and we content ourselves with formulating the conclusion: to develop chaotic trajectories, 'the flow pattern must change with time so that a streamline in one pattern crosses a streamline in a later times' [40]. This condition is illustrated in the blinking vortex flow, introduced by H. ARef. [41],and shown in Fig. 5.23.

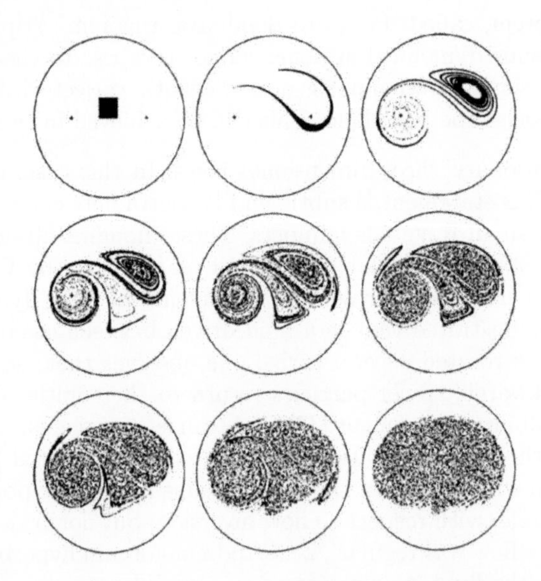

Fig. 5.23: Numerical simulation showing the positions, at different times, in the blinking vortex flow, of particles initially confined in a square. (Reprinted from [41] with permission of Cambridge University Press. Copyright 2022.)

The blinking vortex is a vortex, placed in a tank, that jumps from one position to another in a periodic manner (see Fig. 5.23). Within each period, the vortex develops steady circular streamlines and, thereby, circular trajectories. However, by jumping from one position to the other, fluid trajectories become chaotic. This was shown by H.Aref in 1984 [41], and it came as a surprise, at that time, that abstract concepts could be substantiated in such a simple manner. The paper triggered the discovery of a new domain, called 'chaotic mixing'. This domain provided the conceptual tools to design, one decade later, microfluidic mixers.

5.5.5 Chaotic mixing in time dependent flows with hyperbolic points

It is important to present a less general condition for creating chaotic behaviour, often described in detail in reviews and textbooks, and whose implementation, in microfluidic devices, turns out to be easier. It is based on the temporal displacement of hyperbolic points.

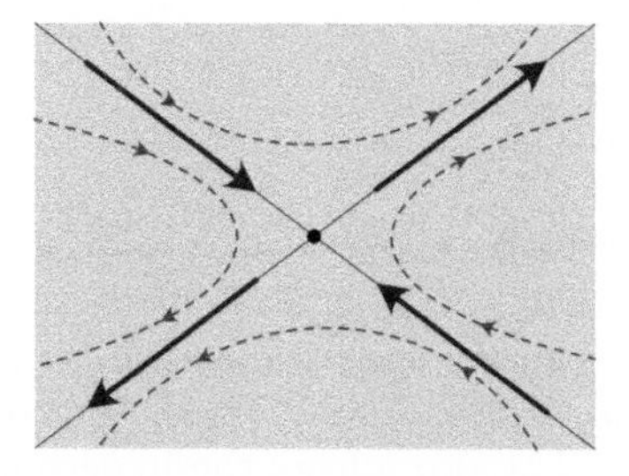

5.24: Hyperbolic point in a two–dimensional flow.

As evoked before, in 2D, a hyperbolic (also called saddle) point is a point where, along one axis, fluid particles converge and along the other, they diverge (see Fig. 5.24). The two axes define separatrices that delineate regions, from where, in a steady regime, fluid particles cannot escape. Close to the hyperbolic point, streamlines are hyperbolas. One key remark is that, in the vicinity of the saddle point, small perturbations can push particles across a separatrix, and expedite it in a different region. Small perturbation, large consequences. This sensitivity is a source of chaos. Poincaré showed that, by periodically perturbing the position of a saddle point, homoclinic orbits (i.e. joining hyperbolic points to themselves), develop complex patterns, which, decades later, were called 'chaotic' [35,36]). The mechanism is subtle and a number of tools of dynamical system theory are needed to describe it. Let us nonetheless attempt to provide hints. Fig. 5.25 (extracted from J. Ottino [28]), shows what happens when a hyperbolic point is forced to move, at constant speed, along a small loop (shown by a dashed line). In this case, the homoclinic orbit develops a complex pattern. Lobes form, whose number grows in time and thicknesses shrinks. In such conditions particles placed on the orbit jump from one region to another, in a manner increasingly sensitive to noise. In this process, after some time, particles initially placed on the orbit pay visits to different flow regions, in an erratic manner. Particles located close to it will manifest similar trends. This is how, though described in an oversimplified manner, chaotic behaviour may develop close to erring saddle points. We refer readers to specialized books (see [26,27,29]) for thorough presentations on the subject.

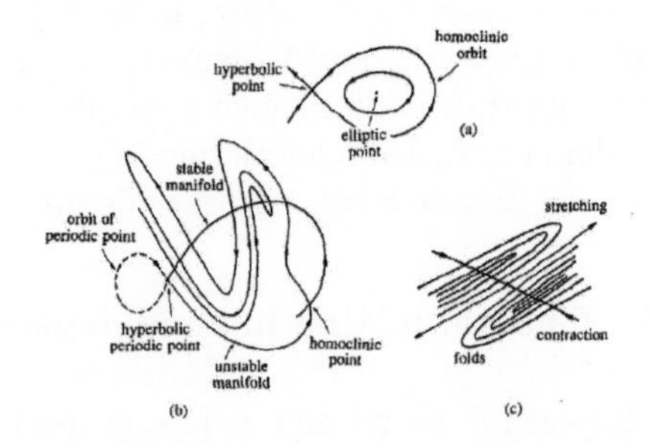

5.25: (A) steady homoclinic orbit, with a fixed hyperbolic point;(B) the hyperbolic point moves along a loop (dashed line);(C) sketch of streaklines produced by the saddle point movement (From J. Ottino [28].)

In practice, modulating in time the positions of hyperbolic points is an efficient strategy for producing chaotic trajectories, and, thereby, mix tracers in reservoirs. Fig. 5.26 shows a numerical example, which is easy to implement (15 lines of Matlab code), that illustrates this feature. We have here a line of periodic counter-rotating vortices defined by the stream function:

$$\Psi = \sin \pi x \cos \pi y$$

where x, y are the plane coordinates. Fig. 5.26(A) shows two eddies, with centres located at $x = 0$ and $x = 1$. The hyperbolic points are located at the walls $y = \pm 0.5$; two of them are denoted by S and S'. In Fig. 5.26(B), the positions, along the x axis, of the two eddy centres are modulated in time (in the simulation, the perturbation is $0.4 \sin(\omega t)$, with $\omega = 0.3$). Indeed, the modulation also affects the hyperbolic points. In such circumstances, fluid particles undergo chaotic behaviour. They visit one eddy then the other, in a random manner, tending to cover the entire space. Interestingly, an island seems to resist.[10] Fig. 5.26(B) thus suggests that modulating hyperbolic points represents an efficient method for mixing fluid particles.

5.6 Mixing in microfluidic devices

5.6.1 How to characterize mixing ?

Complete or full mixing is defined as a state where the concentration field $C(\mathbf{x})$ is homogeneous. But how to characterize mixing when it is not complete? There exist an infinity of measures of mixing. The most widely used, often called 'mixing index', is given by the expression:

$$I = 1 - \left(\frac{\int_V (C(\mathbf{x}) - C_{mean})^2 d\mathbf{x}}{V C_{mean}^2} \right)^{1/2}$$

[10]This is related to KAM theorem, whose description lies beyond the scope of the book (we refer the reader to Refs [26, 27, 29] for presentations)

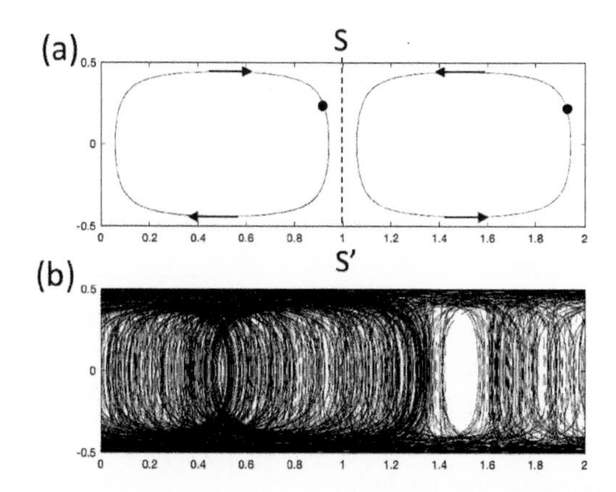

Fig. 5.26: Matlab simulation showing the trajectories of two particles, in (A) steady and (B) modulated counter-rotating vortex flows. S and S' are two saddle points. In both cases, the initial conditions are the same (x = 0.9; y=0.3 and x=1.9, y=0.3 – indicated by dots –).(A) Steady vortex flow: the particles remain on closed streamlines, which reveal the flow structure.(B) Modulated vortex flows: here, the positions of the vortex centers, and thereby those of the hyperbolic points vary periodically in time, along the x axis. In such conditions, the trajectories of the two particles display chaotic behaviour.

in which V is the fluid volume, $d\mathbf{x}$ is the elementary volume of integration, and C_{mean} is the concentration averaged over V. As the system gets more an more mixed, the index approaches unity. In the opposite case, where C is a Dirac function, $I = 0$.

The definition of the mixing index is arbitrary. One could have imagined a criterion based on the concentration gradients, or the derivatives of the concentration gradients, in view of highlighting the presence of strong gradients in the system. Alternatively, one could use an index defined by $I' = 1 - max(C - C_{mean})/C_{mean}$. However, for the sake of conceptual and computational simplicity, and to be able to compare with published data, it is now current to use I. There is no general rule to appreciate whether the mixing is satisfactory or not, it depends on the application, although 90% is often termed, in the literature, as satisfactory, good, excellent, or of high quality.

5.6.2 Mixing index in a box.

Let us take the example of a spot diffusing in a one dimensional box of size $2L$, and calculate the mixing index. We recall the expression of the concentration field for this case (see Eq. (5.9))

$$C(x,t) = C_0 \left(1 + 2 \sum_{1}^{n} \exp(-Dn^2\pi^2 t/L^2) \cos(n\pi x/L) \right) \tag{5.21}$$

in which D is the diffusion coefficient, C_0 is the mean concentration and t is time. In our geometry, the expression of the mixing index is:

$$I = 1 - \left(\frac{\int_{-L}^{L} (C(x,t) - C_0)^2 dx}{2LC_0^2} \right)^{1/2}$$

The mixing index is represented in Fig. 5.27 as a function of the dimensionless time $2tD/L^2$.

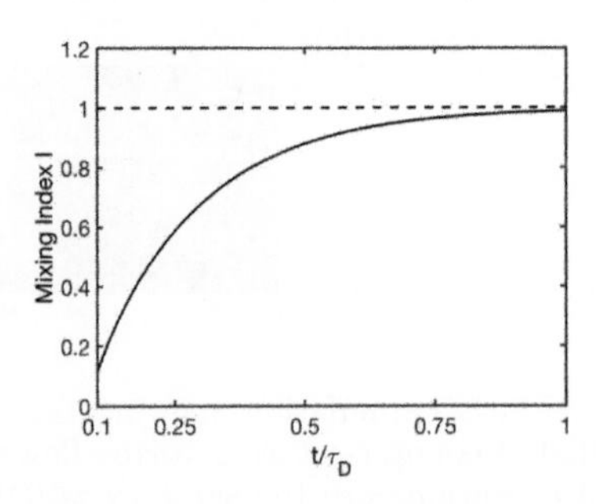

5.27: Evolution of the mixing index with time in a one-dimensional box, in which, at $t = 0$, a small spot of tracer has been injected.

I indeed depends on time. The longer we wait, the more mixed the system is, and the larger I is. At late times, the first term of the series of Eq. (5.21) dominates the others. Thereby, in this range of time, the index factor is approximately given by :

$$I \approx 1 - \exp(-D\pi^2 t/L^2)$$

The approximation shows that complete mixing is approached at an exponential rate, the time constant being $L^2/\pi^2 D$. Thereby, as seen in Fig. 5.27, if we wait a time equal to $\tau_D = l^2/2D$, we obtain a mixing index of 1-exp$(-\pi^2/2) \approx 99.3\%$, which is high. Thus, in this problem, τ_D appears as a time above which mixing is excellent'. Furthermore, although counter–examples may be found, we will consider that, in general, by waiting for a time τ_D, high mixing is achieved.

5.6.3 The problem of the slowness of diffusion.

As suggested above, in the case of pure diffusion, waiting a time equal to l^2/D (where l is a characteristic length and D the diffusion coefficient) guarantees that mixing is achieved. Table 5.2 provides orders of magnitude for such a time.

Sugar in water in a 10 cm high glass	200 days
Fluorescein in water, in a microwell of 1 μm	2.5 ms
Fluorescein in water, in a microwell of 200 μm	100 s
Albumin in a microreactor of 200 μm	2,000 s
20 bp DNA strand in a microreactor of 500 μm	14 h

Table 5.2 Diffusion times for various systems.

The table 5.2 indicates that mixing is rapid in micrometric-sized systems (see the example of the one-micron well). In this case, we can 'laisser-faire' to achieve mixing. This remark applies well to nanofluidics. By contrast, for typical microfluidic reactors, hundreds of μm in size or so, depending of the molecule, the mixing time can reach hours. Moreover, in most situations, fluids move. With speeds of 1 cm/s, translating 100 s into a travel length leads to 1 m. This generates obvious constraints for the design of the device. It thus appeared evident, in the early years of microfluidics, that new ideas for accelerating mixing were needed. A bottleneck had to be resolved. This brings us to the question of micromixing.

5.6.4 A tour in the world of micromixers

When the first (French) edition of this book was written, in 2002, about ten or so micromixers had been invented. Today, more than a hundred exist. Several reviews have been published (see, for instance, [42–44]).

Micro– and milli–mixers can be represented as functions of their Peclet and Reynolds numbers, as was done in Ref. [43] (See Fig.5.28).

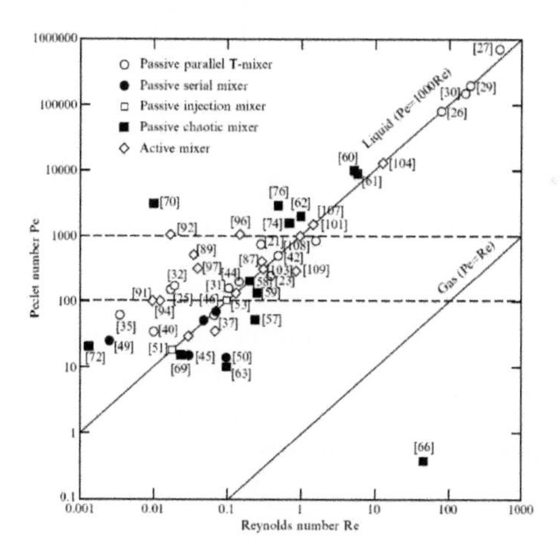

Fig. 5.28: Representation of micromixers and milimixers, for gases and liquids, in a Reynolds–Peclet diagram (Reprinted from Ref. [43], with permission of Institute of Physics Publisher. Copyright 2022).

Peclet and Reynolds numbers are defined, respectively, by $Pe = \frac{Ul}{D}$ and $Re = \frac{Ul}{\nu}$, where U is a characteristic speed, l is a characteristic scale, and D and ν are, respectively, the diffusion constant of the tracer and the kinematic viscosity. The ratio:

$$Sc = \frac{\nu}{D}$$

is the Schmidt number. It is on the order of unity for gas and around 1000 for the liquids. The two lines, one for gas, the other for liquids, drawn on Fig. 5.28, correspond to these values.

Restricting ourselves to liquids, one can distinguish low Reynolds numbers micromixers, working up to $Re = 10$ or so, and moderate Reynolds numbers micromixers or millimixers, which operate at $Re > 10$, and exploit, most of the time, eddies generated by the fluid inertia, as we saw in Chapter 3. These can be called 'inertial micromixers'. In Fig. 5.28, mixers have also been classified in several categories: passive or active, chaotic or not. We now look at particular cases, which will illustrate the different types of mixers published in the literature.

5.6.5 Reducing sizes to speed up mixing

In order to mix, one could think of shrinking the system size so as to decrease the diffusive time, and thereby mix at a faster speed. Fig. 5.29 shows an example of a naive device that would make use of this idea.

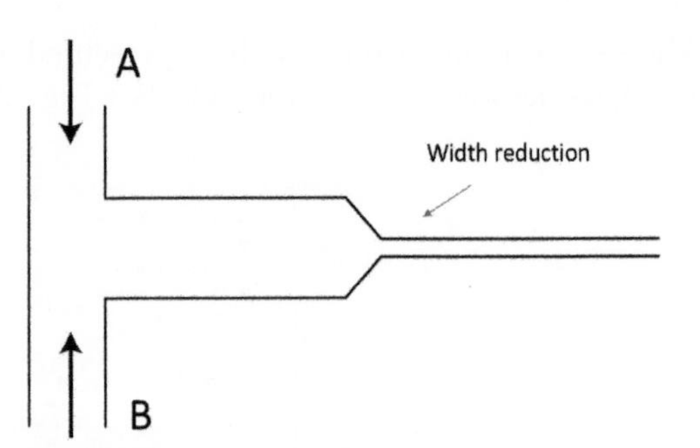

5.29: T–junction followed by a restriction, added to facilitate mixing.

Two fluids, A and B, enter a T-shaped intersection. We previously studied this geometry, called T-sensor. Downstream of the T-junction, assume the cross section of the channel is reduced. For instance, its width w_0 is reduced to $w < w_0$. The corresponding diffusive time τ in this part of the channel is given by the relation :

$$\tau \sim \frac{w^2}{D}$$

This looks easy. Just reduce the size, and the mixing time is reduced by a factor $(w_0/w)^2$. The problem is the mixing length. In the same channel, the two fluids have a speed Q/wh, where Q is the flow rate and h is the channel height, which is the same before and after the restriction. Thus, the length necessary to achieve mixing is :

$$L \sim \frac{Qw}{hD}$$

It increases by a factor (w/w_0). Let us take orders of magnitude: with w and h equal to 100 μm and a flow rate equal to 100 μL/s, L is 100 m. This is far too large. One method is to parallelize the entries. Topological constraints force to escape from the plane, i.e. design a three–dimensional system. An example is shown in Fig. 5.30.

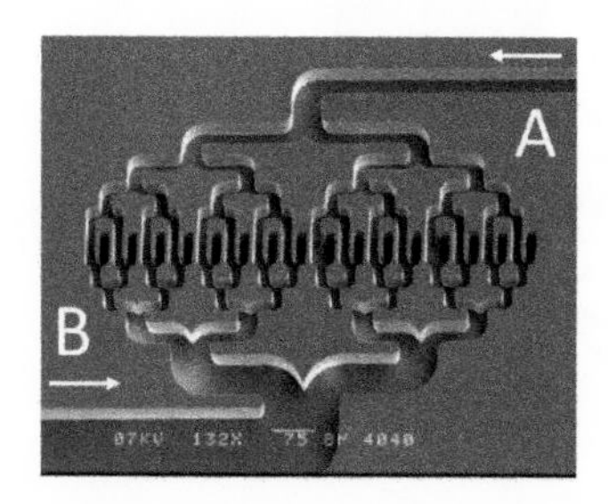

Fig. 5.30: Micromixer built by Upchurch. (Left) SEM image of the micromixer. (Right) Integration of the micromixer at the entry of a microfluidic device.

The two fluids A and B enter from the sides of the left–hand image in Fig. 5.30. Each flow rate is divided into 2^4=16 branches. Branch widths are substantially smaller, a factor or 4 or so. Divided as such, A and B meet in the central part of the left-hand image of Fig. 5.30. They exit normally to the plane of the device (see the tubes of Fig. 5.30, right). Using the above formula, mixing length is decreased 64 times, which is progress.

5.6.6 The herringbone micromixer

The herringbone mixer [45] is shown in Fig. 5.31.

The bottom wall of the channels is structured: oblique grooves, asymmetric with respect to the channel width, are built along it. This is the 'herringbone' structure. They force the flow to develop recirculations along the mean stream, or, said differently, they cause fluid particles to spiral out. Every five grooves, the centres of the two eddies, are displaced. In the frame of reference of the mean flow, the saddle point positions are modulated in time, inducing chaotic behaviour, according to the mechanism mentioned earlier. The following expression for the mixing length L_m was found consistent with the experiments:

$$L_m \approx w \log(Pe)$$

where w is the channel width and Pe is the Peclet number (calculated with w, the average flow and the fluorescein diffusion coefficient). This expression reflects the exponential increase of trajectory separations, discussed in the previous section. In practice, L_m is a few millimers for Peclet numbers on the order of 10^5. [45]. The herringbone

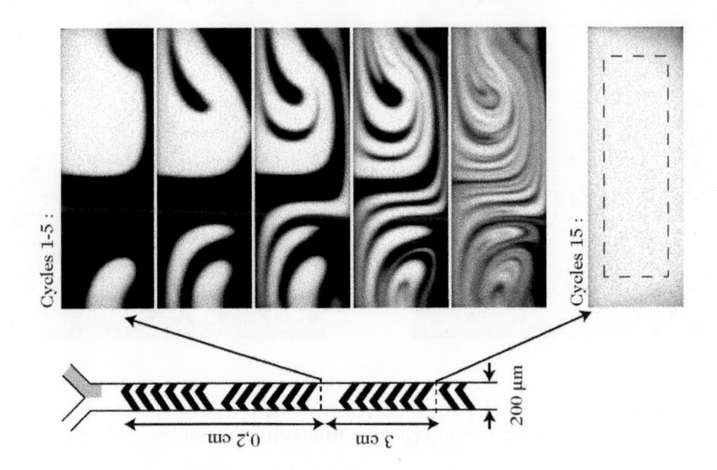

Fig. 5.31: Herringbone micromixer. Structuration of the microchannel bottom wall with grooves, forcing the flow to swirl (see the bottom figure). Mixing of fluorescein in water is obtained after a few microchannel widths. (Reprinted from [45] with permission from American Association for Advanced Science. Copyright 2005.)

mixer, due to its simplicity and efficiency, is frequently used in laboratories. Its limitation, common to practically all micromixers, is the small throughput.

5.6.7 The rotary micromixer

We now consider active mixers. Using polydimethylsiloxande (PDMS) valve technology, S. Quake [46] developed a circular micromixer (see Fig. 5.32). A peristaltic pump, activated by three microvalves (we will describe the technology later), transports DNA and cells in an annular channel, along the curved arrow shown in Fig. 5.32).

The two entities are transported at the mean flow speed, but, in the meantime, velocity gradients disperse the spots along the flow streamlines, We discussed this phenomenon earlier. The tracers form arches, which, over time, because of the confinement, stack onto each other. After a few rotations, thin striations appear. As their widths decreases in time and their number increases, mixing is rapidly achieved. This phenomenon was discussed in detail in Ref. [25]. In practice, mixing is achieved in a few seconds, well before the Taylor dispersion regime establishes.

5.6.8 The cross-channel micromixer

The cross-shaped micromixer is another active mixer. It consists of a principal channel, along which a stationary flow is imposed, which intersects with side channels, along

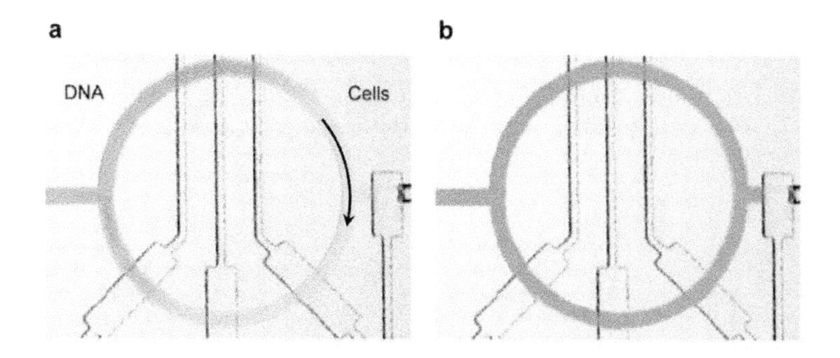

Fig. 5.32: Rotary micromixer: fluids or suspensions are introduced in the annulus and driven by a peristaltic pump, activated by valves (channels transverse to the ring); 5 nl of a micromixer before (a) and after (b) mixing. DNA and cells are mixed in a few seconds. (Reprinted from [46] with permission from IOP. Copyright 2022.)

which a time-periodic flow is imposed (see Fig. 5.33). The velocity field is thus time-dependent, and then, according to the preceding discussion, the development of chaotic trajectories is possible. It has been shown in this system that, at the intersection, a material line is subjected to a succession of folding and stretching [47–51]. This is due to the fact that, by superimposing a time-dependent side flow onto the mean motion, one creates modulated hyperbolic points (located at the junction corners [49]), which induce, as explained in the preceding section, stretching and folding of the material lines.

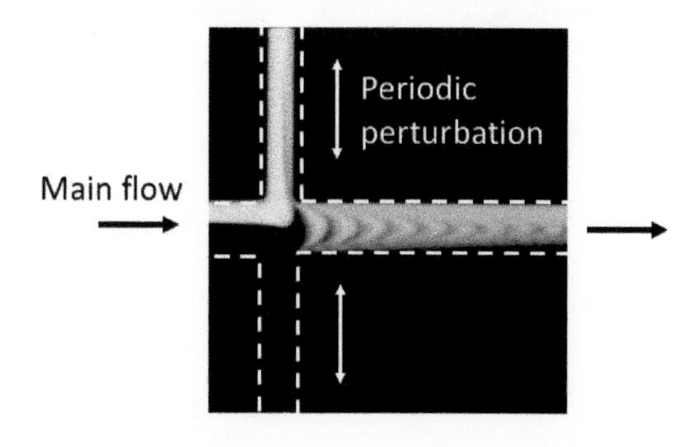

5.33: Cross–channel micromixer: fluids (water and fluorescein solution) penetrate the junction from the left, and are subjected to the time periodic action of a side flow, activated by a pump [49].

In practice, depending on the flow conditions, stretching and folding intervene 1– 3 times in the intersection. A phenomenon of resonance has been uncovered, for which, in certain conditions, lobes develop at the intersection and allow us to achieve high mixing indices [50,51]. Fig. 5.33 illustrates this feature. The length of mixing L_m is, in practice, several w, where w is the channel width. The range of conditions for which mixing is achieved can be broadened by placing several cross-channels in series [49,52]).

5.6.9 Mixing in droplets

As we saw in Chapter 4, as droplets move downstream, they generate internal recirculations [53]. The detail of the corresponding mixing process is shown schematically in Fig. 5.34 [54].

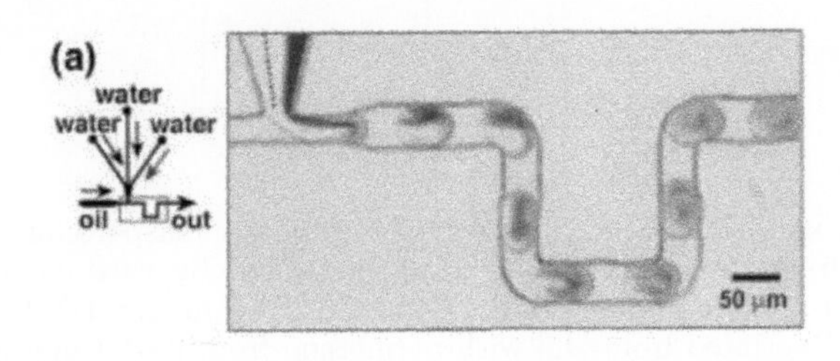

5.34: Mixing in droplets. (Reprinted from [54, 55] with permission from AIP Publishing. Copyright 2022.)

After the T-junction, the dye concentrations, in each half of the droplet, are homogeneous and thus a strong gradient exists across the two regions. The role of the turnings is to vary, in time, the recirculation pattern, so as to develop, in a process discussed in the preceding section, stretchings and foldings of the material lines. Fig. 5.34 shows that after two turnings, striations appear. As the turnings are repeated, chaotic behaviour takes place, leading to mixing. This process, modelled as a baker's transform in [55], allows mixing to occur 'for free'. The process is crucial for biological and chemical applications. Allowing droplets to be treated as chemical microreactors, it has played a major role in the development of droplet–based microfluidics.

5.6.10 Mixing at moderate Reynolds numbers

Mixing with Dean vortices. We saw in Chapter 3 that when a flow is driven in a curved channel, Dean vortices develop, with speeds increasing with the square of the flow rate. These vortices have been exploited to mix fluids. An example is shown in Fig. 5.35 [56].

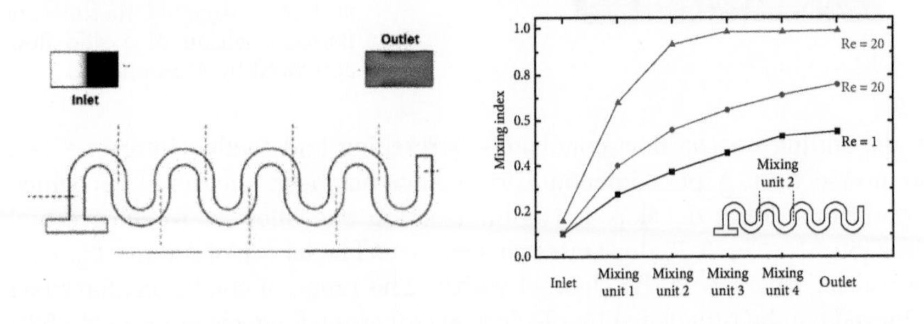

Fig. 5.35: Dean mixer in a serpentine. (Right) Mixing index along the serpentine, at different Reynolds numbers [56]. (Creative Commons Attribution (CC BY) licence.)

In this case, the curved channel, by developing Dean recirculations, mixes the two dyes injected at the entry. Fig. 5.35 shows that, after four turns, for a Reynolds number of 100, the mixing index is almost equal to unity. At smaller Reynolds numbers, Dean vortices are weak and mixing is poor. Dean micromixers thus operate at moderate and large Reynolds numbers, below turbulence onset. In practice, a range of Reynolds numbers ranging between 50 and 500 is frequently used. This range favours high throughput and, in the meantime, keeps the flow under control. However, mixing frequently occurs *before* the fluids penetrate in the curved part of the system. This is because, since the Reynolds is high or moderately high, eddies naturally appear at the entry. This indeed depends on the design of the entry channels. Eddies may also nucleate at corners, when the serpentine has a crenel shape. High mixing indexes are favoured by the presence of temporal instabilities, triggering the onset of chaotic trajectories, as shown recently [61].

Mixing in T junctions at moderately large Reynolds numbers. At moderately large Reynolds numbers, typically above 20-100, microfluidic flows enter a domain in which eddies naturally appear: close to corners, in diverging sections (see Chapter 3), or around curved boundaries. These eddies may, in turn, trigger temporal instabilities. There are various contributions on the subject, focusing, most often, on the flow structures and their connections to mixing. They are periodically reviewed (see, for instance, [62]).

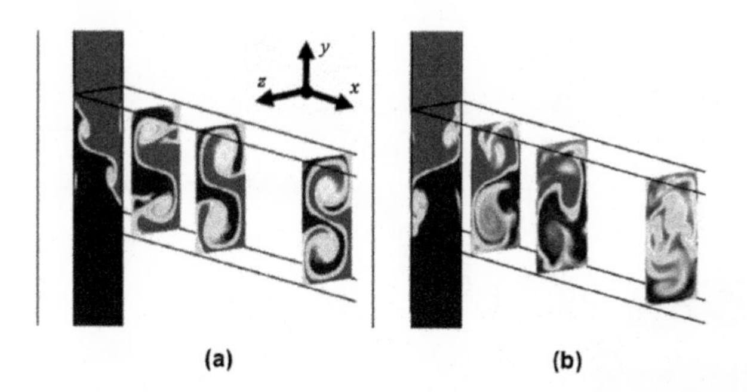

Fig. 5.36: Fluids introduced in a T–junction and developing eddies. (a) Re=186; (b) Re = 600. (Reprinted from Ref. [63], with permission of Elsevier. Copyright 2022.)

In the example we take here, two miscible fluids are injected in a T–junction and are further collected in a rectangular channel. Numerical simulations show that eddies develop right at the entry (see Fig. 5.36). These eddies, being induced by the curvature of the streamlines, are similar to Dean vortices. Here, the Reynolds number Re is defined by $\frac{Ud}{\nu}$ in which U is the 'superficial' speed (i.e. the flow rate of each branch divided by the cross section area), d is the hydraulic diameter, and ν is the kinematic viscosity of the (identical) fluids. For Re comprised between ≈ 20 and 150, two stationary symmetric vortices appear at the entry. Above 150, the flow pattern undergoes a bifurcation towards asymmetric eddies. Above 400, a new regime takes place, in which these vortices become time–dependent. Simulations show that in such conditions, substantial mixing between the two fluids takes place at a few hydraulic diameters from the entry (see Fig. 5.36 (B)). The simulation was found to be in good agreement with the experiment [64]. Here again, the modulation of the saddle point is probably at the origin of mixing[11].

5.7 Four applications of transport of matter in microfluidics

5.7.1 Example 1: measurement of fast reaction kinetics in Y–junctions

In principle, fast reaction kinetics can be measured in T– or Y–junctions. A flow driven at speed U, in such a system, can resolve kinetics on the order of δ/U, where δ is the spatial resolution. With $\delta = 10\mu$m, and U=10 cm/s, a kinetics of 0.1 ms can be resolved. This is two orders of magnitude faster than conventional systems, such as the stop-flow', which consists in pushing, with a syringe, two reagents in a tank, as fast as possible, and stopping brutally. Such systems have temporal resolutions on the order of 20 ms.

The interpretation of the chemical phenomena taking place downstream of T or Y junction is nonetheless not straightforward to analyse [19, 65, 66]. Fig. 5.37(A) shows an example of the fluorescence field produced by two reagents introduced in a Y–junction.

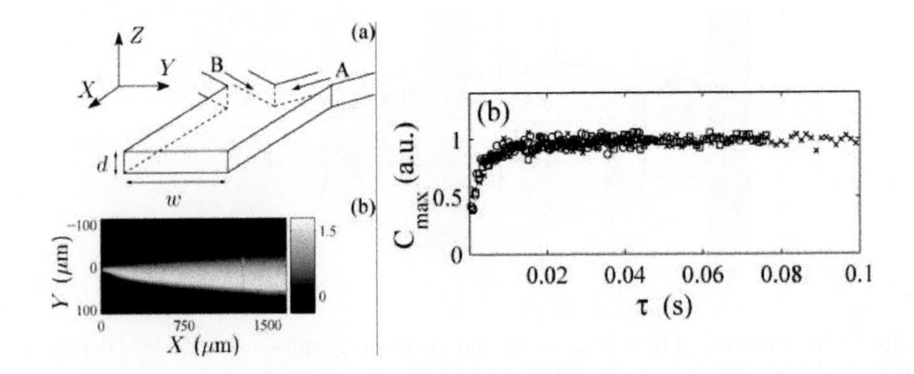

Fig. 5.37: Measurement of the kinetics of a protonation reaction, in which a fluorescent dye (acidic coumarin) is produced. (Upper left) Design of the microfluidic device. (Lower left) Typical stationary field of fluorescence measured in the collecting channel. (Right) Plot of the maximum fluorescence intensity as a function of the streamwise coordinate, converted in time by the relation $\tau = X/U$ (see text) [66].

The solutions driven in the Y-junction of Fig. 5.37 are acetronitrile solutions, one containing basic coumarin and the other triflic acid. The reaction produces acidic

[11]This is an hypothesis. The point is not discussed in the cited paper.

coumarin, which is fluorescent. Fig. 5.37 shows the field developed by the reaction. Owing to the kinetics of the reaction, which depends on the local concentrations of the reagents, the fluorescent field is asymmetric. The concentration fields of the reactants are symmetric (they have the error function shapes described earlier), but the product concentration profile is not. A plot of the fluorescence maximum as a function of time $\tau = X/U$, where X is the downstream coordinate, and U is the mean speed, is shown on the right of the figure. The graph shows that, with microfluidics, the kinetics of fast reactions can be measured. In Ref. [66], a detailed theoretical/numerical analysis was performed.

5.7.2 Example 2 :measurement of fast reaction kinetics in droplets and screening of chemical reactions

A droplet-based method for measuring reaction kinetics was proposed in 2003 [53]. As shown above, in droplets, because of the internal recirculations, mixing time are small. For example, with droplets moving at 10 cm/s in a 100 μm side square channel, the mixing times are on the order of 1 ms. Fast reaction kinetics are thereby accessible to the measurement. The demonstration is shown in Fig. 5.38 [53].

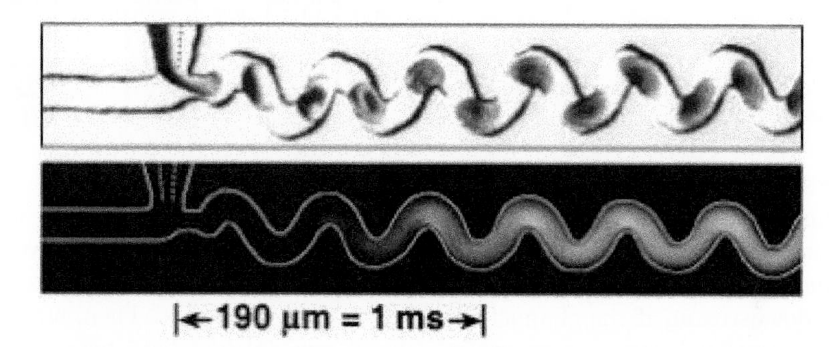

Fig. 5.38: The upper figure shows the droplets formed in the system. In the lower figure Fluo4 and $CaCl_2$ solutions are injected in the T-junction. The increase in the fluorescence signal shows the development of the complexation reaction as droplets are moving downstream. (Reprinted from [53] with permission of Wiley and Sons Eds. Copyright 2022.)

In the experiment, Fluo–4 and $CaCl_2$ solutions are injected in the T–junction. They form, with a separation buffer, the content of the droplets. Fluo–4 is weakly fluorescent, but, in the presence of calcium ions, it forms a highly fluorescent complex. With speeds on the order of tens of centimetres per second (see Fig. 5.38), it was possible to resolve submillisecond kinetics [53].

The same geometry was used to mix chemical reagents, and screen the characteristics of the products at various concentrations ratios [53]. This screening was achieved by varying the flow rates. In this manner, the amounts of reagent could be varied in each droplet. Droplets are emitted at kHz and the screening is fast – orders of magnitude faster than any robotic equipment –. In the meantime, the quantity of reagents used in each reaction is orders of magnitude smaller than traditional technique: a droplet encloses, typically, one nanoltre, while a well, in a 96 or 384 well plate, uses one millilitre or so. The work prompted new ideas that today, are harnessed in a number of successful biotechnological companies.

5.7.3 Example 3: shaping concentration fields

With microfluidics, it is possible to shape concentration fields [67]. The principle is shown in Fig. 5.39.

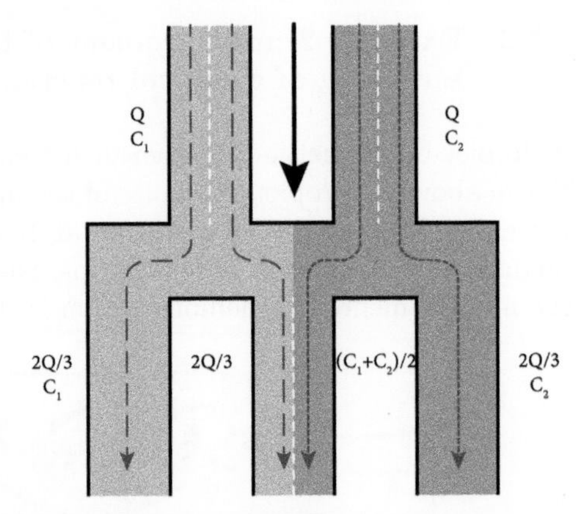

5.39: Principle of the concentration gradient generator.

Two dyes, of concentration C_1 and C_2 are introduced at the top of Fig. 5.39. Four flow streamlines are sketched on the figure. Due to the large Peclet numbers at work, and the short travel times, dye 1 is transported directly in the left channel, dye 2 in the right one, and both of them penetrate, in equal proportions, in the central channel. Further downstream, if the channel is sufficient long, they mix. Then, starting with two concentration levels, one generates a third one, intermediate between the two. Thus, if the two dyes are red and blue, we will create an additional colour, violet. By repeating the process, more colours can be created. The network of Fig. 7.23 shows that nine colours (i.e. nine distinct levels of solute concentration) can be obtained.

In the experiment, the nine concentration levels were recombined in a collecting channel, in order to produce a constant concentration gradient (See Fig.7.23). This system was used in chemotaxis studies: the migration of neutrophils, placed at the bottom of the collecting channel, and subjected to nutrient gradients was analysed [68]. The 10-fold log levels solute distributor [69], presented in Chapter 3, was inspired by this work.

5.7.4 Example 4: Lipid Nano Particles production in inertial micromixers

Lipid Nano Particles (LPN), sketched in Fig. 5.41 (A), are complex nanoparticles, 50 – 100 nm in diameter, formed through a self-assembly process, including, according

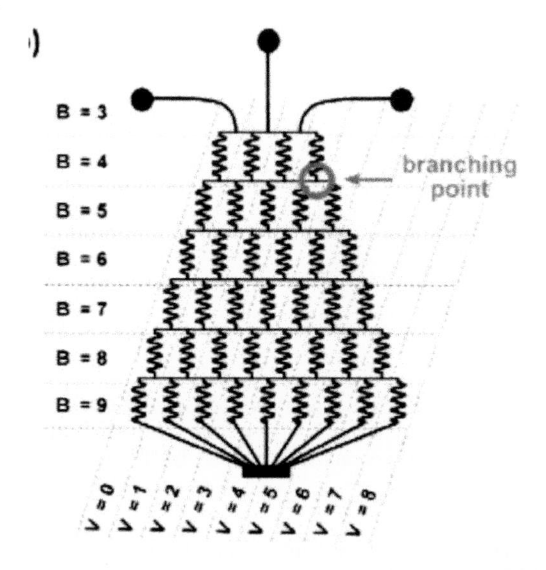

5.40: Image taken from [68], in which a nine–channel gradient generator is demonstrated.

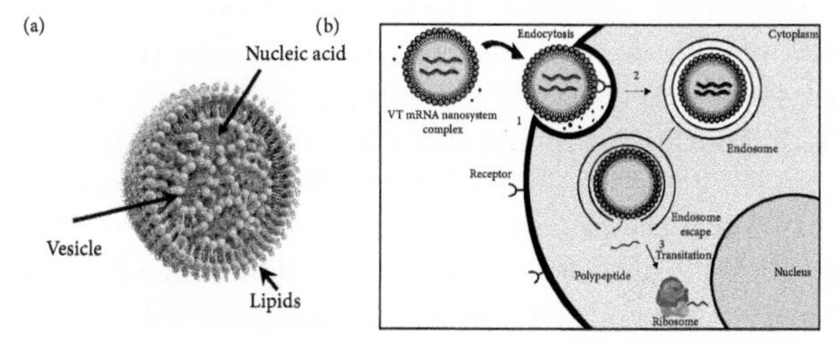

Fig. 5.41: A: LNPs structure: a ionizable lipiidic shell encapsulates vesicles, that encapsulate nucleic acids. B - Nucleid Acid delivery into the cell, including: endocytose, endosome formation, endosome escape and, for the case of mRNA, protein synthesis in the ER (endoplasmic reticulum) and its finalization, prior to excocytose, in the Golgi apparatus (not shown)

to Ref. [57], three steps: discoidal cluster formation, aggregation of clusters into larger patches, and vesicle formation. LNPs are used as vectors for nucleic acids. They migrate in the tissue, attach to the cells, and penetrate into them, thanks to their affinity with the cell membrane. Having penetrated the cell, LNPs deliver their passengers, which are designed to interact with the cell machinery (see Fig. 5.41(A)) . Nucleic acids carried by LNPs can be DNA, mRNA, or siRNA. LNPs have been highlighted during the COVID crisis. They vectorized mRNA into cells and made possible mRNA–based vaccination [58].

How to create LNPs ? In order to assemble the elements forming the LNPs, mixing is required. In large mixers, because of turbulence, size polydispersivity is large and encapsulation efficiency is low. Post–processing steps, such as filtration, extrusion and centrifugation, are needed to produce functional LNPs in this manner. Using microfluidic mixers, in particular inertial milli/micromixers, in view of reaching high throughputs, is an option [59–61]. Fig. 5.42 shows a four–ring micromixer, producing functional LNPs, under flow rates on the order of several tens of ml/min.

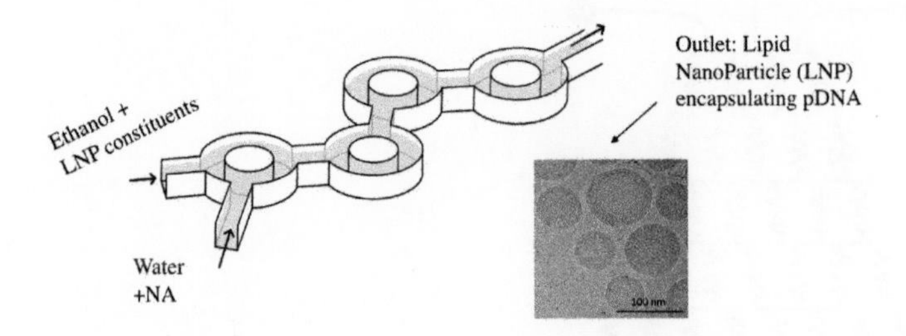

Fig. 5.42: Three–dimensional view of the device, in which the aqueous phase containing the nucleic acid (plasmid DNA (pDNA)) diluted in an acidic buffer is injected on one side, and an ethanol-lipid solution on the other. The mixing of the two constituents lead to the formation of LNPs. The process is well controlled [61].

In Fig. 5.42, a three dimensional view of the four-ring system is shown. The aqueous phase containing the nucleic acid diluted in an acidic buffer is introduced in one branch, and ethanol phase containing the lipids in the other. As they mix together, lipids, exposed to water, form vesicles, encapsulating their partners, in a process called''self assembly'. LNPs thus form. An image of a LNP, obtained in the system of Fig. 5.42, 100 nm in size, transporting plasmid DNA (pDNA) with an excellent encapsulation yield, is shown [61]. With the progress made in the domain over the recent years, and the numerous applications of nucleic acid vectorization to the fields of vaccination and therapeutics, it seems obvious that inertial micromixers will be increasingly used in the forthcoming years.

5.8 Transport of matter across interfaces

5.8.1 Basics

Transport across interfaces occurs in many systems: for instance, osmotic membranes, liquid-liquid extraction, or droplets. The subject is well described in classical textbooks [70, 113]. When two solutions 1 and 2, incorporating solutes of concentrations C_1 and C_2, respectively, are separated by a permeable interface, equilibrium takes place if the following condition is satisfied

$$K = C_1/C_2$$

in which K is the partition coefficient. If the concentration ratio is not equal to K, then fluxes develop across the interface, in order to restore equilibrium. In the chemical engineering literature, a convenient notion – the mass transfer coefficient (we call it here h_m) – is often used to quantify the process. It is schematized in Fig. 5.43..

Here a volume, assumed to be perfectly mixed at all times, is filled with a solution, that includes a solute of concentration C_{body}. Outside, the concentration, also perfectly

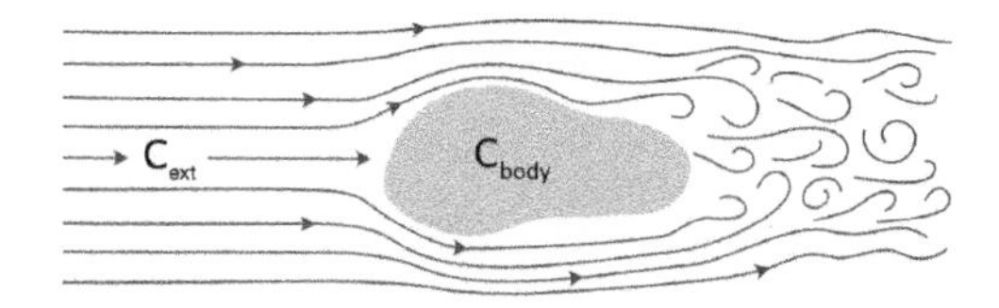

5.43: Body exchanging matter with a fluid. The mass transfer coefficient is a convenient tool allowing the process to be characterized without solving the equations of the problem (see text).

mixed, is equal to C_{ext}. Although different proposals exist in the literature, we consider here the following definition of the mass transfer coefficient [70]:

$$h_m = \frac{J}{A(C_{body} - KC_{ext}}$$

in which J is the mass flux across the exchange area A and K is the partition coefficient. We saw that Fick's law impose fluxes proportional to the concentration gradient. We may thus expect that, at low Reynolds numbers, and in regimes where diffusion prevails, h_m will be proportional to the inverse of the system size. By reducing system sizes, we thus accelerate the transport, and reduce the time needed to achieve processes based on mass transfers. This is where microfluidics can be advantageous.

5.8.2 Microfluidic liquid–liquid extraction

Liquid–liquid extraction in channels. The idea was harnessed in liquid-liquid extraction systems, schematized in Fig. 5.44 [71]. In this work, the wall wettability of the glass microchannel is textured, so that one side is hydrophilic and the other hydrophobic. In such conditions, a stable countercurrent flow of an organic phase could be driven along an aqueous phase, itself driven in the opposite direction. In this system, extraction of a cobalt complex from toluene into water could be achieved in a few seconds, with an efficiency (the ratio of the extracted phase concentration over its initial concentration) of up to 98% [71].

Droplet–based liquid–liquid extraction. This time, extraction is performed with microfluidic droplets. As shown in Fig. 5.45, droplets are formed in a hydrodynamic focusing junction [72]. The droplets are composed of water or water/glycerol mixtures and the external phase is octan-1-ol, to which fluorescein was added. The choice of ethanol was motivated by its high partition coefficient, with respect to a number of solutes, including fluorescein and rhodamine. Droplets move downstream, at a speed of 10 mm/s.

5.44: A stable flow of toluene is driven along an aqueous phase, itself driven in the opposite direction. The configuration is called 'counterflow'. In this system, extraction of a cobalt complex from toluene to water could be achieved in a few seconds with a high efficiency.

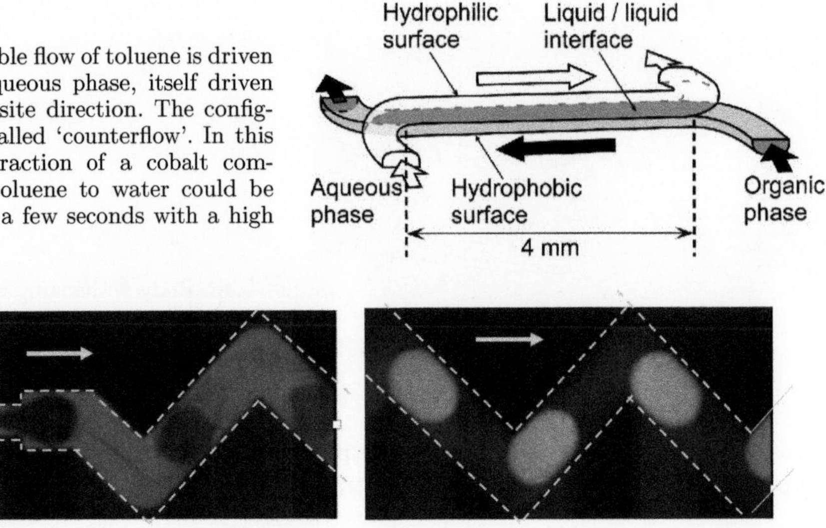

Fig. 5.45: Droplets are composed of water/glycerol mixtures and the external phase is octan-1-ol, to which fluorescein is added. Their mean diameter is 195 μm. Channel height h is 95 μm and width 205 μm. Droplets move downstream, i.e. towards the right, at speeds of 10 mm/s. The left figure shows the system, just after droplet formation, and the right figure, further downstream, at 1 cm from the entry. .

The experiment showed that, after a few seconds, almost all fluorescein is transferred from the external phase (octanol) to the aqueous droplets (See Fig. 5.45, right). In Ref. [72], this was called extraction, considering fluorescein is extracted' from octanol. The authors also investigated the inverse case, i.e. rhodamin transferred from the droplet *to* the external phase. This case can be called 'delivery'. In all cases, they found that, depending on the flow rate, the extraction times vary between 0.2 and 0.8 τ_D, where τ_D is the diffusive time h^2/D, where D is the diffusion constant of the solute in octanol (the other diffusion constants being comparable) and h the channel height. The dependence of the extraction time with the flow speed, is linked to the presence of eddies inside and outside the droplets, as we saw in Chap 3. Using a constant mass transfer coefficient, in this case, would be erroneous.

5.8.3 Bubbles and droplet dissolution: the Epstein-Plesset theory

A bubble immersed in an undersaturated liquid dissolves. As sketched in Fig. 5.46, gas around the bubble diffuses away, depleting the interfacial region, forcing the bubble to transfer gas in the liquid, in order to maintain the interface at equilibrium. In this process, the bubble shrinks. But even in saturated and oversaturated liquids, bubbles may shrink, owing to the action of capillarity, as will be seen later. Since diffusion and interfaces are involved in these situations, one may think that miniaturization will considerably affect the behaviour of the systems. We may expect, for example, that small bubbles will dissolve faster, because mass transfers are accelerated by miniaturization. Depending on the case, this acceleration may be problematic or advantageous.

The problem of bubble dissolution, in a fluid at rest, was described theoretically, in a comprehensive and concise manner, by Epstein and Plesset [73], leading to the so-called Epstein-Plesset theory. Fig. 5.46 sketches the geometry of the problem.

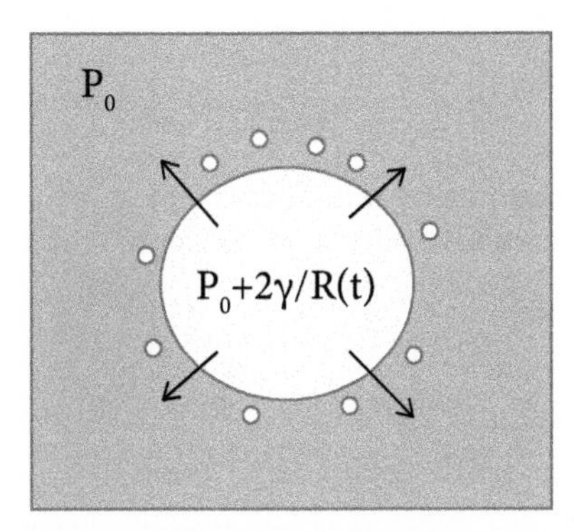

5.46: Epstein–Plesset theory. In non–saturated liquids, gas diffuses in the liquid phase to maintain equilibrium at the interface, dictated by Henry's law.

In the liquid phase, gas concentration $C(r,t)$ (where r is the radial coordinate and t the time), is governed by diffusion equation. In spherical coordinates, taking angular invariances into account, the equation reads:

$$\frac{\partial C}{\partial t} = D\Delta C = D\ \frac{1}{r^2}\frac{\partial C}{\partial r}(r^2\frac{\partial C}{\partial r}) \tag{5.22}$$

in which D is the diffusion coefficient of gas in liquid. Here we consider two cases (we use D. Lohse's review [79] for the presentation).

Large bubbles. .

For large bubbles, capillarity can be neglected. In such conditions, the boundary conditions are:

$$C(r,0) = C_\infty.\ \text{and}\ C(R,t) = C_s$$

where C_∞ is the gas concentration far from the bubble, C_s is the saturation concentration, also called solubility, and $R(t)$ is the bubble radius. C_s is the maximum gas concentration that a liquid, at equilibrium, can solubilize. It depends on pressure and temperature. The warmer the liquid, the smaller the solubility. This dependence explains why, when a glass of water warms up, bubbles spontaneously nucleate at the surface. Forming bubbles is a way, for the liquid to degas, i.e. to decrease its gas content so as to match a smaller C_s.

Let us return to Eq. 5.22. In the limit where the process is slow, $\frac{\partial C}{\partial t}$ can be neglected, and the solution to Eq. 5.22, together with the boundary conditons, read:

$$C(r) = C_\infty + (C_s - C_i)\frac{R}{r}$$

Except when $C_\infty = C_s$, a flux $J = D(\frac{\partial C}{\partial r})_{r=R}$ develops across the interface, in one direction or another, depending on the sign of $C_\infty - C_s$. An important parameter here is the oversaturation, defined by

$$\zeta = \frac{C_\infty}{C_s} - 1$$

ζ can be positive (oversaturated solution) or negative (undersaturated solution). From the previous relations, one can determine the mass exchange rate across the interface. It is given by

$$\frac{dm}{dt} = 4\pi R^2 D(\frac{\partial C}{\partial r})_{r=R} = 4\pi R D \zeta C_s \tag{5.23}$$

in which $m = \frac{4}{3}\pi R^3 \rho_G$, in which ρ_G is the gas density, is the bubble mass. We can establish the following relation:

$$\frac{dR}{dt} = \frac{D}{\rho_G R}(C_\infty - C_s) = \zeta \frac{D}{\rho_G R} C_s$$

After integration, we find:

$$R(t)^2 = R(0)^2 + \frac{2D\zeta C_s}{\rho_G}\, t \tag{5.24}$$

where $R(0)$ is the initial bubble radius. When $\zeta < 0$, the bubble shrinks and when $\zeta > 0$, it grows. In the former case, the bubble lifetime, in a pure liquid ($\zeta = -1$), is given by:

$$\tau = \frac{\rho_G R_0^2}{2DC_s} \tag{5.25}$$

Table 5.3 provides a few values of C_s and τ, for a bubble of 100 μm in radius and three gas/liquid systems.

The formulas we just established have interesting consequences for microfluidics: when bubbles are trapped in a microfluidic device, it is often difficult to remove them, because, most of the time, as explained in Chapter 4, they are anchored by the roughness of the wall surface, trapped by wetting heterogeneities, or captured in dead zones. A strategy currently used in the laboratories is to evacuate the liquid, before filling. Then, according to the theory, trapped bubbles will dissolve. However, theory also tells us that this strategy is not efficient with air, because, according to Table 5.3, significant waiting times would be needed. For instance, for an air bubble, 100 μm in

Gas	Medium	Solubility C*(w/w)	$\tau(s)$
CO_2	Water	1.7 10^{-3}	6
N_2	Water	10^{-5}	1200
O_2	PFC	~2 10^{-4}	60

Table 5.3 Solubilities of various gases in liquids, and dissolution times estimated for a bubble of 100 μm in radius, and pure liquids (see Eq.5.25). The relation between C^* and C_s is $C_s = \rho_L C^*$, where ρ_L is the liquid density

radius, one must wait 20 min before it disappears. By using carbon dioxide, whose solubility is much higher, this time is reduced to 6 s. This is why devices are often pre-filled with this gas. In the field of microfluidics, researchers easily understand that solubility favours rapid dissolution, but they rarely pay attention to Epstein–Plesset theory.

Small bubbles. .

The calculation we made neglected surface tension. As noted, it applies to large bubbles. When a large bubble shrinks, at some point, it becomes small and capillarity ceases to be negligible. To describe the shrinking process fully, we thus need to take capillarity into account. The problem was solved by Epstein and Plesset [73]. The calculations are elaborate, and we give here a summary of the results, using again D. Lohse's review [79].

With capillarity, the situation changes: pressure P_g inside the bubble is above that in the liquid. We have:

$$P_g = P_0 + \frac{2\gamma}{R}$$

in which P_0 is the ambient pressure and γ the surface tension. This is Laplace's law, which we discussed in Chapter 4. The law is interesting, because it provides a criterion for distinguishing between what we called 'small' and 'large' bubbles. From the formula, we may consider that 'small' bubbles have radii below $\frac{P_0}{2\gamma}$ and large bubbles have radii larger than this. For clean water and air, the cross-over between the two cases, i.e. $\frac{P_0}{2\gamma}$, is 1.4 μm.

The gas concentration at the interface, on the liquid side, is given by Henry's law:

$$C(R,t) = \frac{C_S}{P_0} P_g$$

With this, calculations similar to those carried out previously allow us to determine the temporal evolution of R. We will concentrate on the case where capillarity dominates the problem, i.e. $R << \frac{P_0}{2\gamma}$ A calculation carried out in this limit leads to the formula:

$$\frac{dR}{dt} = -\frac{3DC_s}{2\rho_G R}$$

The bubble shrinks, according to a law given by:

$$R^2(t) = R^2(0) - \frac{3DC_s}{\rho_G}\,t$$

Now, the life time of the bubble is as follows:

$$\tau = \frac{\rho_G R^2(0)}{3DC_s}$$

Note that, despite the fact that the regime is dominated by capillarity, the bubble lifetime is independent of surface tension. For a nitrogen bubble measuring 1 μm in radius, in water, the survival time is 1 ms. Micrometric bubbles dissolve extremely fast.

As noted, the full determination of the bubble dynamics, for arbitrary values of ζ and $R(0)$, including saturation and capillarity, was performed by Epstein and Plesset [73]. In all cases, it was found that for $\zeta < 0$, the bubble always shrinks while for $\zeta > 0$, it may shrink or grow, depending on the parameters of the problem. A stationary state exists, for which the oversaturation ζ compensates the action of capillarity, but it is unstable [79]. The theory, applied to nanobubbles, concludes that in practice, bubbles of that size always disappear, in 1 ms or so [79].

5.8.4 The long lifetime of surface nanobubbles

According to Epstein–Plesset theory, one must be fast to see submicrometric bubbles. They dissolve so rapidly that a standard camera (with a frame rate of 100 images/s) coupled to a high magnification microscope would be too slow to image them. Early observations, made in 2000–2001 by three groups [75–77] indicated that in oversaturated water and over hydrophobic surfaces, nanobubbles survive for hours. An example of long–lived nanobubbles [77], observed with the atomic force microscopy (AFM), is shown in Fig. 5.47 (A).

The observations made by the three groups challenged our understanding of the liquid/vapour interfaces, and, for this reason, they attracted much interest. However, for years, the situation remained confusing. The existence of long–lived nanobubbles was confirmed by a number of investigators and rejected by others.[12] Eventually, based on theory, numerics, along with considerable experimental effort, the community came to

[12] In the confusion, some companies proposed to sell *bulk* nanobubbles in bottles, arguing that they enhance the growth of plants, improve the life of oysters, and reduce the occurrence of heart or cancer problems. What we can say today, is that there is no indication that, if these nanobubbles exist, they are not trapped by impurities, and should be seen as surface nanobubbles [80]. Whether they can heal something is another subject.

A

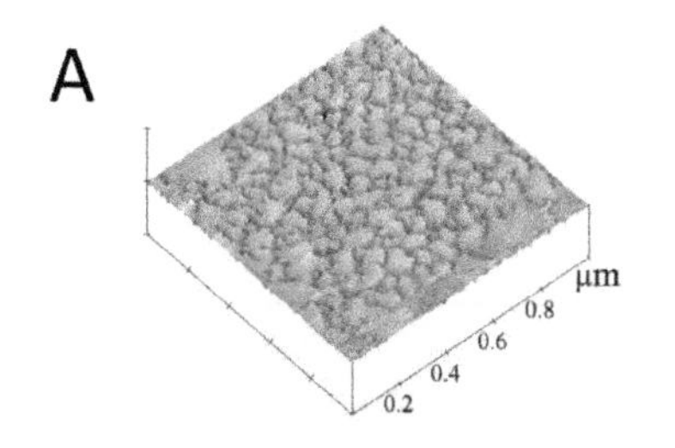

B

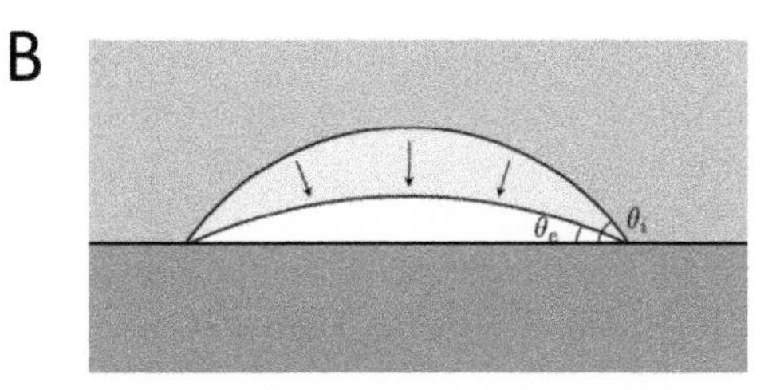

Fig. 5.47: (A) AFM image of a single large surface nanobubble on a hydrophobized gold surface. (Reprinted from [77], with permission of the American Physical Society. Copyright 2022.) (B) Shrinking process of a pinned surface nanobubble with footprint diameter L and initial contact angle $\theta_i > \theta_e$ towards the equilibrium contact angle θ_e [74].

the conclusion that these bubbles exist [78, 79]. There is a diversity of cases, but, in most of them, long–lived nanobubbles are several tens of nanometres in size, contain hundreds of molecules, and have semi-spherical shapes, with contact angles substantially more acute than Young angle.

How can we reconcile these observations with Epstein–Plesset theory? One notes first that Epstein-Plesset theory concerns bulk bubbles, not surface bubbles. The point raised by [74] is that nanobubbles are stabilized by the pinning of their contact line at the surface. As we explained in Chapter 4, when wetting is partial, roughness and wetting heterogeneity generate pinning. Pinned nanobubbles, as they shrink, decrease their contact angle, because they keep the same footprint (see Fig. 5.47 (B)). To understand nanobubble stability, we may reproduce the argument of Ref. [74].[13] The kinetics of the bubble mass M derives from an equation similar to Eq(5.23). It is given by:

$$\frac{dM}{dt} = -\frac{\pi}{2}LD\left((P_0 + \frac{4\gamma\sin\theta}{L})\frac{c_s}{P_0} - c_\infty\right)f(\theta) \sim K(\frac{L_c}{L}\sin\theta - \zeta)$$

in which $f(\theta)$ is a geometric function characterizing the bubble geometry [74, 81], and K is a factor. The key point is that, for $\zeta > 0$, and for a certain range of L, this equation admits a stationary solution, characterized by a contact angle θ_e given by:

$$\sin\theta_e = \zeta\frac{L}{L_c}$$

in which $L_c = 4\frac{\gamma}{P_0} \approx 2.84\ \mu$m. At equilibrium, and for bubbles smaller than L_c/ζ, Laplace pressure counteracts the gas influx induced by the oversaturation. Large surface bubbles, i.e. larger than L_c/ζ cannot reach an equilibrium: they swell and detach from the surface, by buoyancy. The equilibrium, found for small bubbles, is stable. Should the bubble swell, Laplace pressure increases, gas crosses the interface, and the bubble shrinks. A similar reasoning can be made for negative volume perturbations.

[13]Amazingly, the work extended a theory made by Y. Popov, concerning the coffee ring effect. [81]

In all cases, the system is stable. The theory also explains that θ_e is uncorrelated with the Young angle. The theory was well–supported numerically [82]. As a whole, one may conclude that nanobubbles can survive. Interestingly, the theory resolved a fifteen–year puzzle.[14]

Microfluidic contrast agents.
It is possible to delay bubble dissolution by coating them with a polymerized shell and filling them with a heavy gas (for instance, perfluorohexane C_6F_{14}). The shell being porous, the bubble still dissolves, but the process takes a much longer time, tens of minutes or so, in water, for 5 μm diameters. These bubbles, possessing a high echogenicity, can be used as ultrasonic contrast agents (UCAs). With an echograph, it is possible to visualize the bubbles and see where in the body they travel, and at what speed. These bubbles are used to analyse, for instance, the presence of leaks at the level of the heart valves. In the USA each year, this technology is used on 30,000 patients. It is possible to image different parts of the body without contrast agents, but, in this case, the image quality is reduced, and, sometimes, crucial details are impossible to detect. It has been shown that, with microfluidics, one can create highly monodisperse contrast agents [83,100]. An example is shown in Fig. 5.48 [100].

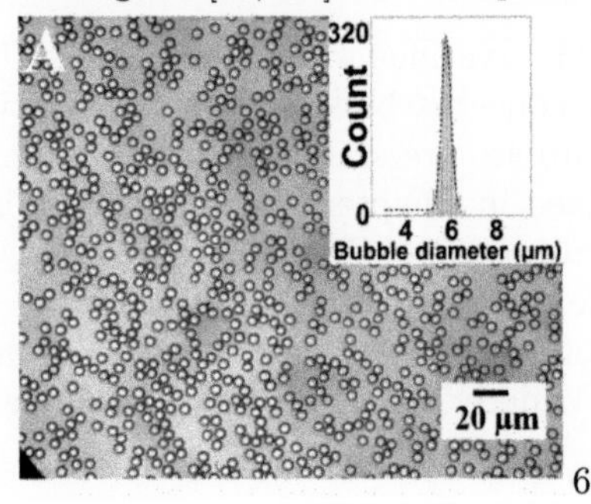

Fig. 5.48: Microfluidic contrast agents, coated with polyvinyl alcohol (PVA), along with their size distribution [100]. (Courtesy of Ugur Soysal and Pedro Azevedo.)

.
Monodisperse contrast agents are advantageous for obtaining more intense acoustic responses (all bubbles resonate and produce strong signals). They are interesting for drug delivery applications (the optimisation of the delivery process, based on bubble bursting, is sensitive to the size), and for performing non invasive pressure measurements (bubbles emit echos depending on the local pressure and the size) [84–86, 100]. At the moment, microfluidic monodisperse UCAs are awaiting commercialization [100]. Note that the coupling between microfluidics and acoustics has been explored long ago, in the years 2003–2010, in particular for particle sorting [24,87].

[14]The theory seems simple. Then why did we have to wait for fifteen years to formulate it ? This is hows science proceeds. Situations may be blurred by experimental confusion, induced by the considerable difficulties experiments had to face. Also, elaborating the correct vision of a complex problem, involving unsettled questions, takes time. In a private communication, D. Lohse told that he discovered the argument during a Taipeh–Amsterdam flight. One may perhaps infer that flying above continents stimulates inspiration.

5.9 Particles and microfluidics

5.9.1 How do small particles diffuse close to a wall?

We saw that small particles diffuse with a diffusion coeffient given by the Stokes–Einstein law. We recall here the expression of the law:

$$D = \frac{kT}{6\pi R\mu}$$

This law was established in an infinite medium. Close to a wall, the movements of the particle are slowed down, and the diffusion constants decrease. This is called hindered diffusion'. How can we describe the phenomenon? Faxen in 1922 [88] and Brenner in 1961 [89] carried out the calculation. The coefficient of diffusion is no longer a single number, but a diagonal tensor, with two components, one parallel to the wall, $D_{//}$ [88] and the other normal to the wall [89,90], $D_{\perp}$. The corresponding formulas are:

$$D_{//} \approx (1 - 9r/16z + r^3/8z^3 - 45r^4/256z^4 - r^5/16z^5)D$$

and:

$$D_{\perp} \approx \frac{6(z-r)^2 + 2r(z-r)}{6(z-r)^2 + 9r(z-r) + 2r^2} D$$

in which r is the particle radius, z the distance of the particle centre to the wall, and D is the bulk diffusion coefficient.

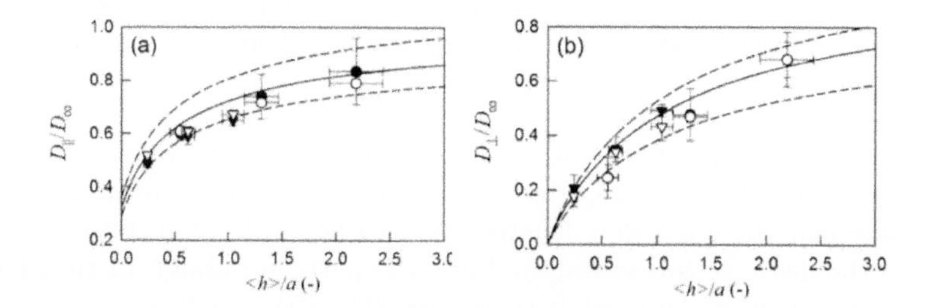

Fig. 5.49: Hindered diffusion correction factor $D_{//}/D$, $D_{\perp}/D$ vs nondimensional gap size between particles and glass surface $(z-r)/r$, where r is the particle radius and z is the centre distance to the wall. D is the diffusion coefficient measured in the bulk. The full line corresponds to the theoretical models discussed in the text. (Reprinted from [91], with permission of AIP. Copyright 2022.)

Figure 5.49 shows measurements, based on evanescent waves, of the two diffusion coefficients as a function of the distance to the wall $(z-r)/r$, in which z, as we said, is the distance of the particle centre to the wall, i.e. its 'altitude' [91]. Similar measurements were obtained by Ref. [92]. As we approach the wall, $D_{//}/D$ decays to a value close to the theoretical expectation, i.e. 0.32, while $D_\perp$ decays to zero. The effect of the wall is thus important at distances comparable to the particle radius.

5.9.2 Particles deposition at low Reynolds numbers

We just saw how walls affect the diffusion constants of particles. More generally, close to a wall, and at small Reynolds numbers, the particle is subjected to a number of effects, sketched in Fig. 6.37.

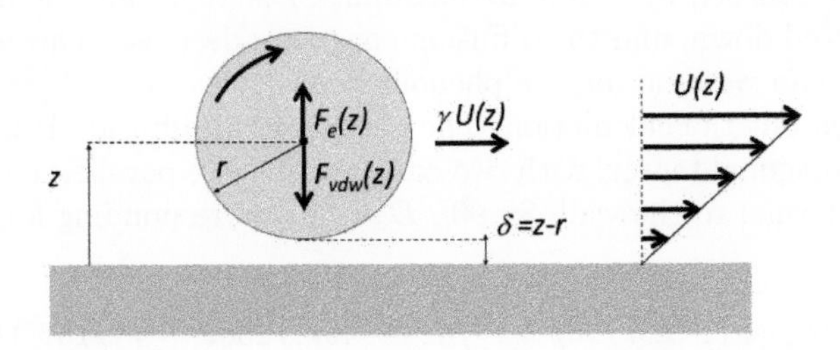

Fig. 5.50: Spherical particle moving close to a wall. Van der Waals force $F_{vdw}(z)$, repulsive electrostatic force $F_e(z)$, and viscous drag $\gamma U(z)$ are represented.

The effects include brownian agitation (that we analysed earlier), viscous forces (due to particle displacement and rotation), Van der Waals forces (attracting the particles towards the wall (see Chapter 2), and electrostatic forces (due to the presence of surface charges, as will be seen in Chapter 6). Depending on the conditions, these effects carry the particles away from the wall or bring them to it, onto which, in most situations, they stick.

The problem looks complicated, but recently, a comprehensive analysis of the different cases has been carried out [93–95]. Here, we consider the case where electrostatic forces are negligible and the particle is located close to the wall, at distances $z - r << r$ (where z is the distance to the wall and r is the particle radius). In such conditions, the equations governing the particle trajectory $x(t), z(t)$, are [93]:

$$\dot{x}(t) \approx \gamma U(z) + f_x(t) \tag{5.26}$$

$$\dot{z}(t) = f_z(t) + \frac{dD_\perp}{dz} + \frac{D_\perp}{kT} F_{vdWz} \tag{5.27}$$

in which

$$F_{vdWz} = -\frac{Ar}{6(z-r)^2} \tag{5.28}$$

where A is the Hamaker constant of the wall/solution/particle system (see Chapter 2).[15] γ is a constant (on the order of 0.7), modelling the action of the shear on the particle speed [93], and U_r is the local velocity, i.e. the flow speed calculated at a distance $z = r$ from the wall. $f_x(t), f_z(t)$ are the components of the Brownian forcing, discussed earlier. We further focus on the case where the particle are so large (typically above 1 μm) that Brownian motion can be neglected.[16] This regime is called 'Van der Waals 1' in Ref. [94]. In this case, the equations simplify into the following:[17]

$$\dot{x}(t) \approx \gamma U_r \tag{5.29}$$

$$\dot{z}(t) \approx -\frac{A}{6kT}\frac{D}{(z-r)} \tag{5.30}$$

On solving Eqs. (5.29) – (5.30), one finds:

$$(z(t)-r)^2 = (z_0 - r)^2 - \frac{ADt}{3kT} \text{ and } x = \gamma U_r t$$

where z_0 is the particle position for $t = 0$, assuming that initially, the particle is located at $x = 0$. As sketched in Fig. 5.51, particles, attracted by Van der Waals forces, falls towards the walls while being transported by the flow. Particles injected at an altitude $z_0 < z_c$, with $z_c - r = \delta_c = \sqrt{\frac{ADL}{3\gamma kTU_r}}$, are caught by the channel, while the others escape. This is the deposition mechanism.

5.51: In the Van der Waals 1 regime, the non–Brownian particle, injected at the 'altitude' z_0, is transported by the flow and, in the meantime, attracted by the wall. For a given channel length L, below a critical altitude $z_0 = z_c$, the trajectory hits the wall, and the particle is captured. Above z_c, the particle escapes the channel.

In practice, particles are injected in the system, at a rate $J = \frac{\varphi Q}{v_P}$, in which φ is the volumetric concentration, v_p is their volume, and Q is the volumetric flow rate. J is the number of particles that penetrate, per unit of time, the channel. Its unit is T^{-1}. Using the previous results, it is not difficult to show that the number of particles $N_A(t)$, deposited at time t on the lower channel wall, is equal to:

$$N_A(t) \approx \frac{\delta_c}{h} J\, t \tag{5.31}$$

[15] We take here the expression for the plane-plane case. This is appropriate because the particle is close to the walls, at distances so small that its surface looks planar.

[16] It is considered, in colloidal science, that particles with diameters larger than 1 μm smooth out the effect of thermal agitation.

[17] In practice, for plastic material, $A/kT \sim 1$, so that, close to the wall, the term $\frac{dD_\perp}{dz}$ is negligible.

with:

$$\frac{\delta_c}{r} = S = \left(\frac{A}{3\gamma kT}\xi_L\right)^{1/2} \quad \text{with } \xi_L = \frac{LD}{U_r r^2} \tag{5.32}$$

$N_A(t)$ is the number of particles that are collected, at time t, by the lower wall, in a channel of length L. Replacing L with x, $Na(t)$ becomes $Na(x,t)$, i.e. the number of particles collected by the same channel at time t, but up to a distance x from the entry. With that, we can define the particle density $p(x,t)$ by the following relation:

$$p(x) = \frac{\partial N_A}{\partial x} = \frac{\delta_0}{4h^{3/2}} J x^{-1/2}\, t \tag{5.33}$$

$p(x,t)$ decreases along the channel. This is the so-called 'retention profile', systematically observed when particles are injected in narrow channels, for instance, in porous media. At all time, the maximum particle density is located at the entry of the channel, i.e. in the framework of our analysis, for $x = h$. Eqs. 5.31 and 5.32 also show that the larger the Hamaker constant A is, the stronger the deposition rate is. This implies, for instance, that deposition of polystyrene particles, as used in Ref. [93], is stronger with hydrophobic walls than with hydrophilic walls.

The experiments of Ref. [93], conducted with high salt concentrations for screening the electrostatic forces, and polystyrene particles, 5 μm in diameter, has confirmed these predictions, along with the scalings given by Eqs. (5.31), (5.32) and (5.33). Figure 5.52 shows an example of comparison between theory and experiment. From this work, proposal was made to measure the Hamaker constant [93].

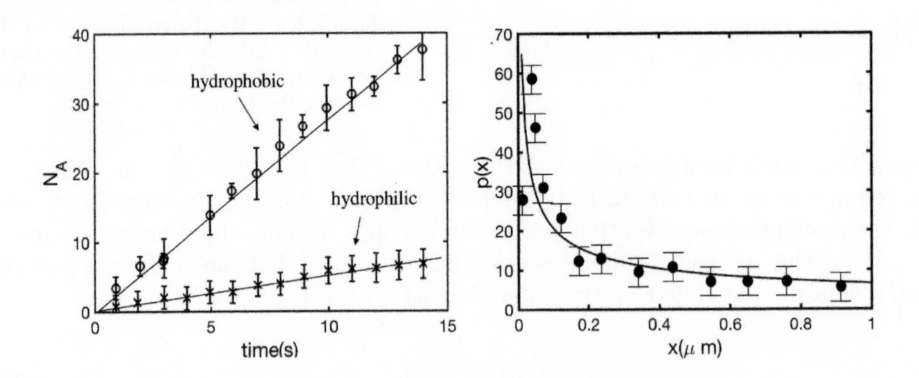

Fig. 5.52: (Left) Number of particles collected by the lower channel wall, as a function of time, for hydrophilic and hydrophobic walls. The lines correspond to Eq. (5.31). (Right) Probability of finding a particle collected by the wall, at a given time, at distance x from the entry. The line is $p(x) \sim x^{-1/2}$. This plot is called 'retention' profile by the filtration community [93].

5.9.3 Clogging

Often, the first impression that one may have, on looking at a microfluidic device, is that channels are so small that they will always clog. Consequently, they will never be used industrially. It turns out that this is wrong: today, as we saw in Chapter 1, hundreds of millions of microfluidic devices are shipped every year. Still, it remains true that clogging is an issue. Any impurity of size comparable to the channel transverse dimensions, or a fraction of it, often present in systems of practical interest, or even generated by the device itself (for instance, PDMS debris), may close channels partially and eventually clog the system. However, over the years, progress has been made and today, many microfluidic devices are used in the industry in a reliable manner, without clogging. In most cases, clogging can be avoided by working with filtered liquids and, most importantly, placing filters with submicrometric pores in line with the device.

How does clogging occur? Fig. 5.53 shows an example.

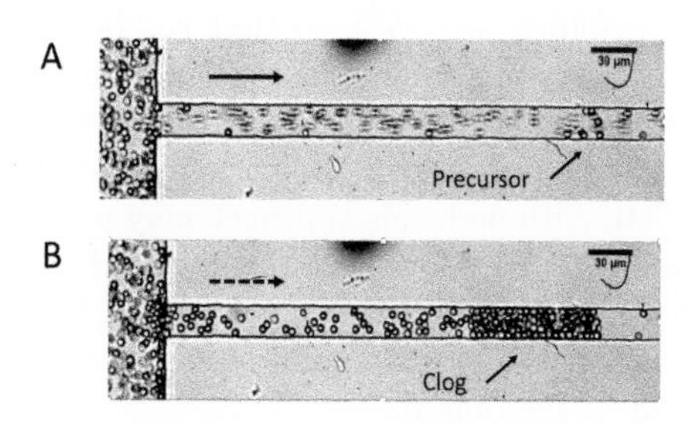

5.53: Clogging in a PDMS microchannel.(A) Several particles deposit at the walls, forming a precursor that partially obstructs the channel.(B) After arches have formed across the channel, the clog appears.The dashed arrow represents the small flow traversing the porous clog, bringing new particles that increase its size. In this experiment, speeds are fractions of mm/s, particles are 5 μm in diameter, and the clog forms in a few seconds. (Courtesy of C. Cejas (2018).)

In Fig. 5.53 (A), particles first deposit at the walls, then accumulate locally, forming arches across the channel, which eventually clogs it (Fig. 5.53 (B)). Today, there is no established description of the phenomenon, and consequently no formula that could characterize the clogging conditions. It is nonetheless possible, using Eq. 5.31, to suggest a criterion. Let us assume that clogging takes place when several particles are adsorbed at the same place, forming a cluster, that develops arches and provokes clogging. Let us further assume that this event will occur first close to the channel entry, i.e. for $x \sim h$ (because p is maximal there). A cluster can be associated to a local full coverage, i.e. $p(x)$ on the order of $\frac{h}{r^2}$.[18] Based on this, using Eq. (5.33), we find the following expression of the clogging time:

$$t_c \sim \frac{h^3}{r^2 \delta_0 J}$$

Applied to the experimental conditions, of Fig. 5.53, we find seconds, which is consistent with the observations. The formula indicates that the smaller the Hamaker

[18]For a channel of width h, this event will occur with a probabiliy $\frac{r}{h}p(x)$. We can also say that, at a distance h from the entry, one must have full coverage, i.e. for a channel of width h, $N_A(h,t) \sim (\frac{h}{r})^2$; then, $p \sim N_A(h)/h \sim) \frac{h}{r^2}$

constant, the larger the clogging time, which agrees with the experiment. This implies that is preferable to use hydrophilic channels to transport polystyrene particles (rather than hydrophobic ones).

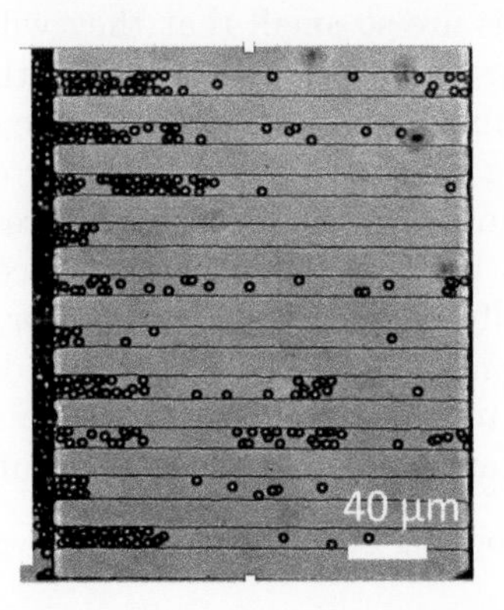

5.54: Clogging in an array of square PDMS microchannels, 20 μm side, in which 5μm particles have been introduced, by driving a flow from left to right. The image is taken a few seconds after the particles have penetrated the channel. Most of the clogs appear close to the channel entries. (Courtesy by C.Cujas (2018).)

One can also show that t_c is proportional to h (since $\delta_0 \sim h^2$), so that the larger h is, the later the channel clogging. From this remark, one may infer that in a channel including restrictions, clogging will most probably occur in the narrowest parts of the channel. The reasoning we present for the Van der Waals regime 1 can be extended to the other deposition regimes analysed in Ref. [94]. Although this approach may provide interesting clues on the clogging conditions, in particular a link with the material properties, it is based on hypotheses and simplifications that need to be confirmed. Other approaches have been proposed [96–99], emphasizing kinetic and statistical aspects.[19] To summarize, much remains to be understood in the domain.

In practice, if neither surface properties, nor dimensions can be tuned, what can be done? The first idea is filtering. Filtering prior to injecting the (liquid) sample in the system through submicrometric pores (200 nm pore sizes are often used), along with care in the microfabrication process, are rules, respected by microfluidic laboratories, to ensure long–term functioning of microfuidic devices. Another approach consists, for instance, in introducing anti-clogging systems, such as the 'rivers' shown in Fig. 5.55.

Here, small bubbles (5 μm in diameter) are produced at several kHz. Gas is introduced in a central channel, and liquids in side channels. To avoid clogging, two large channels, called 'rivers', are added on the sides. Most of the liquid phase (water) circulates in the rivers, only a small part of it penetrate into the flow focusing device. Rivers sweep away dust particles that could enter the system. In the filtration language, this is called 'tangential filtering'. The liquid can be recycled.

For microfluidic systems dedicated to handle particles, the problem is more delicate, because particle *must* circulate in the system. Once the clogs appear, it is difficult, in practice to remove them, and in most cases, the device is lost. The most obvious rule to avoid the problem is to enlarge the channels and minimize the attractive interactions between the particles and the walls. Another approach is to use droplets to transport

[19]These aspects have been overlooked in the model described above

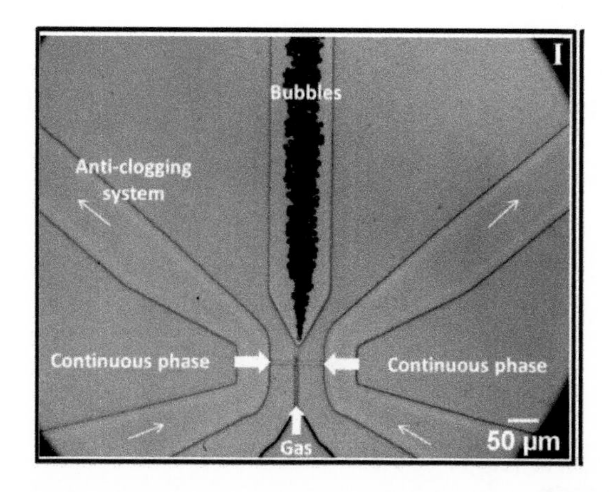

5.55: Anti–clogging system allowing us to produce bubbles without clogging: air is injected in a channel, in the lower part of the figure, and liquid (water) is injected on the sides, through two small channels (barely visible). These small channels are connected to a 'river', i.e. a much larger channel, which carries most the liquid phase. This system represents a tangential filtering configuration [100]. (Courtesy of U. Soysal and P. Nieckele.)

the particles or to work at large speeds, which may be impractical in a number of cases.

5.9.4 Three methods of sorting particles

The H filter. The H filter was invented in the late 1990s [101]. The principle is shown in Fig. 5.56.

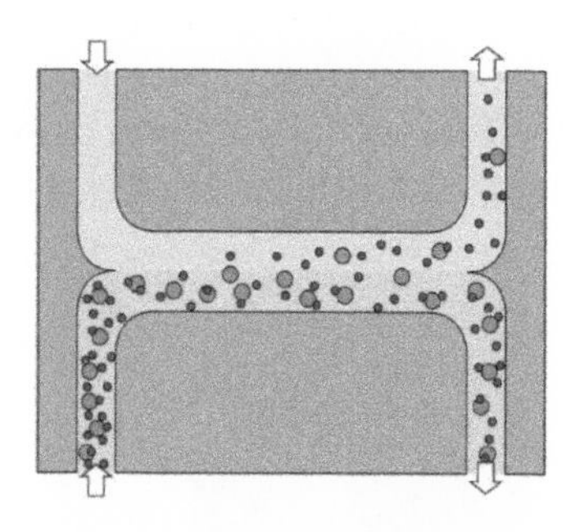

5.56: The H–Filter [101]: two miscible fluids, one including particles, and the other not, form an interface across which particles diffuse, at a speed depending on their sizes. In this way, particles can be sorted in function of their sizes.

The incoming fluids are miscible and their flow rates are equal. Therefore, as we saw in Chapter 3, a diffuse interface establishes in the system. Particles are injected in the left part of Fig. 5.56, in the lower branch. Large particles follow the streamlines. They travel in the central channel, bifurcate to the right, and exit. Small particles diffuse, and consequently, they visit different streamlines during their journey. Part of them cross the interface and jump in the upper branch of the H–filter. By repeating the operation, it is possible, in theory, to sort particles with a large yield. In practice, in order the diffusive process to be efficient, the central channel of the H–filter must be long, and speeds must be low. As a consequence, throughputs are low. Therefore, although conceptually appealing, over the years, other methods have been preferred .

Sorting by confinement (particle flow fractionation (PFF). Different sorters use the effect of the confinement to place particles of different sizes on different streamlines. This gave rise to the acronym PFF' (particle flow fractionation) [102]. An example is shown in Fig. 5.57.

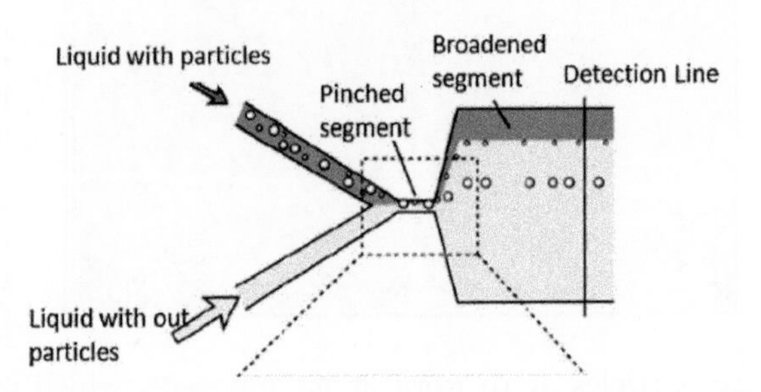

5.57: System enabling particle separation.

By passing trough the restriction, large particles are forced to circulate along the centerline, even though they were injected sideward, as in Fig. 5.57. Further downstream, these particles will stay on the same streamline and, consequently, travel distantly from the walls. By contrast, small particles, also injected sideward, will stay close to the wall in the collecting channel. Thereby separation according to size is achieved. Since diffusion is irrelevant, the device can be operated at fast speeds, as long as the flow does not develop detachment eddies induced by the divergent channel. This constraint restricts the operating speed. In practice, non–Brownian particles and cells, i.e. with sizes above 1 μm, can be efficiently sorted with this method.

The pillar array. The idea was proposed in 2004 [103], giving rise to the acronym DLD (deterministic lateral displacement). The principle is shown in Fig. 5.58 (borrowed from Ref. [104]):

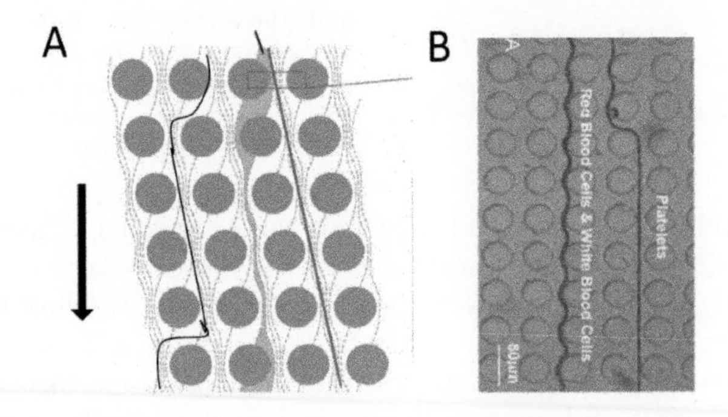

Fig. 5.58: Deterministic lateral displacement (DLD) [104]. Large particles are conveyed along oblique trajectories, while small particles can pass from one row to the other, transported by transverse fluid streamlines.(A) Inclined array.(B) 'Vertical array', in which cells (red and white blood cells) are separated from platelets [104]. (Creative Commons licence.)

Large particles, whose centres are maintained at some distance from the disks, are forced to follow oblique trajectories, into which they are continually displaced at each successive pillar (this justifies the acronym). By contrast, a fraction of small particles follow fluid streamlines passing from one row to another one [104]. Thanks to this process, separation between small and large particles can be achieved. Being based on hydrodynamic transport, separation is much faster than diffusion. Fig. 5.58 (B) shows that the principle works for separating blood cells from platelets. The idea also works for separating long DNA fragments, but with limited resolution.

5.10 Particles in inertial regimes

Although the Segre-Silberg effect was discovered in 1961 [105], it came as a surprise, in the microfluidic community, that it was possible to reorder particles in a microchannel, just by increasing the Reynolds number. The rediscovery of the Segre-Silberg effect, made by Di Carlo et al. in 2006 [106], stimulated the growth of a new topic, called 'inertial microfluidics' . In fact, as we noted in Chapter 3, the term 'inertial microfluidics' encompasses a much broader domain. This concerns moderate Reynolds number flows, driven in inertial regimes, i.e. regimes in which inertia forces come into play, in microfluidic or millifluidic systems, incorporating or not particles or interfaces.

We thus present the behaviour of particles transported in flows driven at moderate Reynolds numbers, i.e. typically 10–100. Above these numbers, time-dependent phenomena develop, giving rise, as we previously saw, to chaotic behaviour. In such regimes, it is impossible to reorder anything. We restrict ourselves to rigid particles. Soft particles, such as blood cells, although important for the applications, are nonetheless too complex to be discussed here. For such particles, membrane elasticity along with deformability generate a plethora of effects that are too extensive to present here.

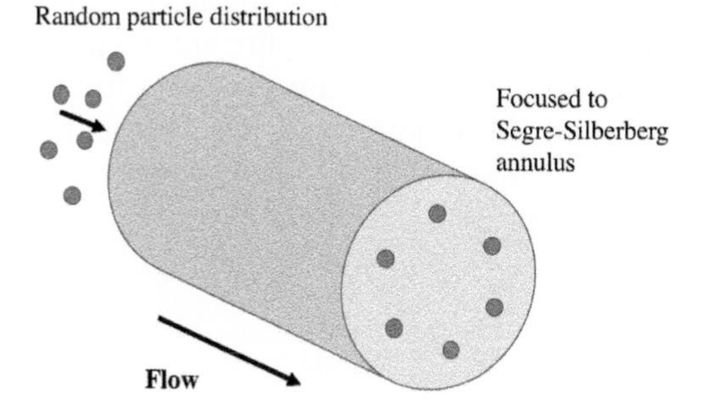

5.59: From Ref. [107]: in a cylindrical pipe, at moderate Reynolds numbers, randomly distributed particles focus to an annulus located between inside the pipe.

The Segre–Silberberg effect [105] is sketched in Fig. 5.59. In dilute suspensions of neutrally buoyant particles, driven in a tube, particles tend to collect on an annulus, whose radius is $0.6R$. Why this is the case? The subject is rich and subtle. To make a long story short, one may first note that, in inertial microfluidics, particles are subjected to viscous drag forces and inertial lift forces. The drag forces drive particles along the flow streamlines, whereas inertia forces drive particles across the flow streamlines. There are several inertial forces, induced by the particle rotation, their speed deficit, with respect to the flow, the wake generated by this deficit, and the coupling of all these effects to the wall. In straight channels, the most important forces are a lift directed away from the wall and a shear gradient lift directed down the shear gradient, i.e. towards the wall (Saffman force). The combination of the two forces gives rise to equilibrium positions towards which particles migrate as they travel downstream. In channels of rectangular cross-section, the equilibrium positions are surfaces parallel to the walls. Ref. [108] provides detailed analyses of these forces. In practice, one needs long channels to observe the Segre–Silberberg effect.

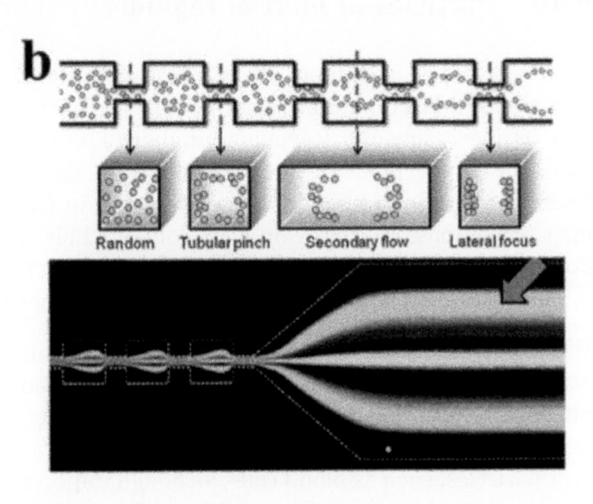

5.60: Inertial separation in a multi-orifice microchannel according to the size-dependent lateral migration termed as multi–orifice flow fractionation [108–110]. Such images are obtained for Reynolds numbers ranging between 20 and 100. (Reprinted from Ref. [108] with permission of the Royal Chemical Society. Copyright 2022.)

Over the years, the effect has been exploited in different geometries to reorder or sort particles and cells. We mentioned an example, in Chapter 3, in which Dean vortices, probably coupled to Segre-Silberberg effect, allowed to isolate Circulating Tumor Cells (CTC) [111]. Another example of the reordering effect which is quite spectacular was demonstrated in Refs. [109, 110]; this is shown in Fig. 5.60 (images taken from Ref. [108]). Some of these systems are commercialized. Their advantages are their simplicity and high throughput.

5.11 Adsorption

5.11.1 Generalities

Due to the presence of attractive forces and thermal motion, molecules located near a wall may be captured by the surface for some time (adsorption) until a thermal fluctuation re-expedites it in the bulk (desorption). The field of adsorption is vast (see, for instance, Ref. [112, 113]), we give here an extremely brief account, motivated by the fact that microfluidics exacerbates the role of the surfaces; therefore, it makes sense, in the context of this book, to introduce a few notions. Adsorption concerns surfaces, while absorption concerns volumes. A simple illustration of the difference is shown in Fig. 5.61.

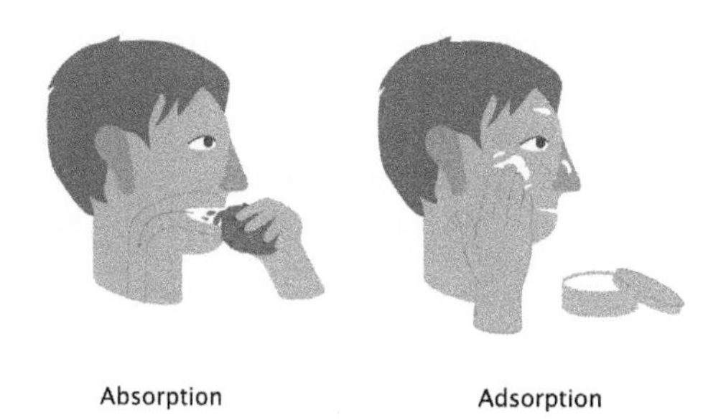

5.61: Difference between adsorption and absorption. (from L. Dehove.)

Adsorption/desorption concerns molecules, macromolecules, and clusters of molecules, small enough to be sensitive to thermal fluctuations. For larger objects (colloids, particles), the terms used most often are sticking, adhesion or deposition. Direct evidence of adsorption is brought by various methods, such as atomic force microscopy (AFM), spectroscopy, or electronic microscopy. An example based on AFM is shown in Fig. 5.62 [114].

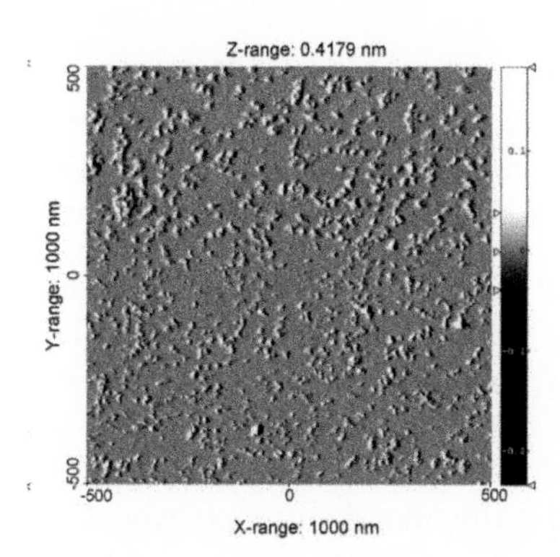

5.62: AFM image of adsorbed albumin molecules on a flat surface. (Reprinted from [114] with permission of Elsevier. Copyright 2022.)

There exist two types of adsorption: physisorption, where the adsorbate (i.e. the adsorbed molecules) is retained by intermolecular forces, and chemisorption, where the adsorbate undergoes a chemical reaction to reside, for some time, or irreversibly, at the interface. One example of physisorption concerns the already mentioned amphiphilic molecules, and one example of chemisorption is oxygen molecules adsorbing and oxidizing a metal surface. The former process is reversible, and the latter process is generally not. Microfabrication often uses one or the other of these two mechanisms for depositing films or layers on silicon wafers or glass plates. We will see examples of this in Chapter 7.

Adsorption and desorption phenomena are crucial for microfluidics. Adsorption can be favourable or unfavourable. A favourable example, is the extraction of nucleic acids in membranes, a technique invented in the 1990s by Boom [115]. Here, by varying the salt concentration, it is possible to change the sign of the forces developed at silicon membrane surfaces, so as to adsorb (capture) or desorb (release) RNA. An unfavourable example is non–specific adsorption (NSA) at the microchannel walls, that often leads to the loss of the biological sample before it can be processed, or, in an electrochromatographic context, can lead to undesirable electrosmotic flows. In order to block NSA, an albumin or PEG coating, prior to sample injection, is currently used.

5.11.2 Langmuir isotherm

In 1916, Langmuir proposed a theory of adsorption [70,113,116–118]. It is schematically represented in Fig. 5.63.

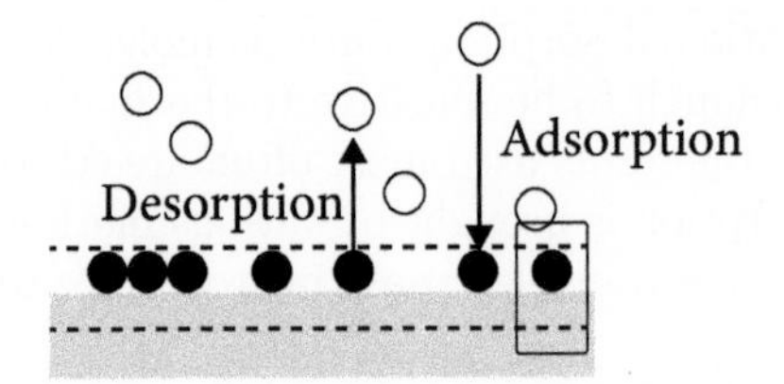

5.63: Langmuir model of adsorption, in which a single layer exchanges with the bulk, through adsorption/desorption events.

Molecules adsorb on the surface for a certain interval of time. Even though they are energetically 'happy', they do not stay anchored to the surface. Subjected to thermal fluctuations, they desorb. In stationary regimes, one may assume that an equilibrium holds between adsorption and desorption fluxes, towards and from the surface.

The theory of Langmuir relates the flux J_1 of molecules adsorbing on the surface and, for liquids, their bulk concentration C. For gas, C is replaced by partial pressure. Let us assume a liquid. In Langmuir theory, one single molecular layer is adsorbed: molecules are retained on the bare areas of the surface, but are warded off by areas that are already covered. The flux J_1 of incident molecules is proportional to the local concentration C and to the area uncovered by the adsorbed monolayer. Calling ϕ the fraction of surface covered by the adsorbate, we can write:

$$J_1 = k_1 C(1 - \phi)$$

where k_1 is a coefficient. On the other hand, the flux of desorbed molecules, represented by J_2, is proportional to the covered area:

$$J_2 = k_2 \phi$$

where k_2 is a second coefficient. At equilibrium, the two fluxes are equal. We deduce:

$$\phi = \frac{KC}{1+KC}$$

where $K = k_1/k_2$ is often called the Langmuir constant. The quantity of material adsorbed at the surface, as a function of the bulk concentration C, is thus:

$$F = \frac{F_0 KC}{1+KC}$$

where F_0 is a constant. This is the Langmuir isotherm.[20]. It is called 'isotherm' to underline that, in the theory, the temperature is maintained constant.[21] At high concentrations, ϕ tends to unity. In this limit, the surface is entirely covered by a monomolecular film.

Langmuir theory is generally found in good consistency with the experiment. One example is shown in Fig. 5.64, for protein adsorption on silica surfaces [119]. The numbers on the axes are typical: a few milligram per m^2 at ambient temperature for full surface coverage, at concentrations on the order of 10 g/l.[22]

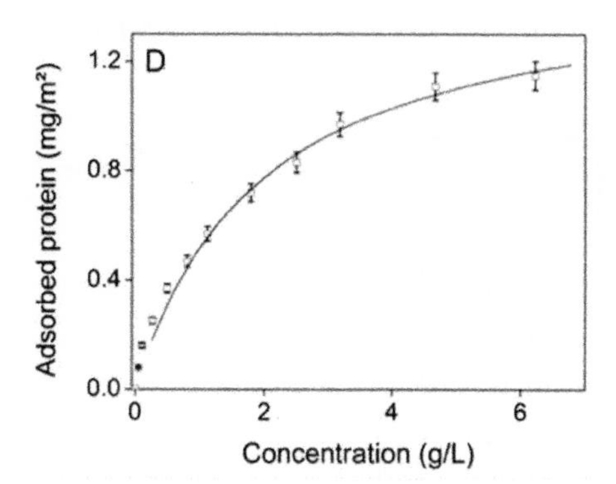

5.64: Comparison, in liquid, between Langmuir isotherm theory and experiment (protein adsorption on silica surfaces.) [119] (Creative Commons licence.)

Other demonstrations of Langmuir's law, using thermodynamics or statistical physics, have been established. The main limitations of this law are that a single layer is assumed and interactions inside it are neglected. For these reasons, and in an attempt to generalize the Langmuir isotherm, different laws have been proposed over the years, such as the Brunauer–Emmett–Teller (BET) isotherm [120]. There also exist purely phenomenological relations, such as Freundlich's law. This law, which reads $F = KC^a$, where a and K are empirical constants, is widely used. It was written in 1909, by Herbert Freundlich [121], seven years before Langmuir.

[20] For gases, the same law applies, but C is replaced by partial pressure

[21] Heat is released in the adsorption process and absorbed by desorption. In the framework of Langmuir theory, the temperature is held constant during the entire processes. Otherwise, the temperature would evolve unboundedly, and the model would be unphysical.

[22] It is often noted, in the field, that the excellent agreement between Langmuir theory – with a free parameter – and the experiment does not demonstrate that theory is valid. It just shows that the theory provides a possible description of the adsorption phenomenon, consistent with the data.

It is usual, in the field of chromatography, to distinguish between two cases of adsorption

- If the function $F = f(C)$ is concave, adsorption is said 'favourable'. Concentration fluxes being the derivative of the curve, it is large at the origin, i.e. at same C. Thus, in the domain of chromatography, it will be possible to detect traces of substances in a sample by capturing them on a surface and, after elution, proceed to their analysis.
- If $F = f(C)$ is convex, the situation is opposite: fluxes are small at small C and adsorption is said 'unfavourable'.

5.12 Chromatography

5.12.1 Principle of chromatography

Chromatography is described in a number of textbooks [122–126] papers and lectures (see, for instance, [127–129]).

The general principle (for column chromatography), is represented in Fig. 5.65.

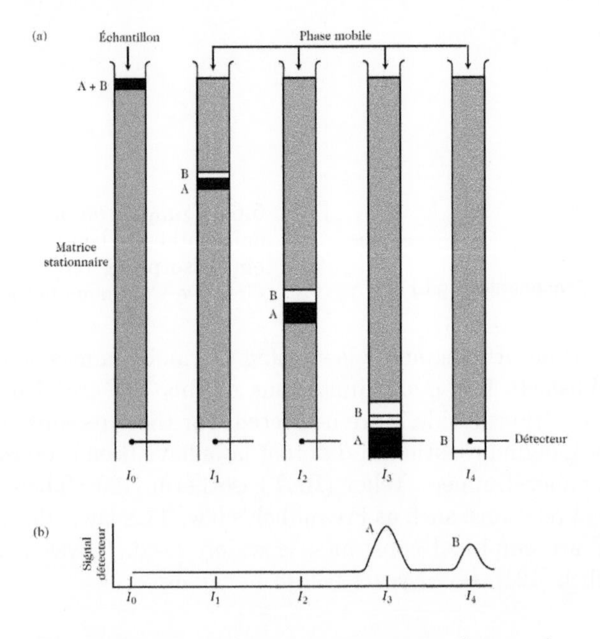

Fig. 5.65: Principle of chromatography (see text).

The column is filled with a porous medium, beads, or a gel, forming the stationary phase (the matrix). The mobile phase is the solution (i.e. the solute and the solvent) that circulates through the column. The sample is deposited at the top of the column; it separates into components as it travels through it.

During its journey, the sample is subjected to adsorption/desorption processes, which depends on the affinities of the molecular species contained in the sample and the matrix. The transport velocity will thereby vary from one species to another. In this way, distinct bands will appear progressively along the column over time, each band containing species with the same partition coefficient. As time goes by, a detector placed at the bottom of the column records a temporal signal (see Fig. 5.66).

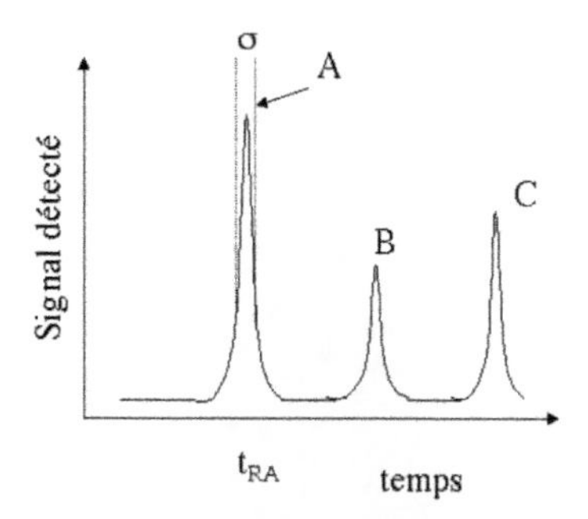

5.66: Typical temporal signal obtained by the detector.

The signal is made up of a series of peaks. It allows the measurement of the time of passage of each band in before the detector (called the 'retention time'). From this are deduced the migration velocities, the corresponding partition coefficients, and, exploiting the calibration data or other associated information, the different components of the sample. The chromatographic signal can also, from the measurement of the area under each peak, determine the concentrations for each component. It is remarkable that the method's efficacy is based on the amplification of small differences: small disparities in the partition coefficients of different constituents of the sample were effectively amplified by repeated passages through the grains forming the matrix.

Once the bands have migrated to the base of the column, we proceed to the detection phase. In chromatographic systems, this problem is crucial, especially when the component in question exists only in trace amounts (this is frequently the case in biiology and the environmental sciences). Nevertheless, there are a wide variety of detectors available that use for example fluorescence (an extremely sensitive method), thermal conductivity, electrical conductivity, photon adsorption, etc.

Let us indicate a few orders of magnitude: for liquid phase chromatography (LPC), columns are typically 20 cm long and 4 mm in diameter, with particles of 20 μm and applied pressures increasing to several bars or tens of bars. The quantities being analysed are on the order of the milligram, and the efficacy varies between 100 and 1 000 theoretical plates. The typical minimum detection for this type of column is on the order of the picogram. The retention time obviously depends on several parameters, but to illustrate, we can say that they are typically on the order of ten or several tens of minutes. The versatility and the richness of chromatographic methods allows us to believe that it is possible to find a solution to whatever analytical problem may arise.

5.12.2 Different types of chromatography

The main types. Without entering into the details, the most common methods of liquid-phase chromatography are described below (see Fig. 5.67).

• **Adsorption chromatography:** : the sample components interact differently with the matrix of the column, leading to different travel times.

• **Partition chromatography:** the column contains a liquid film covering each sphere making up the stationary phase. Certain components of the sample dissolve in this film.

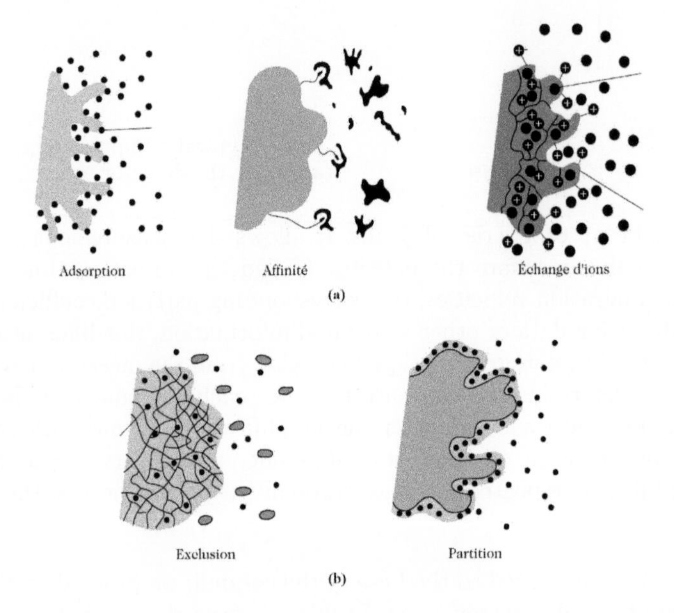

Fig. 5.67: Types of chromatographies.

• **Ion-exchange chromatography:** the matrix is made up of charged spheres that retard the displacement along the column of ions of the opposite sign. The most common matrices are gels of cellulose. The effect of the retardation of the displacement can be controlled and optimized by modifying the pH of the solution.

• **Exclusion chromatography:** the matrix is made up of porous spheres. Small molecules frequently visit the interior of these spheres by diffusion. and thereby progress along the column more slowly than large molecules do: this is the effect of separation by exclusion. The spheres used are, for example, agarose beads, which possess a large diversity of porosities.

• **Affinity chromatography:** in this case, the matrix shows high affinity for certain components of the solute (for example, antibodies specific to antigens). The molecule is captured by the matrix, and is further eluted by circulating a solvent.

High performance liquid chromatography (HPLC). Over the years, the reduction of the grain sizes, along with the augmentation of the column length, have considerably improved the performances of the columns. This gave birth to a new acronym HPLC (high performance liquid chromatography). HPLC is not a specific type of chromatography, but a refinement of preceding techniques, resulting in highly resolved separations. Typically, the matrices of HPLC systems are made of pores between 1 and 5 μm. Columns are long and operates at high pressure (hundreds of bars). With this type of system, as will be shown in the next section, hundreds of thousands of theoretical plates can be obtained.

5.12.3 Performances of the columns

The characterization of a separation process is based on a number of quantities [123–125,129]. The first one is the retention time t_R, defined in Fig. 5.66. This is the time a component of the sample takes to traverse the column and pass before the detector. The second quantity is the standard deviation of the peak corresponding to a given component (Fig. 5.66). It is advantageous to reduce this quantity as much as possible. This quantity is notated as σ. The column efficiency is characterized by a number N, the number of theoretical plates[23]. This number is defined by the following expression:

$$N = \frac{t_R^2}{\sigma^2} \tag{5.34}$$

The larger N is, the higher the column efficiency is. A related quantity is the plate height, defined by.

$$H = \frac{L}{N} \tag{5.35}$$

The smaller H is, the higher the efficiency is. Over the course of time, the different bands (corresponding to each species present in the sample) spreads. A well–accepted model, obtained on a semi–empirical basis, is due to J. J. van Deemter (1956) [130]:

$$H = A + \frac{B}{U} + CU$$

in which A, B and C are empirical constants. Physically, A represents the effect of heterogeneities across the column, B is the effect of longitudinal dispersion, that spreads the solute band, according to mechanisms we presented earlier, and C covers different adsorption phenomena at the pore scale. A thorough discussion of the van Deemter equation, along with other models can be found in C. F. Poole [125]. Chromatography optimizes H by reducing A, B and C, and working in a region close to the minimum

[23] A word borrowed from the domain of distillation

of the function $H(U)$. In HPLC, pores or bead sizes are in the micrometric range, and arguments and experiments indicate that, for the speeds currently used in the domain (a few mm/s), the first term, i.e. constant A, is dominant [125]. Although this term deserves discussion, one may have a rough idea of its amplitude by taking d, the pore size. We thus have:

$$H \sim d \text{ and } N \sim \frac{L}{d} \tag{5.36}$$

The plate number thus can reach high levels, 10^5 or so, for columns of several centimetres size. An example [131] with a number of plates of 10^6 is shown in Fig. 5.68.

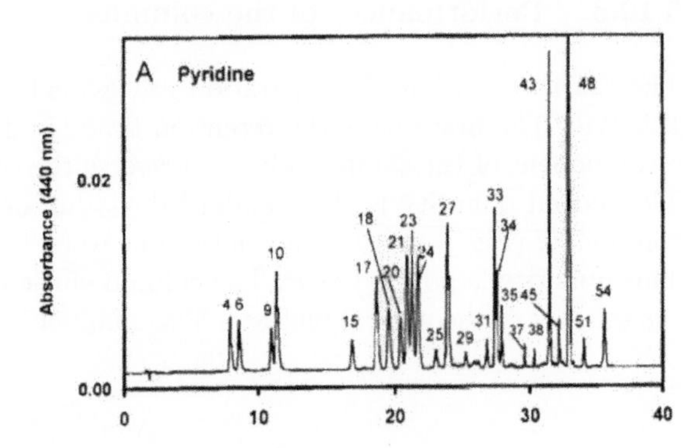

5.68: An example of HPLC with $N \approx 10^6$ [131]. (Creative Commons licence.)

5.12.4 First miniaturization of a gas chromatographic column: Terry (1975)

As noted in the introduction, the first miniaturized chromatographic column was made in 1975 [132]. Fig. 5.69 again shows the device. The system, made in silicon, included an injection valve and a separation column, 1.5 m long. Detection was based on a thermal sensor, fabricated on an external silicon wafer, and fixed to the chip. The design can be understood by using the notions we previously introduced. Here, the chromatography is based on adsorption. The carrier gas and the sample adsorb to the walls with a different kinetics, so that , at the end of the column, one recovers two separate peaks. The column must be long, in order to obtain a high efficiency, and it was challenging to pack a channel 1.5 m long in a miniaturized device. The valve must be small and quick, to work with small samples, favouring resolution. It was fabricated on a separate wafer. The number of plates we could estimate was on the order of several tens of thousands. The authors emphasized the small quantity of carrier gas used to perform the separation and the speed (less than 10 s) [132].

As discussed by Reyes *et al.* [127], this work is historically important, but it did not generate a surge of activity in what we now call microfluidics. At that time, neither researchers in biology, physics, or chemistry expressed interest in the invention. In the 1980s, the visibility of microfluidics was extremely modest.[24]

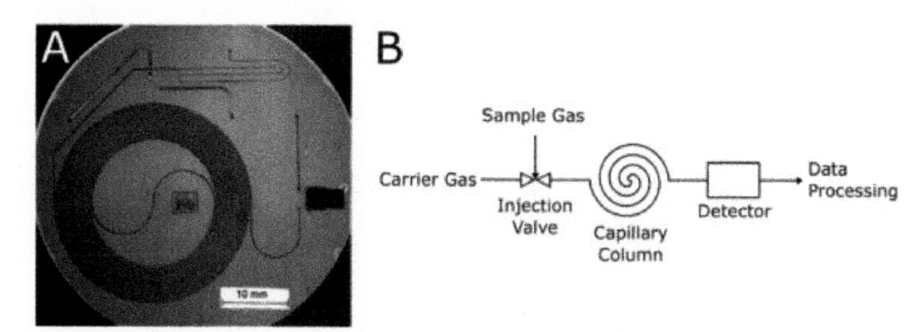

Fig. 5.69: Gas Chromatographer described by Terry et al [132]. The device consists of a 1.5 m long, spiral channel, etchec in silicon, with input (top right) and output (right) for gas sample Channel dimensions are 40 μm depth, 200 μm wide. Flow within the device is controlled with a valve(top left) before the coil; the thermal detector is located on the right of the device. B - Shematic showing the components housed on the silicon. wafer (Reprinted from Ref. [132] with permission of IEEE. Copyright 2022).

Today, miniaturized gas chromatographs are commercially successful. They serve different industries, for instance oil industry interested in on-site analyses, and for which the compactness of the system, coupled to the quality of the analyses, is unique. However, this is a niche. The start-up of microfluidics, which addresses much broader applications, began twenty years later, around 1990, with the coupling of microfabrication technology and capillary electrophoresis [133]. We will develop the subject in Chapter 6.

5.12.5 Is it worth miniaturizing chromatographic columns ?

Why miniaturizing chromatographic columns? Four arguments can be advanced:

- Sample volumes are reduced.
- Compactness is increased. The system becomes transportable.
- The cost is lower.
- By parallelizing several columns, throughput can be increased.

These advantages must, however, be taken alongside a certain number of limitations:

- In conventional HPLC, matrix particles or gel/porous pores are micron-size. As previously explained, these small sizes, coupled to the substantial length of the columns, allow high efficiencies to be reached. By contrast, transverse dimensions of microfluidic channels are, typically, 10 μm, while device lengths are a few centimetres long. This leads to a loss of four orders of magnitude in terms of efficiency, compared to conventional HPLC.

[24]The word 'microfluidics' appeared later, in the 1990s (see Chapter 1).

- To improve the situation, a number of investigators proposed to integrate gels or particles on the device, or reduce the channel dimensions. Even with these improvements, a substantial loss of performances persisted, in comparison with traditional HPLC.

In conclusion, miniaturization of chromatographic columns can be interesting for practical reasons (this is the case for gas chromatographers), but performances are inherently limited. In Chapter 6, we will see that miniaturization becomes extremely advantageous when electrokinetics is used.

5.13 Thermal transport by conduction

5.13.1 Conduction of heat in gases, liquids, and solids

Heat-conduction phenomena have been studied a great deal over the last century; excellent books exist and we will give an extremely brief account here of the domain.

- For gases, the origin of thermal conduction stems from the random movement of molecules. For systems much larger than the mean free path λ (which, as the reader may recall, is about a hundred nanometers at ambient temperature), kinetic theory gives the following estimate for the thermal diffusivity κ:

$$\kappa = \frac{K}{\rho C_{\mathrm{p}}} \sim \lambda c, \tag{5.37}$$

  where K is the thermal conductivity, ρ is the gas density, C_{p} is the heat capacity, and c is the sound speed. This expression is valid for perfect gases. For systems that are about the same size or smaller than the mean free path λ, molecules collide more often the walls than their partners, and the expression of thermal diffusivity must be amended. More generally, the hypotheses leading to the establishment of the macroscopic laws of heat transport (such as Fourier's law) must be re-examined. This problematic is similar to that concerning the validity of the Navier–Stokes equations in small systems, discussed in Chapter 3.

- For solids and liquids, it is necessary to distinguish between metals and insulators (see Kittel [135] for a thorough presentation of the subject). For insulators, heat is transported by the internal modes of vibration in the crystal lattice [134]. These modes are quantized, and one can treat them as quasi-particles known as phonons (quasi indicates that they do not have an autonomous existence independent of the crystal lattice that supports them). These quasi-particles undergo collisions, producing an erratic agitation that resembles that of gas molecules. Just as for gases, a mean free path λ_{p} can be defined. It ranges between 1 and 100 nm at ambient temperature.

  For insulators larger than λ_{p}, the thermal conductivity K_{p} can be estimated in a manner similar to the kinetic theory of gases. We obtain:

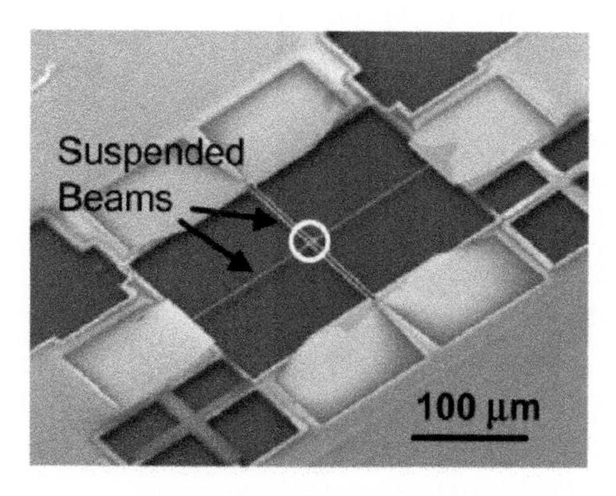

Fig. 5.70: Calorimetry experiment on a 40-nm wide silicon wire. The wire was used to study deviations from Fourier's law of heat conduction, which makes up part of macroscopic heat transport theory. (Reprinted with permission from Cahill *et al.* [134]. Copyright 2003, American Institute of Physics.)

$$K_{\mathrm{p}} \approx \frac{1}{3}\rho C_{\mathrm{ph}}\lambda_{\mathrm{p}}u_{\mathrm{p}}, \tag{5.38}$$

where C_{ph} is the heat capacity of phonons and u_{p} is the speed of sound. For systems smaller than λ_{p}, the calculation is more complicated. It would take us far beyond the scope of this book.

- For pure metals, electrons are the principal thermal carriers. At ambient temperature, their mean free paths and speeds are on the order of 100 nm and 10^3 km/s, respectively. Although electrons have tiny mass, owing to their speeds, the kinetic energy exchanged during electron-phonon collisions is greater than that exchanged between phonons, which is why they are the principal carriers of heat in metals. Just as for insulators, the conduction of heat in nanometric metallic systems is currently an active area of research. One example is an experiment carried out with wires, a few nanometres in diameter (see Fig. 5.70) [134].

To conclude, macroscopic theories are no longer valid at scales below a mean free path, i.e. roughly 100 nm or so in solids and gases. These scales lie on the lower edge of the microfluidic domain (see Chapter 2) and we will not address them. In smaller systems (pertaining to the nanofluidic range), the investigation of thermal phenomena, and their coupling to flow phenomena, is a domain of active research.

5.13.2 Fourier's Law

As observed above, the mean free paths of the particles ensuring heat conduction (molecules, phonons, or electrons) are much smaller than the size of the systems we

System	K ($\mathrm{W\,m^{-1}K^{-1}}$)	κ ($\mathrm{m^2/s}$) $\times 10^{-6}$	C_p(kJ/kg/K)
Water	0.6	0.14	4.2
Silicon	150	80	0.70
Glass	1.4	0.34	0.83
PDMS	1.5	0.1	1.4
Copper	401	113	0.38
Air	2.41 10^{-2}	22	1

Table 5.4 Thermal conductivities, diffusivities, and specific heats for a selection of materials.

will consider. The heat fluxes $\mathbf{q}$ will thus be related, as in ordinary size systems, to the temperature gradients ∇T by Fourier's law, which reads as follows:

$$\mathbf{q} = -K\nabla T \tag{5.39}$$

where K is the thermal conductivity. The units of K are $\mathrm{W\,m^{-1}\,K^{-1}}$. An important quantity, altready introduced, is the thermal diffusivity, denoted by κ. We recall its definition:

$$\kappa = \frac{K}{\rho c_{\mathrm{v}}}, \tag{5.40}$$

where ρ is the density of the fluid and c_{p} is its specific heat. The unit of thermal diffusivity is the same as for the kinematic and mass diffusivities: m^2/s. A few values of K and κ, along with C_p for different liquids, solids and a gas (air) are given in Table 5.4:

Consistently with common knowledge, metals and semi-conductors are excellent conductors. Gas is an insulator and materials, such as pure water and glass, are in between. This hierarchy reflects the existence of different physical mechanisms for the conduction of heat, as discussed in the preceding section.

5.13.3 The heat diffusion equation

The heat equation can be derived in the same manner as the diffusion equation for mass transport, and we refer the reader to textbooks such as [136] for its derivation. With a constant thermal conductivity K, the heat diffusion equation reads:

$$\rho C_{\mathrm{p}} \frac{\partial T}{\partial t} = K\Delta T + s(\mathbf{x}, \mathbf{t}), \tag{5.41}$$

where $T(\mathbf{x},\ \mathbf{t})$ is the local temperature ρ is the fluid density, and $s(\mathbf{x}, \mathbf{t})$ (in $\mathrm{Wm^{-3}}$), is a rate of heat production or absorption per unit of volume, due to a source or a sink. s can be positive or negative. For instance, if the fluid is conducting and an electrical current is applied through it, we will have $s = \frac{E^2}{\rho_E}$, in which ρ_E is the electric resistivity, and E the applied electric field. If the fluid develops an exothermal reaction, s will be the enthalpy produced by this reaction, per unit of volume and unit of time.

Another form of the heat diffusion equation, frequently used, is:

$$\frac{\partial T}{\partial t} = \kappa \Delta T + \frac{s}{\rho C_{\mathrm{p}}} \tag{5.42}$$

With a temperature T_w and a flux q_w at the walls, the boundary conditions are:

$$T = T_w \text{ or } K\frac{\partial T}{\partial z} = q_w \tag{5.43}$$

These boundary conditions satisfy temperature and thermal flux continuity.

5.13.4 Three examples

Example 1: Temperature field in a sandwich. To begin, let us say that a slab of height h, whose top and bottom, located at $z = \pm\frac{h}{2}$ (where z is the coordinate normal to the slab), have temperatures equal, respectively, to T_2 and T_1, Eqs. 5.41 reduce, in the steady state, to $\frac{\partial^2 T}{\partial z^2} = 0$. Thereby, in such a geometry, the temperature profile $T(z)$, reads:

$$T(z) = \frac{1}{2}(T_1 + (T_2) + (T_2 - T_1)\frac{z}{h}$$

Let us now consider, instead of one slab, three slabs, represented in Fig. 5.71. The lower face of the system is cooled down at T_0 and the upper one is kept at T_a ('a' for atmosphere). We examine two cases, often encountered in microfluidics:

a stack of glass, water, and PDMS and

b stack of glass, air, and PDMS.

With our notations, the boundary conditions, at the slab interfaces, reads as follows:

$$T_1 = T_2, \quad K_1\frac{\partial T_1}{\partial z} = K_2\frac{\partial T_2}{\partial z} \tag{5.44}$$

With this, it is possible to calculate, in each slab, the temperature profiles. We do no write the expression of the temperature profiles, but show their shapes in Fig. 5.71.

Depending whether the medium layer is air or water, we have different temperature profiles: in case (a), PDMS and glass have similar conductivities, but water is 2.5 less conductive (see Table 5.4). We thus have a higher slope in the water channel. In case (b), air has a much smaller conductivity than water. Consequently, the temperature slope, in the microchannel, compared to water, is much steeper, while inside PDMS and glass, temperatures are almost uniform. In this system, air plays the role of a

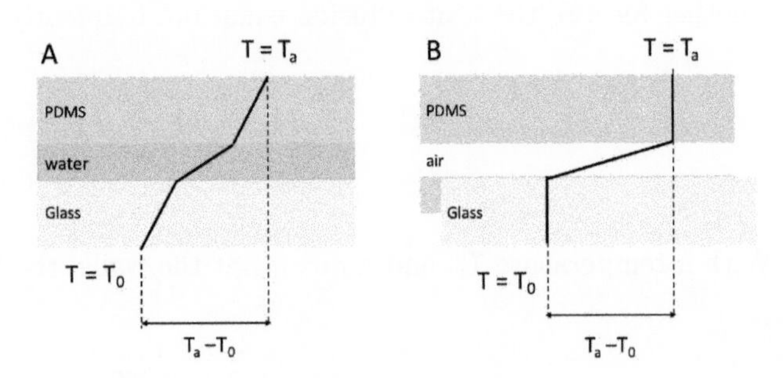

5.71: Temperature profiles in a three–layer system, submitted to the same temperature difference $T_a - T_0 > 0$: (A) PDMS–water–glass; (B) PDMS–air–glass.

thermal insulator, while PDMS and glass thermalize at the temperatures to which they are exposed.

Example 2: Temperature field in a microchannel with internal heating. Now, we consider the microchannel represented in Fig. 5.72 and assume that the liquid is heated by a thermal source s.

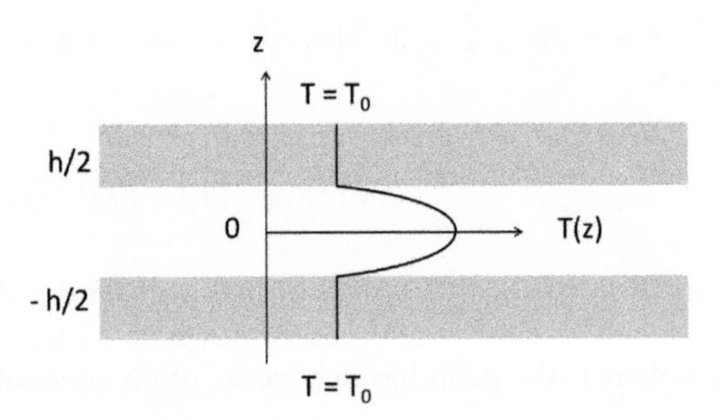

5.72: Temperature profile in a microcanal filled with a motionless fluid, subjected to internal heating, characterized by a source s.

The two plates that confine the liquid are located at $\pm h/2$ and are held at temperature T_w. In the problem, temperature T only depends on z. Therefore, the equation governing the temperature profile, in the fluid, and for the steady state, reads:

$$0 = K \frac{\partial^2 T}{\partial z^2} + s \tag{5.45}$$

The thermal boundary conditions for walls maintained at fixed temperature T_{w} are:

$$z = \pm \frac{h}{2}, \quad T = T_{\mathrm{w}} \tag{5.46}$$

After integrating, we find:

$$T(z) = T_{\rm w} + \frac{s}{K}\left(\frac{h^2}{8} - \frac{z^2}{2}\right)$$

The maximum heating is located at $z = 0$. The temperature difference $\Delta T_{\rm max}$, between the channel centre and the walls, is given by:

$$\Delta T_{\rm max} = \frac{sh^2}{8K}. \qquad (5.47)$$

Just as the scaling laws predicted, ΔT_{max} increases with the square of the transverse dimension of the canal. We can thus expect, in miniaturized system, and for standard heat sources, an extremely limited increase (or a decrease if s is negative) of the fluid temperature. Let us take two examples:

1 – **Joule heating in electrolytes.** . Let us consider an electrolyte of resistivity ρ_E enclosed in a microchannel of length L, subjected to a voltage V. An example is an electrophoresis separation channel. We will have $s = \frac{V^2}{\rho_E L^2}$. With $V = 1$ kV, $\rho_E \sim 1$ Ohm-m, h=10 μm, $L = 1$ cm, we find $\Delta T \approx 10^{-1}$ °C, a small quantity. The channel can thus easily be maintained at constant temperature. By contrast, standard electrophoresis gels, for DNA separation, are several millimetres in height, and 10 cm long. The same calculation leads to an increase in temperature by tens of °C. Heating represents the main source of resolution limitation and low speed of the technique.

2 – **Exothermal reactions.** In exothermal reactions, $s = QU$, where Q is the heat per unit of volume produced by the reaction and U is the flow speed. Rough orders of magnitude, for reactions producing 100 kJ/mole, indicate that, in microsystems similar to the preceding example, temperature increases does not exceed 1°. However, for industrial systems, designed for high throughput and therefore, having higher dimensions, typically millimetres in size, and larger speeds, typically cm/s, the increase can be several or several tens of °C. Heat must be extracted to keep the temperature constant. This brings us to the topic of microexchangers, which we will return to later.

Example 3: Plate suddenly raised at temperature T_w. In a fluid initially held at temperature T_∞, a thin plate, located at $x = 0$ is suddenly heated at temperature T_w. For instance, a hot knife we would have quickly immersed in a cold liquid to cool it down. The plate has a thickness e. The situation is sketched in Fig. 5.13.4.

In the limit of small e, we can model the initial condition by the expression:

$$t = 0, T(x, 0) = e(T_w - T_\infty)\delta(x)$$

in which $\delta(x)$ is the Dirac function, that we saw earlier.

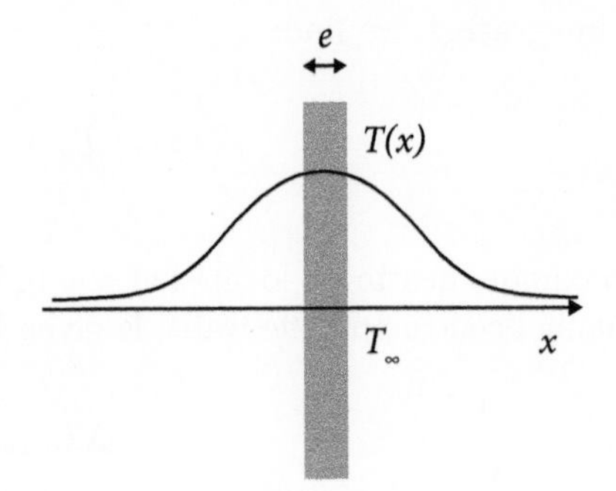

5.73: A thin hot plate is suddenly raised at temperature T_w. The solution to the problem, in the limit where e tends to zero, is a gaussian

The system is governed by the following equation (see Eq. (5.42)):

$$\frac{\partial T}{\partial t} = \kappa \Delta T \tag{5.48}$$

The solution is Gaussian. Its expression is:

$$T(x,t) = T_\infty + \frac{e(T_w - T_\infty)}{\sqrt{2\pi\sigma^2}} \exp\left(-\frac{x^2}{2\sigma^2}\right) \tag{5.49}$$

with $\sigma = \sqrt{2\kappa t}$. The plane cools down and eventually reaches the bath temperature T_∞. To have a sense of the orders of magnitude, suppose you plunge the knife in cold water, held at $T_\infty \approx 10°$.Taking $e = 3$ mm, $\kappa = 0.14$ m^2/s, and $T_w = 90°$C, the calculation shows that you need a time of ≈ 8 s to bring it back to a temperature of 30°C. At this temperature, at least from a thermal prospective, the knife no longer causes any harm.

Note that the solution (5.49) can be found in the Landau-Lifchitz textbook on fluid mechanics [137]. In this reference, the general problem of the temporal evolution, in a fluid at rest, of a temperature field $T_0(x)$ is solved. The case we considered here is also discussed in Ref. [137].

5.14 Convection-diffusion heat equation and properties

5.14.1 The notion of heat transfer coefficient

A quantity central in the engineering literature, is the thermal exchange coefficient, denoted by h. This coefficient, illustrated on Fig. 5.74., was invented by Newton. It is similar to the mass exchange coefficient', introduced earlier. Take a body of uniform temperature T_{body}, immersed in a flow, whose upstream temperature is T_∞. Depending on the sign of $\Delta T = T_{body} - T_\infty$, heat flux Q is oriented towards the body or the fluid.

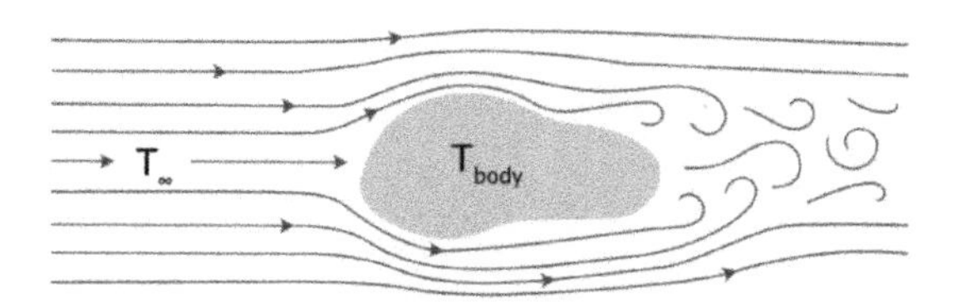

Fig. 5.74: Definition of the heat transfer coefficient.

Newton assumed a relation of proportionality between Q and ΔT. The coefficient h is defined by the relation:

$$h = \frac{Q}{A\Delta T},$$

where A is the surface body area, is known as the heat transfer coefficient. One example is a radiator functioning at fixed temperature heating up a room: in this case, the temperature difference is defined as the difference between the radiator and room temperatures, A is the area of the radiator and Q is the thermal power output. The units of the coefficient h are $\mathrm{W\,m^{-2}\,K^{-1}}$. This incorporates all the mechanisms (natural convection, forced convection, radiation, conduction) participating in the thermal exchange. The higher h is, the greater the thermal exchange is. Table 5.5 shows some values of h, useful for characterizing the performances of common systems.

Type of transfer	Fluid or process	h ($\mathrm{W\,m^{-2}\,K^{-1}}$)
Natural convection	Gas	5 – 30
	Water	100 –1,000
Forced convection	Gas	10 – 300
	Water	300 – 12 000
Phase transition (water)	boiling	3,000 – 60,000
	condensation	5,000 – 110,000

Table 5.5 Typical values of the thermal exchange coefficient h.

.

Heat transfer coefficient is a clever engineering concept: it allows us to have orders of magnitude of heat exchanges without solving any equation, and, moreover, by considering it as a constant, to establish simplified equations. For instance, how long a time is needed to cool down the body of Fig. 5.74? With h constant, the heat transfer between the body and its environment is governed by the following equation:

$$mC_V \frac{dT}{dt} = hA(T - T_\infty)$$

in which m is the body mass, and C_V is its specific heat. The solution to this equation is:

$$T(t) = T_\infty + (T_0 - T_\infty) \exp(-\frac{t}{\tau})$$

where T_0 it the initial body temperature, and

$$\tau = \frac{mC_V}{hA}$$

is a time characterizing the dynamics of the process (cooling or heating). With h= 100 $\mathrm{W\,m^{-2}\,K^{-1}}$, it takes 100 seconds to cool down a body of centimetric size, and a fraction of second or so for an object 10 μm in size.[25] Indeed, in many cases, temperature is not uniform and it is erroneous to assume that h is time – and temperature – independant. Still, taking h as constant, allows us, with the help of Table 5.5, to have a sense of the orders of magnitude involved in a given system, which is extremely useful.

So far, we discussed global heat transfer coefficients, i.e. averaged out over the body surface area. There is also a local version of h. In this case, we have:

$$q = h(T - T_\infty)$$

in which q is the local heat flux density crossing the interface, normally to it, and T is the local interfacial temperature. h, then, is the *local* heat transfer coefficient. This relation will allow us to define thermal boundary conditions.

5.14.2 Convective heat transfer

Convection–diffusion equation. The convection-diffusion equation can be derived from the heat diffusion equation, by placing ourselves in a Lagrangian frame of reference. The reader may refer to Batchelor's and Landau–Lifschitz textbooks for a detailed derivation of this equation [136, 137]. We obtain:

$$\rho C_p \frac{DT}{Dt} = \rho C_p \left(\frac{\partial T}{\partial t} + \mathbf{u} \nabla T \right) = K \Delta T + s(\mathbf{x}, \mathbf{t}), \tag{5.50}$$

in which, s includes external heat sources or sinks and the heat generated by viscous dissipation. As in the preceding section, depending whether a temperature T_w or a flux q_w is fixed at the walls, the boundary conditions are:

[25] Here, for the sake of simplicity, we took the same h for the two cases. We will see later that h increases with the inverse of the system size.

$$T = T_w \text{ or } K\frac{\partial T}{\partial z} = q_w \tag{5.51}$$

When $q_w = 0$, these conditions can be expressed in a more compact form by using the local heat transfer coefficient h :

$$K\frac{\partial T}{\partial z} = h(T - T_w)$$

The preceding boundary conditions correspond to $h = \infty$ and $h = 0$, respectively.

These equations, coupled to the boundary conditions, encompass a considerable variety of phenomena: natural convection, forced convection, heat diffusion, and heating due to the presence of heat sources. In the framework of miniaturized systems, this diversity can be reduced by exploiting the scaling laws presented in Chapter 2. Phenomena like internal heating with viscosity, buoyancy (giving rise to convection), and radiation emitted by the fluid, are negligible in microfluidics, in the most common situations. These estimates will allow us to concentrate on specific situations, relevant to microfluidics.

The Biot number. The Biot number Bi characterizes heat transfers at the boundaries. It is currently used in the engineering literature. Its definition is:

$$Bi = \frac{hl}{K}$$

where l is a characteristic scale. In miniaturized system, since h depends on the system size, it is difficult to estimate, in general, an order of magnitude for the Biot number.

Thermal Peclet, Nusselt, and Prandtl numbers. Let us first define the thermal Peclet number. In a system of size l subjected to a temperature difference ΔT, with a flow U, heat transfers, characterized by Q, take place. An example is the body immersed in a cold (or hot) fluid moving around it, which we represented in Fig. 5.74. In such a situation, one can write:

$$Q = f(\kappa, U, l, \Delta T, \rho, c_p).$$

We have here 7 variables and 4 dimensions. We can thus write this equation in the form:

$$\frac{Q}{Kl\Delta T} = f\left(\frac{Ul}{\kappa}, \frac{c_p \Delta T}{U^2}\right).$$

At small ΔT, we may neglect the second term in the parenthesis of the right hand size. Under this approximation, we obtain:

$$\frac{Q}{Kl\Delta T} \approx f\left(\frac{Ul}{\kappa}\right),$$

which can be written in the form:

$$Nu = f(Pe_{\text{th}}),$$

where Pe_{th} is the thermal Peclet number, defined by:

$$Pe_{\text{th}} = \frac{Ul}{\kappa},$$

and Nu is the Nusselt number defined by:

$$Nu = \frac{Q}{Kl\Delta T}$$

.

The thermal Peclet number can also be introduced by using Eqs. (5.50). The approach is similar to that leading to the definition of the Reynolds and (massic) Peclet numbers. It provides the physical following physical interpretation of Pe_{th}:

$$Pe_{th} = \frac{\text{Transport by convection}}{\text{Transport by diffusion}}$$

At small thermal Peclet numbers, heat is mostly transported by the materials (solid parts and fluids taken at rest) and at large Peclet numbers, mostly by the flow. Heat exchangers, which use flows to enhance thermal transport, must obviously operate at large Peclet numbers. The Nusselt number has a long history in thermo-fluidics. It allows to quantify, in a dimensionless form, the heat exchanges developed in a system. In the literature, it is often derived from the heat transfer coefficient h, using the formula:

$$Nu = \frac{hl}{K}$$

How large is the thermal Peclet number? As the velocity varies with l (at fixed pressure, see Chap 1), the thermal Peclet number varies as l^2: this suggests that it is small in microsystems, which might tempt us to neglect the flow contribution to the heat transport. This remark is illustrated by Fig. 5.75.

In these experiments, a heat source is placed on the right of the figure, on the side of a microchannel, along which a flow is driven. The Reynolds numbers vary from 7 to 91. The local heat source (a resistance) develops a thermal field, which crosses the flow. The Peclet number, in this case vary from 1.1 to 13. One sees that, for a Peclet number of 1.1, the thermal field is unaffected by the flow. At the largest speed, where the thermal Peclet number reaches a value of 13, an effect of entrainment of heat by the flow is visible.

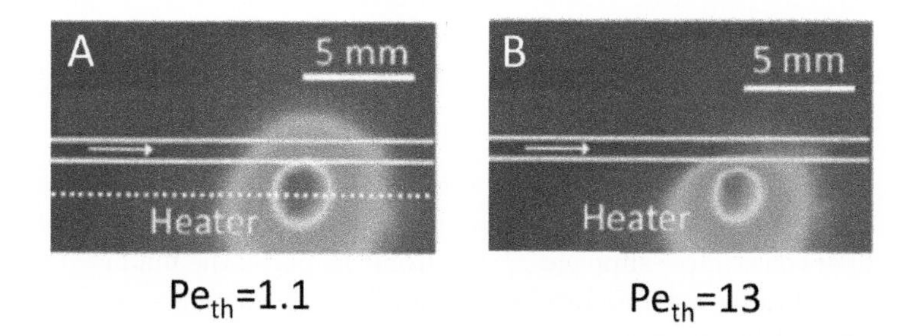

Fig. 5.75: Effect of the thermal Peclet number, shown by placing a heat source close to a microfluidic channel. At the highest Peclet number, heat is transported by the flow, erasing the thermal field. (A)10 μl/min. (B) 120 μl/min. (Reprinted from [138], with permission of Royal Chemical Society. Copyright 2022.)

In most cases, microfluidic devices, dedicated to chemical or biological analysis, operate at small thermal Peclet numbers, leaving the thermal field unaffected by the presence of a flow. By contrast, microfluidic heat exchangers must interact with the thermal field, and thereby operate at high thermal Peclet numbers. The problem is that, in order to obtain significant Peclet numbers, we need to operate at high or moderately large Reynolds numbers. In effect, the ratio of the two numbers is:

$$\frac{Pe_{th}}{Re} = \frac{\nu}{\kappa} = Pr$$

in which Pr is the Prandtl number. For usual gases, this number is close to unity. For water, most currently used in heat exchangers, because of its low viscosity and high capacity, the Prandtl number is seven. It is thus impossible, with water, to reach large Peclet numbers and, in the meantime, work at small Reynolds numbers. This raises a constraint on the design of micro heat exchangers. In practice, in order to achieve Reynolds numbers up to 1000, typical channel transverse dimensions are 100 μm - 1 mm and velocities are on the order of m/s. In such conditions, the thermal Peclet numbers range between 10 and 100.

One may conclude that, in practice, in efficient micro heat exchangers, thermal Peclet and Reynolds numbers must be substantial. In fact, these systems operate in inertial regimes. This represents a major difference between microfluidic heat-exchangers and lab-on-a-chips. In practice, the two communities, although sharing the same microfabrication toolbox, most often work along separate pathways.

5.15 Heat transfer in the presence of a flow in microsystems

5.15.1 The laminar thermal boundary layer

When a flow is driven over a surface maintained at a temperature different from the upstream fluid, a thermal boundary layer develops. There is a vast literature on the subject, which analyses heat transfers in laminar and turbulent regimes, examines the coupling between velocity and thermal fields, and calculate heat transfers and exchanges coefficients. Here, we restrict ourselves to an elementary situation, whose analytics is simple and which is relevant to microfluidics.

The geometry we consider is shown in Fig. 5.76. The plate is maintained at fixed temperature T_w. The velocity field is supposed uniform, which can be justified at small Prandtl numbers or at large slippages.[26] Far from the plate, the fluid is at temperature T_∞. In such conditions, a temperature field develops. Assuming stationarity, the field $T(x, z)$ is governed by the equation:

$$U\frac{\partial T}{\partial x} = \kappa\frac{\partial^2 T}{\partial z^2} \tag{5.52}$$

where U is the fluid velocity The boundary conditions are:

$$T(x, 0) = T_w \text{ and } T(x, \infty) = T_\infty. \tag{5.53}$$

Equation (5.52) is a diffusion equation. We already solved this type of equation. The solution is:

$$T(x, z) = T_w + (T_\infty - T_w)\,\mathrm{erf}(\frac{z\sqrt{U}}{2\sqrt{\kappa x}})$$

A boundary layer, sketched in Fig. 5.76 develops. Its width δ_{BL} is given by:

$$\delta_{BL}(x) = \sqrt{\frac{2\kappa x}{U}} \tag{5.54}$$

Thus, the thickness of the thermal boundary layer grows with the square root of the distance from the edge of the plate.

In terms of orders of magnitude, taking x =100 μm, $\kappa = 0.14\ 10^{-6}$ m^2/s (for water), and $U_0 = 1$ mm/s, one finds $\delta_{BL} \approx 170\mu$m. The boundary layer grows rapidly in the cross-stream direction. We will return to this point in the next subsection.

[26] At small Prandtl numbers, the hydrodynamic boundary layer is much thinner than the thermal boundary layer and, from the viewpoint of the thermal field, it can be neglected.

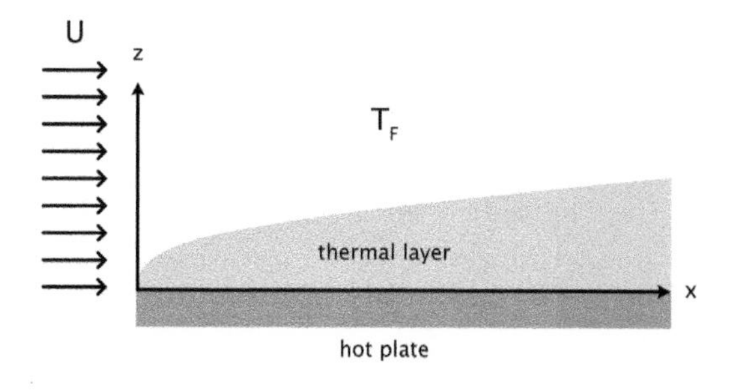

5.76: Thermal boundary layer over a plate, with a uniform flow U.

5.15.2 How long does it take to thermalize a fluid entering a microdevice?

To answer this question, it is convenient to define the thermal length L_{e} as the distance beyond which the thermal boundary layer 'hits' the walls, i.e. its width equals the microchannel height. This image, although approximate and schematic, leads us to consider that, beyond L_{e}, the fluid is thermalized at the device temperature, forgetting its thermal history. The expression of this length can be established by using the definition of the thermal boundary layer thickness δ_{BL} (see above). By imposing $\delta_{BL} = h/2$, we obtain:

$$L_{\mathrm{e}} \approx \frac{1}{8}\frac{Uh^2}{\kappa}$$

Another way to obtain the same result is to note that the temperature field takes a time equal to $h^2/8\kappa$ to diffuse more than half the height of the canal. During this time, the field is transported by the flow, at a distance equal to $Uh^2/8\kappa$. This leads to the expression of L_{e}.

The thermal length L_{e} can be rewritten in the following form:

$$\frac{L_{\mathrm{e}}}{h} \approx \frac{1}{8} Pe_{\mathrm{th}}$$

where the thermal Peclet number is defined by:

$$Pe_{\mathrm{th}} = \frac{Uh}{\kappa}$$

By taking the same numbers in the previous subsection, with $h = 100$ μm, one finds $L_{\mathrm{e}} \approx 9$ μm. In practice, with current fluids, typical speeds and channel sizes, L_{e} does not exceed the height of the canal. Indeed, these estimates neglect entry effects related to flow momentum. In fact, at small Reynolds numbers, the entry length for the establishment of a Poiseuille flow profile is also on the order of h (this point is not discussed in this book). From this estimate, we may confirm the conclusion we just drew out for the temperature field.

5.15.3 Application: microfluidic polymerase chain reaction (PCR)

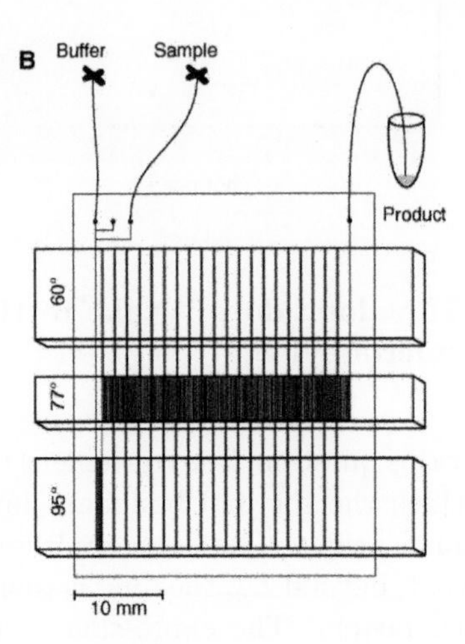

5.77: Fast PCR: device performing a 20-cycle PCR amplification of a 176-base pair fragment in 90 seconds [139]. Traditional technique takes 2 h for achieving the same task.

Small thermal lengths allow the time required for amplifying nucleic acids to be shortened [139]. Fig. 5.77 shows a system in which polymerase chain reaction (PCR) is performed along a microfluidic serpentine [139]. Driven by a syringe pump, the DNA sample visits different temperature zones on a glass plate. For performing PCR, three temperatures are needed: 60°C (annealing), 70°C (extension), and 95°C (denaturation). With microfluidics, a 20-cycle PCR amplification of a base pair fragment can be achieved in a pair of minutes, to be compared to 2 h in non miniaturized setups [139]. The key feature here is the rapid thermalization of the biological mix (i.e. primers, sample, enzymes, nucleotides, buffer) as it is transported downstream. In practice, the continuous flow format is not used today because of hydrodynamic dispersion, which tends, as explained earlier, to dilute the sample along the serpentine. We will see, in Chapter 6, that this effect can be suppressed by using electrokinetics.

5.15.4 Heat transfer in microchannel flows far from the entry (developed region)

We consider a microchannel, subjected to a heat flux q_0 applied normally to the walls. The geometry is shown in Fig. 5.78.

To model the situation, we consider that the fluid is confined between two infinite planes separated by h, with the flow developing along x. The Poiseuille profile, which characterizes the flow, is given by the expression:

$$U(z) = U_0(1 - 4z^2/h^2),$$

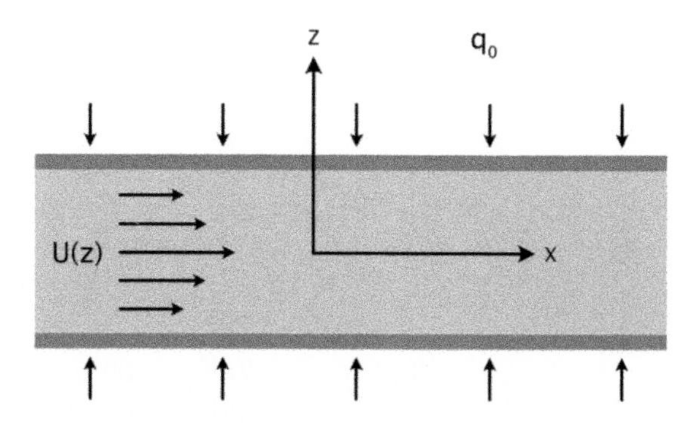

Fig. 5.78: Flow in the presence of a fixed thermal flux q_0.

where U_0 is a speed and z is the co-ordinate normal to the planes. The temperature field, assumed stationary, only depends on x and z. In such conditions, the convection-diffusion equation reads:

$$U_0(1 - 4z^2/h^2)\frac{\partial T}{\partial x} = \kappa\frac{\partial^2 T}{\partial z^2}. \tag{5.55}$$

The thermal boundary conditions at the walls $z = \pm\frac{h}{2}$ are given by the equations:

$$z = \pm\frac{h}{2}, \quad K\frac{\partial T}{\partial z} = \mp q_0, \tag{5.56}$$

where K is the fluid thermal conductivity. We look for a solution of the form:

$$T(x, z) = Ax + F(z),$$

where A and $F(z)$ are unknown quantities. By identification, we find the following expression for the function $F(z)$:

$$F(z) = \frac{AU_0}{\kappa}\left(\frac{z^2}{2} - \frac{z^4}{3h^2}\right) + B,$$

where B is a constant. Then, considering the boundary conditions (5.56), we obtain the relations:

$$F(z) = \frac{3q_0 h}{K}\left(\frac{z^2}{2h^2} - \frac{z^4}{3h^4}\right) + B, \tag{5.57}$$

Now we must determine B. For this purpose, we introduce a temperature T_b, called the 'bulk' temperature, or the 'mixing cup' temperature, which will be later taken for the calculation of the coefficient of heat transfer h_c (c' for channel, to avoid confusion with the height h). By definition, T_b is:

$$T_b = \frac{\int_{-h/2}^{h/2} U(z)T(x,z)\mathrm{d}z}{\int_{-h/2}^{h/2} U(z)\mathrm{d}z}$$

This temperature is extremely important. It represents the temperature of a sample, taken out of the channel, and mixed in a cup'.[27] By identification, after some calculation, one finds the following formula for the temperature field:

$$T(x,z) - T_b(x) = \frac{3q_0 h}{K}\left(\frac{z^2}{h^2} - \frac{2z^4}{3h^4}\right) - \frac{39}{280}\frac{q_0 h}{K} \tag{5.58}$$

Let us calculate global values. The temperature difference ΔT along the microchannel is equal to:

$$\Delta T = Ax = \frac{3q_0}{\rho C_\mathrm{p} h U_0} x \tag{5.59}$$

Then the exit temperature T_e is related to the inlet temperature T_i by the relation:

$$T_\mathrm{e} = \frac{3q_0}{\rho C_\mathrm{p} h U_0} L + T_\mathrm{i}. \tag{5.60}$$

The following observations can be made about this formula:

- The smaller the height, the larger the temperature difference between the entry and the exit. This behaviour is intuitive.
- The higher the speed, the smaller the temperature difference between the entry and the exit. According to Eq. (5.60), should the fluid be at rest, the exit temperature would be infinite. To suppress this unphysical behaviour, one must incorporate, in the analysis, thermal inertia, represented by the term $\rho C_\mathrm{p} \frac{DT}{Dt}$. With this term, the temperature is no more infinite, but increases with time. Here, the presence of the flow, however small it is, allows a stationary regime to be achieved.

Let us now calculate the heat transfer coefficient h_c. One obtains:[28]

[27] This sampling temperature is not the average, along z, of the fluid temperature in the channel, which could be tempting to take instead of T_b. The reason is that when we take a fluid sample out of the system, the fast regions are more represented than the slow ones. This fact is not accounted for in spatial averaging.

[28] The factor 4 in the expression of h_c comes from that fact atht, by convention, the hydraulic diameter $2h$ is used as the characteristic scale, and the total heat flux across the walls is $2q_0$

$$h_c = \frac{4q_0}{T_w - T_b} = \frac{140K}{17h} \tag{5.61}$$

in which T_w is the wall temperature, i.e. the temperature at $z = \pm h/2$. The result deserves four remarks:

1. There is an apparent paradox: the heat transfer coefficient, being independent of flow speed, would be the same if the fluid were at rest. To understand this behaviour, one must note that, according to Eq. (5.58), temperature gradients across the channel (then along z) do not depend on x or the flow speed. Therefore, temperature differences, between the wall and a reference such as T_b, only depend on the channel height, the fluid thermal conductivity, and the heat flux at the wall. Because of that, the heat transfer coefficient h_c, which quantifies the heat fluxes normal to the channel wall, is flow independent, however strange it may look at a first glance. This does not mean that we can stop the flow: in this problem, the flow has two functions: the first is to maintain a stationary state in which a permanent heat transfer across the wall can develop. Without flow, no steady state. The second function is to keep the fluid temperature low (see below).

2. The Nusselt number is $Nu = h_c h/K$. Its value is $\frac{140}{17}$, a number difficult to guess without carrying out the full and quite lengthy calculation. In a tube, a similar calculation leads to a Nusselt number equal to $\frac{48}{11}$.

3. If the heat transfer coefficient does not depend on the flow rate, why are large thermal Peclet numbers and large Reynolds number used? In fact, heat must be transported along the flow, and released outside the system, in order to avoid working at elevated temperatures, leading to boiling. To satisfy this condition, Peclet numbers must be large. To stress this point, one may rewrite Eq. (5.60) in the form: $T_e = \frac{3q_0 L}{K} Pe_{th}^{-1} + T_{\rm i}$, in which $Pe_{th} = \frac{U_0 h}{\kappa}$. The formula shows that large Pe_{th} prevent temperatures to reach high levels. In terms of order of magnitude, let us take, with water, $U_0 = 10$ mm/s, $q_0 = 1\ kW/m^2$, $L = 0.1$ m and $h = 100\mu$m. We obtain: $Pe_{th} \approx 7$, $\frac{3q_0}{K} \approx 167°$ and consequently, $T_e - T_i \approx 23°$. Working with a thermal Peclet number of 0.1 would lead to a temperature elevation of 1670°, i.e. above the melting temperature of silicon.

4. Result 5.61 shows that, as announced previously, the exchange coefficient is inversely proportional to the characteristic size of the system, i.e. the channel height. In the system we discuss here, we can produce thermal exchange coefficients on the order of 10^5- 10^6W m^{-2} K^{-1}. This is much larger than with non miniaturized systems (see Table 5.74)).

5.15.5 Why do bird legs not freeze in winter?

Small bird legs are several millimetres in diameter (see Fig. 5.79). A naive reasoning, based on thermal diffusion, would lead us to conclude that they should freeze in winter.

However, they do not. The explanation is provided on the right–hand part of the figure.

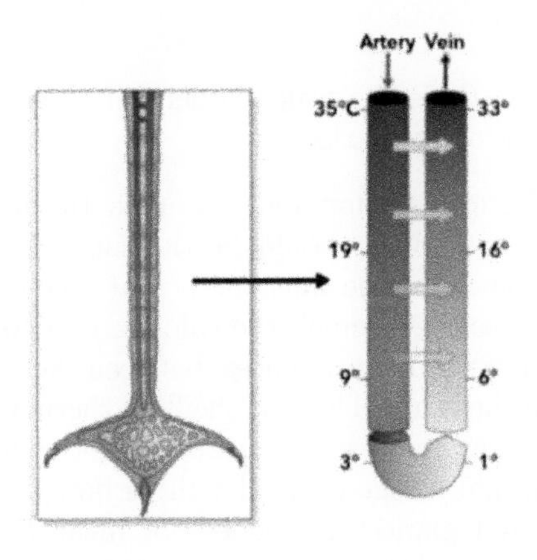

Fig. 5.79: Small bird in the snow. Why do its legs not freeze ? The explanation, based on the right–hand figure, is given in the text.

In the legs, veins and arteries are located close to each other. Consequently, due their small sizes, they develop substantial thermal exchanges between them. Therefore, the hot channel (35°C) can keep the other at above-zero temperatures, avoiding freezing. This mechanism also holds for larger birds, such as ducks, in which veins and arteries are also placed close to each other. Moreover, they are placed distantly from the outer surface. In this manner, evolution has helped the small and medium–sized birds to survive in winter.

5.16 Evaporation and drying

5.16.1 Generalities

When a liquid/vapour system is heated, the liquid is transformed into gas until a new equilibrium state is reached. This is evaporation. Evaporation occurs when the saturation pressure is larger than the partial pressure but smaller than the ambient pressure. This is the case of a puddle evaporating under the sun. When no equilibrium is found, the liquid disappears or, if it contains non–volatile components, precipitates. This is drying. Boiling is different from evaporation. It occurs when the saturation pressure is larger than both. This is the case when water is heated above 100°C. We focus here on evaporation.
Let us consider the heated droplet shown in Fig. 5.80.

The droplet being hotter, the vapour concentration close to the interface will be larger than in the far region, where the temperature is kept cold. Thereby, a diffusive profile

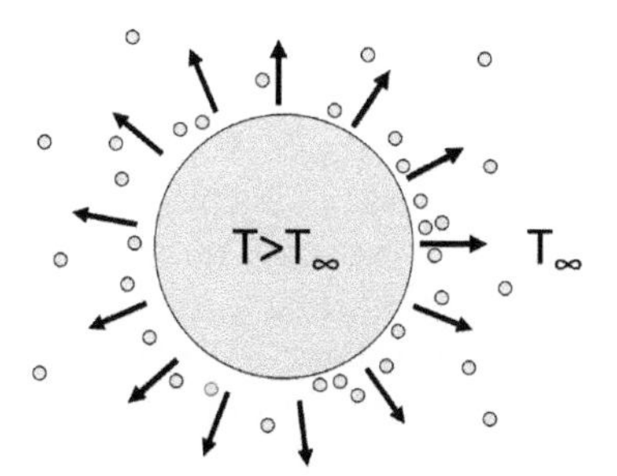

5.80: Sketch of the evaporation of a hot droplet in a cold gas. A flux of matter develops from regions rich in vapour, close to the interface, towards regions poor in vapor, far from the interface. In this process, the droplet shrinks.

develops, that maintains a flux of matter from regions rich in vapour, close to the interface, towards regions poorer in vapour, far from it. In this process, the droplet shrinks.

The phenomenon looks simple, but in practice, it is complex [140]. In many cases, it involves capillarity, hydrodynamics and heat transfer, that couple together to induce convection, Marangoni flows [141], crystallization, or precipitation.[29] Efforts have been expended for decades to investigate these situations. Some aspects have been reviewed [140, 144, 145].

In microfluidics, evaporation is an important phenomenon. It can be detrimental. For instance, small samples exposed to atmosphere can evaporate fast and get lost. It can also lead to the realization of new structures, impossible to create in non–miniaturized systems. An example shown in Fig. 5.81 is monodisperse micrometric solid foams [147] (see also [150]). These foams are obtained from an assembly of microfluidic bubbles, for which the continuous phase, a chitosan solution, was polymerized, cross-linked, and dried.[30] This ability to create new objects, by controlling drying, opens opportunities in material science. In the case of the foams, applications to photonic band gap materials and stem cell expansion were proposed [147–149].

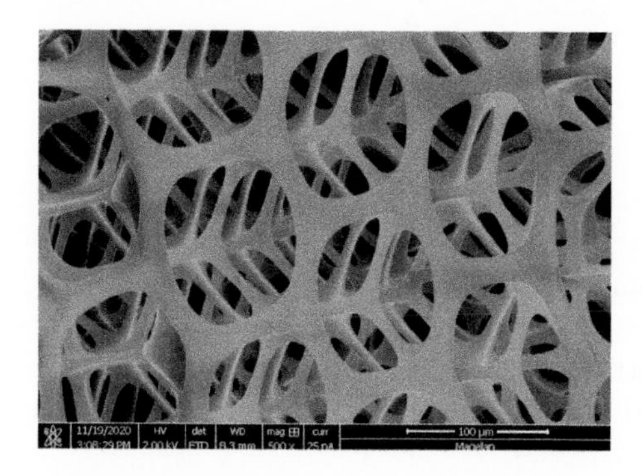

5.81: Scanning electron microscopy (SEM) image of a solid structure obtained by cross-linking an assembly of monodisperse microfluidic bubbles. In this system, the liquid phase, a chitosan solution, was polymerized, cross-linked and dried [147]. (Courtesy of M. Russo, I. Maimouni.)

[29]In addition, surprising phenomena have been reported, such as a temperature jump at the interface, in apparent violation with thermodynamics [143]

[30]During the drying process, the thin films located between each bubble burst and disappear, leaving the pores open.

5.16.2 How long does it take for an isolated spherical droplet to dry?

The first and simplest theory of droplet evaporation is due to Maxwell [151]. He succeeded in solving the equations of the problem, by assuming vacuum, no internal convection, and no latent heat. Some assumptions are probably questionable – in particular the neglect of latent heat, which cools down the droplet as it evaporates, and giving rise to Stefan flows [141, 142, 152] –. Maxwell theory nonetheless provides an important starting point, which turns out, in many cases of practical interest, to be extremely useful for interpreting the experiments. Maxwell theory shares features with the Epstein-Plesset theory of dissolving bubbles [73], mentioned previously. The calculation is lengthy but not difficult and the interested reader will find it described in Ref. [144]. The main outcome of the Maxwell model is the determination of the evolution of the droplet radius $a(t)$, which is given by a formula similar to Epstein-Plesset theory (see Eq. (5.24)).

$$a^2 = a_0^2 - \frac{8DMp_{sat}}{\rho_L RT} t \tag{5.62}$$

in which a_0 is the droplet initial value, D is the diffusion constant of the liquid vapour in the gas medium, M is the molar mass of the liquid, ρ_L is its density, R is the gas constant and T_L is the liquid temperature.[31] The theory compares well with the experiment. An example is shown in Fig. 5.82, in which droplets were levitated electrostatically.

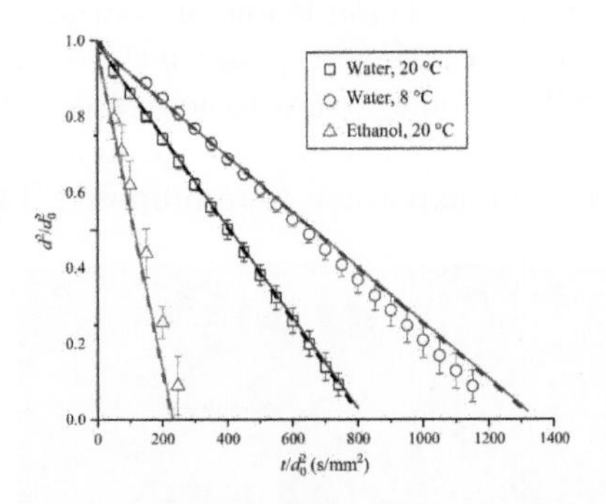

5.82: Radius evolution, with time, of water and ethanol microdroplets at different temperatures and 0% relative humidity. Symbols: experimental; solid lines: Stefan-Fuchs model; dashed lines: Maxwell model (Reprinted from [154], with permission of Springer Nature. Copyright 2022.)

The authors also compared the measurements to the Stefan-Fuchs model [153], in which Stefan flows are taken into account. It turned out that in the experimental conditions that were considered, there is little difference between the two models. This suggests that latent heat can be neglected in this case. From $a(t)$, one can calculate the lifetime τ of the droplet:

[31] Maxwell and Epstein-Plesset formulas have the same form. This can be seen by replacing c_{sat} by $p_{sat}M/RT$ (using the law of perfect gases).

$$\tau = \frac{a_0^2 \rho_L RT}{8DMp_{sat}} \tag{5.63}$$

This time decreases with the droplet size to the square, indicating that the smaller the droplet, the shorter its lifetime. In practice, 1 μm diameter droplets survive for 1 ms, and 100 μm droplets live for 10 s.

5.16.3 How long does it take for a surface droplet to dry?

When the droplet is deposited on a plane substrate, we can anticipate, from reading Chapter 4, that complications will arise, due to the presence of a contact angle, roughness, and wetting heterogeneities. In fact, two situations can be defined [155]: (a) the contact angle is constant or (b) the droplet radius is constant (see Fig. 5.83).

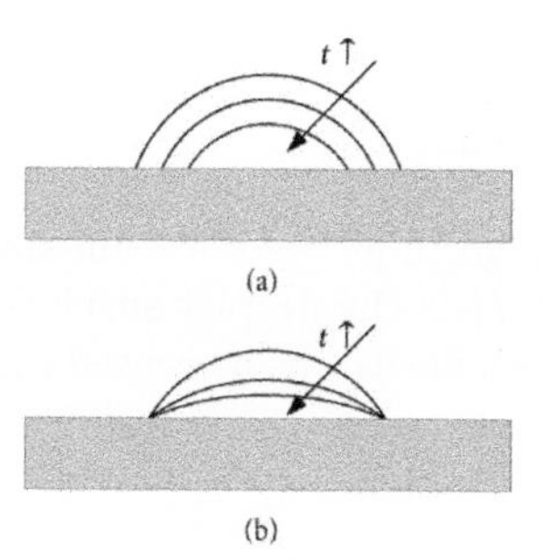

5.83: Two situations for the evaporation of a droplet on a substrate: (A) free contact line. Contact angle is locked at the Young value, as the droplet evaporates.(B) Droplet pinned on the surface, with a constant footprint and a free contact angle.

Typically, contact angles are constant on high energy surfaces, and they vary on low energy surfaces, because, again, of contact line pinning. Note that our droplet can decide to jump from one state to the other during the evaporation process, or get pinned only partially, developing a footprint evolving in a complicated manner.

Let us restrict ourselves to case (b), i.e. a variable contact angle and a constant footprint, characterized by radius a_0. Different authors succeeded in solving the problem, within slightly different frameworks of approximation [156–160]. These theories essentially agree with each other. We use Ref. [160] which provides a simple approximate formula for the temporal decay of the droplet mass:

$$m(t) \approx m_0 - 4D(1-H)c_{sat}a_0\, t \tag{5.64}$$

in which m is the droplet mass at time t, m_0 is its initial value, H is the relative humidity, D is the diffusion coefficient of liquid vapor in the ambient gas, c_{sat} is the saturation gas concentration far from the droplet and, again, a_0 is the footprint radius. The evolution of $a(t)$, given by Eq. (5.64), and compared to the experiment, is shown in Fig. 5.84.

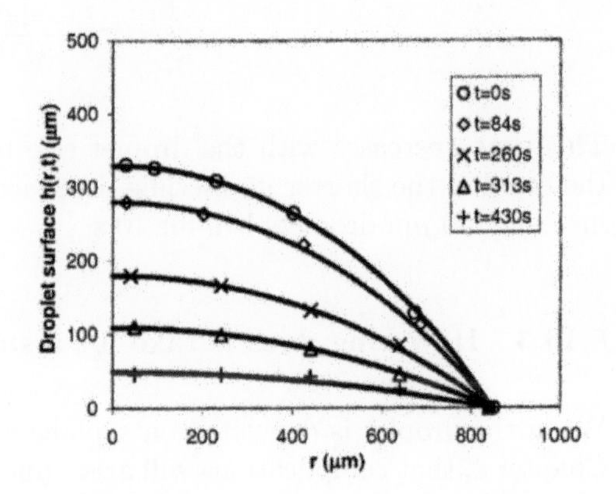

5.84: Measured droplet profiles at different times, for water evaporating from a droplet of initial radius $a_0 = 0.85$ mm and initial height 0.329 mm. The symbols show the locations of fluorescent particles on the droplet surface, and the lines are the fittings of circular arcs to these data. (Reprinted from [160] with permission of the American Chemical Society. Copyright 2022.)

Fig. 5.84 shows that the Hu and Larson formula reproduces well the experimental observations. From Eq. (5.64), the droplet's lifetime τ is:

$$\tau \approx \frac{m_0}{4D(1-H)c_{sat}a_0}$$

Here again, it is instructive to estimate orders of magnitude.[32] For a 1 μm-sized droplet in a dry environment – so, roughly, a volume of 1 $fl.$ – the droplet survives for 1 ms. For 100 μm-sized droplets, the life time is 10 s. This time can be increased considerably by raising H.

5.17 Microexchangers for electronic components

5.17.1 The considerable needs of cooling of data centres

Since the invention of the integrated circuit, the course of the integration of transistors onto Central Processing Unit(CPU) has proceeded with impressive speed. In 1971, the operating units of computers contained around 2,000 transistors. In 2002, this number had multiplied by 25,000, approaching 60 million transistors for the 3 GHz Pentium IV. At the time of writing (2022), 50 billion transistors are currently integrated on microprocessors.

This evolution is accompanied by an increase in the number of commutations and, consequently, a raise in the quantity of heat produced by the unit. It is critical to extract this heat for the transistors to function at a stable temperature, between 80° C and 95°C.

Today, in the USA alone, data centers consume 24 TWh of electricity and 100 billion litres of water to cool them down [161]. This represents the residential needs of a

[32]Maxwell and Hu-Larson formula have the same form. This can be seen by replacing c_{sat} by $p_{sat}M/RT$ and developing m_0. The two formulas therefore provide, for the droplet lifetime, comparable orders of magnitude.

city of the size of Philadelphia. In 2027, the needs will account for 31% of Ireland's electricity [161].

Traditionally, processors are cooled down with radiators. These radiators are fabricated with a good thermal conductor (copper), ventilated, or traversed by cooling channels. In some case, the entire CPU is immersed in an insulating fluid. The seminal work of Tuckerman and Pease [162] showed that the usage of microcanals placed directly on or in good thermal contact with the processor allows to reach exchange coefficients much higher than traditional methods.[33] This is where micro and millifluidics come on the stage.

5.17.2 Old method

The traditional method consists, thus, in installing a radiator cooled by forced or natural convection (i.e. with or without a fan). Nature has used this solution to cool down the stegosaurus. This dinosaur, a peace-loving herbivorous animal, shown on Fig. 5.85, grew large plates on its back, to thermoregulate its body and, in the meantime, scare predators.

5.85: A traditional radiator, possessing cooling fins increasing the surface area of exchange with the cooling source (air). The animal (stegosaurus) bears the radiator on his back. (iStock image, Credit MR1805 Bochum, Germany.)

In all cases, including our stegosaurus, the heat flux q extracted by the radiator is given by:

$$q = \frac{P}{A} = h_c \Delta T \tag{5.65}$$

where h_c is the exchange coefficient, ΔT is the temperature difference between the component and the exterior, P is the power produced by the component, and A is the exchange surface. For a radiator cooled by natural convection, the exchange coefficient is on the order of ten or so $\mathrm{W\,m^{-2}\,K^{-1}}$ (see Table 5.5). In such conditions,

[33] We may quote, from Tuckerman and Pease (1983 Symposium on VLSI Technology): 'The widely held belief that a VISI circuit (VLSIC) must not use more than a few watts of power due to thermal limitations has led to major constraints on the performance of VLSI systems. There is, however, no fundamental justification for this belief.

with a produced heat q is on the order of 10^3 W m^{-2}, it is possible to maintain the microprocessor at a temperature between 80 and 90°C. The usage of fans, by improving h up to an order of magnitude, allows us to work at larger q. However, for heat densities currently required in data centres, such radiators are no longer able to thermoregulate the system.

5.17.3 Optimizing microexchangers

We recall that in microchannels, the thermal exchange coefficient h_c is given, in term of order of magnitude, by the relation $h_c \sim h^{-1}$, in which h is the channel height (see Eq. (5.61)). By miniaturizing the system, higher exchange coefficients can thus be obtained. In terms of orders of magnitude, for water in a microchannel, 100 μm high, one achieves coefficients h_c on the order of 10^5 W m^{-2} K^{-1}. As noted, this represents an improvement of two orders of magnitude compared to the passive radiator. This point was stressed by Tuckerman and Pease [162]. In their experiment, they achieved a heat transfer coefficient on the order of 10^5 W m^{-2} K^{-1}, so that they could extract a power density equal to 7.9 MW/m^2, keeping the temperature rise below 71°C.

A simplified view of a microexchanger, of the type discussed by Tuckerman and Pease, is formed by stacking microchannels (see Fig. 5.86).

The channels have a width w, a height h, and a total length L. They are placed in parallel, along a serpentine (not represented), located just beneath, and in excellent thermal contact with a CPU, idealized by a rectangular plate. We assume that the solution obtained in Section 4.14.4 can be transposed, in terms of order of magnitude, to our case. The processor produces a thermal power P, and the heat flux q that the microexchanger extracts is:

$$q \sim \frac{P}{wL}$$

In Section 4.14.4, we calculated the temperature differences developing in the fluid, along the streamwise direction (ΔT_x) and the cross-stream direction (ΔT_z). We found:

$$\Delta T_z \sim \frac{q_0 h}{K} \quad \text{and} \quad \Delta T_x \sim \frac{q_0 L}{\rho U C_{\mathrm{p}} b} \tag{5.66}$$

in which U is the flow speed, ρ is the fluid density, and C_{p} is the thermal capacity.

ΔT_z increases with h, while ΔT_x decreases with it. In an optimal microexchanger, the two temperature rises must be minimized, in order to work in thermally acceptable conditions. The optimal conditions, obtained by equating ΔT_z and ΔT_x, are defined by the following expression:

$$Pe_{\mathrm{th}} = \frac{U_0 h}{\kappa} = \frac{L}{h} \text{ and then } h \sim \sqrt{\frac{L\kappa}{U}} \tag{5.67}$$

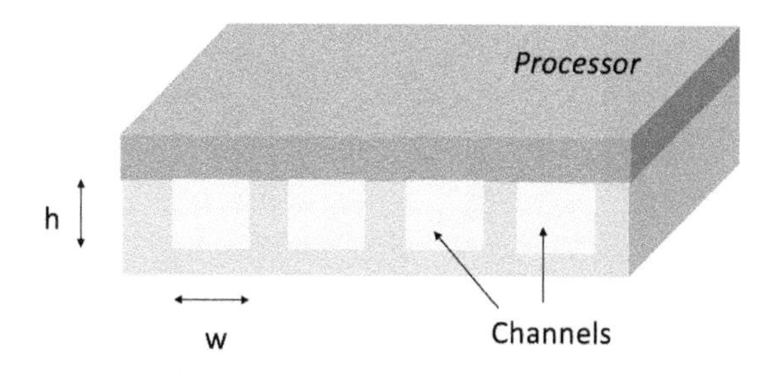

5.86: Example of an oversimplified microexchanger using several microchannels in parallel, placed along a serpentine.

Ideally, it would be interesting to increase U, so as to decrease h and increase the thermal exchange coefficient. However, in this approach, the power of the pump assuring fluid circulation must be raised. Let us recall the formula for the pressure drop along a microchannel, established in Chapter 3:

$$\Delta P \sim \frac{\mu L U}{h^2} \tag{5.68}$$

where μ is the fluid viscosity. Increasing U and, in the meantime, decreasing h implies increase ΔP. This brings us to the question of the pump. Excellent pumps are needed to drive, in a compact design, flows at substantial speeds in narrow channels. In the early days of microfluidics, electro–osmotic pumps were proposed [163]. Today, the subject is an active area of research.

References

[1] R. Brown, *Edinburg New Philosophical Journal*, 358 (July – Sept. 1828).
[2] Y. Pomeau, J. Piasecki, *CRAS* (2017).
[3] J. Perrin, *Les atomes* (1914).
[4] A. Einstein, *Ann. Phys. (Berlin)*, **17**, 549 (1905).
[5] P. Langevin, *Comptes Rendu de l'Academie des Sciences*, vol. **146** , 530 (1908).
[6] M. von Smoluchowski, *Ann. der Phys.*, **21**, 756 (1906).
[7] W. Sutherland, *Philos. Mag.*, **9**, 781 (1905).
[8] P. Levy, *Theorie de l'Addition des Variables Aleatoires*, ed. Gauthler Villars, Pans (1954).
[9] A. Y. Khimtchme, P Levy, *C. R. Acad. Sci. (Paris)*, **202**, 374 (1936).
[10] J.-P. Bouchaud, A. Georges, *Physics Reports*, **195**, 127 (1990).
[11] B. Duplantier, *Seminaire Poincaré*, **1**, 155 (2005).
[12] J. H. Van't Hoff, *Kongliga Svenska Vetenskaps-Academiens Handlingar*, Stockholm, **21**, 1 (1884).
[13] W. B. Russel, D. A. Saville, W .R. Showalter, *Colloidal Dispersions*, Cambridge University Press (1989).
[14] P. J. A. Kenis, R. F. Ismagilov, G. M. Whitesides, *Science*, **285**, 83 (1999).
[15] A. Kamholz, B. Weigl, B. Finlayson, P. Yager, *Anal. Chem.*, **71**, 5340 (1999).
[16] A. Kamholz, P. Yager, *Biophys. J.*, **80**, 155 (2001).
[17] A. Kamholz, P. Yager, *Sens. Actuators B*, **82**, 117 (2002).
[18] R. Ismagilov, A. Strooke, P. Kenis, G. Whitesides, H. Stone,*Appl. Phys. Lett.*,**76**, 2376 (2000).
[19] S. Takayama, J. Cooper McDonald, E. Ostuni, M. N. Liang, P. J. A. Kenis, R. F. Ismagilov, G. M. Whitesides, *Proc. Nat. Acad. Sci. USA*, **96**, 5545 (1999).
[20] P. B. Rhines, W. R. Young, *J. Fluid. Mech.*, **133**, 133 (1983).
[21] Private communication.
[22] G. I. Taylor, *Proc.Roy.Soc.*, A219, 186 (1953).
[23] R. Aris, *Proc. Roy. Soc. Lond.* **A 235**, 67 (1956).
[24] H. Bruus, *Theoretical Microfluidics*, Oxford University Press (2007).
[25] A. Adjari, N. Bontoux, H. A. Stone, *Anal. Chem*, **78**, 387 (2006).
[26] P. Bergé, Y. Pomeau, C. Vidal, *Order in chaos*, J. Wiley & Sons, New York (1986).
[27] J. Ottino, *The kinematics of mixing : stretching, chaos and transport*, Cambridge University Press (1989).
[28] J. Ottino, *A. Rev. Fluid. Mech.*, **22**, 207 (1990).
[29] A. Lichtenberg, M. Lieberman, *Regular and stochastic motion* (2013).
[30] E. W. Lorenz, *Journal of the Atmospheric Sciences*, **20**, 130 (1963).

[31] J. H. Poincaré, *Acta Mathematica*, **13**, 1 (1890).
[32] P. Grassberger, I. Procaccia, *The Theory of Chaotic Attractors*, Springer, New York, NY. (2004).
[33] D. Ruelle, F. Takens, *Communications in Mathematical Physics*, **20**, 167 (1971).
[34] D. Ruelle, *Predictability*, AAAS 139th Meeting (1972).
[35] S. Smale, *Bulletin of the American Mathematical Society*, **73**, 747 (1967).
[36] V. K. Melnikov, *Trans. Moscow Math.*, **12**, 1 (1963).
[37] https://www.youtube.com/watch?v=rA6T83FKuE8.
[38] S. W. Jones, O. M. Thomas, H. Aref, *Journal of Fluid Mechanics*, **209** , 335 (1989).
[39] S. Wiggins, J. M. Ottino, *Phil. Trans. R. Soc. Lond. A*, 362, 937 (2004).
[40] J. Ottino, *Scientific American*, 56 (1989).
[41] H. Aref, *J. Fluid Mech.*, **143**, 1 (1984).
[42] V. Hessel, H. Löwe, F. Schönfeld, *Chem. Eng. Sci.*, **60**, 2479 (2005).
[43] N. T. Nguyen, Z. Wu, *J. Micromech. Microeng.*, **15**, R1 (2005).
[44] G. Cai, L. Xue, H. Zang, J. Lin, *Micromachines*, **8**, 274 (2017).
[45] A. D. Stroock, S. K. W Dertinger, A. Ajdari, I. Mezić, H. A. Stone, G. M. Whitesides, *Science* , **295**, 647 (2002).
[46] J. W. Hong, Y. Chen, W. F. Anderson, S. R. Quake *J. Phys.: Condens. Matter* , **18**, S691(2006).
[47] Y–K. Lee, P. Tabeling, C. Shih, C. M. Ho, *SME Int. Mech. Eng. Congress and Exposition*, **1905**, 505 (2000).
[48] Y–K. Lee, J. Deval, P. Tabeling, C. M. Ho, *Microreaction Technology*. Springer, Berlin, 185 (2001).
[49] X. Niu, Y–K. Lee, *J. Micromech. Microeng.*, **13**, 454 (2003).
[50] F. Okkels, P. Tabeling, *Phys. Rev. Lett.*, **92**, 38301 (2004).
[51] C .N. Baroud, F. Okkels, L. Ménétrier, P. Tabeling, *Physical Review E*, **67**, 060104 (2003).
[52] F. Bottausci, I. Mezić, C. D. Meinhart, C. Cardonne, *Phil. Trans. R. Soc. Lond. A*, **362**,1001 (2004).
[53] H. Song, D. L. Chen, R. F. Ismagilov, *Ang. Chem. Intl. Ed.*, **45**, 7336 (2000).
[54] H. Song, M. R. Bringeer, J. D.Tice, C. J. Gerdts,R. F. Ismagilov, *App. Phys. Lett.*, **1**, 4664 (2003).
[55] J. Wang, J. Wang, L. Feng, T. Lin, *RSC Advances.*, **126**, 104138 (2015).
[56] J. Clark, M. Kaufman, P. S. Fodor, *Micromachines*, **9**, 107 (2018).
[57] G. Gompper, *J. Chem. Phys*, **125**, 164908 (2006).
[58] R. Cross, *Chem. Eng. News*, **99**, 16 (2021).
[59] M. J. W. Evers, J. A. Kulkarni, R. der Meel, et al. *Small Methods*, **2**, 1700375 (2018).
[60] M. Maeki et al. *PLoS One*, **12**, e0187962 (2017).
[61] M. Ripoll, E. Martin, M. Enota, O. Robbe, C. Rapisarda, M.C. Nicolai, A. Deliot, P. Tabeling, J-R. Authelin, M. Nakach, P. Wils, *Sci. Reports*, **12**, 9483 (2022).
[62] Q. Zhao, D. Yuan, J. Zhan, W. Li, *Micromachines*, **11**, 461 (2020).

[63] A. Minakov, V. Rudyak, A. Dekterev, A. Gavrilov, *Int. J. Heat. Fluid. Flow*, **43**, 161 (2013).
[64] M. Hoffmann, M. Schluter, N. Rabiger, *Chem. Eng. Sci.*, **61**, 2968 (2006).
[65] C. Baroud, L. Menetrier, F. Okkels, P. Tabeling, *Phys. RevE*, **67**, 60104 (2003).
[66] J-B. Salmon, C. Dubrocq, P. Tabeling, S. Charier, D. Alcor, L. Jullien, F. Ferrage, *Anal. Chem.*, **77**, 3417 (2005).
[67] N. L. Lean, S. W. Dertinger, D. T. Chiu, I. S. Choi, A. D. Stroock, G. M. Whitesides, *Langmuir*, **16**, 8311 (2000).
[68] N. L. Leon, H. Baskaran, S. W. Dertinger, G. M. Whitesides, L. Van de Water, M. Toner, *Nature Biotech.*, **20**, 826 (2002).
[69] K. W. Oh, K. Lee, B. Ahn, E. P. Furlani, *Lab on a Chip*, **12**, 515 (2012).
[70] E.L.Cussler, *Diffusion*, Cambridge University Press (2009).
[71] A. Aota, M. Nonaka, A. Hibara, T. Kitamori, *Angewandte Chemie*, **119**, 896 (2007).
[72] P. Mary, V. Studer, P. Tabeling, *Anal. Chem.*, **80**, 2680 (2008).
[73] P. S. Epstein, M. S. Plesset, *J. Chem. Phys.*, **18**, 1505 (1950).
[74] D. Lohse, X. Zhang, *Phys. Rev. E*, **91**, 031003 (2015).
[75] N. Ishida., T. Inoue, M. Miyahara, K. Higashitani, *Langmuir*, **16**, 6377 (2000).
[76] S. T. Lou, Z. Q. Ouyang, Y. Zhang, X. J. Li, J. Hu, M. Q. Li, F. J. Yang, *J. Vac. Sci. Technol. B*, **18**, 2573 (2000).
[77] J. W. G Tyrrell, P. Attard, *Phys. Rev. Lett.*, **87**, 176104 (2001).
[78] M. Alheshibri, J. Qian, M. Jehannin, V. S. J. Craig *Langmuir*, **32**, 11086 (2016).
[79] D. Lohse, X. Wang, *Reviews of Modern Physics*, **87**, 981 (2015).
[80] Private communication.
[81] Y. Popov, *Phys. Rev. E* , **71**, 036313 (2005).
[82] Y. Liu, X. Zhang, *J. Chem. Phys.*, **141**, 134702 (2014).
[83] T.Segers, N.de Jong, M. Versluis, *The Journal of the Acoustical Society*, **140**, 2506 (2016).
[84] S. Hernot, A. L. Klibanov, *Advanced Drug Delivery Reviews*, **601153**, (2008).
[85] T. Segers, P. Kruizinga, M. P. Kok, G. Lajoinie, N. de Jong, M. Versluis, *Ultrasound in Medicine & Biology*, **44**,1482 (2018).
[86] A. J. Dixon, J. Li, J. M. R. Rickel, A. L. Klibanov, Z. Zuo, J. A. Hossack, *Ann. Biomed. Eng.*, **47**, 1012 (2017).
[87] T. Laurell, F. Petersson, A. Nilsson, *Cham. Soc. Reviews*, **36**, 492 (2007).
[88] H. Faxen, *Ann. Phys.*, **373**, 89 (1922).
[89] H. Brenner, *Chem. Eng. Sci.*, **16**, 242 (1961).
[90] A. J. Goldman, R. G. Cox, H. Brenner, *Chem. Eng. Sci.*, **22**, 637 (1967).
[91] Y. Kazoe, M. Yoda, *Appl. Phys. Lett.*, **99**, 124104 (2011).
[92] P. Huang, K. S. Breuer, *Phys. Rev. E*, **76**, 046307 (2007).
[93] C. M. Cejas, F. Monti, M. Truchet, J-P Burnouf, P. Tabeling, *Langmuir*, **33**, 6471 (2017).
[94] C. M. Cejas, F. Monti, M. Truchet, J-P. Burnouf, P. Tabeling, *Phys. Rev. E*, **98**, 62606 (2018).
[95] T. Porto Santos, R. Lopes Cunha, P. Tabeling, C. M. Cejas, *Physical Chemistry Chemical Physics*, **22**, 17236 (2020).

[96] H. M. Wyss, D. L. Blair, J. F. Morris, H. A. Stone, D. A. Weitz, *Phys. Rev. E*, **74**, 061402 (2006).
[97] A. Sauret, E. C. Barney, A. Perro, E. Villermaux, H. A. Stone, E. Dressaire, *Appl. Phys. Lett.*, **105**, 074101 (2014).
[98] E. Dressaire, A. Sauret, *Soft Matter*, **13**, 37 (2017).
[99] A. Sauret, K. Somszor, E. Villermaux, E. Dressaire, *Phys.Rev. Fluids*, **3**, 104301 (2018).
[100] U. Soysal, P. N. Azevedo, F. Bureau, A. Aubry, M. S. Carvalho, A. C. S. N. Pessoa, G. Lucimara, O. Couture, A. Tourin, M. Fink, P.Tabeling, *Med Physics*, **48**, 1484 (2022).
[101] B. H. Weigl, P. Yager, *Science*, **283**, 346 (1999).
[102] P. Sajeesh, A. K. Sen, *Microfluiidics Nanofluidics*, **17**, 1 (2014).
[103] L.R. Huang, E.C. Cox, R. H. Austin, J. C. Sturm, *Science*, **304** 987 (2004).
[104] J. McGrath, M. Jimenez, H. Bridle, *Lab on a Chip*, **14**, 4139 (2014).
[105] G. Segre, A. Silberberg, *Nature*, **189**, 209 (1961).
[106] D. Di Carlo, D. Irimia, R. G. Tompkins, M. Toner, *Proc. Natl. Acad. Sci.*, **104**, 18892 (2006).
[107] D. Di Carlo, *Lab on a Chip*, **9**, 3038 (2009).
[108] J. Zhang et al., *Lab on a Chip*, **16**, 10 (2016).
[109] J.-S. Park, H.-I. Jung, *Anal. Chem.*, **81**, 8280 (2009).
[110] T. S. Sim, K. Kwon, J. C. Park, J.-G. Lee, H.-I. Jung, *Lab Chip*, **11**, 93 (2011).
[111] M. E . Warkiani, B .L. Khoo, L. Wu, A. K. Ping Tay, A. A. S Bhagat, J. Han, C.T. Lim, *Nature Protocols*, **11**, 134 (2016).
[112] D. M. Ruthven, *Principles of Adsorption and Adsorption Processes*, John Wiley & Sons (1984).
[113] P. W. Atkins, *Physical chemistry*, Oxford University Press (1994).
[114] S. Demanèchea, JP Chapel, L. Jocteur Monrozier, H. Quiquampoix*Colloids and Surfaces B: Biointerfaces*, **70**, 226 (2009).
[115] R. C. J. A Boom, C. J. Sol, M. M., Salimans, C. L. Jansen, P. M. Wertheim-van Dillen, J. P. M. E. Van der Noordaa, *Journal of clinical microbiology*, **28**, 495 (1990).
[116] I. Langmuir, *Part I: the Research Laboratory of the General Electric Company*, 2221 (1916).
[117] I. Langmuir, *Part II: the Research Laboratory of the General Electric Company*, 2221 (1918).
[118] I. Langmuir, *Journal of the American Chemical Society*, **40**, 13 (1918).
[119] C. Mathe, S. Devineau, J.–C. Aude, G. Lagniel, S. Chédin, V. Legros, M. H. Mathon, J-P. Renault, S. Pin, Y. Boulard, J. Labarre, *PloS one*, **8**, e81346 (2013).
[120] S. Brunauer, P. H Emmett, E. Teller, *Journal of the American Chemical Society*, **60**, 309 (1938).
[121] H. Freundlich, *Akademische Verlagsgesellschaft* (1909).
[122] D. A. Skoog, D. M. West, F. J. Holler, *Fundamentals of Analytical Chemistry*, De Boeck University (1996).
[123] R. P. W. Scott, *Techniques and Practices of Chromatography*, M. Dekker (1995).
[124] Y. Kazakevich, H. McNair, *Basic Liquid Chromatography*, hplc.chem.shu.edu/NEW/HPLC_B

[125] C. F. Poole, *The Essence of Chromatography*, Elsevier (2003).
[126] A fundamental work explaining chromatographic methods used for the identification of proteins is *Molecular Biology of the Cell*, by B. Alberts, D. Bray, J. Lewis, M. Raff, K. Roberts, J. Watson (1994).
[127] D. Reyes, D. Iossifidis, P-A. Auroux, A. Manz, *Anal. Chem.*, **74**, 2623 (2002).
[128] P. A. Auroux, D. Iossifids, D. Reyes, A. Manz, *Anal. Chem.*, **74**, 2637 (2002).
[129] S. Descroix, *Cours Les Houches* (2009).
[130] J. J. van Deemter, F. J. Zuiderweg, A. Klinkenberg, *Chem. Eng. Sci.*, **5**, 271, (1956).
[131] M. Zapata, F. Rodriguez, J. L. Garrido *Marine. Ecol.Prog.*, **189**, 29, (2000).
[132] S. C. Terry, J. H. Jerman, J. B. Angell, *IEEE Trans. Elec. Dev.*, **26**, 1880 (1979).
[133] A. Manz, N. Graber, H. Widmer, *Sens. Actuators*, **B1**, 244 (1990).
[134] D. Cahill, W. Ford, K. Goodson, G. Mahan, A. Marjumdar, H. Maris, R. Merlin, S. Philipot, *J. Appl. Phys.*, **93**, 2, 793 (2003).
[135] C. Kittel, *Introduction to Solid State Physics*, Wiley, New York (1953).
[136] G. Batchelor, *Introduction to Fluid Dynamics*, Cambridge University Press.
[137] L. Landau, V. Lifschitz, *Mécanique Des Fluides*, Mir, (1967).
[138] P. Yi, R. A. Awang, W. S. T. Rowe, K. Kalantar-Zadeh, K. Khoshmanesh, *Lab on a Chip*, **14**, 3419 (2014).
[139] MU Kopp, A. J. de Mello, A. Manz *Science*, **280** 1046 (1998).
[140] H. Y. Erbil, *Advances in Colloid and Interface Science*, **170**, 67 (2012).
[141] E. Sultan, A. Boudaoud, M. B. Amar, *J. Fluid. Mech.*, **543**, 183 (2005).
[142] I. Bilimi, *J. of Thermal Science and Technology*, **40**, 309 (2020).
[143] G. Fang, C. A. Ward, *Phys. Rev. E*, **59**, 417 (1999).
[144] R. Hołyst, D. Litniewski, K. Jakubczyk, M. Kolwas, K. Kowalski, S. Migacz, S. Palesa, M. Zientara, *Rep. Prog. Phys.*, **76**, 034601(2013).
[145] S.Tarafdar, Y. Y. Tarasevich, M. Choudhury, T. Dutta, D. Zang, *Advances in Condensed Matter Physics*, **2018**, 1 (2018).
[146] J. W. van Honschoten, N. Brunetsa, N. R. Tas, *Chem. Soc. Rev.*, **39**, 1096 (2010).
[147] I. Maimouni, M. Morvaridi, M. Russo, G. Lui, K. Morozov, J. Cossy, M. Florescu, M. Labousse, P. Tabeling, *ACS Applied Materials & Interfaces*, **12**, 3261 (2020).
[148] J. Ricouvier, P. Tabeling, P. Yazhgur, *PNAS*, **116** (19) 9202 (2019).
[149] M. A. Klatt, P. J. Steinhardt, S. Torquato, *Proc. Natl. Acad. Sci.*, **116**, 23480 (2019).
[150] S. Andrieux, W. Drenckhan, C. Stubenrauch, *Polymer*, **126**, 425 (2017).
[151] J. C. Maxwell, *Collected Scientific Papers*, Cambridge, p. 625 (1890).
[152] Z. Pan, J. A. Weibel, S. V. Garimella, *International Journal of Heat and Mass Transfer*, **152**, 119524 (2020).
[153] N. A. Fuchs, *Evaporation and droplet growth in gaseous media*, Elsevier (2013).
[154] M. Ordoubadi, F. K. A Gregson, O. Melhem, D. Barona, R. E. H. Miles, D. D. Sa, S. Gracin et al. *Pharm Res*, **36**, 100 (2019).
[155] F. Parisse, C. Allain, *Journal de Physique II*, **6**, 1111–1119 (1996).
[156] N. N. Lebedev, *Special Functions and Their Application*, Prentice-Hall: Englewood Cliffs, New Jersey (1965).
[157] R. G. Picknett, R. J. Bexon, *J. Colloid Interface Sci.*, **61**, 336 (1977).

[158] R. G. Deegan, O. Bakajin, T. F. Dupont, G. Huber, S. R. Nagel, T. A. Witten, *Nature*, **389**, 827 (1997).
[159] R. D. Deegan, O. Bakajin, T. F. Dupont, G. Huber, S. R. Nagel, T. A. Witten, *Phys. Rev. E*, **62**, 217 (1999).
[160] H. Hu, R. G. Larson, *J. Phys. Chem. B*, **106**, 1334 (2002).
[161] R. van Erp, R. Soleimanzadeh, L. Nela, G. Kampitsis, E. Matiol, *Nature*, **585**, 211 (2020).
[162] D. B. Tuckerman, R. F. Pease, *IEEE Electron Device Lett.*, **EDL-2**, 5, 126 (1981).
[163] L. Jiang, J. Mikklsen, J. Koo, L. Zhang, D. Huber, S. Yao, A. Bari, P. Zhou, J. Santiago, T. Kenny, K. E. Goodson, *Proceeding of Thermal Challenges in Next Generation Electronic Systems*, Thermes, Santa Fe, New Mexico, USA, 133 (2002).

6

Electrokinetics

6.1 Introduction

In the years 1980s, stimulated by the surge of micro–electromechanical systems (MEMS) technology, and in the hope of creating performing, miniaturized, electro-mechanical actuators, researchers worked at integrating, in microsystems, electrodes with extremely small separations. In this manner, they could produce high electric fields, and thus large electromotive forces. The idea came to a small group, based in Europe, to exploit these large fields for separating ions, sorting particles and driving electrolytes. In 1992, Manz and Widmer [1] invented the first miniaturized electrokinetic system of separation based on MEMS technology. They demonstrated that, by working in microchannels, ions can be separated in seconds, with excellent efficiency, while traditional methods needed heavy equipment and hours to perform the same task. The concept was appealing, and, in a visionary impetus, it gave rise to the notion of lab on a chip, i.e. a device performing all operations carried out in laboratories, but in a centimetric system, using small amounts of reagents and short times.[1] Microfluidics was born.

6.2 Basic notions of electrostatics of macroscopic media

6.2.1 Governing equations

Here, we briefly review some basic results of electrostatics. For further reading, we recommend two excellent textbooks [2, 4]. In the field of electrostatics, by definition, there is no magnetic field and Maxwell equations, in a vacuum, reduce to:

$$\text{div } \epsilon_0 \boldsymbol{E} = \rho_e \tag{6.1}$$

$$\mathbf{curl\ E} = 0 \tag{6.2}$$

in which $\mathbf{E}$ is the electric field, ρ_e is the charge density (C/m^3), and $\epsilon_0 = 8.8410^{-12}$ F/m is the vacuum permittivity. The first equation leads to the Gauss theorem (the

[1] We may quote a sentence in the conclusion of Ref. [18]: Such systems could lead to laboratories on a chip that offer rapid, sophisticated analyses in a mobile package that is free to leave the laboratory.'

flux of $\mathbf{E}$ across a surface enclosing a volume is equal to the charge contained in it) and the second one to electric field circulation along a closed line equal to zero. From Eq. (6.2), one introduces an electric potential ϕ given by:

$$\mathbf{E} = -\nabla\phi$$

Combining this with Eq. (6.1), on obtains ::

$$\Delta\phi = -\frac{\rho_e}{\epsilon_0} \tag{6.3}$$

In purely dielectric media, i.e. with zero conductivities, and in the presence of an electric field, charges of opposite sign, electrons or protons, separate, so that dipoles, aligned along the field line, emerge from the background. On the other hand, the permanent molecular dipoles (for instance, water) reorient and align along the field direction. All these molecular contributions sum up to develop a macroscopic polarization P (see below). The phenomenon is illustrated in Fig. 6.1.

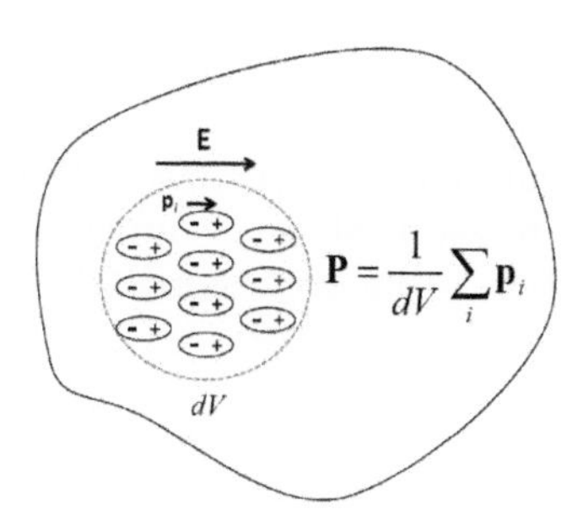

6.1: A dielectric volume δV, subjected to an electric field $\mathbf{E}$, polarizes. Microscopic dipoles $\mathbf{p}_i$ form, giving rise to a local, macroscopic, polarization vector $\mathbf{P}$, which sums up all of them.

Determining the consequence of the presence of a polarization field is nothing but straightforward. First, let us recall that for a dipole formed by two opposite charges q, separated by a vector $\mathbf{d_i}$, the polarization vector is equal to $q\mathbf{d_i}$. In vacuum, and at distances large compared to d_i, this dipole develops, at location $\mathbf{r}$ (with the origin taken at the dipole centre), an electric potential $\phi_i(\mathbf{r})$, given by:

$$\phi_i = \frac{\mathbf{p_i}.\mathbf{r}}{4\pi\epsilon_0 r^3} \tag{6.4}$$

In a dielectric medium, a polarization vector par unit of volume, $\mathbf{P}$, builds up by summing up, in the elementary volume δV, all the molecular dipoles i induced by the presence of the local electric field (see Fig. 6.1). Without electric field, no charge separates, then no polarization appears. $\mathbf{P}$ is therefore a function of $\mathbf{E}$. In linear, isotropic materials, which will be considered all along the chapter, $\mathbf{P} = \epsilon_0\chi\mathbf{E}$, where χ is the dielectric susceptibility.

It has been a mathematical achievement to demonstrate that, for these materials, the effect of polarization is fully taken into account by replacing, in Maxwell equations, and in the boundary conditions, the vacuum permittivity ϵ_0 by the medium permittivity $\epsilon = \epsilon_0(\chi + 1)$. In other words, the effect of polarization is to replace the displacement field $\epsilon_0\mathbf{E}$ by $\epsilon_0\mathbf{E} + \mathbf{P}$. We will admit this result, which is not straightforward to demonstrate, and is well–presented in textbooks.[2] Often, a physical picture is offered, in which a dielectric material is shown to be equivalent to a non–dielectric material with a charge distribution at its surface, called 'bound charges'. This representation allows us to visualize the effect of the polarization, but it is not useful for calculating the electrostatic field, because the charge distribution is, in fact, most often, an unknown of the problem.

In microfluidic systems, where oils, poydimethylsiloxane (PDMS), plastic, and glass are used, the dielectric materials are isotropic and linear, and ϵ is typically a few times the vacuum permittivity. Water is extremely polarizable: its permittivity is 80 ϵ_0 (at 20°C).

6.2.2 Boundary conditions

Having stated that, formally, the effect of polarization is absorbed in ϵ, called 'dielectric constant of the medium', it becomes possible to establish, in a simple manner, the boundary conditions. At the frontier between two media of permittivities ϵ_1, and ϵ_2, with E_1 on one side and E_2 on the other, one has (see Fig. 6.2):

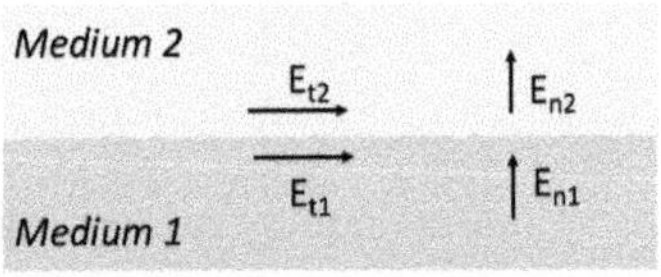

6.2: Boundary conditions on the electric field at the frontier between two dielectric materials.

$$E_{t1} = E_{t2} \text{ and. } \epsilon_2 E_{n2} - \epsilon_1 E_{n1} = \sigma_S \tag{6.5}$$

in which σ_S is the surface charge and t and n represents the tangential and normal coordinates, respectively, with respect to the interface. The condition on the electrical potential is the continuity at the interface, along with a relation, similar to that shown in Eq. (6.5) on the normal derivatives. In conducting materials, charges can move towards the interfaces, building up surface charges. But here, we restrict ourselves to pure dielectric material, in which charges are fixed, so that, in most cases, σ_S will be taken as equal to zero.

[2]And also, at the time this book is written (February 2021), Wikipedia

6.2.3 Electric field in dielectric slabs

In microfluidic systems, electric fields are produced by electrodes integrated in the device. Obviously, in the exceedingly important case of water, in order to avoid electrochemical interactions, electrodes cannot touch the fluid. Then, an insulating layer must be placed between it and the electrodes, as sketched in Fig. 6.3. How does the electric field behave in such a situation? Is it screened by this layer? The question can be answered by using Eqs. (6.1), (6.2), (with $\rho_e = 0$), together with the boundary conditions (6.5) (with $\sigma_S = 0$).

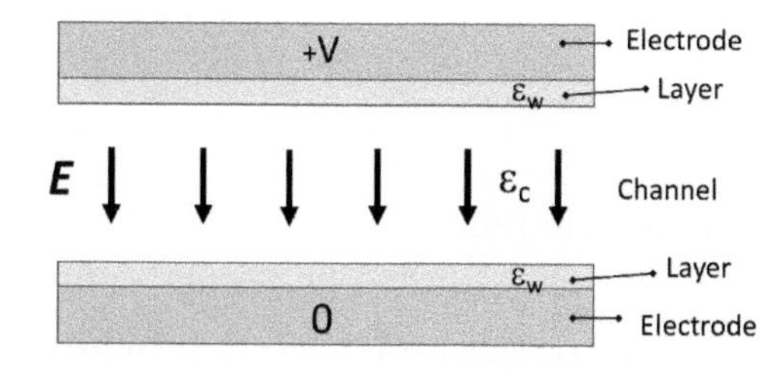

6.3: Two electrodes, held at a potential difference V, developing an electric field $\mathbf{E}$, across a channel of permittivity ϵ_c, separated from the electrodes by a layer of permittivity ϵ_w.

One finds that the electric field E, oriented normally to the electrodes, is constant in each domain, i.e. in the layers and the channel (see Eq. 6.1). Applying further the boundary conditions (6.5), one finds, for the channel region:

$$E_c = \frac{\epsilon_w V}{h_w \epsilon_w + h \epsilon_c} \tag{6.6}$$

The calculation shows that if the working fluid is more polarizable than the walls, for instance, water for the liquid and PDMS for the walls, the electric field is screened by them. It is somehow lost in the walls. In the case where the walls have comparable or larger permittivities, than the working fluid, there is no screening effect. In such cases, an intense electric field can be applied within the channel. To fix orders of magnitudes, with PDMS channels of 100 μm high, a layer of 10 μm and silicone oil as the working fluid, by applying tensions of 300 V across the pair of electrodes, electric fields of 27 kV/cm can be imposed inside the fluid. Note that this level approaches the electrical breakdown. In air, for instance, the electrical breakdown is 30 kV/cm; in silicone oil, it ranges, roughly, between 30 and 100 kV/cm. There is not much difficulty, in microfluidic systems, to work with high electric fields.

The microfluidic device shown in Fig. 6.4 provides an interesting example, in which Eq. 6.6 can be used.

Here, water droplets are driven in oil. An electric field is produced by two electrodes, integrated in the PDMS walls. Owing to the dielectric constants of the channel and the walls are comparable, one may estimate that droplets will see' an electric field on the order of V/d, where d is the distance between the two electrodes. Should the emulsion be direct (oil droplets in water), the application of a voltage would be inefficient. The

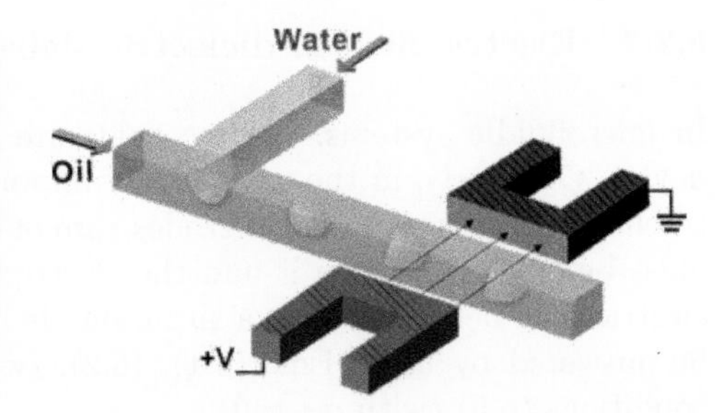

6.4: Microfluidic sorter of aqueous droplets, transported in an oil phase. The electric field inside the droplets is substantial and it is possible to exploit it to deflect their trajectories and sort them. Should the continuous phase be water instead of oil, the electric field inside the droplets would be small and it would be impossible to perturbe their trajectories.

electric field would develop in the continuous phase and would be extremely weak in the droplet.

6.2.4 Dielectric sphere in an uniform electric field

Here we consider the important problem of a dielectric sphere of radius R, permittivity ϵ_s immersed in a medium of permittivity ϵ_m, and subjected to a uniform electric field $\mathbf{E_0}$, applied along z axis. Here again, the medium is a pure dielectric material, and there is no permanent volumetric nor surface charge. Therefore, both ρ_e and σ_S are zero. Coordinates (r,θ) are shown in Fig. 6.5.

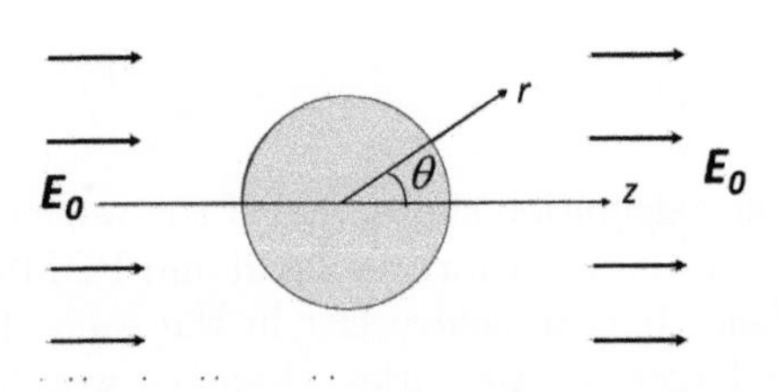

6.5: Sphere immersed in a uniform electric field, showing the spherical coordinate system, with radius r and polar angle θ as the variables (the system is independent of the azimuthal angle).

The problem can be exactly solved, using a spherical system of coordinates and noting that the solution is independent of the azimuthal coordinate.(see, for instance, Ref. [2]). Quantities depend only on r, the distance from the sphere centre, and θ, the polar angle. One finds, for the potential ϕ:

$$r < R : \phi = -E_0(1 + K_0)r \cos\theta. \tag{6.7}$$

$$r > R : \phi = -E_0(r + K_0 \frac{R^3}{r^2}) \cos\theta \tag{6.8}$$

in which:

$$K_0 = \frac{1 - \epsilon_r}{\epsilon_r + 2}$$

and $\epsilon_r = \frac{\epsilon_s}{\epsilon_m}$. K_0 is the equivalent of *minus* the Clausius–Mossoti factor for steady fields. It plays a crucial role in the phenomenon of dielectrophoresis, as will be seen later. Let us write the expressions, still in spherical coordinates, of the electric field:

$$r < R : E_r = E_0(1 + K_0)\cos\theta \text{ and } E_\theta = -E_0(1 + K_0)\sin\theta \tag{6.9}$$

$$r > R : E_r = E_0(1 - 2K_0\frac{R^3}{r^3})\cos\theta \text{ and } E_\theta = -E_0(1 + K_0\frac{R^3}{r^3})\sin\theta \tag{6.10}$$

Expression (6.9) includes two terms: the first term is the applied field, while the second is the field developed by the sphere. This field has a pure dipolar structure (see Eq. (6.4)), with a dipole moment **p**, located at $r = 0$, aligned along $\mathbf{E_0}$, and equal to:

$$\mathbf{p} = -4\pi\epsilon_1 K_0 R^3 \mathbf{E_0}$$

For the external medium, as far as the electric field is concerned, the dielectric sphere is equivalent to a dipole of moment **p**, located at the particle centre. Equation (6.9) tells us that, in the sphere, the electric field is constant, and oriented along x. Its value is $E_0(1 - K_0)$. Thus, for a highly dielectric sphere, the electric field inside the sphere is zero, while in the opposite case, it is $\frac{3}{2}E_0$, i.e. larger than the applied field. The field patterns are shown in Fig. 6.6, in two extreme cases: (a) a sphere much more polarizable ($\epsilon_r >> 1$) and (b) one much less polarizable than the medium ($\epsilon_r << 1$).

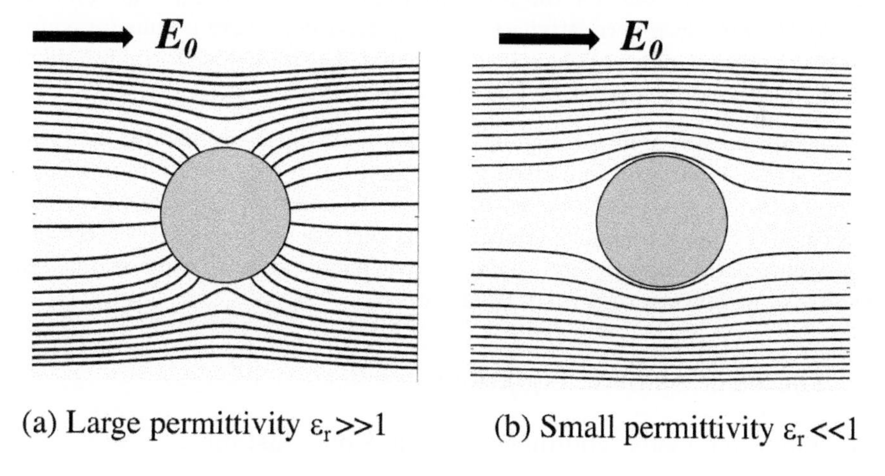

Fig. 6.6: A dielectric sphere immersed in a uniform electric field. The sphere deforms the electric field lines. (A) Sphere of large permittivity (compared to the medium).(B) Sphere of small permittivity.

In case (b), the field lines enter the sphere normally to its surface, as if they were attracted by it. In case (b), the field lines tend to be tangential to the sphere, as if they seek to avoid it.

Is it possible to understand why this is the case? Let us write again the boundary conditions across two dielectric media:

$$E_{tm} = E_{ts} \text{ and. } \epsilon_m E_{nm} - \epsilon_s E_{ns} = 0 \tag{6.11}$$

In the case of large permittivity (case (a)), the boundary conditions impose $E_{ns} \approx 0$, at the interface, inside the sphere. This relation suggests that $\mathbf{E} = \mathbf{0}$ inside the sphere, and then $E_{tm} = 0$ at the interface. This leads to electric field lines normal to the interface, as shown in Fig. 6.6 (A). In the case of a sphere of low dielectric constant (case (b)), Eq. (6.11) implies $E_{nm} \approx 0$, so that the electric field is, in the medium, tangential to the sphere. This explains the pattern shown in Fig. 6.6 (B).

6.2.5 The analogy between dielectric and conducting systems

Conducting materials are physically different from dielectric materials. In the former, ions or electrons are mobile, while in the latter, they are fixed. Nonetheless, there exists an analogy between conducting systems, in which the conductivity σ is constant (for example, homogeneous metals), and homogeneous dielectric systems: if we neglect charge diffusion, the relation $\mathbf{J} = \sigma \mathbf{E}$, in which $\mathbf{J}$ is the electric current density, holds. Electric and current lines are therefore undistinguishable. Restricting ourselves to the steady–state, the boundary conditions are, for the normal component of the electric field, $\sigma_1 E_{n1} = \sigma_2 E_{n2}$ and, for the tangential one, $E_{t1} = E_{t2}$. They are analogous to Eqs (6.5), if we replace permittivities by conductivities. There is thus an equivalence between the steady–states of electric and current density fields. Formally, solving the equations for the former case provides the solution for the latter.

This analogy applies well for system composed of different metallic materials, for instance, metallic particles immersed in mercury, but is not a very interesting system in practice. It does not apply to systems of metallic particles suspended in electrolytes. In this case, because ions, which transport the current, cannot penetrate the particles, charge accumulation occurs, and the analogy, which is about steady–states, ceases to be valid. We will return to this situation later.

6.3 The electrokinetic equations

6.3.1 Charge density, conductivity, diffusion constant,...

We concentrate on electrolytes. Before proceeding to the establishment of the governing equations, it is useful to define several quantities.

Charge density: the charge density, denoted by ρ_e is the electrostatic charge per unit of volume. It is related to the ionic mass concentration C by the relation $\rho_e = \frac{q}{m} C$, where q is the ion charge and m is its mass.

Ionic mobility: In the presence of an electric field $\mathbf{E}$, an ion migrates at speed $\mathbf{v}$, according to the law: $\mathbf{v} = \mu_e \mathbf{E}$, in which μ_e is the ionic mobility. This mobility can be estimated by using the Stokes-Einstein law. One finds:

$$\mu_e = \frac{q}{6\pi\mu a} = \frac{qD}{kT}$$

in which μ is the fluid viscosity, a is the ionic radius (or ionic complex radius) and D is the diffusion coefficient. a includes the presence of a hydration shell bound to it, and moves with it. For instance, for sodium ions, for which the hydration shell is formed with three water molecules, a is 0.25 nm, i.e. 2.5 times its ionic radius. For potassium ions, there is no hydration shell, and a is equal to the ionic radius (0.14 nm).

Ionic conductivity: The relation between the mobility μ_e and the conductivity σ reads:

$$\sigma = \rho_e \mu_e = \rho_e \frac{qD}{kT}$$

.The important fact here is that, unlike with metals where conductivities are constant, σ is proportional to the ionic density. We will see later that this dependence has considerable implications on the electrokinetics of the electrolyte.

6.3.2 The electrokinetic equations for a single ion species

Electrolytes include several species of ions, positive and negative, of different masses and charges. However, here, for simplicity, we assume that the ions contained in the solution have the same mass m, diffusion constant D, and charge q. This electrolyte could be called 'mono-ionic'. Later we will return to realistic cases, where several types of ions, with different characteristics, compose the solution. At this stage, introducing several species would make the equations opaque.

Within the electrolyte, electric currents develop, due to ion displacements, develop. Their origin, schematized in Fig. 6.13, is threefold:

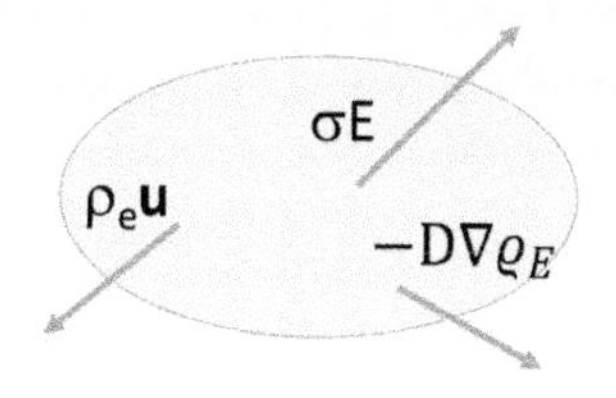

6.7: Three origins of the electric current in electrolytes. With the vector orientations shown on the figure, Eq. (6.13) tells us that the charge density ρ_e will decrease.

- Diffusion. This process is analogous to the ordinary diffusion process, and is governed by Fick's law, which we saw in Chapter 5. Let us recall the expression of this law:

$$\boldsymbol{J}_{\mathrm{m}} = -D\nabla C$$

where $\boldsymbol{J}_\mathrm{m}$ is the mass current (i.e. the mass crossing a surface element per unit time) and C is the mass concentration. Dividing the two sides of the equation by q/m, we obtain the relation:

$$\boldsymbol{J}_\mathrm{D} = -D\nabla\rho_\mathrm{e} \tag{6.12}$$

where $\boldsymbol{J}_\mathrm{D}$ is the current with a diffusive origin, and ρ_e is the charge density, i.e. the electric charge per unit volume. Indeed, in our model, ρ_e is everywhere positive or negative, depending on the sign of q.

- Transport by the flow $\mathbf{u}$. The current induced by this effect is written:

$$\boldsymbol{J}_\mathrm{T} = \rho_\mathrm{e}\boldsymbol{u},$$

where $\boldsymbol{u}$ is the fluid velocity. This term is the equivalent of the flux of matter uC in Chapter 5.

- Migration in the presence of an electric field $\mathbf{E}$. The corresponding flux is given by Ohm's law:

$$\boldsymbol{J}_\mathrm{e} = \sigma\boldsymbol{E}$$

where σ is the ionic conductivity.

Adding up these three contributions, we arrive at the equation:

$$\boldsymbol{J} = -D\nabla\rho_\mathrm{e} + \rho_\mathrm{e}\boldsymbol{u} + \sigma\boldsymbol{E}. \tag{6.13}$$

The relation is valid for electrolytes, where, as noted above, currents are conveyed by ions. In liquid metals (such as mercury), where currents are transported by electrons, the physics is different, and the first two terms are absent. In practice, as noted above, electrolytes include several types of ions, and several equations of this type should be written, one for each species. We will come to this point later.

The equation for the conservation of the electric charge is similar to the conservation of mass:

$$\frac{\partial\rho_\mathrm{e}}{\partial t} + \mathrm{div}\,\boldsymbol{J} = 0 \tag{6.14}$$

As in the preceding section, Maxwell equations constrain the electric field with two relations:

$$\text{div } \epsilon \boldsymbol{E} = \rho_{\text{e}} \text{ and } \textbf{curl E} = 0 \tag{6.15}$$

in which ϵ is the homogeneous dielectric permittivity of the medium. ϵ incorporates, in a single parameter, the polarization of the medium induced by the electric field.

Equations (6.13), (6.14), and (6.15) must be completed by the Navier-Stokes equation, to which Coulomb forces are added. Coulomb forces apply to the ions, but on the scale of the fluid particle, owing to fast momentum exchange, these ions drag the neutral elements. Consequently, Coulomb forces, weighted by the ionic density ρ_{e}, apply to the fluid element as a whole. With ϵ homogeneous, there are no other forces to consider. The flow equation thus read;

$$\rho \frac{D\boldsymbol{u}}{Dt} = -\nabla p + \mu \Delta \boldsymbol{u} + \rho_{\text{e}} \boldsymbol{E} \tag{6.16}$$

where $\boldsymbol{u}$ is the fluid velocity. In the context of microfluidics, as emphasized in Chapter 3, we can eliminate inertial terms, and restrict ourselves to the following form:

$$-\nabla p + \mu \Delta \boldsymbol{u} + \rho_{\text{e}} \boldsymbol{E} = 0 \tag{6.17}$$

Equations (6.13), (6.15), and (6.17) define the system of equations governing electrolyte flows. They must be completed by boundary conditions. Compared to the preceding section, where the media were not conducting and therefore charges did not move, here the situation is more complex. Electrical currents flow, covering interfaces with charges, which in turn affect the electric field distribution. The boundary conditions thus must include several relations, including speed, electric fields, currents, charges. We will not develop the subject here, but content ourselves, in the next sections, with considering elementary situations for which the boundary conditions take a simple form.

Because of the dependence of the electric conductivity on the ionic density, the equations are non–linear. This feature gives rise to complex phenomena, such as instabilities, chaotic behaviour, and turbulence. We will not discuss this subject, but restrict ourselves to the description of important cases, pertinent to microfluidics.

6.4 The electrical double layer

6.4.1 Origin of the surface charges

The appearance of charges at interfaces is a general phenomenon. Two major causes exist.

Ionization or dissociation of surface groups A glass immersed in an aqueous solution becomes negatively charged, because the silane terminals SIOH, located at the surface of the glass, are deprotonated, i.e. they loose their H^+. The electrical potential associated with these charges at a pH of 7 is on the order of -100 mV. The process is general: many chemical groups can dissociate in water, conferring the surface with a net charge. Examples are carboxylic and sulphate groups.

Ion adsorption As we saw in Chapter 3, ions can develop an affinity with the surface and get adsorbed, in a process similar to Langmuir adsorption, described in Chapter 5. An important example for microfluidics is hydroxyl group on hydrophobic surfaces. The phenomenon, unexplained until the year 2000, is due to the enhanced autolysis of water on these surfaces (see [5]).

6.4.2 Gouy Chapman and Stern models

General structure:. The study of electrokinetics close to charged surfaces has a long history. The first model was proposed by Helmholtz [6], in 1879. Immersed in an electrolyte, charged surfaces adsorb a fixed layer of counter-ions that neutralizes the charge. This model had drawbacks, in particular a capacitance independent of the ionic concentration, which contradicted the experiment. We will return to this point later. Also, physically, it was not clear why, in the presence of thermal agitation, all counter–ions would condensate or adsorb at the walls. This led Gouy (1910) [7] and Chapman (1913) [8] to propose another model where, unlike the previous one, all counter–ions, whatever their distance to the wall, are mobile. Later, a two-layer model, combining the two preceding ones, was proposed by Stern [9], in 1924. The first layer, which is fixed (see Fig. 6.8), is called 'Stern layer'. At its edge, there is a plane, called 'slip plane', associated to a potential ζ, beyond which ions are mobile (see the dashed line). This is the 'diffuse layer'. The model is called 'double layer' (DL or EDL for electrical double layer), because it is composed of two layers: one fixed and the other diffuse.

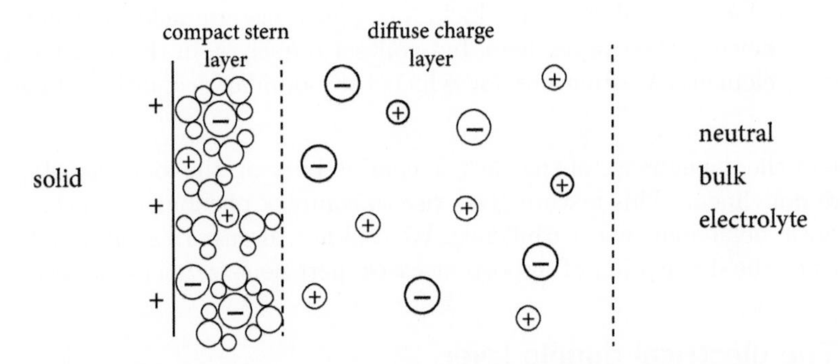

Fig. 6.8: Stern model: a fixed layer, composed of a featureless accumulation molecules, charged or not, and a diffuse layer, in which ions are mobile. There is a plane, called slip plane', associated to a potential ζ, whose location is not precisely defined, supposed to mark the separation between the two layers.

Structure of the Stern layer. For a long time, the structure of the Stern layer remained inaccessible to measurements. Theoretical considerations led researchers to think that its content is a dense mixture of molecules, ions, and counter-ions, as sketched in Fig. 6.8. Its thickness is on the order of the Bjerrum length, which represents the separation distance between two elementary charges storing an electrostatic cohesion energy equal to kT. We may suggest that the Stern layer, storing such an energy, will form a condensed phase. The expression of the Bjerrum length is:

$$\lambda_\mathrm{B} = \frac{e^2}{4\pi\varepsilon\, kT}$$

where e is the elementary charge. For water at room temperature, $\lambda_\mathrm{B} \approx 0.7$ nm.

In recent years, thanks to progress in spectroscopy, along with indications provided by numerical modeling (see [10], for instance), a number of information has become accessible. In the work of Ref. [11], performed with a glass surface exposed to an NaCl solution, the authors concluded that the chemical content of the Stern layer includes a first layer of hydrated sodium ions, linked to the surface, above which hydrating water molecules, forming either part of the first layer, or a second or third layer, in contact with the fluid. Fig. 6.9 shows measurements carried out on this system.

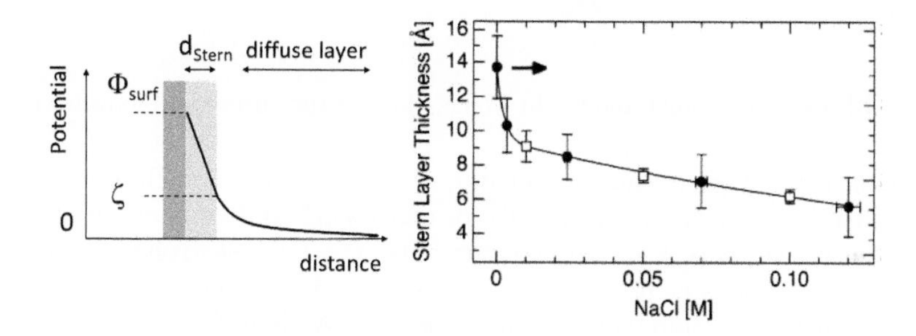

Fig. 6.9: A variety of measurements characterizing the Stern layer: (Left) Quantities characterizing the Stern layer. (Right) Evolution of the Stern layer thickness, in function of the NaCl concentration. (Reprinted from [11] with permission of Wiley and Sons. Copyright 2022.)

Different quantities could be measured: the surface potential, Φ_S, inferred from X–ray photoelectron spectroscopy, the charge surface, measured by Fourier transform Infrared (FTIR) spectroscopy, and the ζ potential, inferred from electrophoretic measurements. Φ_{surf} is on the order of -300 mV, i.e. one order of magnitude larger than ζ. There is a strong drop across the Stern layer, leading to electric fields on the order of 300 MV/m, well above the breakdown field.[3] From the knowledge of the electrical potentials, and that of the surface charge, the thickness of the Stern layer d_{Stern} can be estimated. It is represented in Fig. 6.9 (right). d_{Stern} was found on the order of a few Angstroms, in agreement with estimates based on Bjerrum length. There is a decrease of d_{stern}, with the NaCl concentration, consistent with the evolution of the ζ potential with this parameter. Much remains to be learned in this area.

[3]The breakdown field in common materials is below 10 MV/m.

6.4.3 The electric field in the diffuse layer

Let thus concentrate on the diffuse layer, and take z as the coordinate normal to the wall (see Fig. 6.10). In the presence of a surface charge, the mobile ions located close to the surface feel an attractive forces that drive them towards the wall, while thermal agitation works at suppressing concentration gradients. The analysis of this situation can be carried out by using statistical or thermodynamical arguments (as currently done in textbooks). Here we investigate the problem with the electrokinetic equations presented previously.

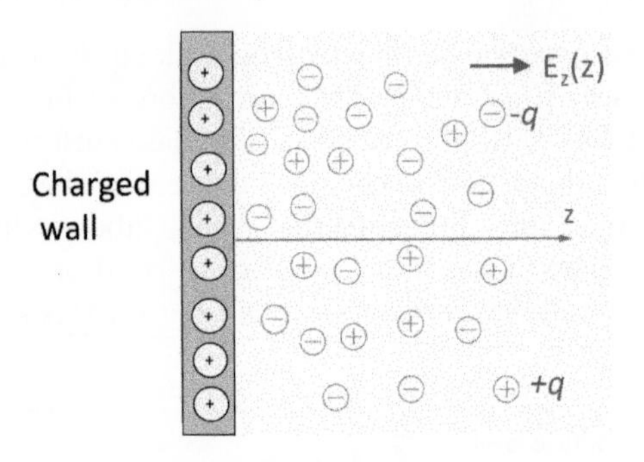

Fig. 6.10: Gouy-Chapman model, in which, close to the surface, all ions are mobile.

Let us assume, as Debye and Huckel did, that the solution includes two types of ions, one positive, of charge $+q$ ($q > 0$), of density ρ_e^+ and conductivity σ^+ and the other of charge $-q$, density ρ_e^- and conductivity σ^-. This electrolyte is called binary symmetric'. The flow is at rest and quantities depend on z only. This implies that $\mathbf{E}$ and $\mathbf{J}$ have one single component, oriented along z. We look for a stationary solution. Equation 6.14 imposes $J_z = Cst$. We assume here that the wall is insulating, implying the constant is zero. Based on this, Eq. 6.13 simplifies into:

$$0 = \sigma^+ E_z - D^+ \frac{\mathrm{d}\rho_e^+}{\mathrm{d}z} \text{ and } 0 = \sigma^- E_z - D^- \frac{\mathrm{d}\rho_e^-}{\mathrm{d}z}$$

Conductivity σ depends on charge densities, and we need to take this dependence into account to go further. This is done by using the expressions shown in the last section:

$$\sigma^+ = \rho_e^+ \frac{qD^+}{kT} \text{ and } \sigma^- = -\rho_e^- \frac{qD^-}{kT}$$

With this relation, and using the electrical potential ϕ defined by:

$$E_z = -\frac{d\phi}{dz}$$

the previous equation can be rewritten as follows:

$$D^{+}(-\rho_{e}^{+}\frac{q}{kT}\frac{d\phi}{dz} - \frac{\mathrm{d}\rho_{e}^{+}}{\mathrm{d}z}) = 0 \tag{6.18}$$

$$D^{-}(\rho_{e}^{-}\frac{q}{kT}\frac{d\phi}{dz} - \frac{\mathrm{d}\rho_{e}^{-}}{\mathrm{d}z}) = 0 \tag{6.19}$$

On integrating them, and taking into consideration that, far from the wall, $\rho_e^{\pm}$ tend to $\pm\rho_\infty$, one obtains the following relations between the ionic densities and the potential:

$$\rho_{e}^{+} = \rho_{\infty}\exp(-\frac{q\phi}{kT}) \tag{6.20}$$

$$\rho_{e}^{-} = -\rho_{\infty}\exp(\frac{q\phi}{kT}) \tag{6.21}$$

These expressions are the Boltzmann distributions of the ionic charges.

With this, using $\rho_e = \rho_e^{+} + \rho_e^{-}$, adding up Eqs. (6.20) and (6.21), one obtains:

$$\rho_e = -2\rho_\infty \sinh(\frac{q}{kT}\phi) \tag{6.22}$$

and using

$$\frac{d^2\phi}{dz^2} = -\frac{\rho_e}{\epsilon}$$

we get:

$$\frac{d^2\phi}{dz^2} = 2\frac{\rho_\infty}{\epsilon}\sinh(\frac{q}{kT}\phi)$$

This equation is called the Poisson-Boltzmann equation. Together with the boundary conditions $\phi = \zeta$ at $z = 0$ and $\phi = 0$ at $z = \infty$ (neutrality condition), it can be solved. The solution is:

$$\phi = \frac{4kT}{q}\tanh^{-1}(\tanh(q\zeta/4kT)exp(-z/\lambda_D)) \tag{6.23}$$

where λ_{D} is the Debye length, defined by the relation:

$$\lambda_D = \sqrt{\frac{\epsilon kT}{2\rho_\infty q}} \tag{6.24}$$

It is timely to introduce a parameter, called κ, equal to the inverse of the Debye length. Its definition is thus:

$$\kappa = 1/\lambda_D$$

This parameter is often used in the literature. It facilitates the writing of the equations. We will also use it.

A most useful approximation is the Debye approximation, which consists in assuming that $\frac{q\phi}{kT}$ is small. Physically, this condition means that the potential energy $q\phi$, which tends to concentrate the counter-ions close to the wall, is weaker than the thermal energy kT, which tends to disperse them. In this context, one may expect almost flat counter–ions density profiles and therefore small deviations of ρ_e from its bulk values. With $\frac{q\phi}{kT}$ small, Eq. (6.23) becomes:

$$\frac{d^2\phi}{dz^2} \approx \frac{2\rho_\infty q}{\epsilon kT}\phi \tag{6.25}$$

The solution is:

$$\phi(z) = \zeta e^{-z/\lambda_\mathrm{D}}, \tag{6.26}$$

where ζ and λ_D have the same meaning as before. Figure 6.11 (A) represents the evolution of the potential $\phi(z)$, obtained in the framework of the Debye approximation.

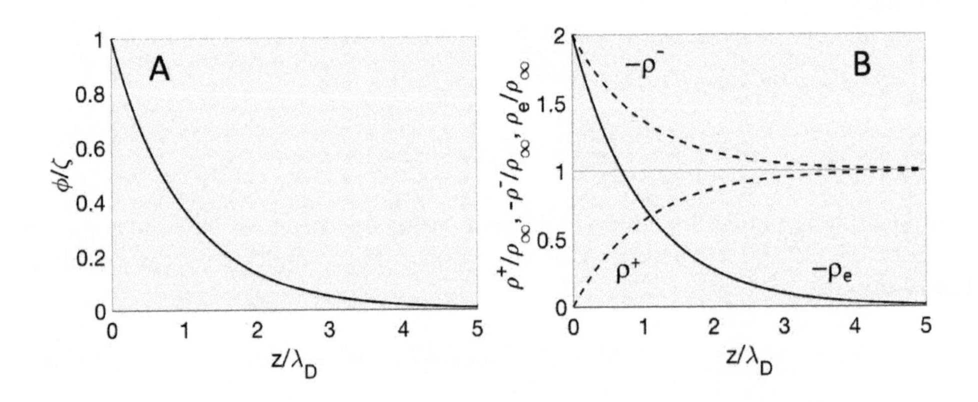

Fig. 6.11: (A) Electrical potential profile $\phi(z)/\zeta$ in the Debye layer, obtained in the framework of the Debye approximation.(B) Density charge profiles, $\rho_+, -\rho_-$, and $-\rho_e$, with the same approximation, with $\Theta = 1$ (see text).

The potential decays over a scale on the order of λ_D. Above 2 λ_D, the potential and the electric field strength are less than 10% of their values at the wall. This is called, in the jargon of the domain, 'Debye screening' λ_D is thus a crucial parameter. In the literature, it is often expressed in terms of massic rather than ionic concentrations, as we did. We can thus reformulate Eq. (6.24) in terms of C_∞, the concentration of ions in mole per unit of volume.

$$\lambda_D = \sqrt{\frac{\epsilon k T}{2 N_A C_\infty q^2}} \tag{6.27}$$

in which N_A is the Avogadro number. A practical formula, for monovalent ions (i.e. $q = e$, the electron charge) and binary electrolytes, is

$$\lambda_D \approx 0.304 nm/\sqrt{C_\infty}$$

in which C_∞ is the concentration in moles per litre, more commonly used in physico-chemistry. For instance, for an electrolyte of concentration of 1 mM/l, the Debye length is 9.6 nm.

Two remarks can be made:

- In practice, it is difficult to obtain Debye lengths much larger than several hundreds of nanometres. In the limiting case of pure water, C_∞ is 10^{-7} M/l; therefore, λ_D is approximately 0.96 μm. This is the highest Debye length that can be obtained in aqueous solutions. However, the unavoidable presence of dissolved ions delivered by walls in contact with the fluid raises the ionic concentration and consequently decreases the Debye length.
- It is not difficult, on the other hand, to obtain a Debye length comparable to the molecular scale. With a concentration of 1 M/l, the Debye length is 0.3 nm. So, what happens, physically? Surprising observations have been made recently [12]. As the Debye length decreases below the ionic size, a new regime occurs, where the screening length, i.e. the distance within which the electrolyte 'feels' the ζ potential, drastically increases with the concentration, reaching values orders of magnitude larger than λ_D. Much remains to be learned on the subject.

From Eqs. (6.26), (6.20), and (6.21), along with $\rho_e = \rho_e^+ + \rho_e^-$, one obtains the expressions of the charge densities:

$$\rho_+ = \rho_\infty(1 - \Theta \exp(-\frac{z}{\lambda_D})), \rho_- = -\rho_\infty(1 + \Theta \exp(-\frac{z}{\lambda_D}))$$

$$\rho_e(z) = -2\Theta\rho_\infty \exp(-z/\lambda_{\mathrm{D}})$$

in which Θ is defined by:

$$\Theta = \frac{q\zeta}{kT}$$

Θ is an important dimensionless number. It could be called dimensionless thermal voltage, but has no official name. $\Theta << 1$ defines the condition for which the Debye approximation can be made. The opposite case, $\Theta >> 1$, defines the domain of nonlinear electrokinetics, in which a rich variety of interesting phenomena exists.

The above equations confirm a statement made previously, i.e. Θ being small, the deviation of the positive and negative charges, from their bulk values $\pm\rho_\infty$, are small. Fig. 6.11 (B) represents the positive and negative charge density profiles, along with the total density charge $-\rho_e$, in the case Θ=1, i.e. in the case where the wall charge, and then the potential ζ, are positive. Even though Θ is slightly too large, in amplitude, for the Debye approximation to strictly apply, Fig. 6.11 (B) provides an acceptable representation of the structure of the charge density profiles. We see that the total charge ρ_e is negative. Its maximum value, at the wall, in this particular case, is $-2\rho_\infty$.

The present analysis was found to be in agreement with Israelachvili's experiments [13], described in Chapter 2, and, later, with molecular simulations. In 2008, the theoretical expression of ϕ could be compared more directly with the experiment (see Fig. 6.12) [14]. The difficulty was to resolve the Debye layer, i.e. scales in the range 5-100 nm. This was achieved by using total internal reflection fluorescence (TIRF) microscopy technique.

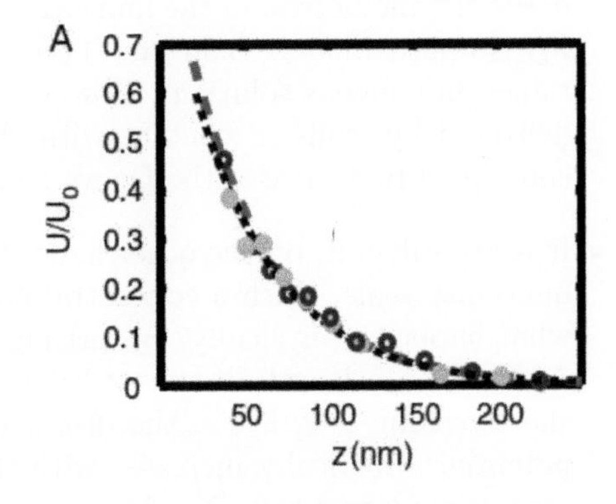

6.12: Potential profile $\phi(z)$ inferred from the measurement of distributions of nanoparticles close to a glass wall, in a microfluidic channel. The TIRF technique was used to resolve sub-100 nm scales and obtain profiles along z. The dashed line is Eq. (6.23) in which λ_D and the group $q\zeta/4kT$ are free parameters [14].

The potential was inferred from the measurement of distributions of nanoparticles suspended in the fluid, and located close to the wall. The particles, negatively charged, are repelled by the negatively charged wall, and their distributions, given by Boltzmann theory, depend on the potential $\phi(z)$. In this manner, the potential could be measured, providing an additional confirmation of the theory (see Fig. 6.12).

6.4.4 Divergence of the charge density

Let us calculate the ionic density ρ_e at the edge of the Stern layer, i.e. for z=0. Using Eq. (6.22), we obtain:

$$\rho_e(z) = -2\rho_\infty \sinh(\Theta) \tag{6.28}$$

At the wall, there is an amplification, by a factor $2\sinh\Theta$ of the ionic bulk concentration. This factor can reach several orders of magnitude. With $\zeta = -120$ mV, the amplification factor is equal to 54. Therefore, if we take an electrolyte with an ionic concentration of 1 M/l, the concentration on the slip plane would reach three ions per nanometre. With such a concentration, potassium ions would be close–packed, and sodium ions, with their hydration shell, would overlap. This example shows that, for a given concentration, the magnitude of Θ and thus ζ cannot be arbitrary large.

Let us make the argument more precise: for a monovalent electrolyte, the close-packing limit is given by:

$$\rho_e \sim \frac{e}{a^3}$$

in which a is an ion size. Using Eq. (6.28), the corresponding zeta-potential limit ζ_{max} is equal to:

$$\zeta_{max} \sim \frac{kT}{e} \ln \frac{e}{a^3 \rho_\infty}$$

kT/e is the thermal potential, equal to 25 mV. The multiplication by the logarithmic term does not change this order of magnitude. In practice, for a millimolar solution of monovalent salt, such as NaCl, the maximum zeta potential values is on the order of 100 mV. This argument provides an explanation why, in practice, the ζ potentials range typically, for non–metallic materials, between -100 and +100 mV.

6.4.5 Debye layers in microchannels

So far we have considered one isolated surface immersed in a liquid. The calculation can easily be extended to the case of two walls, forming, for instance, a shallow channel, again by using Debye approximation. One finds:

$$\phi(z) = \zeta \frac{\cosh(\kappa z)}{\cosh(\kappa h/2)} \tag{6.29}$$

$$\rho_e(z) = -2\rho_\infty \frac{q\zeta}{kT} \frac{\cosh(\kappa z)}{\cosh(\kappa h/2)} \tag{6.30}$$

in which h is the distance between the two walls. Several profiles of $\phi(z)$ are shown on Fig. 6.13, for different values of κh.

For $\kappa h = 20$, two thin Debye layers are established close to the walls $z = \pm h/2$. In the main part of the channel, the electric potential, along with the charge density ρ_E, are extremely small. In this case, close to each wall, the Debye profile given by Eq. (6.26) is recovered. As κh decreases, the Debye layers overlap. This situation is interesting, because, in such cases, the whole fluid acquires a charge density. The feature has been used to filter charged particles in function of their sign. In practice, overlap of the Debye layers occurs only in nanofluidic channels, with heights smaller than 100 nm.

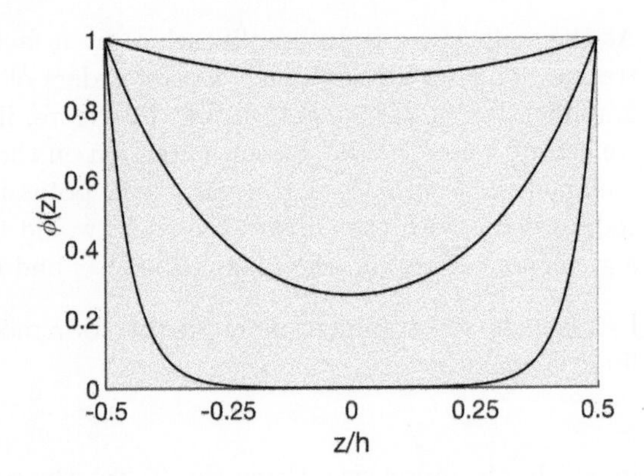

6.13: Potential profiles $\phi(z)$ for three values of κh: 1, 4, and 20.

6.4.6 The Debye layer around a sphere

The mechanism discussed for the case of charged surfaces also holds for charged spheres. The charge attracts counter-ions, which distribute around the sphere in a manner similar to the case of the plane. The situation is sketched in Fig. 6.14.

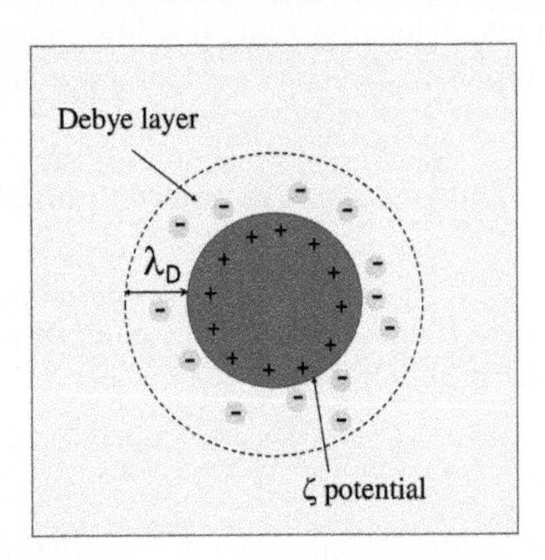

6.14: Sphere immersed in an electrolyte, with a potential ζ at its surface.

We suppose here that the surface of the sphere is held at a potential ζ. This potential gives rise, in the fluid, to a charge distribution, which is governed, in the Debye approximation, by the same equation as for plane surfaces (see Eq. (6.25)) but in spherical coordinates. We thus have:

$$\Delta\phi = \frac{1}{r^2}\frac{d}{dr}(r^2\frac{d\phi}{dr}) = \kappa^2\phi \tag{6.31}$$

This leads to the following expressions:

$$\phi(r) = \zeta \frac{R}{r} \exp(-\kappa(r - R)) \tag{6.32}$$

$$\rho_e(r) = -\epsilon \frac{1}{r^2} (\frac{\partial}{\partial r} r^2 \frac{\partial \phi}{\partial r}) = -\epsilon \kappa^2 \phi \tag{6.33}$$

ζ is linked to total amount of counter–ions $Q = \int_R^\infty 4\pi r^2 \rho_e dr$, generated, in the liquid, by the charged surface, through the relation:

$$\zeta = -\frac{Q}{4\pi\epsilon R(1 + \kappa R)} \tag{6.34}$$

in which, as said before, $\kappa = 1/\lambda_D$ the inverse of the Debye length. The particle thus creates, in the fluid, a charge distribution, whose net charge is Q. As the system is neutral, the charge of the particle should be $-Q$.

Two cases can be singled out:

- $\lambda_D >> R$ ($\kappa R << 1$): In this case the counter-ion density around the sphere is small, and the fluid around it is weakly charged. The field thus extends over substantial distances, repelling partners conveying charges of the same sign, thereby preventing aggregation.
- $\lambda_D << R$ ($\kappa R >> 1$): In this case, a thin charged layer builds up around the sphere, screening the electrical field developed by the surface charge. This situation can be created by adding salt in the solution (with salt, ρ_∞ is increased, and thereby λ_D is decreased (see Eq. (6.24)). In such conditions, particles can approach each other at distances so small that Van der Waals forces come into play, leading to aggregation.

These notions are critical for the stability of colloidal suspensions. Working with $\kappa R << 1$ allows the suspension to be stable, as sketched in Fig. 6.15. In the opposite case, the electric field developed by the particles is screened; clusters form, breaking the suspension (see Fig. 6.15).

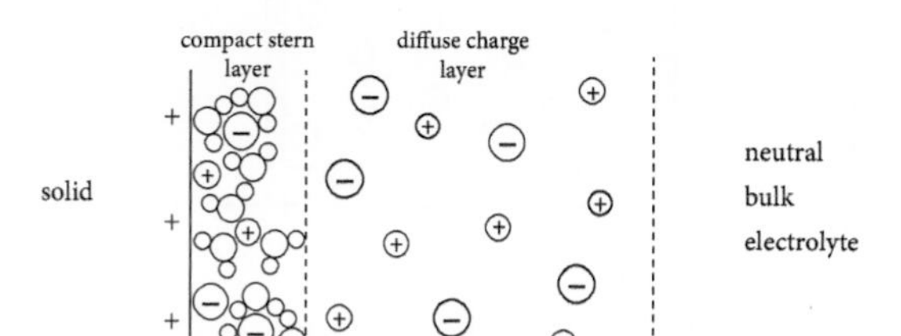

6.15: In the case $\kappa R << 1$, an electric field develops, preventing aggregation, thus ensuring colloidal suspension stability; in the other case ($\kappa R << 1$, the electric field is screened, and aggregation takes place.

Colloidal aggregation plays important roles in different domains, such as paper production, cheese fabrication, and water treatment. A spectacular example, well–connected to the situation that we discuss here, is the formation of deltas. The Mississippi delta, schematized in Fig. 6.16, is the largest delta of the world. Its formation is due to the sedimentation of fine particles, carried by the river's freshwater, which precipitate as they meet the sea water. The sedimentation rate is extremely large: 90 millions tons each year, leading to the continous creation new lands. Hope is raised using this huge quantity of mud to elevate the level of the existing lands, so as to resist the general rise of the sea level.

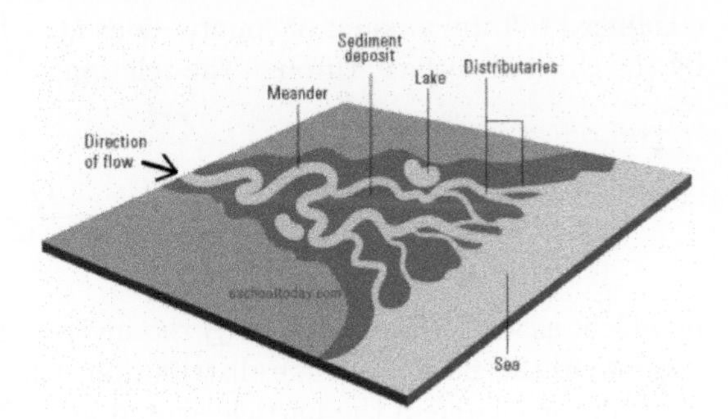

6.16: Schematic representation of the Mississipi delta.

6.4.7 The diffuse layer capacitor

So far, we have adopted a local point of view to describe the electrokinetics of electrolytes. Let us now consider a circuitry viewpoint. Fig. 6.17 shows an electrolyte confined between two blocking electrodes, i.e. two electrodes capable of imposing an electrical voltage, but unable to support a Faradaic current, owing to the fact that no chemical reactions develops at the interface, and no ion penetrates the electrodes. The situation was discussed in depth in [15]. As mentioned in this paper, the naive picture would be that in the device shown in Fig. 6.17, once the voltage is applied, a homogeneous electric field $E = V/L$ develops immediately across the electrolyte.

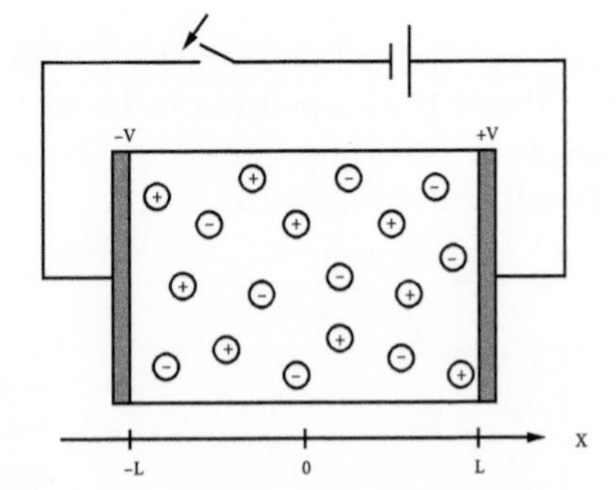

6.17: Circuit including an electrolyte between two blocking electrodes, i.e. that do not support a Faradaic current(inspired from [15]).

In fact, during the process, opposite charges move towards the electrodes and develop double layers at each side.[4] After some time, a steady–state is reached. The process is similar to a capacitor and the name given to it is differential capacitance C_D (taken

[4]We will see later that in such a geometry, electro-osmotic flows develop along the walls parallel to x. We will neglect these flows, or consider that the fluid boundaries, parallel to the x axis, are free surfaces (which is not particularly realistic)

per unit of the electrode exposed area). By differentiating Eq. (6.22), using the relation $d\rho_e \approx C_D \lambda_D d\phi$, along with the definition of λ_D (see Eq. (6.24)), this capacitance, first calculated by Chapman in 1913 [8], is found equal to:

$$C_D = \frac{\epsilon}{\lambda_D} \cosh(\frac{q}{kT}\zeta)$$

An electrolyte confined between two blocking electrodes is thus equivalent to a resistance per unit of electrode area (that of the bulk electrolyte), equal to $\frac{\lambda_D^2 L}{\epsilon D}$ [15], in series with a capacitor. A rigorous analysis, including historical notes and a review of the topics, with far-reaching insights, can be found in Ref. [15].

6.4.8 Dependence of the ζ potential with pH and ion concentration

We shall see later how the potential ζ is measured. The question we address here is how it varies with important parameters, such as the pH and the ionic concentration of the electrolytes. The question was reviewed in [16] and a compilation of data, extracted from the reference, is shown in Fig. 6.18). The data concerns naked silica surfaces and different electrolytes.

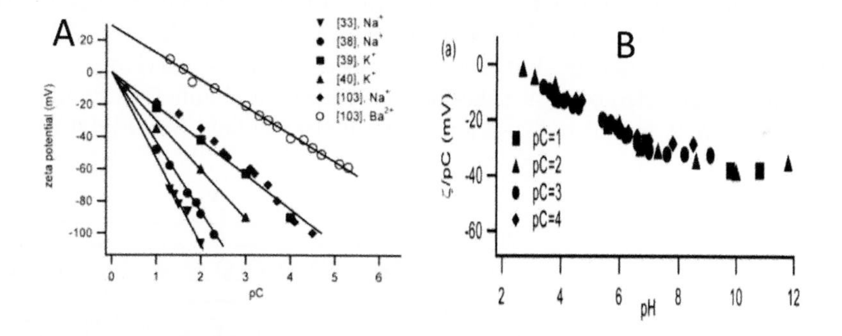

Fig. 6.18: (A) Series of ζ measurements, obtained on silica surfaces, with different electrolytes, as a function of $pC = -log_{10}(C)$, where C is the massic concentration of ions.(B) A compilation of ζ/pC as a function of the pH of the electrolyte. (Reprinted from [16] with permission of Wiley and Sons. Copyright 2022.)

Dependence with the ionic concentration: Fig. 6.18 (A) shows that ζ varies linearly with the logarithm of the bulk concentration C (i.e. the ionic concentration, expressed in M/l), noted by $pc = -\log_{10}(C)$. Why is it the case ? The argument, discussed in [16], is the following: at large Θ, the total charge, per unit of area, Q, stored in the Debye layer, is equal to:

$$Q = \int_0^{\infty} \rho_e(z)dz \approx 2\rho_\infty \lambda_D \exp(\frac{q\zeta}{2kT})$$

Because of the neutrality condition, Q is also the charge stored in the Stern layer. If we assume that the structure of the Stern layer and the charge it contains are independent of the bulk concentration of the electrolyte, Q will not depend on ρ_∞. This heuristic reasoning [16], which needs to be confronted with the knowledge gained recently on the Stern layer, leads to the linear dependence of ζ with $\log \rho_\infty$ (and thus pc), consistent with Fig. 6.18.

Dependence with the pH For the same series of ions as above, Fig. 6.18 (B) shows that the ζ potential decreases as the pH increases. Interestingly, the sign of ζ changes as the pH decreases below 3. This behaviour is inherent to the reaction of deprotonation of silanol groups on the silica surface. The point where $\zeta = 0$ is called isoelectric point. In general, the possibility to invert the sign of ζ potential by varying the electrolyte ionic concentration (for instance, of chaotropic salts) is interesting for capturing, and releasing molecular species such as nucleic acids [17].

6.5 Electro-osmosis

6.5.1 Electro-osmotic flows in channels

Historically, the term *electro-osmosis* was introduced by H. Dutrochet [19], in 1826, after the experiments of F. F. Reuss [20] and Porrett [21], to evoke an analogy between their observations and the phenomenon of osmosis, i.e. the spontaneous movement of solvent molecules through a semi-permeable membrane, induced by a concentration gradient (a historical account of electro-osmosis can be found in [37]). The analogy was abandoned by the author, but the word 'electro-osmosis' survived. The simplest geometry in which electro-osmosis takes place is a straight channel, with charged walls located at $z = \pm h/2$, submitted to a longitudinal electric field (see Fig. 6.19).

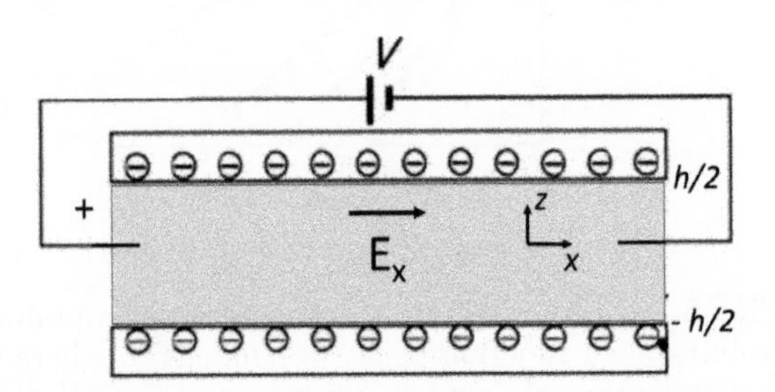

6.19: Geometry of electro-osmosis: a channel with charged walls, submitted to an electric field along x

The electric field is applied in the x direction, with component E_x. This electric field is superimposed on that induced by the walls. The corresponding electrical potential is $\phi_T = -E_x x + \phi(z)$, in which ϕ is the potential induced by the walls. In the previous section, we calculated the density of charges appearing in an electrolyte at rest, confined between two charged walls. When the electrolyte flows in a direction parallel to the walls, and an electric field is applied in the same direction, the calculations remain valid, because the electric field along x does not modify the charge distribution. In the framework of the Debye approximation, ϕ is thus given by the following (see Eq. 6.29):

$$\phi(z) = \zeta \frac{\cosh(\kappa z)}{\cosh(\kappa h/2)} \tag{6.35}$$

The movement of the electrolyte is governed by the following equation:

$$0 = -\nabla p + \mu \Delta \boldsymbol{u} + \rho_e \boldsymbol{E}, \tag{6.36}$$

where p is the pressure, $\boldsymbol{u}$ is the velocity of the fluid (assumed along x), μ is the fluid viscosity, ρ_e, as previously, is the electrical density of ionic charges, and $\boldsymbol{E}$ is the electric field. The projection of Eq. (6.36) along x reads:

$$0 = -\frac{\partial p}{\partial x} + \mu \frac{\partial^2 u}{\partial z^2} - \epsilon E_x \frac{\mathrm{d}^2 \phi}{\mathrm{d}z^2} \tag{6.37}$$

On the other hand, since there is no flow component along z, the projection along this direction reduces to:

$$0 = -\frac{\partial p}{\partial z} + \frac{1}{2} \epsilon \frac{dE_z^2}{dz} \tag{6.38}$$

These equations are associated to the no-slip conditions $u = 0$ at $z = \pm \frac{h}{2}$.

From these equations, one infers:

$$P(x, z) = -\frac{\Delta P}{L} x + \frac{1}{2} \epsilon E_z^2$$

in which L is the channel length and ΔP is the pressure along it. The expression shows that a pressure, $\frac{1}{2}\epsilon E_z^2$, builds up in the channel. This electrostatic pressure has no dynamical consequence. It just pressurizes the fluid, and exert forces on the walls, without producing any motion.

Where does this ΔP come from? A pump can deliver it. However, one can imagine to force $\Delta P = 0$, by imposing the same pressure at the inlet and the outlet of the channel (for instance, leaving them exposed to the ambient atmosphere). Still, there will be a flow, because, in order to balance the electrostatic term in Eq. (6.37), one needs viscous forces, and therefore a flow.

We thus consider two cases.

No applied pressure gradient (pure electro-osmotic flow). Here, we have $\Delta P = 0$. Eq. (6.37) can be integrated and, in the framework of Debye approximation, one obtains (assuming no slippage at the walls):

$$u(z) = \frac{\epsilon E_x}{\mu}(\phi - \zeta) = -\frac{\epsilon\zeta}{\mu} E_x \left(1 - \frac{\cosh(\kappa z)}{\cosh(\kappa h/2)}\right). \tag{6.39}$$

Two profiles, for two values of κh, are shown in Fig. 6.20.

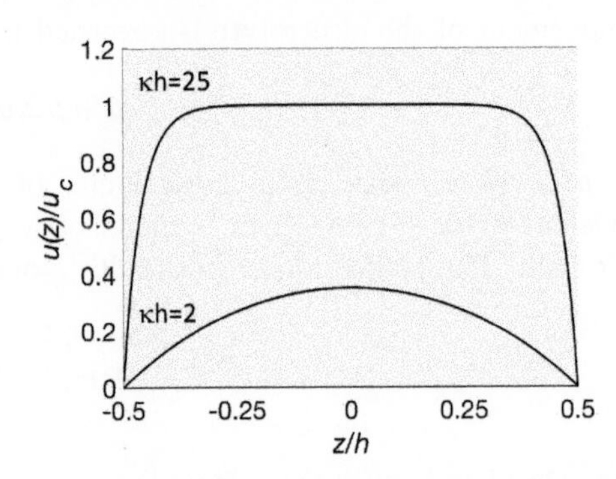

6.20: electro-osmotic profiles for two values of κh: 2 and 25.

At large κh, there is a plateau, which occupies most of the channel space. The plateau speed is:

$$u(z) = u_c = -\frac{\epsilon\zeta}{\mu} E_x. \tag{6.40}$$

This velocity u_c is called the Helmholtz-Smoluchowski velocity. Close to the wall, i.e. within the Debye layer, the velocity decreases to zero so as to match the no-slip boundary conditions. Physically, Debye layers, which are charged, are dragged by the electric field, and in turn mobilize the rest of the flow, in a manner similar to walls moving at speed u_c. Tthis flow exerts an mechanical stress on the walls equal to $\frac{\mu u_c}{\lambda_D}$.

Over the years, a considerable amount of experimental measurements have confirmed Debye Huckel theory. Recently, velocity profiles could be measured directly in the Debye layer in two cases: hydrophilic and hydrophobic walls (see Fig. 6.21) [14].

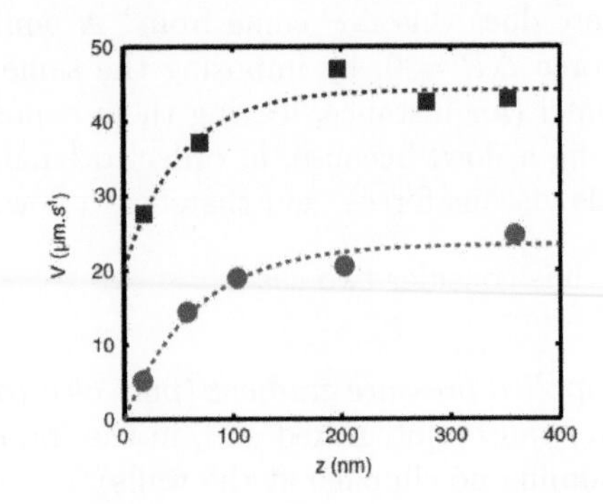

6.21: Electro-osmotic flows close to hydrophilic glass (disks) and hydrophobic OTS (octadecyltrichlorosilane) (squares). The imposed electric field is 500 V/m. Dashed lines are fits to the theoretical predictions. The velocity profile in the hydrophobic case reveals slippage amplification [14].

In the hydrophilic case, for which there is no slippage at the wall, good agreement was found with the theory (see disks and dashed line in Fig. 6.21).

In the literature, an important notion is the electro-osmotic mobility μ_{EOF}. It is defined by the expression:

$$u_c = \mu_{\mathrm{EOF}} E_x$$

Using Eq. (6.40), we obtain:

$$\mu_{\mathrm{EOF}} = -\frac{\epsilon\zeta}{\mu}$$

In terms of order of magnitude, with an electric field of 100 kV/m, and a ζ potential of 100 mV, the Helmholtz–Smolukowski speed is on the order of 1 cm/s, which is substantial. One remarkable consequence of Eq. (6.40) is that the speed is independent of the channel size, as long as the Debye layers are thin. Miniaturization does not cost anything, in terms of flow speed. This appeared, in the early times of microfluidics, with an excess of optimism, as a definitive solution to the problem of liquid pumping at the microscale. We will return to this question later.

Electro-osmotic flow with walls of different ζ potentials. It often happens that microchannel walls are made in different materials: for instance, a PDMS channel bonded to a glass plate. In such circumstances, there is no reason that the ζ potentials of the walls are the same, and thereby, the solution in Eq. (6.39) that we obtained is no more valid. So, what happens? Intuitively, we expect that with each Debye layer playing the role of moving walls, due to the speeds being different, a Couette flow profile will develop in the central part of the channel. The answer is given by resolving Eqs. 6.37. The solution reads:

$$\phi(z) = \zeta_1 \frac{\sinh \kappa(h/2 - z)}{\sinh \kappa h} + \zeta_2 \frac{\sinh \kappa(h/2 + z)}{\sinh \kappa h}$$

$$u(z) = -\frac{\epsilon}{\mu} E_x \left((\zeta_2 - \zeta_1)\frac{z}{h} + \frac{1}{2}(\zeta_1 + \zeta_2) - \zeta_1 \frac{\sinh \kappa(h/2 - z)}{\sinh \kappa h} - \zeta_2 \frac{\sinh \kappa(h/2 + z)}{\sinh \kappa h} \right)$$

in which ζ_1 and ζ_2 are, respectively, the wall potentials located at $z = -h/2$ and $z = h/2$. The solution $u(z)$, calculated with $-\frac{\epsilon}{\mu}E_x = 1\ V^{-1}$, is shown in Fig. 6.22, for $\kappa h = 25$, $\zeta_1 = 80$ mV and $\zeta_2 = 20$ mV.

As expected, outside the Debye layer, we have a Couette profile. Using asymmetric walls could be viewed as an elegant method for producing Couette flows in a microfluidic channel. On the other hand, as we saw in Chapter 5, Couette flows are associated

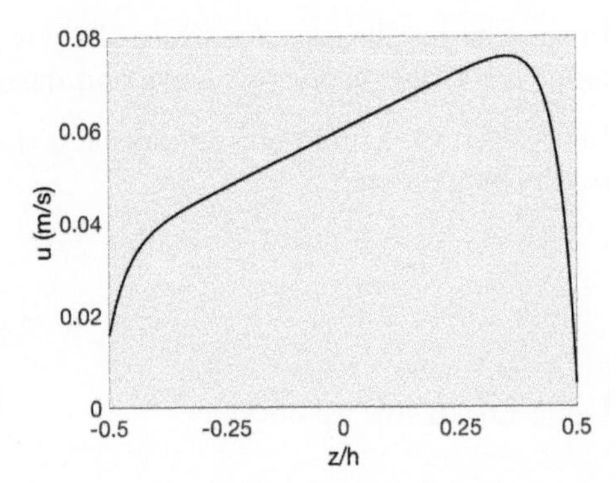

6.22: electro-osmotic profile calculated with $\kappa h = 25$, $-\frac{\epsilon}{\mu}E_x = 1\ V^{-1}$, $\zeta_1 = 80$ mV and $\zeta_2 = 20$ mV.

to large longitudinal dispersion, leading, as we saw in the same chapter, to dilution and loss of the sample transported in the microfluidic device. The plot of Fig. 6.22 can thus also be viewed as a regime to avoid, by equalizing, as much as possible the ζ potentials of the walls.

Pressure gradient combined with electro-osmotic flow. The mixed case corresponds to the situation where walls are identical, and a constant pressure gradient $G = -\mathrm{d}p/\mathrm{d}x$ is applied, or induced (for instance, when the flow develops in a closed cavity). In such conditions, the expression of the flow profile reads :

$$u(z) = \frac{\Delta P h^2}{8\mu L}\left(1 - \left(\frac{2z}{h}\right)^2\right) - \frac{\epsilon\zeta}{\mu}E_x\left(1 - \frac{\cosh(\kappa z)}{\cosh(\kappa h/2)}\right) \tag{6.41}$$

In the central part of the channel, the flow is no longer a plug flow, but a superposition of a parabolic Poiseuille profile and a plateau. The total flow rate is given by the expression:

$$Q = \frac{wh^3}{12\mu L}\Delta P - \frac{\epsilon\zeta wh}{\mu}E_x\left(1 - \frac{2}{\kappa h}\tanh(\kappa h/2)\right)$$

in which ΔP is the applied pressure, L is the channel length, and w is the channel width. Q can be positive or negative. The profile is sketched on Fig. 6.23, in the case where the liquid is confined in a cavity, so that Q=0.

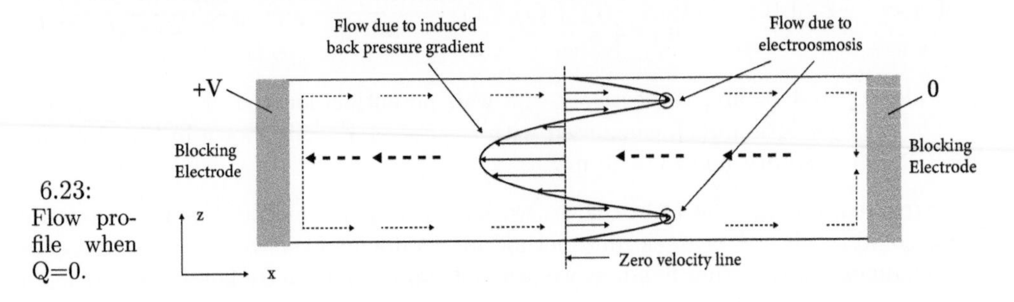

6.23: Flow profile when Q=0.

The figure shows that a flow is created, and recirculates within the cavity. Wrong reasoning would be: by closing the channel, the liquid stays at rest. This would not be true. In fluid dynamics, an external force **f** can be balanced by a pressure gradient, and thus maintain the fluid at rest, only if **curl f** $= 0$. Otherwise, no static equilibrium is possible, and a flow develops. In the present case, the electrostatic force **f** has a x component $f_x = \rho_e(z)E_x$, where $E_x = \frac{\Delta V}{L}$ is constant. The other component, f_z, being equilibrated by a pressure gradient, can be discarded. The curl of **f** has only one component, normal to the plane, equal to $-E_x \frac{\partial \rho_e}{\partial z}$. As this component is non–zero, a flow must develop.

6.5.2 The similarity principle

In an insulating channel, as we saw, the speed **u** is given, at the edge of the Debye layer, by the Helmholtz-Smoluchowski relation (6.40). It is thus proportional to the applied field **E**. This relation, obtained for a straight channel can be generalized to any curved bodies as long as their characteristic scale is much larger than λ_D. This is because, in such a case, the liquid/solid interface looks flat; therefore, we can apply the results obtained for a straight channel. For instance, in Fig. 6.24 this condition would hold if the obstacles developed gently curved shapes, with curvatures much less than κ.

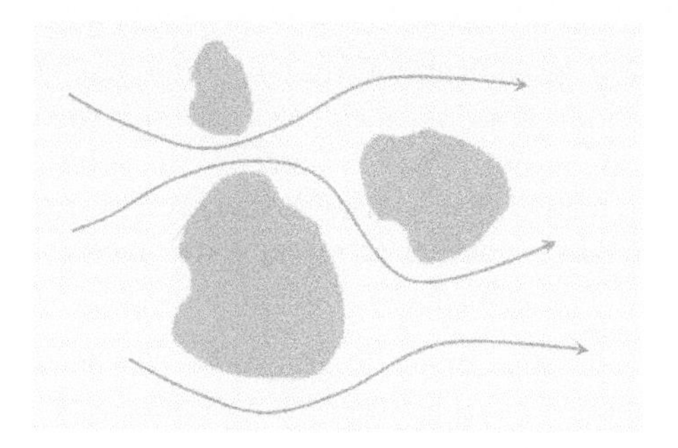

6.24: When Debye layers are small compared to the characteristic size of the objects, and in the absence of pressure gradient, the electric field lines and the flow streamlines are undistinguishable.

We thus have, at the edge of the Debye layer, the following relation:

$$\mathbf{u}(\mathbf{x}) = -\frac{\epsilon\zeta}{\mu}\mathbf{E}(\mathbf{x}). \tag{6.42}$$

The similarity principle consists in considering that, in the absence of imposed pressure gradient, the velocity field given by Eq. (6.42) represents, for infinite spaces, a solution to the problem. This is not just at the edge of the Debye layer, but in the entire outer region, i.e. outside the Debye layers. This statement is justified by two arguments:

- Outside the Debye layers, $\rho_e = 0$, and therefore, the velocity field given by Eq. (6.42), associated to a uniform pressure field (p constant), satisfies the Stokes equations (see Eq. (6.17)).[5]

[5]Because, in a neutral medium, $\boldsymbol{\Delta}\mathbf{E} = -\mathbf{curl}(\mathbf{curl}\mathbf{E}) + \mathbf{grad}\,div(\mathbf{E}) = 0$.

- Inside the Debye layers, the electro-osmotic velocity decreases exponentially to zero. Therefore, the velocity field given by Eq. (6.42), associated to the Debye layers, satisfies the no–slip boundary conditions at the liquid–solid interfaces.

The solution we constructed is thus proportional to the electric field. The flow we found is 'similar' to the electric field, in the same manner as in mathematics, two triangles are similar. However surprising it may look, the flow is potential, as it would be at infinite Reynolds numbers. We met the same paradox, in Chapter 3, in the section dedicated to Hele–Shaw cells. We may infer that, in microfluidics, this type of paradox does not appear infrequently.

The similarity solution develops a uniform flow at infinity. This looks surprising. Hold firmly an insulating object in a tank, apply an electric field and suddenly, the whole fluid is put into motion. This kind of situation is not specific to electro-osmosis. It also occurs in hydrodynamics. Move an infinite plate in a fluid initially at rest. After some time (neglected in the Stokes equation, due to the instantaneity principle, seen in Chapter 3), the fluid, in the entire space, is put into motion. Do the same in a cylindrical geometry. Rotate a cylinder in a fluid, again in an infinite space. Even far from the cylinder, fluid molecules feel a stress induced by the cylinder rotation. In response, they move at the same rotation rate. Eventually, the entire fluid will rotate as a solid object. Thus we are led to the statement that a long spaghetti, plunged in the Atlantic, and rotating at a constant rate, puts the whole ocean into rigid rotation.

These statements are correct, as long as there is no wall. A rotating hair has the capacity to put an ocean in rotation, but, for similar reasons, a wall can entirely stop it. In our case, if we insert a wall in the system, the uniform flow is blocked, the solution can no longer be represented by (6.42) and, consequently, the similarity principle breaks down. In these circumstances, the electro-osmotic flow remains confined in a region of limited extent, around the objects.

In applying the similarity principle, the question that must be addressed is whether the similarity solution is physically admissible, i.e. compatible with the geometry of the system. On might think that the similarity solution is mathematically elegant, but physically unrealistic, because there are always walls somewhere. In fact, this criticism can be circumvented. We will see later that the similarity principle does yield a physically admissible solution, that we will exploit to determine electrophoretic mobilities.

6.5.3 Slippage amplification

A surprising phenomenon exists when the electrolyte slips at the wall. Let us solve again Eq. (6.37), in the framework of the Debye approximation, but now with slippage conditions at the wall, i.e:

$$u(\pm\frac{h}{2}) = \mp b\frac{\partial u}{\partial z}$$

.

where b is the slip length (see Chap 2). We obtain:

$$u(z) = -\frac{\epsilon\zeta}{\mu}E_x\left(1 + b\kappa\tanh(\kappa h/2) - \frac{\cosh(\kappa z)}{\cosh(\kappa h/2)}\right) \tag{6.43}$$

An important feature of Eq. (6.43) is that, for $\kappa h >> 1$, the slip length enters the profile through the dimensionless product $b\kappa$, i.e. b/λ_D. The effect of slippage is thus *not* on the order of b/h, as in Chapter 3, but on the order b/λ_D, which is much larger. The Debye length plays somehow the role of an effective channel height. There is thus an amplification of the manifestation of slippage on the flow, as compared to ordinary' hydrodynamics. We can return to the velocity profiles shown precedently (see Fig. 6.21), and discuss the two cases: hydrophilic (lower curve) and hydrophobic (upper curve) walls. One sees a neat effect of the slippage: it accelerates the plug speed by a factor of 2. By comparing the experimental profiles to the theory, b was found on the order of 40 nm, consistently with the slip length measurements shown in Chapter 3.

6.5.4 Three microfluidic applications or implications of electro-osmosis

Transport of solutes in microchannels. In the case of pure electro-osmotic flows, whose velocity profile is given by Eq. (6.39), solutes are transported along the channel, as in a plug flow. The deviation, due to the presence of Debye layers close to the walls, is on the order of $(\frac{u_c\lambda_D}{D_s})^2$, in which D_s is the solute diffusion coefficient. In microchannels, this is negligible. These considerations have been confirmed experimentally, as shown in Fig. 6.25 [22, 23].

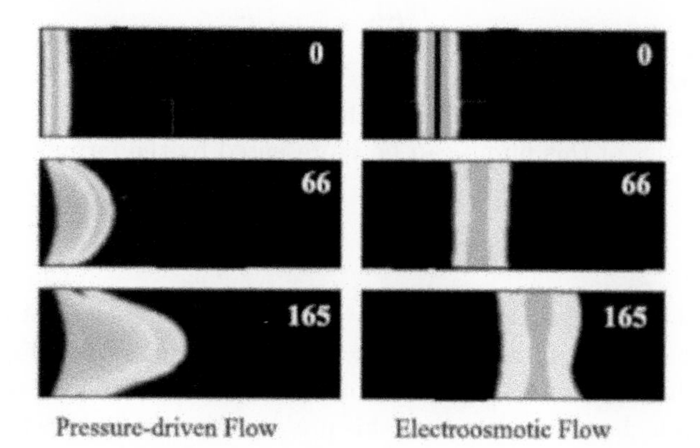

Fig. 6.25: Fluorescence imaging of a fluorescence spot, injected in a microchannel, and tracked at different times (given in seconds), for two cases. (Left) Poiseuille flow (viewed region 100 by 200 μm). (Right) Electro-osmotic flow in a glass capillary (viewed region 75 by 188 μm). (Reprinted from Ref [23] with permission from American Chemical Society. Copyright 2022.)

The figure shows the striking difference between Poiseuille and electro-osmotic flow. Being able to transport solutes with a minimal dispersion is key to the success of electrokinetic separation techniques, as will be seen later.

Electro-osmotic pump. The calculations developed in Section 5.5.1 naturally lead to the notion of electro-osmotic pumps. A general diagram is shown in Fig. 6.26 (A).

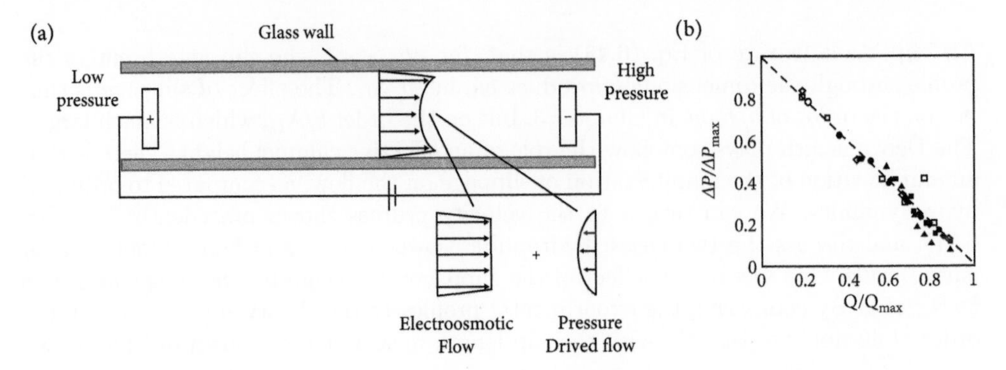

Fig. 6.26: (A) Principle of the electro-osmotic pump. (B) Universal characteristics of the electro-osmotic pumps (Reprinted from [24] with permission from Wiley and Sons. Copyright 2022.)

Two electrodes impose an electric field along a channel $E_x = \Delta V/L$, where ΔV is the voltage drop and L the channel length, in the presence of a pressure gradient $\Delta P/L$. The flow is the combination of a planar Poiseuille and an electro-osmotic flow, as sketched in Fig. 6.26(A). Let us take Eq. (6.41) in the limit $\kappa h >> 1$:

$$Q \approx \frac{wh^3}{12\mu L}\Delta P - \frac{\epsilon\zeta wh}{\mu}E_x \tag{6.44}$$

The pressure gradient results from the existence of a hydrodynamic resistance to which the electrokinetic channel is connected, through which the flow rate is Q established. The characteristics of this pump can be represented by the formula:

$$\frac{Q}{Q_{max}} = 1 - \frac{\Delta P}{\Delta P_{max}}$$

in which $Q_{max} = -\frac{\epsilon\zeta wh}{\mu}E_x$ and $\Delta P_{max} = \frac{12L\epsilon\zeta\Delta V}{h^2}$, independent of the viscosity and the channel length. The formula agrees with the experiment (see Fig. 6.26 (B)) [24]. In practice, typical maximum pressures, obtained with such pumps are on the order of 0.3 bar for 10 μm channels. Note that this pressure rises to 30 bars for 1 μm channels.

6.5.5 Streaming potential

Streaming potential is the opposite of the electro-osmotic pump: instead of imposing an electric field to obtain a flow rate, we impose a flow rate to produce an electric field. The principle is sketched in Fig. 6.27

The electric potential produced in this system can be inferred from Eq. (6.5.5).

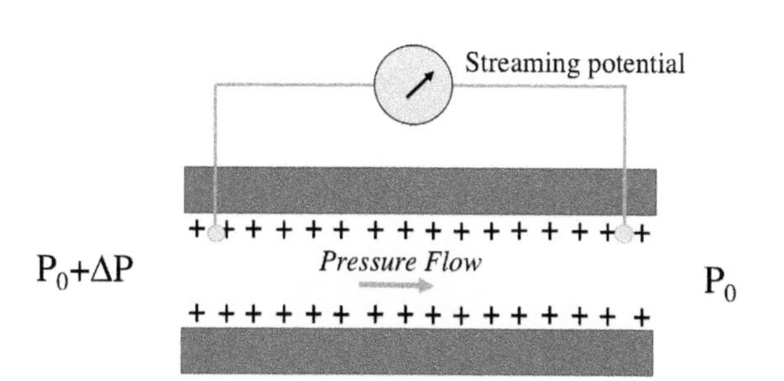

6.27: A pressure difference ΔP generates a flow. This flow, by entraining the positive charges of the Debye layer, gives rise to an electrical current and induces a 'streaming potential'.

$$\frac{\epsilon\zeta wh}{\mu}E_x \approx Q + \frac{wh^3}{12\mu L}\Delta P$$

The flow entrains the charges of the Debye layer, giving rise to an electrical current, called streaming current'. This mechanism has been proposed to produce electricity across membranes made of aligned carbon nanotubes (CNTs) [25, 26]. As we saw in Chapter 3, these membranes exhibit large slippage velocities and thus reduce the hydrodynamic losses. A configuration for producing energy is a semi-permeable membrane separating brine from fresh water, giving rise to an osmotic flow inside the CNTs, which dilute the salt. This flow transports charges and produces a current, which, after collection at electrodes, converts osmotic into electrical energy, with interesting performances [26].

6.6 Electrophoresis

6.6.1 The physics of electrophoresis

Electrophoresis is defined as the phenomenon for which ions, particles, bacteria, macromolecules, or, more generally, *charged* objects, distributed in a solution, move in the presence of an imposed electric field. The phenomenon is characterized by an electrophoretic mobility, defined by the relation

$$\mathbf{U} = \mu_e \mathbf{E}_\infty$$

in which $\mathbf{U}$ is the particle speed and $\mathbf{E}_\infty$ the applied field, i.e. its value far from the particle. The question is: what is the expression of μ_e ? To answer this question, different elements must be taken into account (see Fig. 6.28).

The particle is surrounded by a counter ionic cloud, i.e. a 'ionosphere'. The applied field $\mathbf{E}_\infty$ is deformed by the presence of the particle, giving rise to a spatially dependent field $\mathbf{E}$, This field exerts a force on the ionosphere. A pressure field develops, exerting, in turn, a force on the particle. In the meantime, the 'ionosphere' is put into motion. This flow drags the particle. Figure 6.29 shows an example of a velocity and a pressure field, calculated for a particle as conducting as the fluid, for $\kappa R = 1$. The calculation is based on Henry's solution [29].

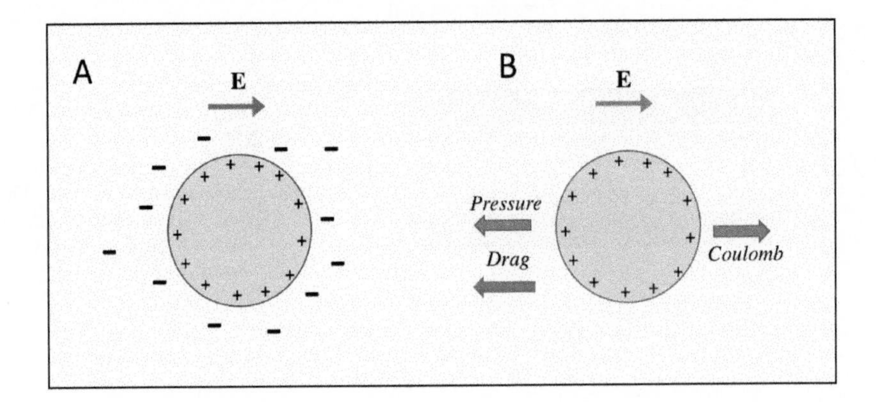

Fig. 6.28: Physics of electrophoresis. (A) The particle, associated to a potential $\zeta > 0$, develops a cloud of counter-ions around it, forming an 'ionosphere'. The system is subjected to an electric field **E**. (B) **E** exerts a force on the cloud. These forces build up a pressure in the fluid and drive a flow, which, in turn, exert a drag on the particle, and a Coulomb force onto the fixed charges of the particle (surface and dipolar)

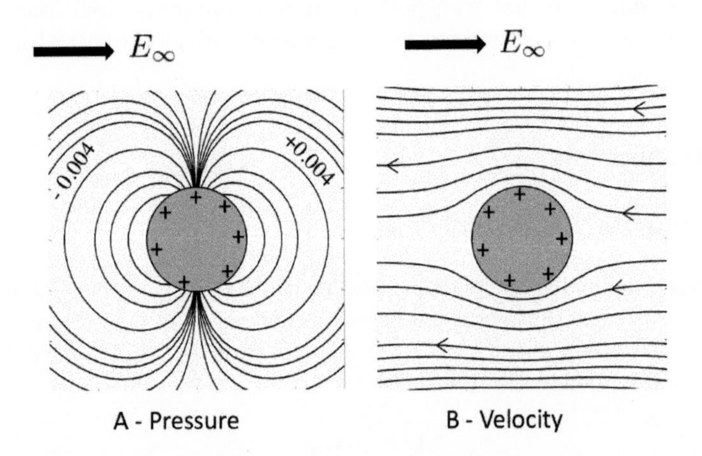

Fig. 6.29: (A) Pressure field developing around a particle, characterized by a positive ζ potential and a conductivity equal to the fluid. Here, $\kappa = 1$. The calculation is based on Henry's solution [29]; the numbers shown on the figure correspond to pressure levels, in dimensionless (arbitrary) units.(B) Streamlines obtained in the same conditions.

The pressure field develops lobes, with negative levels on the left side of the sphere and positive levels on the right (see Fig.6.29(A)). There is therefore a force pushing the particle leftwards. The electrolyte flow works in the same direction: it is directed to the left of the figure, dragging the particle leftwards (see Fig. 6.29(B)). At large distances, this flow decays as the inverse of r. It is thus confined in a region surrounding the sphere. The particle has a fixed charge. It will therefore be subjected to a Coulomb force, pointing to the right of the figure. In principle, in order to solve the problem, one should determine pressure and velocity fields, and, by integrating stresses over the particle surface, determine the forces exerted on it, and thereby, after adding the force due to the fixed particle charge, establish the electrophoretic mobility.

This calculation is not straightforward. In principle, it should take into account that, due to the applied electric field, the ionosphere adopts an elliptic shape, because negative charges, in the bulk, are pulled by the electric field [30,31]. In practice, the effect is significant at large ζ. For small ζ (for instance, on the order of 10 - 20 mV), the cloud deformation can be neglected.

By assuming a symmetric ionosphere, Henry performed, in 1931, the calculation of the electrophoretic mobility, for arbitrary conductivities and arbitrary κ [29]. The calculation takes eight pages. Henry's work solved a controversy between Smoluchowski [27] and Huckel [28], who proposed different formulas for μ_e. Later, Russel, Saville and Schowalter [2] obtained the same result in a simpler manner. We will start by considering two limiting cases:

- (a) $\lambda_D >> R$ (or equivalently $\kappa R << 1$). The Debye layer much thicker than the sphere radius
- (a) $\lambda_D << R$ (or equivalently $\kappa R >> 1$). The Debye layer much thinner than the sphere radius.

We shall moreover focus on the case of insulating spheres. The reason for this is that in electrolytes, ions cannot penetrate the sphere. Therefore, at the sphere interface, the normal component of the ionic current, and therefore that of the electric field, must be zero, as for insulators. Thus, as strange as it looks, the conducting sphere behaves as an insulator. In fact, things are more complicated. When the sphere is conducting, ions, driven by field components pointing normally to the surface, accumulate. This gives rise to transient regimes, where the sphere starts to behave as a conductor, and eventually behaves as an insulator. This will be described later. For insulating particles, there is no component of the applied electric field normal to the surface and therefore, no accumulation.

6.6.2 Case (a): Debye length much larger than the particle ($\kappa R << 1$)

Let us recall the general expression of the potential ϕ and charge density ρ_e around a sphere, induced by a surface charge on the particle, characterized by a potential ζ:

$$\phi(r) = \zeta \frac{R}{r} \exp(-\kappa(r - R)) \tag{6.45}$$

$$\rho_e(r) = -\epsilon \frac{1}{r^2} \frac{\partial}{\partial r} r^2 \frac{\partial \phi}{\partial r} = -\epsilon \kappa^2 \phi \tag{6.46}$$

When κR is small, the charge density $\rho_e(r)$ is small; consequently, the flow and the pressure, generated by the coupling between the applied field and the charge density are small. When κ tends to zero, $\rho_e(r)$ tends to zero. In limit, the electrostatic pressure and viscous stresses become negligible. We are thus left with the Coulomb force $Q\mathbf{E}_\infty$, for which we recall that Q is the fixed particle charge and $\mathbf{E}_\infty$ the applied field, far from the particle. From this result, and on using Stokes relation, the mobility can be determined. This leads to the following formula:

$$\mu_e = \frac{Q}{6\pi R\mu}. \tag{6.47}$$

The expression is called the Huckel formula [28]. Expressed differently, if we use the potential ζ given by the Gauss theorem ($\zeta = \frac{Q}{4\pi\epsilon R}$ when κR tends to zero), we obtain:

$$\mu_e = \frac{2\epsilon\zeta}{3\mu} \tag{6.48}$$

Eqs. (6.55) and (6.48) are valid for particles of arbitrary shapes, as the reasoning suggests.

6.6.3 Case (b): Debye length much smaller than the particle $\kappa R >> 1$

As the Debye length decreases (then κR increases), as in the straight channel, the highest flow gradients tend to concentrate in a thin Debye layer. In fact, at large κR, and in the case of insulating particles, the mobility can be elegantly determined by applying the similarity principle. The principle tells us that outside the Debye layer, in an infinite medium in which no pressure gradient is applied, the flow field is given by the expression:

$$\mathbf{u} = -\frac{\epsilon\zeta}{\mu}\mathbf{E}$$

in which, again, $\mathbf{E}$ is the applied field. This leads to the following formulas:

$$u_r = -\frac{\epsilon\zeta}{\mu}E_\infty(1 - \frac{R^3}{r^3})\cos\theta \text{ and } u_\theta = \frac{\epsilon\zeta}{\mu}E_\infty(1 + \frac{R^3}{2r^3})\sin\theta.$$

This flow is sketched in Fig. 6.30, again in the case $\zeta > 0$.

The flow is potential. It is associated to a constant pressure field, i.e. an absence of pressure gradient. The flow is oriented leftwards, because Smoluchowski speed, at the edge of the Debye layer, points in that direction. A large distance from the sphere, the flow is uniform. Thus, what we have here is a particle at rest, immersed in a uniform flow, oriented from the right to the left. This solution can also be viewed as a particle moving at speed $\mathbf{U} = +\frac{\epsilon\zeta}{\mu}\mathbf{E}$ in a motionless fluid. An example is a particle

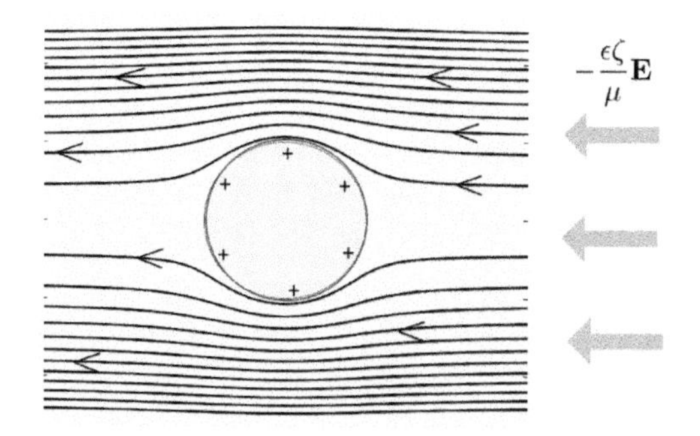

6.30: Flow around the sphere, obtained by applying the similarity principle. The flow is potential, and is associated to a uniform pressure field. This solution corresponds to a particle travelling at speed $\mathbf{U} = +\frac{\epsilon\zeta}{\mu}\mathbf{E}$ in a quiescent fluid.

moving in a large container, in which the fluid is quiescent. The fact that, viewed in this manner, the fluid is at rest at large distance from the sphere warrants that the boundary conditions at infinity are satisfied and, consequently that the similarity solution is physically admissible. Note that the similarity flow, which is potential, is the superimposition of a non–potential Stokes flow generated by the displacement of the particle, and another non–potential flow, produced by the movement of the ionospheric charges, discussed in a previous section.

We thus have succeeded in working out, without calculation, *the* solution to the problem. These considerations allow us to calculate the electrophoretic mobility. It is given by the formula:

$$\mu_e = \frac{\epsilon\zeta}{\mu} \tag{6.49}$$

Eq. (6.49), obtained by Smoluchowski [27], is called 'Smoluchowski approximation' or 'Smoluchowski formula'.

It is possible to examine the problem from the viewpoint of the forces, as Henry did. The exercise is much more complicated than applying the similarity principle. We may content ourselves to check that forces acting on the sphere are equilibrated. With the flow described above, the pressure forces are cancelled, and the remaining stress acting on the particle, integrated over the sphere, is $4\pi R^2\mu\kappa U_S$, in which U_S is the Smolukowski speed. We can check that this force balances Coulomb forces, whose expression is, at large κR, $F_e = QE_\infty \approx 4\pi\epsilon\zeta\kappa R^2$ (see (6.34)).

The arguments we used apply to insulating particles of arbitrary shapes. The Smoluchowski speed thus concerns any object of size l, provided it is insulating and κl is large. If the particle is conducting, things change. For instance, according to Henry's calculations, metallic particles have no electrophoretic mobility.

6.6.4 The Henry function

As noted above, by neglecting the ionosphere ovalization, Henry solved the general case, i.e. κR arbitrary and arbitrary conductivities. Further, assuming the Debye approximation, Henry expressed, for the case of insulating particles, the mobility in terms of a dimensionless function $f(\kappa R)$, defined by:

$$\mu_{\mathrm{e}} = \frac{\epsilon\zeta}{\mu} f(\kappa R) \tag{6.50}$$

where f is given by:

$$f(x) = 1 - \exp(x)(5E_7(x) - 2E_5(x)) \tag{6.51}$$

in which E_n is the exponential integral of order n.[6]) The function is called the 'Henry function'. It is plotted in Fig.6.31.

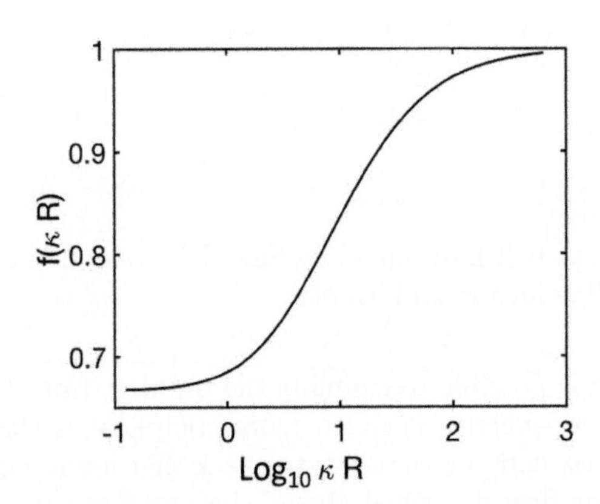

6.31: The Henry function [29].

At small κR, the function is close to 2/3, and at large κR, it tends to unity. The Henry function thus fills the gap between the two limiting cases we considered. In practice, $f(x)$ is currently used in the colloid industry, to infer the ζ potential from the measurement of the electrophoretic mobility. It is often called the 'Henry factor'.

[6] $E_n(x) = \int_1^\infty \frac{\exp(-xt)}{t^n} dt$

Why is the electrical force, exerted by the applied electric field on the particle $Q\mathbf{E}_\infty$?

We considered that the force $\mathbf{F}$ applied on a particle of charge Q immersed, in an uncharged fluid, in an electric field $\mathbf{E}_\infty$, is equal to $Q\mathbf{E}_\infty$. This result is independent of the sphere permittivity, while, as we saw in this Chapter, the electric field pattern strongly depends on it. So where does this independency come from? The result is often presented as straightforward, but it is not. The demonstration can be found in [2]. The starting point is well explained in [3]: in a solid immersed in a fluid, the local force exerted by the applied field is given by the divergence of the Maxwell stress tensor T_{ij}, defined by:

$$T_{ij} = \epsilon(E_i E_j - \frac{1}{2}E_{ii}^2)$$

in which E_i and E_j are the electric field components, *in the fluid* [3]. The force applied on the particle is given by the integral of $DivT_{ij}$ over a volume V around the sphere. Since the fluid around the sphere is uncharged, we may integrate throughout a sphere of large radius containing the particle. Owing to the divergence theorem, this force is also equal to:

$$F = \int_V T_{ij}\, dV = \int_S \epsilon(E_i E_j - \frac{1}{2}E_{ii}^2)\, dS$$

in which S is the surface or the large sphere. Further, by noting that, far from the sphere, the field reduces to $\mathbf{E}_\infty$, the integral can be calculated. In the calculation, on using the Gauss theorem, one obtains the classical result $\mathbf{F} = Q\mathbf{E}_\infty$. The expression of the force is consistent, as it should be, with Coulomb law.

6.6.5 The case of perfectly conducting spheres

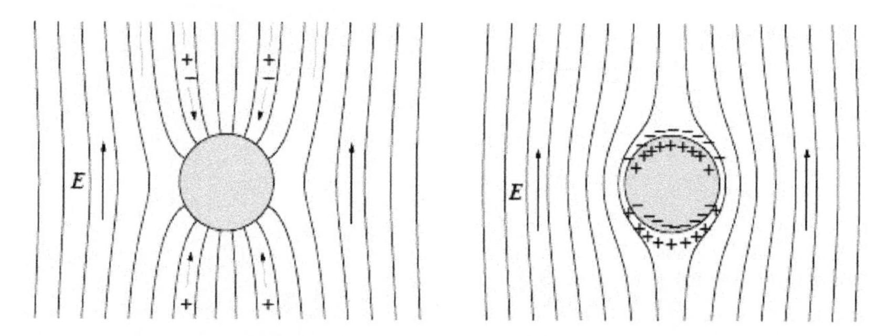

Fig. 6.32: (Left) Electric field lines, just after the electric field E is applied. (Right) After some time, charges accumulate, expelling the field lines from the interior of the particle. In such circumstances, the object behaves as an insulator. (From [32]).

The case of uncharged particles. The case of perfectly conducting spheres is rich and, sometimes, counter-intuitive. We refer here to the work of Ref. [32]. Let us immerse a perfectly conducting sphere in an electric field $\mathbf{E}$. Initially, there is no volumetric nor surface charge, and the sphere permittivity is infinite. Consequently, the electric field pattern will take the form shown in Fig.6.32(B), i.e. with field lines penetrating normally into the sphere and, inside, an electric field equal to zero. In the electrolyte, the solution thus reads:

$$\phi = -Er\cos\theta\left(1 - \frac{R^3}{r^3}\right) \tag{6.52}$$

A ionic current thus develops, conveying charges to the surface. Because ions cannot penetrate into the sphere, they will accumulate. After some time, estimated to λ_D^2/D [32], a charge distribution establishes, expelling the electric field lines (see Fig. 6.32). The positive charges accumulate at the bottom of the particles (the south hemisphere), while the negative charges, moving against the electric field, gather in the north hemisphere. Eventually, a steady–state is reached, for which the accumulation process is complete, implying the current normal to the sphere has vanished. The corresponding electric field line pattern is shown in Fig. 6.32. The electrical field lines being tangential to the surface, the perfectly conducting sphere behaves as an insulator. The expression of the steady–state potential, in the electrolyte, now reads:

$$\phi = -Er\cos\theta\left(1 + \frac{R^3}{2r^3}\right) \tag{6.53}$$

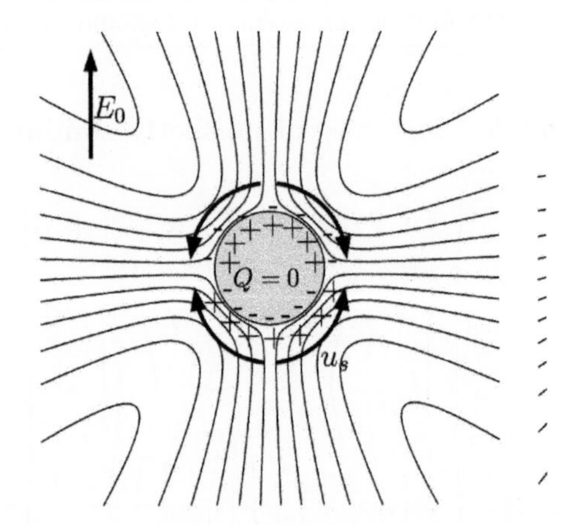

6.33: A metallic particle develops a quadripolar flow around it. (From [32].)

The corresponding ζ potential, $\phi(R)$ is equal to $-\frac{3}{2}Er\cos\theta$. It gives rise to a quadripolar electro-osmotic flow pattern, shown in Fig. 6.33. The electrophoretic mobility is zero, and therefore the particle remains immobile, but the fluid moves around it. The phenomenon gives rise to interesting applications: examples are mixing, particle handling, etc. Just insert a metallic particle in a microwell, with two electrodes, and mixing is achieved. This is conceptually very elegant. In practice however, controlling the oxidation of the particle surface is difficult, so that, in general, mixing takes place only for a limited time.

The case of charged particles with $\kappa R >> 1$. Let us now assume a thin Debye layer, and charge the particle with a surface charge $q_0 = Q/4\pi R^2$, or a zeta potential, $\zeta_0 \approx \frac{q_0}{\epsilon\kappa}$

(see Eq. (6.34)). After the system has reached a steady–state, due to the linearity of the equations, we will have:

$$\zeta = \zeta_0 - \frac{3}{2} E r \cos\theta$$

In the presence of the electric field, whose pattern is shown in Fig. 6.32 (right), the counter–ions in the Debye layer will be dragged and give rise to an electro-osmotic velocity around the sphere. As in the previous case, the similarity principle can be applied, leading to a mobility equal to:

$$\mu_e = \frac{\epsilon \zeta_0}{\mu} \tag{6.54}$$

We thus recover, in this case, the Smoluchowski result for insulating spheres.

6.6.6 The electrophoretic mobility of ions

Ions are much smaller than the Debye length, and, if they are viewed as a sphere of radius a, the results of the preceding sections in the case $\kappa a << 1$ apply, leading to:

$$\mu_e = \frac{Q}{6\pi a \mu}. \tag{6.55}$$

in which a is the hydrodynamic radius'. As mentioned in Chapter 2, depending on the level of hydration, and for small ions, a ranges between 0.1 and 0.5 nm.

6.6.7 The electrophoretic mobility of DNA

With the buffers used in DNA studies or DNA electrophoresis systems, ionic strengths are of the order 10 –100 mM and Debye lengths are on the order of 1–2 nm. These lengths are comparable to the sizes of the molecular entities pertaining to the DNA chain. Therefore, we may use Smoluchowski's formula to estimate their mobilities. Since the ζ potentials for each molecular entity are comparable, they will travel at the same speed. The unfortunate consequence is that the electrophoretic speeds of DNA strands or double strands are independent of their length. In free solutions, no way to 'fractionate' DNA fragments, i.e. to separate them in function of their sizes.[7]

The argument has naturally led the community to use gels. Much work has been done on the subject, in particular during the period 1970–2010. The work has been well reviewed by J. L. Viovy [34] and, more recently, by K. Dorfman [35]. When a DNA chain migrates in a gel, it undergoes different regimes. They are sketched in Fig. 6.34.

[7] This is the argument. Looking in more detail, without gel, end effects and differences in chain flexibilities induce measurable differences in mobilities, between long (above 100 bp) and short chains (10 bp or so) (see [33]). However, the separation efficiencies are much smaller than in gels.

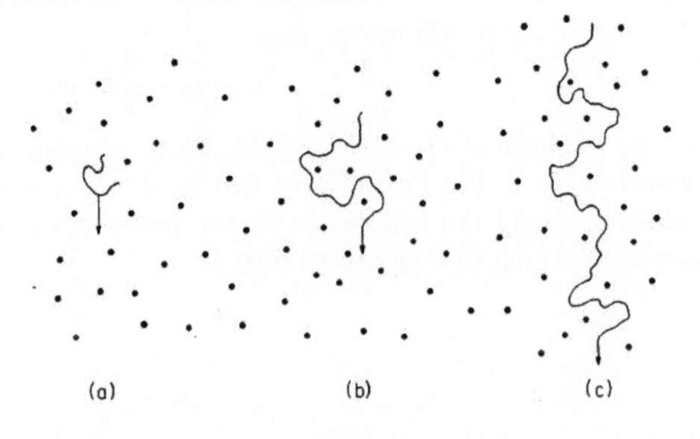

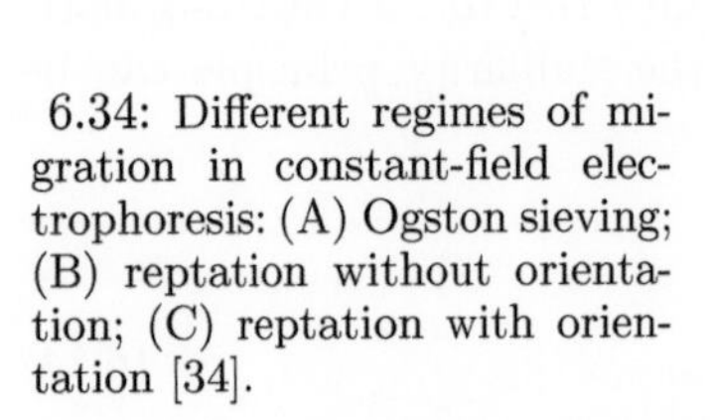
6.34: Different regimes of migration in constant-field electrophoresis: (A) Ogston sieving; (B) reptation without orientation; (C) reptation with orientation [34].

Ogston sieving [37] corresponds to the case where the DNA coil is smaller than the pore (see Fig. 6.34 (A)). Its migration is slowed down by the gel matrix. Reptation, with and without orientation, occurs when the molecule size is larger than the pore (see Fig. 6.34 (B) and (C)). These regimes give rise to a dependence of the mobility with the chain length [34, 35]. They were revealed experimentally in the 1990s (see Fig. 6.35) [36].

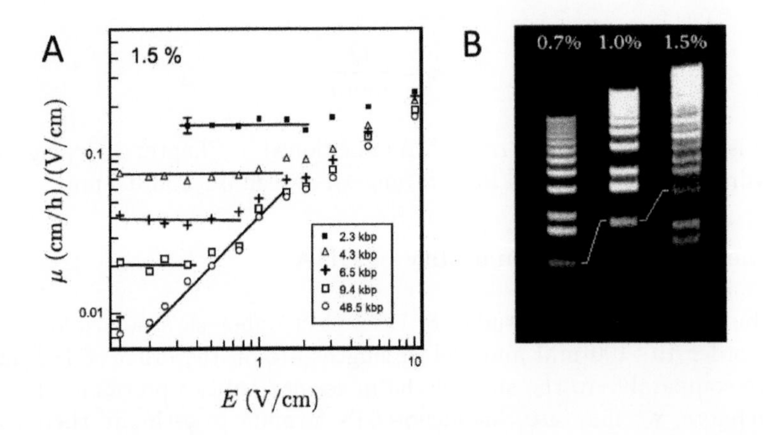

Fig. 6.35: (A) Electrophoretic mobility of different sized DNA in a 1.5% agarose gel as a function of the electric field [36].(B) Typical DNA separation images obtained in agarose gels, for three different gel concentrations. Often (this is not the case here), gels include a reference ladder, showing several DNA fragments with known sizes (Reprinted from Ref [36] with permission of John Wiley and Sons. Copyright 2022.)

Fig.6.35 A shows the evolution of the mobility of DNA duplexes with the electric field, for different DNA lengths. In practice, DNA was visualized with fluorophores, giving rise to an image such as Fig. 6.35 (B), utilizable for measuring mobilities. Within some range of electric field, the mobility depends on the length and is independent of the applied field. In these regimes, fractionation is possible. At larger electric fields, the dependence with the length collapses and no fractionation is possible.

In the years 2000–2010, new systems, attempting to produce gel-like structures with microfabrication techniques, flourished: examples are microfabricated posts with DRIE (see Chapter 7) [40], self assembled functionalized magnetic columns [38], or microchambers forming entropic traps [39]. These systems, which offered alternatives to standard gels, are reviewed in Dorfmann [35].

Electronic Paper

An interesting application of electrophoresis is electronic paper (e–paper). This currently uses hydrocarbon oil droplets, in which pigments of opposite signs are suspended (see Fig. 6.36(A)). Typically, TiO_2 pigments, negatively charged, are used for the white colour. Depending on the orientation of the electric field, black or white pigments migrate at the bottom or the top of the droplet. Observed in reflection, the colour will be either black or white. Droplets are several tens of micrometre in diameter. The electrodes, which impose the electric field, can be addressed individually. In this way, it is possible to display any pattern.

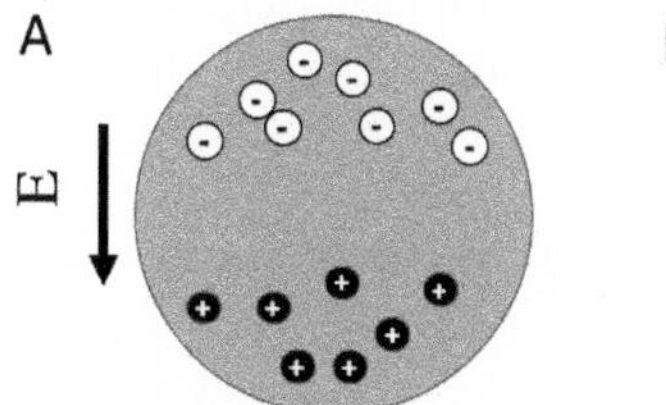

Fig. 6.36: (A) Principle of the electronic paper: under the action of the electric field, white pigments migrate upwards, and black pigments downwards. Seen from above, and in reflection, the droplet looks white. By inverting the electric field, the droplet looks black.(B) An example of an electronic paper, the Kindle reader.

This looks simple. There exist limitations. One of them is the commutation speed, too slow to display movies. E–paper technology possesses important advantages compared to standard electronic display technologies. E–paper consumes much less energy. Also, eye fatigue is reduced. Since 2007, Kindle (Amazon) has sold more than 30 millions e–books.

6.6.8 Particle deposition in microchannels with electrically charged walls

We described, in Chapter 5, based on Ref. [41], how particles, entrained in a channel, deposit on the walls. The analysis can be extended to the case when a surface charge is present, giving rise to a potential ζ, which in turn develops an electric field repelling or attracting the particle, depending on the signs of the ζ potentials. The situation is depicted in Fig. 6.37.

In Chapter 5, we considered the case where Van der Waals forces, drag forces, and Brownian movement are present. Here, we add an electric field, produced by the wall

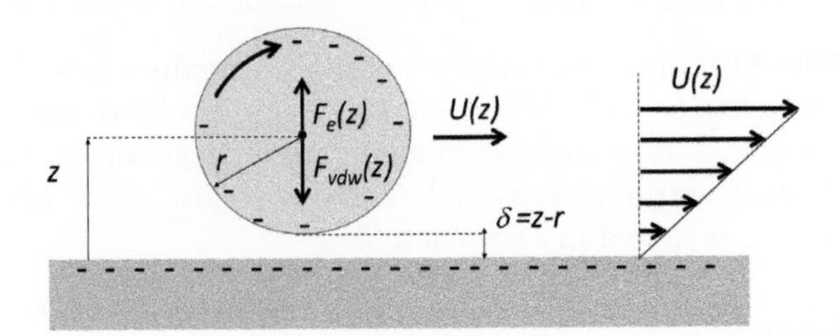

Fig. 6.37: Particle moving in a channel, close to a wall. Channel walls and particles have zeta potentials ζ_w (wall) and ζ_p (particle) [42].

charges. The general situation is rich. It was treated in Ref. [42] and compared with experimental measurements performed in microchannels. Let us take a part of this analysis, by considering the case where particles are non Brownian and the forces developed by the charged walls are much larger than the Van der Waals forces. We also assume that zeta potentials are negative, and thereby, wall and particles repel each other. In such a case, the equations governing the particles trajectories are:

$$\dot{x}(t) \approx \gamma U_r(z) \text{ and } \dot{z}(t) \approx \beta_z \frac{\kappa D}{kT} \chi \exp(-\kappa(z-r) \qquad (6.56)$$

in which $\chi = 4\pi\epsilon\epsilon_0\zeta_w\zeta_p R$

Here, ϵ, ϵ_0 are the dielectric constant of the fluid transporting the particles and the permittivity of vacuum, respectively, and ζ_w and ζ_p are respectively the zeta potentials of the channel wall and the particle. γ is a factor taking particle rotation and shear effect into account and β_z is the hindered diffusion factor (see Chapter 5), D the particle diffusion constant, kT the thermal energy, z the distance of the particle center from the wall, and r the particle radius. κ has been defined in this chapter. From equation 6.56, we infer $\dot{z} > 0$. This implies that particles move away from the wall. Thereby, no particle sticks to the wall. Indeed, when Van der Waals forces are added, the situation changes. Ref [41] showed that, in this case, deposition rates depend on the ratio between electric and Van der Waals forces. Figure 6.38 shows a general diagram, where the complete analysis of the situation, incorporating Brownian, Van der Waals, drag, and electrical forces,is carried out.

The diagram has a form of a shoe. The abscissa is the salt concentration (which controls the Debye length), ξ_L, an inverse of the Peclet number, which controls the ratio of diffusion over advection terms. The larger ξ_L, the more important the diffusion. The ordinate involves the parameter S, which measures the amount of particle, per unit of time, captured by the wall. One learns, from this diagram, that below a ionic strength of 100 mM, no particle deposits in the channel. When the electric field is screened ($C_{salt} > 100mM$), deposition occurs. In this range, depending on ξ_L, two different regimes of deposition occur: at small ξ_L (for instance, with small particles, the regime is Brownian, while, at large ξ_L (for instance, large particles) as we saw in Chapter 5,

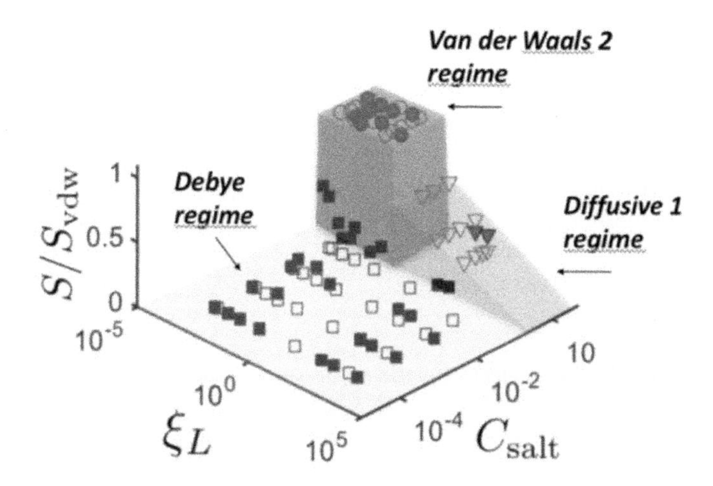

Fig. 6.38: Universal diagram showing the (non-dimensional) rate of deposition S, as a function of the salt concentration and a parameter, ξ_L, representing the ratio between the diffusive and advective terms, for the particle. The points represent measurements performed in microchannels. Below 100 mM, no deposition occurs. In practice, above this concentration deposition occurs, dominated either by Van der Waals forces (Van der Walls 2 regime), or by Brownian motion (Diffusive 1 regime). (From [42].) The various regimes, along with their frontiers, are well accounted for by the theory.

the regime is dominated by Van der Waals forces. The latter regime is the most prone to cloggingl.

6.7 Microfluidic electrokinetic separation

6.7.1 Acronyms

Electrokinetic methods of separation are numerous and the non–specialist may get confused between the different acronyms used in the field. Before going further, it is useful to list the full names of some of them, and the type of interactions they rely on.

6.7.2 Free solution capillary electrophoresis (FSCE)

In free solution capillary electrophoresis (FSCE), the sample is injected in a capillary, in a *free solution*, i.e. without gel, and subjected to an electric field, imposed by two electrodes submerged in external reservoirs (see Fig. 6.39).

As we saw above, the field has two effects: it induces an electro-osmotic flow that displaces the whole fluid as a plug flow, and it separates ions. Those migrate at different velocities, according their charge over mass ratios, as expressed by Eq. 6.55. We have:

Method	Acronym	Interaction
Zone electrophoresis	ZE	Charge
Capillary zone electrophoresis	CZE	Charge
Capillary electrophoresis	CE	Depends on case
Free solution capillary Electrophoresis	FSCE	Charge
Capillary gel electrophoresis	CGE	Sieving
Micellar electrokinetic chromatography	MEKC	Charge+affinity
Micellar electrokinetic capillary chromatography	MECC	Charge+affinity
Microfluidic electrokinetic separation method	MESM	Case dependent
Capillary electro chromatography	CEC	Charge+matrix interation
Isoelectric focussing	CIEF	Isoelectric Point

Table 6.1 Different electrokinetic separation methods, with acronyms

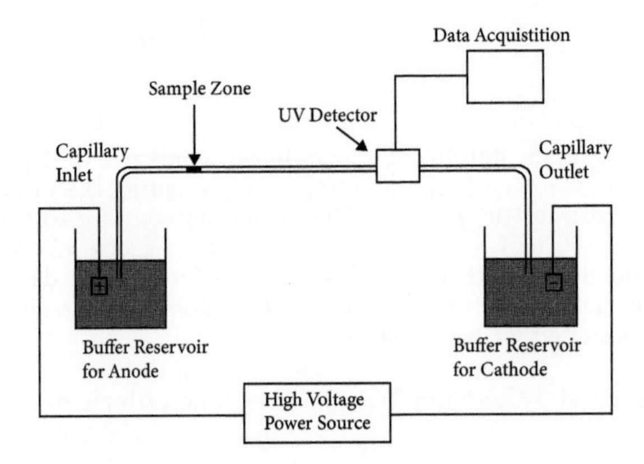

Fig. 6.39: Typical setup of capillary electrophoresis.

$$U_{\mathrm{b}} = U_{\mathrm{EOF}} + U_{\mathrm{q}},$$

where U_{EOF} and U_{q} are the electrosmotic and electrophoretic velocities, respectively.

In CE, the electro-osmotic velocity dominates the electrophoretic one, so as bands of analytes pass under the detector at different instants without overlapping. Detection is more often based on fluorescence, conductivity, or absorption.

An example is shown in Fig. 6.40.

Each peak corresponds to the passing of ions that have the same electrophoretic mobility. From the analysis of Fig. 6.40, and using calibration, we can deduce the migration velocities, and can then identify the different solutes present in the sample. In the case of the figure, the analytes were methylanthracene and perylene. The sharpness of the peaks indicates an excellent resolution of the method.

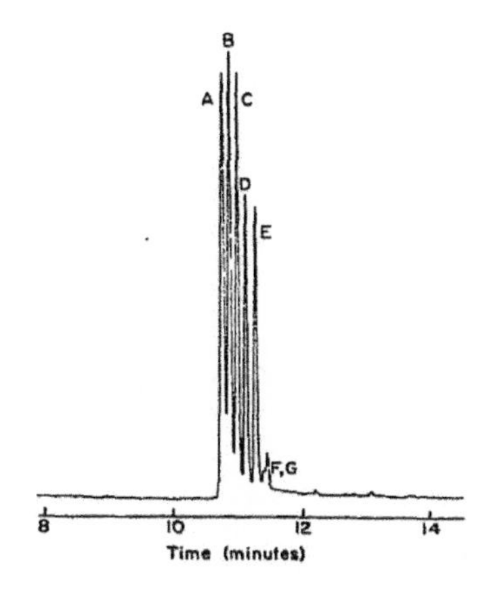

6.40: Early electropherogram of n-alkylamines as fluorescamine derivatives. demonstrating the high resolution of free solution capillary electrophoresis (FSCE). Peaks: A = octyl; B = heptyl; C = hexyl; D = pntyl; E = butyl; F = unknown impurity; G = propyl. Approximately 7 pmoles of each derivative, except for propylamine. (Reprinted from [45], with permission of Elsevier. Copyright 2022.)

6.7.3 The performance of free solution capillary electrophoresis

What are the performances of free solution capillary electrophoresis? The high performances of the technique were pointed by Jorgenson et al. in 1981 [45]. Because solutes are transported by a plug flow, there is no shear and thereby the spreading δX of the spots is controlled solely by molecular diffusion. δX can be estimated by the following expression:

$$\delta X \sim \sqrt{Dt}$$

where D is the diffusion coefficient of the ions forming the spot, and t is the time. At the end of the capillary tube, i.e. at a time equal to L/U_{EOF}, where L is the capillary length, the band has acquired a width W equal to:

$$W \sim \sqrt{\frac{DL}{U_{EOF}}}$$

Then the number of theoretical plates reads:

$$N \sim \frac{U_{EOF}L}{D}$$

The previous formula shows that in order to obtain large number of theoretical plates, it is advantageous to work with strong electric fields and long capillaries. In typical systems, the capillary is a silicon tube with a circular section, from 25 to 50 μm in diameter, and 20 or so cm long. In terms of orders of magnitude, the applied voltages in capillary electrophoresis are of several thousand volts and speeds are on the order of 1 mm/s. Taking $U_{EOF} = 1$ mm/s, $D = 10^{-9}$ m$^2/s$ and L =0.1 m, one finds N $\approx 10^5$, which is a number comparable to the best performing separation techniques, such as high performance liquid chromatography (HPLC), as we saw in Chapter 5.

6.7.4 Miniaturizing free solution capillary electrophoresis

Why miniaturize CE? There are several advantages:

- manipulate small volume samples;
- integration (favouring automation);
- parallelization (allowing high-throughput analysis);
- short analysis times; and
- high performances.

The first three advantages are applicable to miniaturization in general. The fourth advantage is justified in the following manner: we take the expression of the number of theoretical plates given above. We have:

$$N \sim \frac{\mu_{\mathrm{EOF}} E L}{D}$$

where μ_{EOF} is the electro-osmotic mobility, L is the capillary length (the micro-column length), and D is a representative diffusion coefficient of the ions. The formula shows that it is advantageous to work at elevated electric fields. The electric field strength is limited by Joule heating, due to the electric current developing in the electrolyte. We have the following equality, in terms of order of magnitude:

$$Q \sim \sigma_{\mathrm{e}} E^2 L d^2 \sim K \Delta T L$$

where d is the capillary diameter, Q is the heat produced by the Joule effect in the microsystem, K is the conductivity, and ΔT is the temperature difference between the fluid contained in the capillary and the exterior. From this equality, we obtain the following estimate for the electric field :

$$E \sim \sqrt{K \Delta T / \sigma_e} d^{-1}$$

Thus, at fixed ΔT, the smaller d is, the higher E will be. Miniaturization thus allows us to work at an elevated electric field. We are thus led to the following scaling relation:

$$N \sim L/d$$

Even though L is in the decimetre range, the ratio can be large. With the pre-factor in front of this relation(not shown), one can have high N. Thus, miniaturization does not degrade the resolution of the separation process. The scaling of the retention time in the micro–column is given by:

$$t_{\mathrm{r}} \sim L d$$

As the retention time is proportional to miniaturized quantities (L and d), it can be small. In practice, compared to standard columns, working in a microfluidic format allows us to reduce the separation time by orders of magnitude.

6.7.5 The first microchip capillary electrophoresis (MCE) system (1992)

As said in the introduction to this chapter, the first proof of concept of a miniaturized free electrophoretic separation system was obtained by Manz's group, in 1992 [1]. This gave rise to the acronym MCE, which means: microchip capillary electrophoresis. The device is shown in Fig. 6.41.

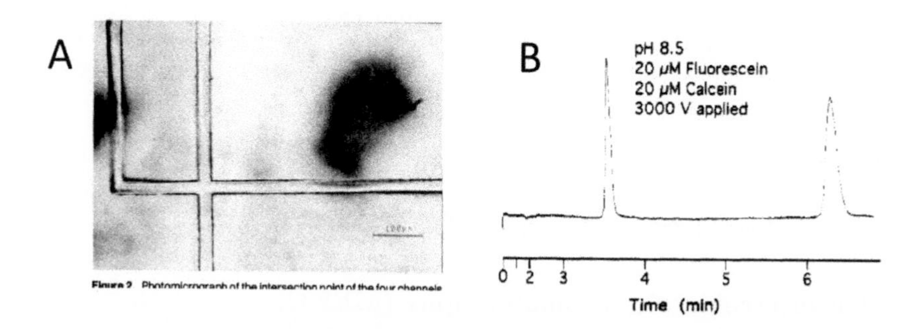

Fig. 6.41: First proof of concept showing that capillary electrophoresis can be performed in a microfluidic channel (1992) [1]. (A) Microscope image of the chip. (B) Electropherogram showing the fast (a few minutes) separation of fluorescein and calcein. (Reprinted from Ref. [1] with permission of the American Chemical Society. Copyright 2022.)

The separation channel, of rectangular cross-section, is 10 cm long, 20 μm high, and 30 μm wide. Compared to standard capillary electrophoresis (CE), the column length is five time shorter. Sample injection has been miniaturized, in order to keep the resolution high [44]. In the system of Ref. [1], the sample injection is performed with the cross-channel shown in Fig. 6.41. The sample is transported in the column by an electro-osmotic flow developing sideway. After the sample has reached the intersection, an electric field is applied along the 'column', i.e. the main channel. A plug flow develops and the sample decomposes into bands. The electropherogram reported in [1] is shown in Fig. 6.41. Plateau numbers rangis between 10^4 and 10^5 were obtained. Developments of this approch can be found in Ref. [43].

In practice, MCE requires an excellent control of the pressures applied at the system entries, in order to avoid back–flows. High hydrodynamic resistances are often used to reduce parasite flows, for instance, induced by capillarity. Such flows disperse the bands and degrade the resolution.

Separating ions in 800 μs. In 1998, Jacobson et al. [47] separated the components of a sample in 800μs (see Fig. 6.42).

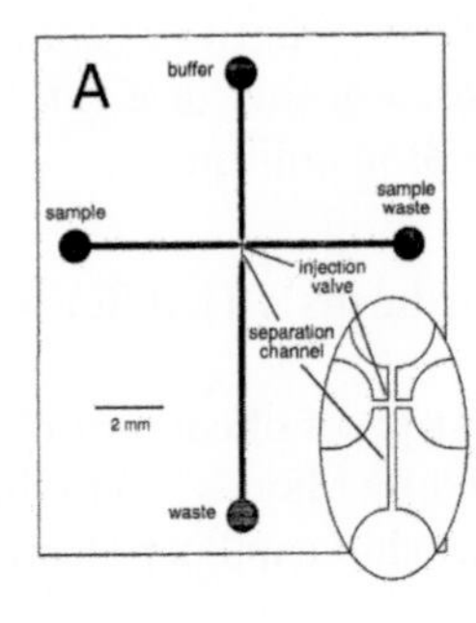

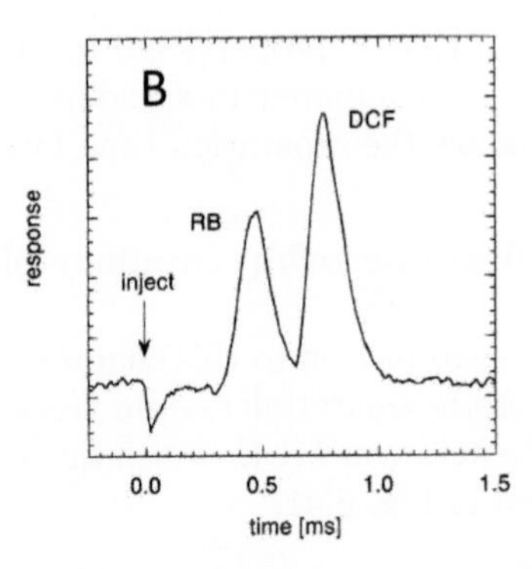

Fig. 6.42: Demonstration that, with microfluidics, samples can be separated in less than one millisecond. (A) Design of the MEC system.(B) Electropherogram demonstrating large speeds of separation. (Reprinted from [47], with permission of the American Chemical Society. Copyright 2022.)

To perform the same task, traditional gel-based methods would take hours or fractions of hour. The success of the demonstration was due to the miniaturization of the device and a careful design of the injection section (see Fig. 6.42) [47].

6.7.6 Micellar electrokinetic chromatography (MEKC)

With CE and MCE, impossible to separate neutral species. This is where micellar electrokinetic chromatography (MEKC) comes on stage. With MEKC, charged and uncharged molecules can be separated, by exploiting, in addition to electrophoretic properties, the affinity of hydrophobic species with micelles Thus, two criteria add up to achieve the separation, providing denser electropherograms. In practice, charged micelles are introduced in the system, by mixing the sample with ionic surfactants, such as Sodium Dodecyl Sulfate (SDS), above the Critical Micellar Concentration (CMC – see Chapter 5 –). Hydrophobic molecules, which hate water, are happy to embark on micelles. They travel together along the column, i.e. the separation channel, at a speed determined by the micelle mobility, with its passengers. Further downstream, the molecule can be detected.

MEKC was carried out in a microdevice, for the first time, in 1995 [48] and 1996 [49]. The electropherogram shown in Fig. 6.43 was obtained in 2000 [50].

In Fig. 6.43, the 19 peaks result from the separation of tetramethylrhodamine-labeled amino acids. Separation was accomplished in 165 s, with an average plate number of 280,000.

6.7.7 Capillary electrochromatography (CEC) and DNA separation

With the preceding techniques, one can separate ions and organic species, but not DNA fragments. A first reason was given previously: the mobility of DNA fragments does not depend on their sizes. Thereby, electrophoresis, whether CE or MEC, cannot

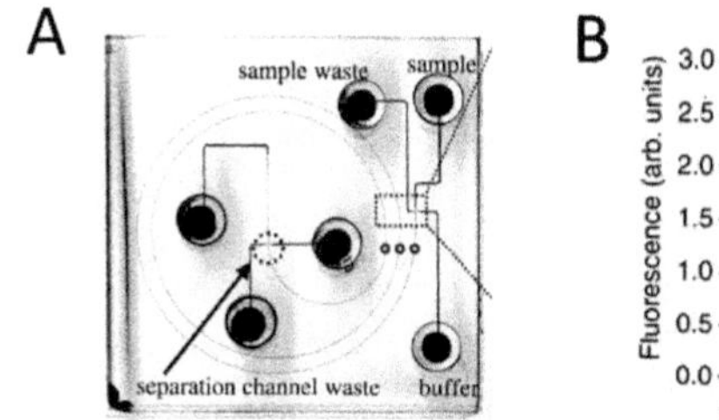

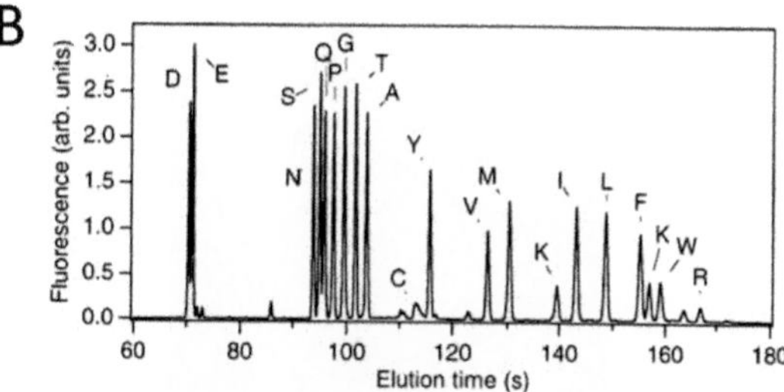

Fig. 6.43: (A) Image of the MEKC microdevice. (B) MEKC separation of 19 TRITC-labelled amino acids in a 10 mM sodium tetraborate/50 mM SDS buffer with 10% (v/v) 2-propanol. The field strength was 770 V/cm, and the detection point was 11.87 cm from the injection cross. The peak locations of the amino acids are indicated by their standard one-letter abbreviations (Reprinted from [50] with permission of the American Chemical Society. Copyright 2022).

fractionate them. The second reason is physico-chemical: DNA, soluble in water, feels 'happy' in aqueous environments. Therefore, it does embark on organic micelles. This excludes MEKC from the list. Therefore, none of the preceding techniques allows us to fractionate DNA.

This is where capillary electro chromatography (CEC) comes on stage. The technique combines chromatography (i.e. separation in a matrix) and capillary electrophoresis. A typical CEC system consists of a capillary filled with a matrix (e.g. silicon particles or a gel) and electrodes imposing a longitudinal electric field. A sample placed at the head of the capillary will be fractionated, because, as discussed earler, during their journey, DNA fragments interact differently with the matrix, and consequently migrate at different speeds.

The first commercial CEC product was launched by Caliper Technologies around the turn of the century [51]. The channels were etched in glass. Detection was based on fluorescence. The electric field was about 1 kV/cm and the column length was a few millimeters. In this device, the separation time took tens of seconds, with a resolution of several thousand theoretical plates. The technology offered important advantages over gel slabs, for which fractionation takes hours. By marking the entry of microfluidics in biology, the Caliper system represents a landmark for the field.

In the meantime, the genome programme was blooming. The genome sequencing was exclusively based on Sanger method [53]. The method used restriction enzymes, fragmenting the genome into small single-stranded pieces. A crucial step was the determination of their sizes. Typically, ten fractionations per kilobase were needed to determine primary sequences [56]. For the human genome, this represented no less than ten millions separations. This number is huge. In this context, researchers invented new CEC configurations, faster, more integrated, more automatized [57], and more parallelized [52]. One example is shown in Fig. 6.44 (B) [55]. DNA samples are placed

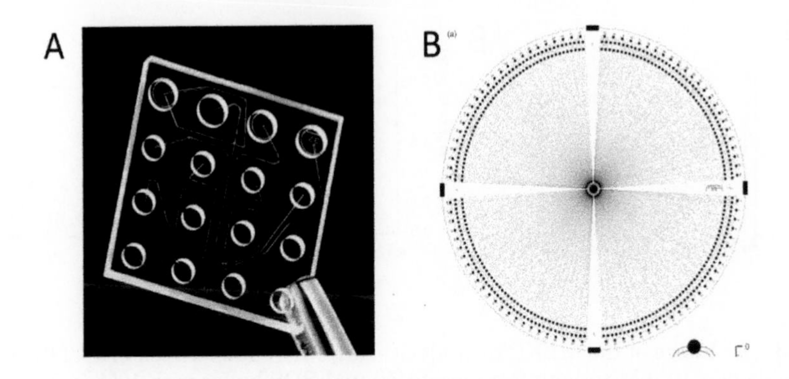

Fig. 6.44: (A) CEC device, commercialized by Caliper Technologies Corp. (Now Agilent Technologies) [51]; (B) 384 well systems for DNA fractionation. (Reprinted from Ref. [55], with permission of the American Chemical Society. Copyright 2022.)

in 384 wells, situated at the periphery of the system. The number of theoretical plates was 4×10^6 and complete analysis of 384 samples took about 6 min.[8]

It turned out that the most popular sequencer, ABI Prism 3700, that sequenced the human genome, did not use this work. To increase the throughput, the machine placed many glass capillaries in parallel. No microfluidic chip was on stage, so to speak. The penetration of microfluidics came later, with the advent of Next Generation Sequencing (NGS). NGS, less accurate than Sanger's method, but much less expensive, no longer needed fractionation. The first NGS machine, the GS20, produced by 454 Sequencing, appeared in 2005. The success of GS20 was due in part to the use of large microfluidic assays. More recent methods [58], such as nanopore sequencing [59], or zeroth mode cavity [60], also made full use of microfluidic or nanofluidic technology. Since then, micro–nanofluidics have become, clearly, a core technology for genome sequencing.

6.8 Dielectrophoresis

In a previous section we calculated the force exerted by a homogeneous electric field on a charged sphere. This led to electrophoresis. When the sphere is neutral, the force is zero. But this is not the end of the story. We saw, in this chapter, that, in a purely dielectric system (not conductive), a sphere subjected to a uniform electric field develops a dipolar field. The corresponding dipolar moment $\mathbf{p}$ is equal to the following (see Eq. (6.45)):

$$\mathbf{p} = 4\pi\epsilon_1 R^3 \left(\frac{\epsilon_r - 1}{\epsilon_r + 2}\right) \mathbf{E}$$

[8]In the extension of this work, architectures with higher throughputs were proposed [54].

A question we may ask is whether the interaction between this dipole and the applied electric field gives rise to a global force. The answer is no, for symmetry reasons. By replacing the sphere by bounded charges at its surface, of opposite signs, and located symmetrically with respect to the sphere centre, one sees that the global Coulomb force exerted on the system is zero. An electric field does not behave like a wind pushing the sphere. However, should the field be non–uniform, there is a possibility that, the symmetry being broken, a net force develops.

Let us investigate this possibility by considering an elementary dipole, shown in Fig. 6.45.

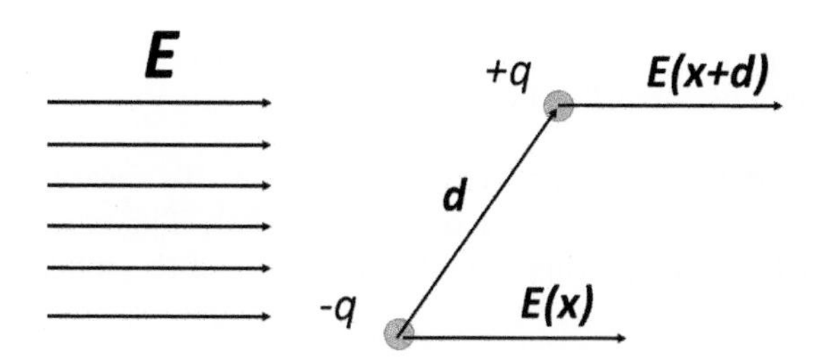

6.45: Dipole immersed in a non homogeneous electric field.

The dipole is formed by two charges of opposite signs, separated by **d**. **E** is the applied field. In the limit where **d** is much smaller than the characteristic scale of variation of **E**, the Coulomb force exerted on the dipole is:

$$\mathbf{F} = q(\mathbf{E}(\mathbf{r}+\mathbf{d}) - \mathbf{qE}(\mathbf{r})) \approx \mathbf{p}\nabla\mathbf{E}$$

in which $\mathbf{p} = q\mathbf{d}$ is the dipolar moment. Applying this result to the case of the sphere, where the corresponding dipolar moment is given by Eq. (6.45), one finds:

$$\mathbf{F} = 2\pi\epsilon_m R^3 \left(\frac{\epsilon_r - 1}{\epsilon_r + 2}\right) \nabla\mathbf{E^2} \tag{6.57}$$

in which ϵ_m is the medium permittivity, and $\epsilon_r = \frac{\epsilon_s}{\epsilon_m}$ is the relative permittivity (in which ϵ_s the sphere permittivity). The formula shows that a force appears when the particle is immersed in an electrical field gradient. This force is called dielectrophoretic [61]. In the literature, it is abbreviated as DEP. The origin of the force is sketched in Fig. 6.46.

When the particle is more polarizable than the medium, the factor $\frac{\epsilon_r-1}{\epsilon_r+2}$is positive, and, according to Eq. (6.57), the force points towards the field *maxima* (see Fig. 6.46 A). In the opposite case, i.e. when the liquid is more polarizable than the particle, the particle is attracted towards the *minima* of the electric field (Fig. 6.46B).

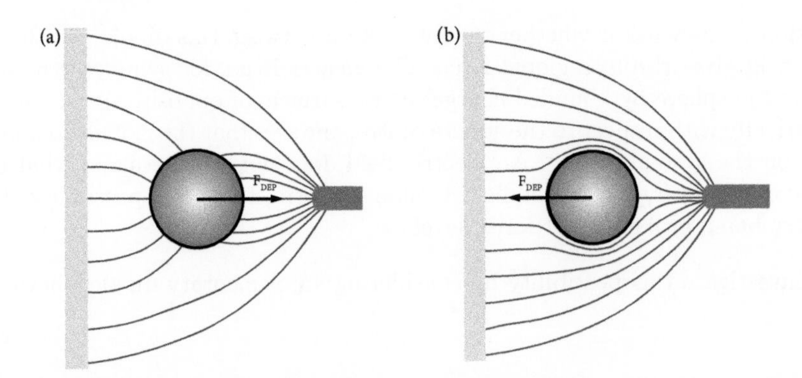

Fig. 6.46: Physics of dielectrophoresis (DEP).(A) Positive DEP: the particle is attracted by field maxima.(B) Negative DEP: the particle is repelled by them.

This conclusion can also be drawn out from an energetic point of view. To create a dipole, one must bring two charges $\pm q$ at a distance $\mathbf{d}$ from each other. The corresponding energy cost is $-q\mathbf{E}.\mathbf{d} = -\mathbf{pE}$. Applying this formula to the case of the sphere leads to:

$$W = -\mathbf{pE} = -4\pi\epsilon_1 R^3 \left(\frac{\epsilon_r - 1}{\epsilon_r + 2}\right) \mathbf{E^2} \tag{6.58}$$

The formula tells us that a particle more polarizable than the environment ($\epsilon_r > 1$) always reduces the electrostatic energy of the system. It will thus be attracted by regions rich in electric field. In the opposite case ($\epsilon_r < 1$), the particle will feel repelling forces.

An important remark is that the force does not depend on the sign of the field. Should the field be oscillatory, the force would keep the same sign because it is proportional to the gradient of the *square* of the electric field.

This remarks calls for addressing the case of an oscillatory electric field, i.e. an electric field given by the expression:

$$\mathbf{E} = \mathbf{E}(\mathbf{x}) \cos(\omega t)$$

in which ω is the pulsation. As is currently done in the literature, we will generalize our presentation to the case where the particles and the medium are both dielectric *and* conducting. The establishment of the expressions of the forces, in such a case, is available in Refs. [67,76]. This consists in introducing the expression of the electric field in Eq. (6.57), averaging out, replacing the permittivities by their complex expressions, and taking the real part. By doing this, the expression of the DEP force $\mathbf{F}$ becomes:

$$\mathbf{F} = 2\pi\epsilon_{rm} R^3 Re(K) \nabla E_{rms}^2 \tag{6.59}$$

where E_{rms}^2 is the average quadratic value of the applied electric field, *Re* denotes the real part, and K is the complex Clausius–Mosotti factor. This factor is defined by the following formula:

$$K = \left(\frac{\epsilon_{\rm p} - \epsilon_{\rm m}}{\epsilon_{\rm p} + 2\epsilon_{\rm m}}\right).$$

in which:

$$\epsilon_m = \epsilon_{rm} - i\frac{\sigma_m}{\omega} \text{ and } \epsilon_p = \epsilon_{rp} - i\frac{\sigma_p}{\omega}$$

where ϵ_{rm}, σ_m and ϵ_{rp}, σ_p are the dielectric constant and electric conductivities of the medium and the particle, respectively. The dielectrophoresis is 'positive' is $Re(K)$ is positive and 'negative' in the opposite case. K varies considerably with the frequency. An example, for water and dielectric particles, is shown in Fig. 6.47.

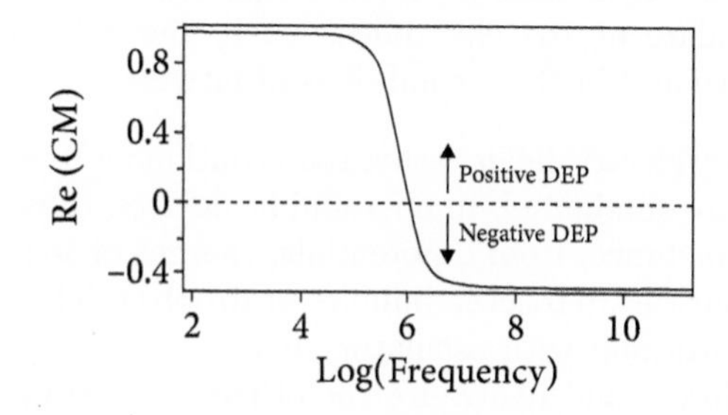

6.47: Typical evolution of the real part of Clausius-Mossotti factor $Re[K(x)]$ versus frequency for dielectric particles in water.

At low frequencies, K is dominated by conductivity effects, and, at large frequencies, by dielectric effects. Fig. 6.47 shows that in the former range, the DEP is positive, while it is negative in the upper range. The cross-over frequency is typically on the order of 1 MHz, again for dielectric particles in water. Varying frequency is a method for changing the sign of the forces.

Let us now analyse the movement of the particles put in motion by DEP forces. In microfluidics, the viscous force acting on a particle of radius R, as we saw in Chapter 3, is given by the Stokes law:

$$\boldsymbol{F}_{\rm v} = 6\pi R\mu\boldsymbol{U},$$

where μ, as usual, is the fluid viscosity, $\mathbf{U}$ the particle speed, and R its radius. Using Eq. (6.59), we obtain:

$$\boldsymbol{U} = \frac{R^2\epsilon_{\rm rm}}{3\mu}{\rm Re}(K(\omega))\nabla E_{\rm rms}^2$$

With an excitation of 1 kV/cm, and a particle of 10 μm in size, the forces and speeds are of the order of tens of pN and μm/s. DEP forces are limited by the electrical breakdown of the material (see Chapter 2). Still, as will be seen later, the forces deployed by DEP technology are large enough to develop interesting functionalities.

An experimental evidence of the DEP forces is shown in Fig. 6.48.

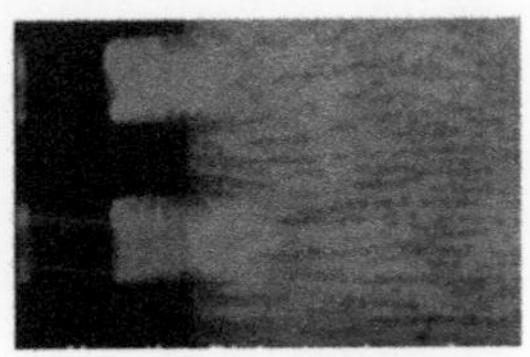

6.48: Laboratory experiment demonstrating the effect of DEP forces on polystyrene particles immersed in water, with electrodes held at 300 v (MMN, J. Goulpeau, 2003.)

The left-hand part of the figure shows polystyrene particles in water, above 1 MHz. They are repelled by the electrodes. On the right-hand part, the frequency is below 1 MHz and the particles tend to accumulate at the electrodes, where the field is maximum. In the first case, DEP is positive and in the second, it is negative.

The important advantage of using oscillatory electric fields rather than continuous ones is that all forces linear in the electric field are eliminated. In microfluidic devices, these forces are ubiquitous. They originate, for instance, from ζ potentials, charges or ions adsorbed at the walls, or conveyed by impurities, particles, bubbles or droplets. They are, in practice, largely uncontrolled. By working with oscillatory fields, we capture the dielectrophoretic part of the electric forces and neutralize the others. Important to note, dielectrophoretic forces are volumetric and not surfacic. More information on dielectrophoresis can be found in Refs. [64, 68, 76, 81].

6.9 Three illustrations/applications of dielectrophoresis

6.9.1 Capturing submicrometric particles

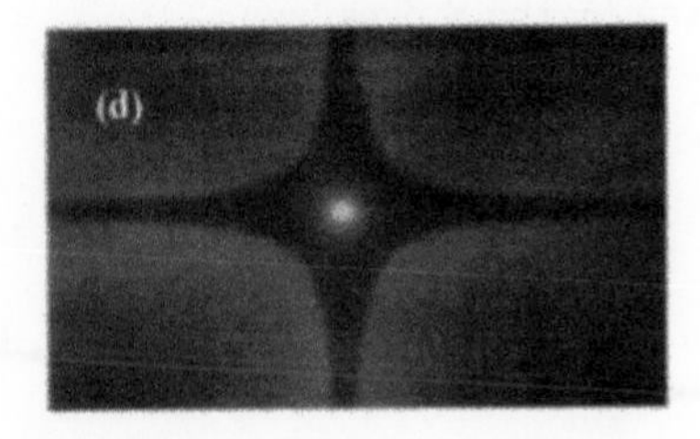

6.49: Dielectrophoretic trap localized between four polynomial electrodes. At the centre, where the electrical field is minimal, a 557 nm particle is captured with negative DEP (Reprinted from Ref. [62] with permission of the American Chemical Society. Copyright 2022.)

An early demonstration of the capacity of dielectrophoresis to trap sub–micrometric particles is shown in Figure 6.49 [62, 63] . The figure shows a 557 nm in diameter particle captured in the region where the electric field, produced by four polynomial electrodes, is minimal [62]. DEP is negative in this case.

6.9.2 Droplet becoming rectangular in an AC electric field

A water droplets is produced at a T-junction and moves rightwards along the main channel (see Fig. 6.50). As it penetrates in the inter-electrode region, the flow is stopped and a voltage V is applied.

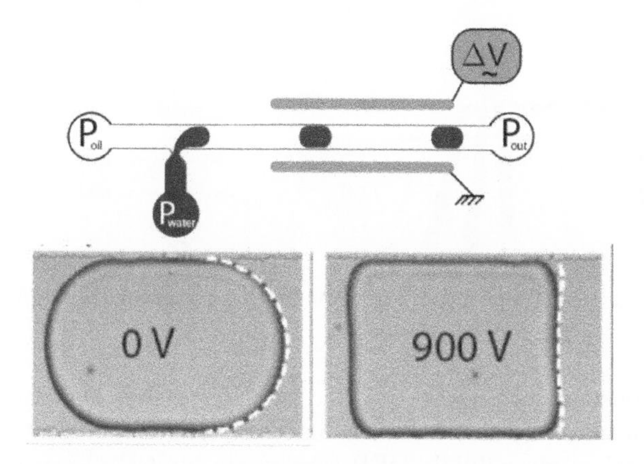

6.50: Water droplet at rest, subjected to an electric field, in a channel 100 μm wide and 30 μm high, filled with oil. (MMN, L. Menetrier Deremble (2010).)

Droplet shapes are shown in Fig. 6.50. At $V = 0$, we find the usual shape of the obstructing droplet, dictated by surface tension and confinement, with a round nose and an elongated part (see Chapter 4). As the voltage is raised to 900 V, the water droplet becomes rectangular. Why this is the case? As often in interface problems, it is more convenient to think in terms of energy rather than forces. As water is polarizable, the droplet will reshape itself to place water, as much as possible, in the highest field regions. It will thus tend to become rectangular.

One may suggest that the effect become important when electrical energy ϵE^2 becomes larger than capillary energy γ/w, in which E is the electric field in the channel, γ is the interfacial tension water/oil, and w is the channel width. A dimensionless number can be constructed:

$$J = \frac{\epsilon w E^2}{\gamma}$$

In the experiment of Fig. 6.50, with w= 100 μm, and E= 9 MV/m, one has $J \approx 1$ indicating that according to the argument just given, above this tension, electric forces control the droplet shape, as observed in the experiment.

6.9.3 Sorting droplets with DEP

Figure 6.51 shows the concept [69]: droplets of different colours (black and white) are driven towards a Y junction. When the electrodes are not polarized, droplets, arriving at the junction, choose the upper branch, because its hydrodynamic resistance, compared to the lower one, is smaller. When an AC electric field is applied, droplets are attracted, by positive DEP forces, towards the electrodes. Their trajectories bent downwards, so that, at the junction, they chose the right branch. Then, by activating the electrodes in function of the droplet colour (identified by a laser), the device operates as a droplet sorter.

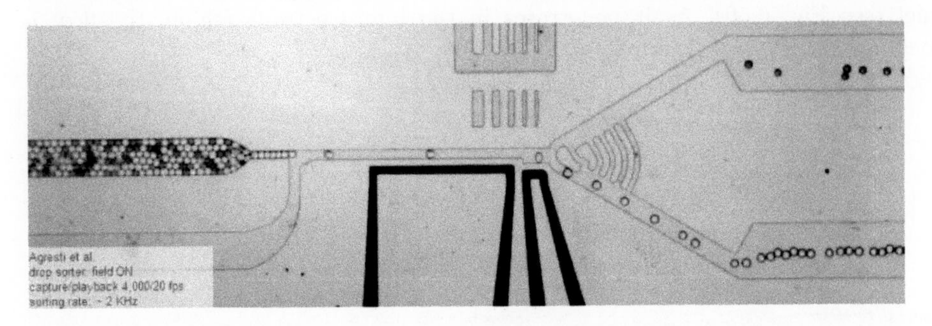

Fig. 6.51: Droplet–based sorting device. (Reprinted from [69] with permission of PNAS. Copyright 2022.)

The sorting rate is typically thousands of droplets per second. To increase this rate, higher electric fields would be needed. This is not possible because of the dielectric rupture limit. Therefore, in practice, microfluidic DEP sorters are limited to several kHz. Although the rate is well below the fluorescence-activated cell sorting (FACS), microfluidic DEP sorters are extremely useful. They were used in Ref [69], to identify new functional mutants of the enzyme horseradish peroxidase. From an initial population of 10^8 enzymes, 100 variants were selected in less than 10 h, using less than 150 μL of total reagent volume. Compared with the state of the art, the microfluidic screening assay represents a 1,000-fold increase in speed and a 1-million-fold reduction in cost [69].

6.10 Electrowetting

Let us consider a conducting droplet (for instance, salted water), polarized at tension V, placed over a thin hydrophobic insulating layer, deposited on a metallic electrode (see Fig. 6.52).[9].

The configuration of Fig. 6.52 is called electrowetting on dielectrics (EWOD). It was popularized by B. Berge [73]. Without an insulating layer, electrochemical reactions develop at the liquid–electrode interface, imposing to work at small voltages. In such circumstances, the droplet does not depart much from its shape at zero voltage. With the insulating layer, as we saw in this chapter, the electric field strength is reduced, but this reduction is outweighed by the possibility to apply large voltages. In practice, by applying voltages ranging between 50 and 1000 V, in the configuration of Fig.

[9]I am much indebted to F. Mugele for the writing of this section

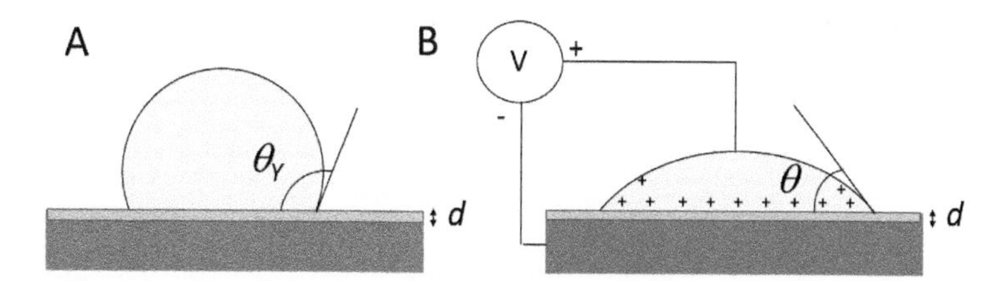

Fig. 6.52: Electrowetting on dielectrics (EWOD): a conducting droplet is placed on an insulating layer of thickness d, below which a metallic electrode represents the ground.(A) No voltage is applied; the contact angle is equal to the Young angle θ_Y.(B) $V > 0$, the contact angle is decreased.

6.52, one significantly modifies the droplet morphology, in a stable and reproducible manner.

Fig. 6.52 defines a capacitor. When a voltage is applied, a charge Q appears at the droplet/insulator interface (see Fig. 6.52). This charge is given by $Q = CVS$, in which S the droplet footprint and $C = \frac{\epsilon}{d}$ is the capacity per unit of area (here d is the layer thickness). In the droplet, counter–ions, attracted by the surface charge, build up a Debye layer at the solid–liquid interface. In the meantime, charges of the same sign populate the liquid–vapor interface, essentially close to the contact line, as will be seen later. The question is: how to describe the situation?

Several descriptions can be found in the literature. Here, we use the description of Refs. [72, 74]. For the sake of simplicity, let us assume that the droplet conductivity is large and, consequently, Debye layers are thin and there is almost no electric field in the liquid bulk. Let us take an energy viewpoint, as we did in in Chapter 4, in the section dedicated to Young equation. In our problem, two energies must be considered: surface and electrostatic energies. Suppose that we move, on Fig. 6.53, the contact line by $dx > 0$ (flooding the surface). This displacement increases the capacity of the system by a quantity, per unit of length, equal to Cdx. This in turn decreases the electrostatic energy E_e of the system, by an amount, still per unit of length, equal to $dE_e = \frac{1}{2}CV^2dx$.[10]) In this displacement, the contact angle changes and the surface energy of the system, E_S, varies by a quantity equal to:

$$dE_S = (\gamma_{SL} - \gamma_{SG} + \gamma \cos\theta)dx = \gamma(\cos\theta - \cos\theta_Y)dx$$

where γ_{SV}, γ_{LS} and γ are the solid-vapor, solid-liquid and liquid-vapour surface tension, respectively, θ_Y is the Young angle, and θ is the contact angle adopted by the

[10]The word 'decrease' is important. Two condensator plates attract each other, because their charges are of opposite sign. Should they be free to move, they would approach each other. By approaching, from infinity, at a distance d, their energy decreases by an amount equal to $\frac{1}{2}CV^2S$ (where $C = \epsilon/d$), like a mass m falling on earth, whose gravitational energy decreases by a quantity equal to mgh, where h is its initial altitude and g the gravity. Similarly, if the capacity, per unit of length, increases by Cdx, the energy will *decrease* by an amount equal to $\frac{1}{2}CV^2dx$.

system when the voltage is applied. As the system is at equilibrium, the sum $dE_e + dE_S$ must be equal to zero. We thus obtain:

$$\cos\theta = \cos\theta_Y + \frac{\epsilon V^2}{2d\gamma} \tag{6.60}$$

This equation is called the Young-Lippman equation (YL) [70]. To summarize, the droplet spreads, as much as it can, to augment the capacitance of the system and then decreases its energy. This decrease is balanced by an increase in surface energy. The outcome is the YL equation.

Then, droplets flatten out to satisfy YL equation. Where do the forces come from? Why does the droplet spread out ? As we said, inside the droplet, counter–ions, attracted by the charges embedded in the dielectrics, build up two charged layers. One at the horizontal surface, and another one, at the liquid–vapour interface, near the contact line, covering only a part of it. Let us assume, as above, that the liquid conductivity is large, and, consequently, these layers are thin. Within each layer, charges repel themselves, and thus work at spreading the droplet over the solid surface. This where the forces originate from. Their calculation requires some technique. The most efficient method for carrying out the calculation is using Maxwell stresses. This was done by Jones, in 2002 [74, 76]. The calculation showed that the corresponding line tension, which pulls the contact line outwards, is $\frac{\epsilon V^2}{2d}$. The result is, as expected, equivalent to the energetical approach. Charges and stresses are localized at the contact line [75,77]. The reason is geometrical: interfaces form a wedge, and, by approaching the apex, each charged layer gets closer to each other. Consequently, the local electric field increases.[11] A singular behaviour takes place there. In fact, calculations show that, at the contact line, electric field and pressure are infinite [75,77]. Still, despite this singular behaviour, the tension (per unit of length) exerted on the contact line, by electrostatic forces, is the one calculated by Jones, i.e. $\frac{\epsilon V^2}{2d}$ [77].

Several remarks can be made:

- Lippman angle is an apparent contact angle. At distances comparable to the insulating layer thickness, i.e. d, the interface bends so as the contact angle, at the surface, its equal to the Young angle [80]. This is sketched in Fig. 6.53

  The reason is that, as shown in Ref. [72, 74], the Maxwell stress is zero at the contact line. Thereby, surface tension forces prevail and the Young angle is recovered. Then, if we look the system at a scale comparable to d, we see that the electric field affects the droplet shape, but not the contact angle, unlike what the term 'electrowetting' might suggest.

- The Debye layer has its own capacitance. It is placed in series with the dielectric/electrode capacitor. Since the Debye layer capacitance is large, it can be neglected. More accurate formulas, taking the Debye layer into account, have been established in the literature [74].

[11] In the calculation, Kang assumed charged layer thicknesses on the order of the Debye length

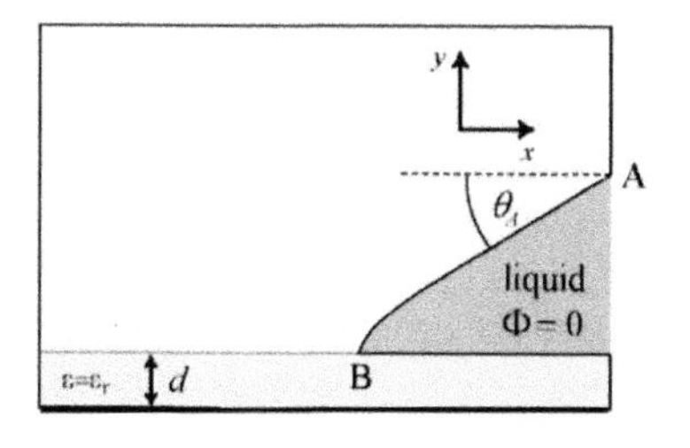

6.53: Droplet shape, close to the solid surface. At point A, the apparent contact angle, characterizing the slope of the liquid interface, is given by the Young-Lippman equation. At point B, i.e. at the surface, the contact angle is θ_Y. (From Ref [74]).)

- Switching between hydrophobic and hydrophilic states takes, typically, 1 ms. From an engineering prospective, this high speed is interesting.

Early measurements of the evolution of the contact angle with the applied tension, carried out with salt water in air, are shown in Fig. 6.54 [75].

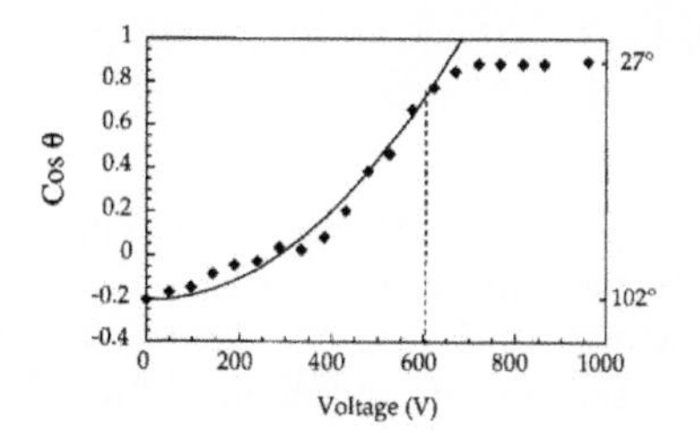

6.54: Early measurement of the cosinus of the contact angle θ as a function of the applied tension V, for a PTFE film in air, 50 μm thick, using salted water. The full line – a parabola – represents Young–Lipmann equation (6.60). (Reprinted from [75] with the permission of Springer Nature. Copyright 2022.)

Fig. 6.54 shows that the contact angle varies from 102°to 27°, as tension V is increased from 0 to 1000 V [75]. The full line – a parabola –represents YL Eq. (6.60). The work showed that theory is in excellent agreement with experiment, in a broad range of voltages. At larger V, the contact angle levels off, a phenomenon called 'contact angle saturation', still not fully understood [74, 79].

Electrowetting is exploited in 'digital droplet technology', also called 'digital microfluidics' (DMF) [81–83]. The concept is shown in Fig. 6.55.

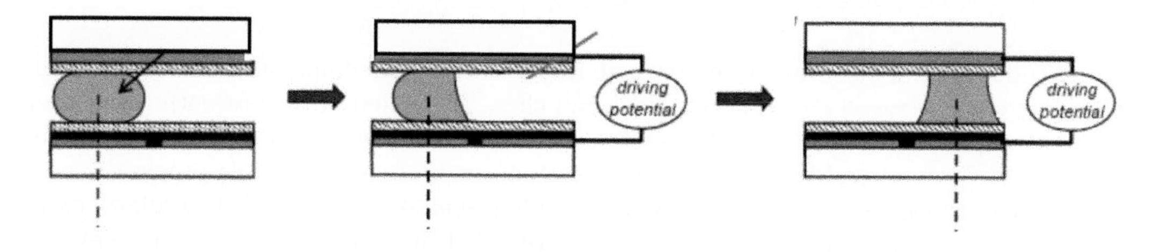

Fig. 6.55: A droplet, initially at rest, feels the action of the right electrode, and, in response, decreases its contact angle. Laplace pressure moves it rightwards. At the end of the journey, it adopts a symmetric shape and returns to rest. The process can be repeated to displace the droplet over an array of electrodes addressable individually.

Let us consider a droplet placed between two plates, with the lower surface patterned with two electrodes. When the electrodes are not polarized, the two contact angles, on the right and on the left are equal, and the droplet stays at rest. When the right electrode is activated, the right contact angle decreases, the interface tends to inverse its curvature, and consequently, due to the action of Laplace pressure, the droplet moves to the right. At the end of its journey, it adopts a symmetric shape and stops. The process can be repeated to displace the droplet on a plane. This is the idea of digital droplet technology.

Fig. 6.56 shows a system where a water droplet is moved at will, over a surface patterned with addressable electrodes [85].

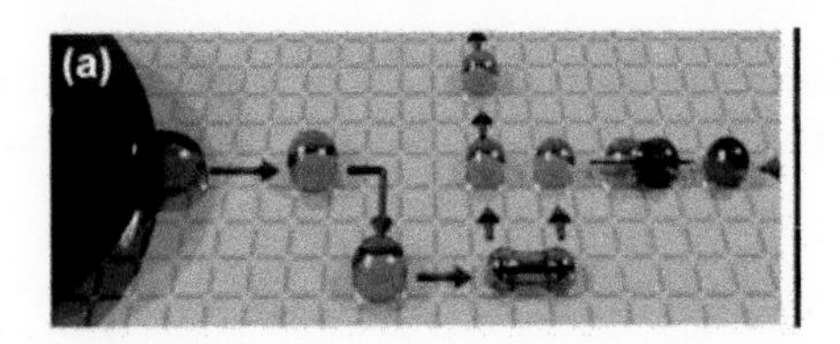

6.56: An EWOD system, in which droplet move on a patterned plate, driven by successive activations of addressable electrodes (Reprinted from [89] with the permission of John Wiley and Sons. Copyright 2022.)

Over the last two decades, several research groups [79, 82, 84–88] have demonstrated impressive realizations: rotating droplets, droplets moving on a substrate in a complicated manner, interacting together, mixing reagents, initiating chemical reactions, cutting droplets. The technology substantiated well the concept of lab on a chip'. This was clearly a groundbreaking technology and some papers promised the advent of a revolution [89].

Several remarks can be made: DMF requires precise microfabrication. Electrodes must be placed close to each other, so as, at any time, droplets feel' the electric field. Typical inter electrode distances range between 5 and 10 μm. With voltages on the order of 100 V, the local electric field is 10 kV/cm. Therefore, DMF does not operate far from dielectric rupture. In fact, today, this constraint does not raise technological nor economical issue. The real practical issue is the long term stability of the coatings over thousands (or millions for displays) of actuation cycles, and for bio-applications, fouling onto the hydrophobic coatings [90].

Today, digital microfluidics remains an important branch of microfluidics, but its commercial success is still concentrated on niches. The situation has recently been analyzed [91]. Commercial EWOD systems are used in two areas: optical (e.g. liquid lenses – see above –) and biomedical, the important example being sample preparation for DNA sequencing.[12] By automatizing sample preparation steps, EWOD allows us to complete sample preparation in ~30 min instead of the 4–5 hours of manual operation. This example illustrates the impressive capabilities of the technology.

[12]Although, today, other systems are preferred

Lenses controlled by electrowetting

Electrowetting is used in optics to control the focus of liquid lenses. The principle, pioneered by Berge [73], is shown in Fig. 6.57. An oil layer, forming a concave lens, is confined in a system filled with water. The oil/water contact line touches an electrode. By varying the voltage, the contact angle changes, as does the focal length. This approach allows for faster focusing. It also makes the lens less susceptible to shock. Mechanical autofocus systems achieve focus in around 500 milliseconds. Fluidic lenses are expected to achieve this in only a couple of milliseconds. The technology may thus substantially impact the mobile phone market.

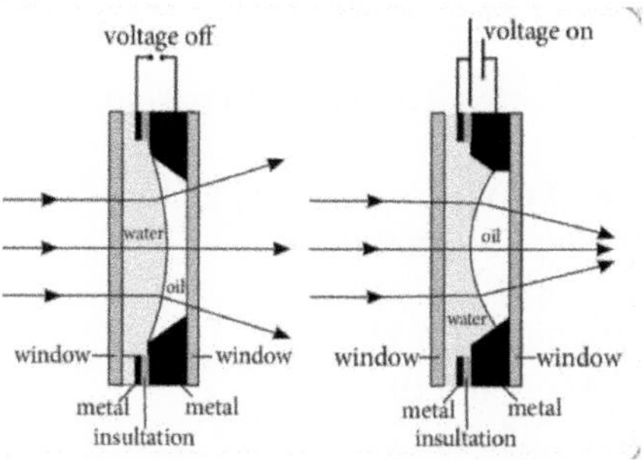

Fig. 6.57: Principle of a liquid lens, controlled by electrowetting. This type of lens can be used on mobile phones.

References

[1] J. Harisson, A. Manz, Z. Fan, H. Ludi, H. M. Widmer, *Anal. Chem.*, **64**, 1926 (1992).

[2] W. B. Russel, D. A Saville, W. R.Schowalter, *Colloidal Dispersions*, Cambridge University Press (1989).

[3] L. Landau, E. Lifshitz, *Electrodynamics of continous media*, Pergamon, New-York,(1984).

[4] R. J. Hunter, *Fundation of Colloidal Science*, Oxford University Press (2001).

[5] J. K. Beattie *Lab on a Chip*, **6**, 1409 (2006).

[6] H.Helmholtz, *Ann. Phys.*, 337 (1879).

[7] M. G. Gouy, *J. Phys. Radium*, 457 (1910).

[8] D. L. Chapman, *Philos.Mag.*, **6**, (1913).

[9] O. Stern, *Z. Elektrochem.*, **30**, 508 (1924).

[10] I. Siretanu, D. Ebeling, M. P.Anderson, S. L. Svane Sipp, A. Philipse, M. C. Stuart, D. van den Ende, F. Mugele *Sci. Reports.*, **4**, 4956 (2014).

[11] M. Brown, A. Goel, Z. Abbas *Angew. Chem.*, **128**, 3854 (2016).

[12] A. M. Smith, S. Perkin, *Phys.Rev.Lett.*, **119**, 26002 (2017).

[13] J. N. Israelachvili, *Intermolecular Surface Forces*, Academic Press (2011).

[14] C. I. Bouzigues, P. Tabeling, L .Bocquet *Physical Review Letters*, **101**, 114503 (2008).

[15] M. Z. Bazant, K. Thornton, A. Ajdari *Physical Review E*, **70**, 021506 (2004).

[16] B. J. Kirby, E. F. Hasselbrink Jr, *Electrophoresis*, **25**, 187 (2004).

[17] R. Boom, C. J Sol, M. M Salimans, C. L Jansen, P. M. Wertheim-van Dillen, J van der Noordaa, *J. Clin. Microbiol.*, **28**, 495 (1990).

[18] C. J. C.Biscombe, *Angew. Chem. Int. Ed.*, **56**, 8338 (2017).

[19] H. Dutrochet, *L'Agent immediat du mouvement vital devoile dans sa nature et dans son mode d'action, chez les vegetaux et chez les animaux*, Dentu, Paris (1826).

[20] F. F. Reuss, *Mem. Soc. Imp. Nat. Moscou*, 2, 327 (1809).

[21] R. Porrett Jr, *Ann. Philos.*, **8**, 74 (1816).

[22] S. Wereley, *ME517, Lecture 36, 'Microfluidic Diagnostics'* (2000).

[23] Paul, M. G. Garguilo, D. J. Rakestraw, *Anal. Chem.*, **70**, 2459 (1998).

[24] P. Wang, Z. Chen, H. C. Chang, *Sensors and Actuators B*, **B 79**, 107 (2001).

[25] B. J. Hinds, N. Chopra, T. Rantell, R. Andrews, V, Gavalas, L.G. Bachas, *Science*, **303**, 62 (2004).

[26] A. Siria, P. Poncharal, A. L Biance, R. Fulcrand, X. Blase, S. T. Purcell, L. Bocquet, *Nature*, **494**, 455 (2013).

[27] M. Smoluchowski, *Bull. Acad. Sci. Cracovie*, 182 (1903).

[28] E. Huckel, *Physik. Z.* **25**, 204 (1924).
[29] D. C. Henry, *Proc. Roy. Soc. Lond.*, **A133**, 106 (1931).
[30] J. Th. G. Overbeek, *Theorie der electrophorese,* PhD. thesis, Utrecht University, H. J. Paris, Amsterdam (1941).
[31] A. S. Jayaraman, E. Klaseboer, D. Y. C Chan, *Journal of Colloid and Interface Science*, **553**, 845 (2019).
[32] T. M. Squires, M. Z. Bazant, *Jour. Fluid. Mech.*, **509**, 217 (2004).
[33] E. Stellwagen, N. C. Stellwagen*Electrophoresis*, **23**, 2794 (2002).
[34] J. L. Viovy, *Rev. Mod. Phys.*, **72**, 813 (2000).
[35] K. D. Dorfman *Rev. Mod. Phys.* **82**, 2903 (2010).
[36] Heller, C., T. Duke,J.-L. Viovy, *Biopolymers*,**34**, 249 (1994).
[37] A.G. Ogston, *Trans. Faraday Soc.*, **54**, 1754.(1958).
[38] PS Doyle, J Bibette, A Bancaud, JL Viovy, *Science*, **295** (5563), 2237 (2002).
[39] J. Han, H. G. Craighead, *Science*, **288**, 1026 (2000).
[40] N. Kaji, Y. Tezuka, Y. Takamura, M. Ueda, T. Nishimoto, H. Nakanishi, Y. Horiike, Y. Baba, *Anal. Chem.*, **76**, 15 (2004).
[41] C. M. Cejas, F Monti, M. Truchet, J-P. Burnouf, P. Tabeling, *Langmuir*, **33**, 6471 (2017).
[42] C. M. Cejas, F. Monti, M. Truchet, J-P. Burnouf, P. Tabeling, *PRE*, **98**, 62606 (2018).
[43] D. Harrison, K. Fluri, K. Seiler, Z. Fan, C. Effenhauser, A. Manz, *Science*, **261**, 895 (1993).
[44] S. Jacobson, R. Hergenroder, L. Koutny, J. Ramsey, *Anal. Chem.*, **66**, 2369 (1994).
[45] J. W. Jorgenson, K. D. Lukacs, *Journal of Chomatography*, **218**, 209 (1981).
[46] J.W. Jorgenson, K.D. Lukacs, *Anal. Chem.*, **53**, 1298 (1981).
[47] S. Jacobson, C. Culbertson, J. Daler, J. Ramsey, *Anal. Chem.*, **70**, 3476 (1998).
[48] A. W. Moore, Jr, S. C. Jacobson, J. M. Ramsey, *Anal. Chem.*, **67** 4184 (1995).
[49] F. Von Heeren, E. Verpoorte, A. Manz, W. Thormann, *Anal. Chem.*, **68**, 2044 (1996).
[50] C. T. Culbertson, S. Jacobson, J. M. Ramsey, *Anal. Chem.*, **72**, 5814 (2000).
[51] Company website: www.agilent.com.
[52] A. T. Wooleyn, R. A. Mathies, *Proc. Nati. Acad. Sci.* **91**, 11348 (1994).
[53] F. Sanger, S. Nicklen, A. R. Coulson, *Proc. Natl. Acad. Sci.*, **74**, 5463 (1977).
[54] R. G. Blazej, P. Kumaresan, R. Mathies, *Proc. Nat. Acad. Sci.*, **103**, 7240 (2006).
[55] C. Emrich, H. Tian, I. Medintz, R. Mathies, *Anal. Chem.*, **74**, 5076 (2002).
[56] C. Venter, *Nucleic Acids Res*, **35**, 6227 (2007).
[57] M. A. Burns et al., *Proc. Natl. Acad. Sci.*, **93**, 5556 (1996).
[58] M. L. Metzker, L. Michael, *Nature reviews genetics*, **11**, 31 (2010).
[59] D. Branton, D. W. Deamer, A. Marziali, H. Bayley, S. A. Benner, et al., *Nat. Biotech.*, **26**, 1146 (2008).
[60] M. J. Levene, J. Korlach, S. W. Turner, M. Foquet, H. G. Craighead, W. W. Webb, *Science*, **299**, 682 (2003).
[61] H. A. Pohl, *Dielectrophoresis: The behaviour of Neutral Matter in Nonuniform Electric Fields*, Cambridge University Press (1978).
[62] G. Green, H. Morgan, *J. Phys. Chem.*, **103**, 41 (1999).

[63] G. Green, H. Morgan, *J. Bioechem. Biophys. Methods*, **35**, 89 (1997).
[64] A. Ramos, H. Morgan, N. Green, A. Castellanos, *J. Phys. D: Appl. Phys.*, **31**, 2205 (1998).
[65] M. Washizu, T. Jones, *J. Electrostatic.*, **37**, 121 (1994).
[66] T. Jones, *Electromechanics of Particles*, Cambridge University Press, New York City (1995).
[67] H. Girault, *Electrochimie physique et analytique*, Polytechniques and Universitaires Romandes Presses.
[68] C. Barbaros, D. Li, *Electrophoresis*, **32**, 2410 (2011).
[69] J. J. Agrestia, E. Antipov, A. R .Abate, K. Ahna, A. C. Rowata, J-C Baret, M. Marquez, A. M. Klibanovc, A. D. Griffiths, D. A. Weitz *PNAS*, **107**, 6551 (2010).
[70] G. Lippmann, *Ann. Chim. Phys.*, **5**, 494 (1875).
[71] F. Mugele, J-C. Baret *J. Phys.: Condens. Matter*, **17**, R705 (2005).
[72] F. Mugele, *Cours les Houches* (2015).
[73] B. Berge, *C. R. Acad. Sci. II*, **317**, 157 (1993).
[74] F. Mugele, J-C. Baret *J. Phys.: Condens. Matter*, **17**, R705 (2005).
[75] M. Vallet, M. Vallade, B. Berge, *Eur. Phys. J. B.*, **11**, 583 (1999).
[76] T. B. Jones, *Langmuir*, **18**, 4437 (2002).
[77] K. H. Kang *Langmuir*, **18**, 10318 (2002).
[78] J. Buehrle, S. Herminghaus, F. Mugele, *Phys. Rev. Lett.*, **91**, 086101 (2003).
[79] W. C. Nelson, C. J. Kim. *Journal of Adhesion Science and Technology*, **26**, 1747 (2012).
[80] F. Mugele, J. Buehrle, *J. Phys.: Condens. Matter*, **19**, 375112 (2007).
[81] M. Washizu, *IEEE Trans. Ind. Appl.*, **34**, 732 (1998).
[82] M. G. Pollack, R. B. Fair, A. D. Shenderov, *Appl. Phys. Lett.*, **77**, 1725 (2000).
[83] J. Lee, H. Moon, J. Fowler, T. Schoellhammer, C.-J. Kim, *Sensors Actuators A*, **95**, 259 (2002).
[84] B. Hadwen, G. R. Broder, D. Morganti, A. Jacobs, C. Brown, J. R. Hector, Y. Kubota, H. Morgan, *Lab Chip*, **12**, 3305 (2012).
[85] K. Choi, A. H. C Ng, R. Fobel, A. R Wheeler, *Annual Review of Analytical Chemistry*, **5**, 413 (2012).
[86] Y. Fouillet, D. Jary, A. G. Brachet, J. Berthier, R. Blervaque, et al., *ASME 4th International Conference on Nanochannels, Microchannels, and Minichannels*, Limerick, Ireland (2006).
[87] P. Dubois, G. Marchand, Y. Fouillet, J. Berthier, T. Douki, F. Hassine, S. Gmouh, M. Vaultier, *Anal. Chem.*, **78**, 4909 (2006).
[88] S. K. Cho, H. Moon, C.-J. Kim, *Microelectromech. Syst.*, **12**, 70 (2003).
[89] M. Abdelgawad, A. R. Wheeler, *Advanced Materials*, **21**, 920 (2009).
[90] F. Mugele, private communication (2022).
[91] J. Li, C. J Kim, *Critical Review, Lab Chip*, **20**, 1705 (2020).

7

An introduction to microfabrication

7.1 Introduction

In the chapter, we will briefly describe the most common microfabrication techniques used in microfluidics today. We will start with silicon technology. The reasons are the followings: the possibilities offered by this technology are considerable, it represents an immense resource of knowledge, and it is used in many commercial devices. Additionally, up until the turn of the century, microfluidics was primarily a silicon or glass-based technology. In a second part of the chapter, we will describe soft technology, which is currently most frequently used in the laboratories. We may give tribute for this to G. Whitesides, whose papers, published in 1998, induced a change of paradigm in the field [1]. Before the advent of soft technology, many scientists believed that microfluidics could be extremely useful in biology and chemistry. However, they were hindered by the clean room barrier. G. Whitesides played a crucial role in lowering this barrier and opening a box of innovations. We will devote significant attention to this technology in this chapter.

This chapter does not intend to be exhaustive. For more detail, the reader can refer to books, such as *Fundamentals of Microfabrication* [2], *Handbook of Microlithography, Micromachining and Microfabrication* [3], or *Fundamental and Applications of Microfluidics* [4].

7.2 Current situation of microtechnologies

7.2.1 Hard, plastic and soft technologies

From a material prospective, one can classify the microfluidic technologies into several categories: hard materials (silicon, glass), plastic materials (COC, PMMA, etc.) and soft materials (PDMS, PUV, hydrogels, etc.).

Hard materials: these include silicon and glass. The technologies are based on etching/lithography/deposition. The typical dimensions associated to hard materials range between 20 nm and 500 μm.

Plastic materials: these materials allow channels to be created with sizes that mainly range between 20 μm and several milimiters, i.e. embracing micro– and millifluidic scales. A large variety of materials are used, with different surface properties, e.g.. hydrophobic and hydrophilic. Examples are polymethylmethacrylate (PMMA), Cyclic olefin copolymer (COC) or SU8, a photoresist. Plastic devices are made using different technologies, such as micromilling, laser ablation, or injection. Plastics are extremely important for the industrialization of microfluidic devices, due to their low cost. With the development of 3D printing, they are being increasingly used in research laboratories.

Soft materials: these materials are mostly used for creating channels (or other structures) between 0.5 and 500 μm of characteristic sizes. PDMS is by far the most used one.

7.2.2 Tolerances

Figure 7.1 represents the dependence of the relative tolerance with the size of the objects to be created.

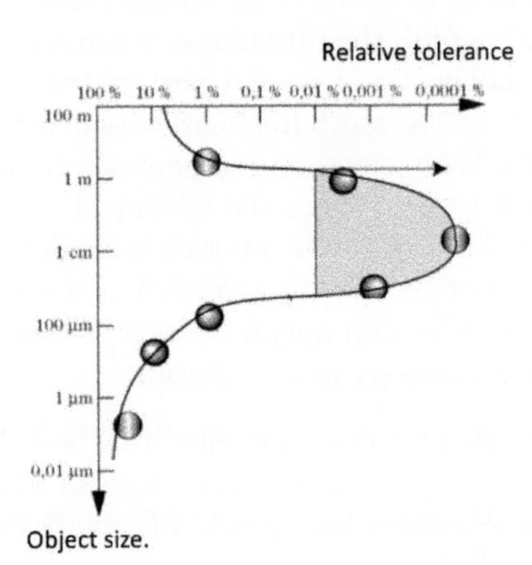

7.1: Relative tolerance at which several objects, characterized by their dimensions, are fabricated.

A number of comments can be made:

- In the construction domain, houses are built with a tolerance of 1%. For instance, for a vertical wall, a tolerance of 8 mm for a height of 1 m is recommended by the agencies. This represents a relative uncertainty of 0.8 %.
- Traditional machining (milling machine, lathe, etc.) creates objects between 1 mm and a few decimetres. The technique has a precision on the order of tens of μm. The relative tolerance is thus high, up to 10^{-6} (upper grey area of Fig. 7.1). Over the last decades, the resolution has been improved. The extreme case is the Robonano machine, which creates metallic objects with a precision of tens of nanometres [5] (see Fig. 7.2 (A)). The machine has five degrees of freedom (3 in translation, 2 in axial rotation). Its displacement precision is of 1 nm, and its angular precision is $(10^{-5})^{\circ}$. Fig. 7.2 (B) shows a Boudda micromask made with such a machine.

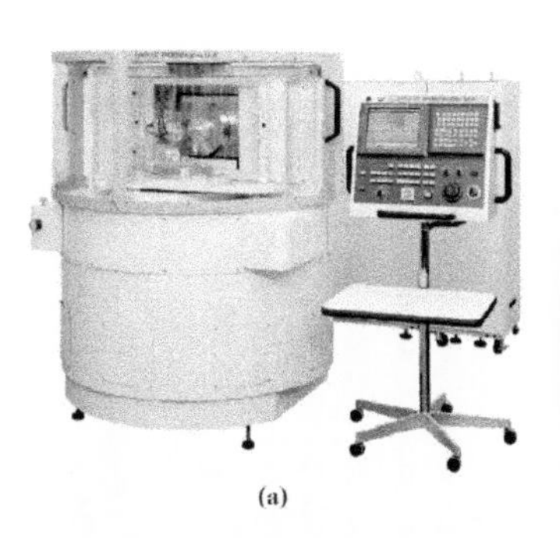

(a)

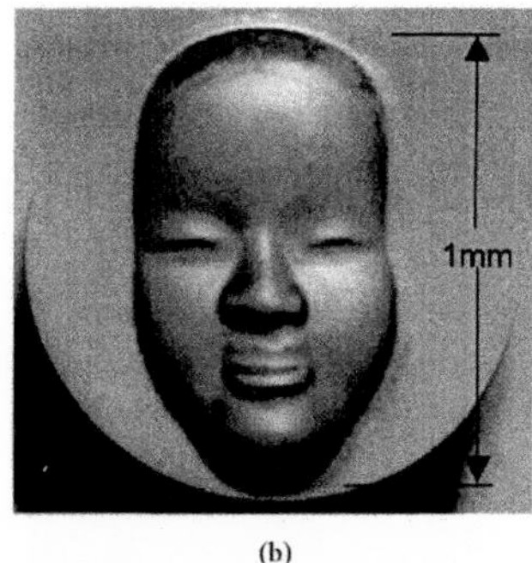

(b)

Fig. 7.2: (A) Machine ROBOnano-Ui. It has 5 degrees of freedom (3 in translation, 2 in axial rotation). The displacement precision is of 1 nm, and its angular precision is of $(10^{-5})°$. (B) Bouddha micromask fabricated with Robonano. (photos by Pr Takenchi, University of Electro-Communications, Osaka.)

Robonano has not penetrated the laboratories for obvious reasons: it can only be used with metals, it is extremely expensive and slow. On may appreciate these machines as great technological feats, the crowning achievement of traditional methods, but with very limited usage in practice.

- The domain of microfabrication involves scales spanning between one micrometre and a millimetre. The absolute tolerance is 1–10 μm and the relative tolerance 1–10 percents. It is amazing to note that microfluidics is less accurate than masonry.
- The fabrication of nanometric objects (between 1 and 100 nm) embraces a large variety of techniques (silicon technology, carbon nanotubes (CNTs), etc.). They are associated to high absolute precision, and relative tolerances typically low or extremely low.

7.3 The environment of microfabrication

7.3.1 The environment of microfabrication

Dust particles erring in laboratories have sizes ranging, roughly, from a few tens nanometres to tens of micrometres. They wander in the air and deposit onto surfaces. In a standard environment, there exist millions of dust particles per feet cube. This dust does not perturb human activity, but, considering the scales used in microfabrication, they may jeopardize the fabrication of microfluidic devices. Figure 7.3 shows a pollen dust particle stuck on a microelectronic circuit. It is not difficult to imagine that such particles can clog microchannels or affect the surface properties of the material. Microelectronics faced this problem for years, until Willlis Whitefield from SANDIA Laboratories invented, in 1962, the concept of 'clean room', i.e. isolated spaces permanently cleaned by air fluxes. In clean rooms, air circulates from the ceiling to the floor, where it is extracted. This configuration forces the particles produced by the users – the most abundant origin – downwards, preventing benches, equipments, etc.. from dusting. This invention represented a milestone in the field.

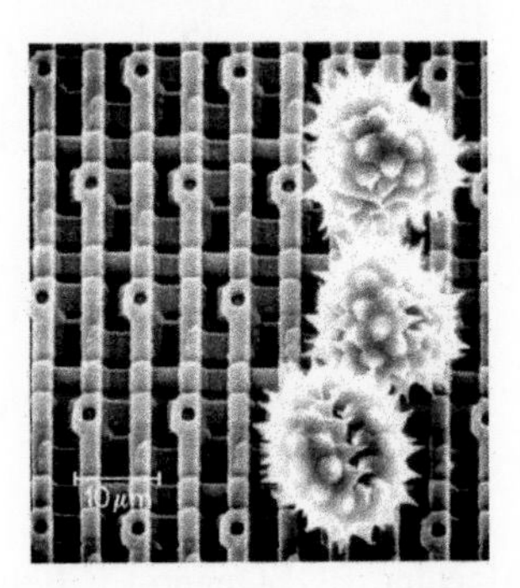

7.3: Pollen on a microelectronic circuit [2].

What is a clean room? It is an environment thermally regulated (around 20 °C), in which humidity is controlled, and which is permanently traversed by filtered air fluxes. The quality of the clean room is measured by counting the number of particles of a size less than 4 μm contained in one cubic inch[1]. Typical clean room classes for MEMS microfabrication are 1000 to 10000. Now, more and more frequently, ISO classification is preferred to the notion of clean room classes. A clean room of class 1 is ISO3, of class 1000 is ISO6, etc.. Inside laminar hoods, classes are typically between 10 and 100. In microelectronic industry, it is not rare to find large rooms classed at 10 or 1 to reduce failure rates. For a class of 1000 to 10000, it is necessary to wear specialized clothing, to cover hair, and to put on gloves and shoe–covers. These precautions are not necessary for rooms of higher classes (i.e. less 'clean').

An example of microfluidic facilities, located in the Pierre GIlles de Gennes Institute in Paris, includes a 200 m^2 'grey room' (class 10000) and a 100 m^2 clean room (class 1000). It is shown in Fig. 7.4.

7.3.2 Using or not using clean rooms for microfluidics

Figure 7.5 shows two extreme cases: a sophisticated submicrometric device (Fig. 7.5 (A)) and a crude millichannel (Fig. 7.5 (B)). In the first case, the device would be impossible to create outside a clean room. In the other case, a photocopyer and a modest amount of PDMS were enough to create the milli–channel.[2] The cross-section of the canal is not geometrically well defined and the walls are rough. Nonetheless, the device was functional.

Microfluidics stands between these two extremes: sophisticated technologies, most often associated to high costs and long processes, and low technologies, most often associated to low costs and speed.

[1]Recall that an inch equals 2.56 cm.

[2]The idea was proposed in Ref. [8].

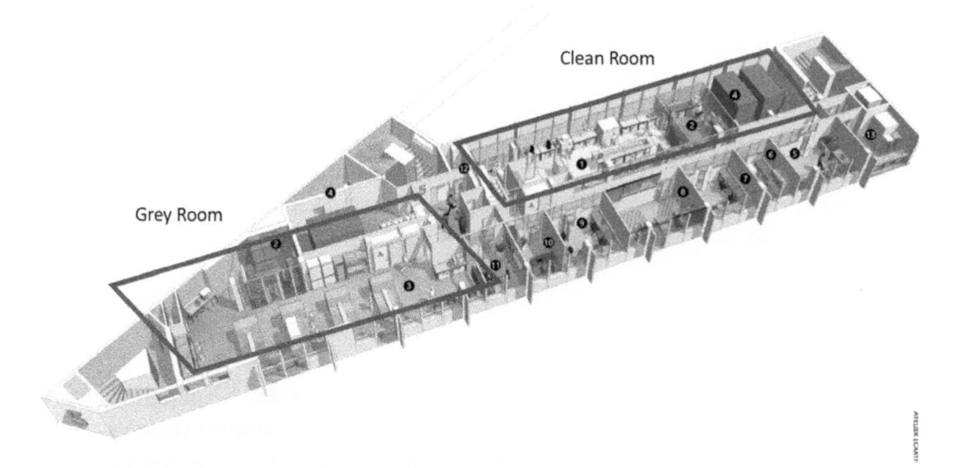

Fig. 7.4: Institut Pierre Gilles de Gennes (IPGG) platform, in Paris, including a 200 m^2 grey room (class 10000) and a 100 m^2 clean room (class 1000). (Drawn by Ecart Fixe.) The grey room includes oxygen plasmas, nanoimprint apparatus, ovens, benches, an actinid room and the clean room an electronic microscope, profilometers, deposition equipment, aligner, spin coater, ellipsometer, and ovens. .

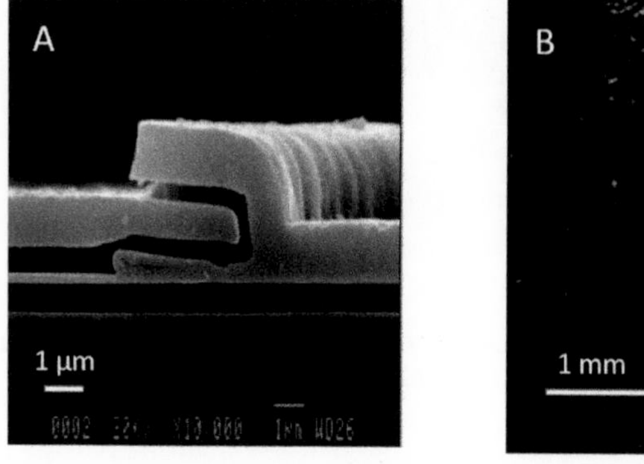

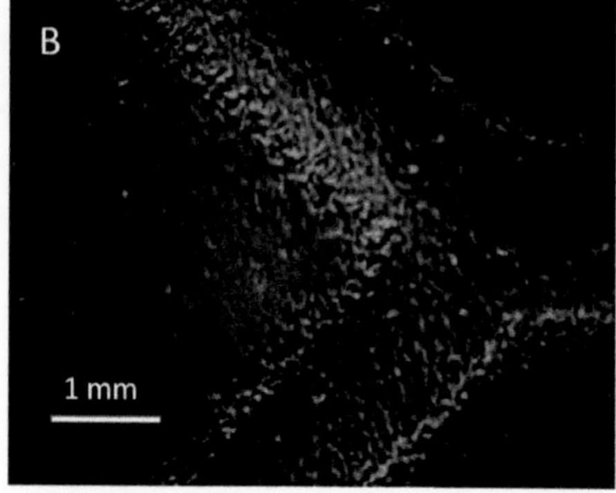

Fig. 7.5: (A) Submicrometric structure(Courtesy of P. Joseph). (B) Channel made with a photocopyer (From T. Jues, R. Prunieres, E. Brunet, P. Tabeling (2002).)

7.4 Photolithography

7.4.1 Introduction

A process playing a central role in microfabrication, is photolithography. The term photolithography' includes the word 'litho', which means stone'. The term refers to the first photolithography, made on a limestone, by Alois Senefelder, in 1798 [37].

We will see that photolithograpy uses collimated sources. Today, different types of sources are available: X-ray, electron, and photon. In practice, microfluidics uses optical lithography, with wavelengths between 200 and 450nm. Using X-rays has been done, giving rise to a technology called 'lithographie, galvanoformung, abformung' (LIGA). LIGA conveyed the hope that, by scaling down microfluidic devices, new valuable systems could be created, as in microelectronics. Despite its potential usefulness, it has not been widely used due to the challenges posed by its complexity. Electron lithography is essentially used for making precision masks. Lithography in the optical range, widespread in microfluidic laboratories, will be the focus of the chapter.

7.4.2 Photolithography masks

Masks are of two types (see Fig. 7.6): high resolution (plates of quartz on which deposits of chrome forming a pattern are made) and low resolution (5 μm or so, made by ink printing on a plastic sheet, most currently used in the labs, owing to its low price and availability). High precision masks are made by using electron beams with a precision on the order of a fraction of a micrometre. We will not describe the technique in this book.

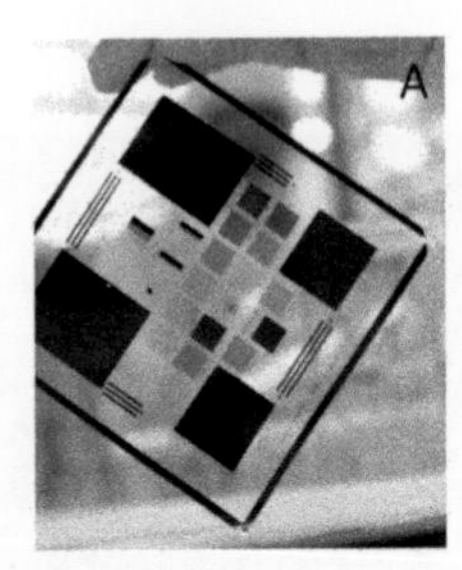

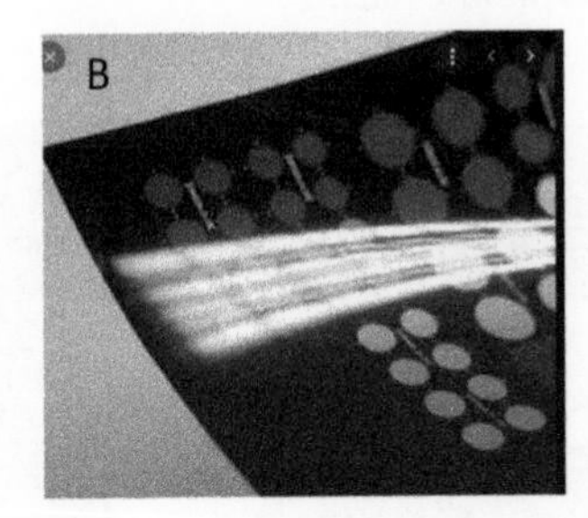

Fig. 7.6: (A) High resolution mask, made by depositing chromium on quartz (submicrometric resolution). (B) Low resolution mask (5 μm or so), using a transparent sheet.

7.4.3 Deposition of the photoresist

The photosensitive polymer is deposited in a thin layer on a solid substrate of silicon or glass, usually a wafer. The deposit is made using a spincoater, as depicted in Fig. 7.7. This system consists of a disk, that maintains the wafer on it by evacuating air. The wafer rotates at rotation speeds ranging between 1000 rpm and 10000 rpm. While the resist spreads on the substrate, the solvent evaporates, and the film thickness saturates. The material thus formed resembles a soft solid. It is remarkable that the thickness h of a film obtained in this way is extremely uniform (to about 20 nm). The evolution of the thickness h of the film is described in [10]. It is governed by the following equation:

$$dh/dt = -2Kh^3 - E$$

in which t is time, $K = \frac{\omega^2}{3\nu}$, where ω is the angular velocity, ν is the kinematic viscosities, and E is the evaporation rate. Two regimes can be singled out, with a cross–over time $\tau = 2\pi/3^{3/2}(2E^2K)^{-1/3}$. At small t/τ, the evaporation is negligible, and the solution spreads, under the action of the centrifugal forces. At late times, evaporation comes into play, the solution concentrates, its viscosity increases, and h levels off. Eventually, the limiting height is given by the following expression:

$$h = kC\left(\frac{\mu}{\omega^2}\right)^{1/3}$$

where C is the initial concentration of polymers in solution, and k is a constant (currently on the order of 0.8). In practice, it takes seconds or tens of seconds to reach the equilibrium thickness.

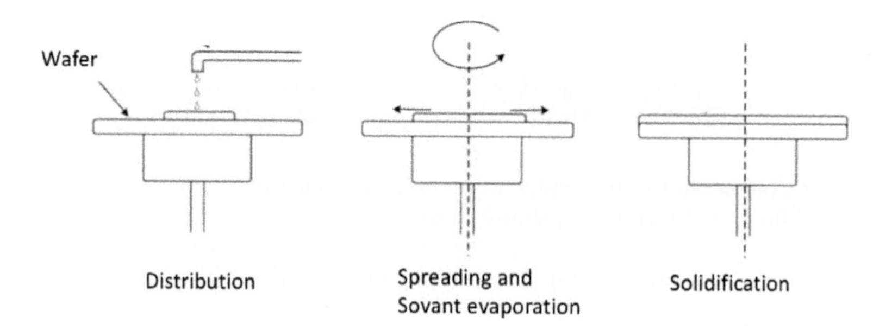

Fig. 7.7: Deposition of the resist on a spincoater).

Depending on the polymer used, film thicknesses vary between a fraction of a micrometre to 200 μm. Once the spreading is completed, the resist still typically consists of about 15 % solvent. If this solvent is not removed, cracks or fissures appear once the film is completely reticulated. To eliminate the solvent completely, the resin is heated slightly (at 70 °C for a few minutes).

7.4.4 Exposure

After spin coating, the photosensitive polymer and its substrate are placed in an aligner. The film is thus exposed to a luminous flux produced by a source crossing the photolithographic mask. The source is most often a mercury vapour lamp that delivers substantial luminescent power (10 to 20 mW), in wavelengths between 300 and 450 nm. [3] The luminous flux initiates physico-chemical reactions in the polymer, which modify its solubility. There are two types of resists: positive and negative.

[3] g- and i-lines are used at 436 nm and 365 nm, respectively.

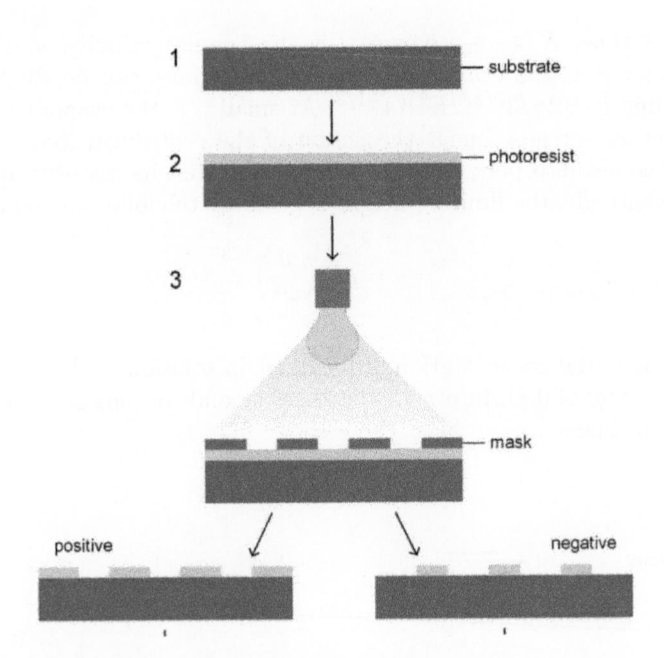

Fig. 7.8: Principle of photolithography. In practice, in lithographic machines (called aligners), beams are collimated, which is not shown here.

- **Positive resist:** the insolated zones become soluble (in some solvent), and the others remain insoluble.
- **Negative resist:** the insolated zones become insoluble and the others remain soluble.

Thus, for positive resists, insolation across a photolithographic mask defines the zones that will, after development (that is, after immersion in a solvent), form holes in the resist film. For negative photoresists, this is the opposite. The resist must be sufficiently transparent to allow the illumination of the whole thickness of the deposited layer, and sufficiently sensitive for the light to develop the appropriate chemistry. In practice, the thickness of a positive resist films does not exceed 10 – 20 μm. The invention of SU8 – a negative photoresist.–, by IBM, in the 1990s, constituted a small revolution in the field. With SU8, it is possible to deposit thick films (hundreds of micrometres).

Several phenomena alter the resolution of the photolithography. One is diffraction at the edges of the mask openings. As in microscopy, this effect limits the resolution to a quantity close to the diffraction limit, i.e. λ/NA, in which NA is the numerical aperture of the collimating lens. In practice, resolutions are on the order of 1 μm.

Another limitation is linked to the penumbra effect: as shown in Fig. 7.9, a collimated beam crossing a mask, and projected onto a screen, produces lit and dark regions. In between, there is a penumbra. Another phenomenon limiting the precision is diffraction. The size δ of the diffraction zone is given by the following formula:

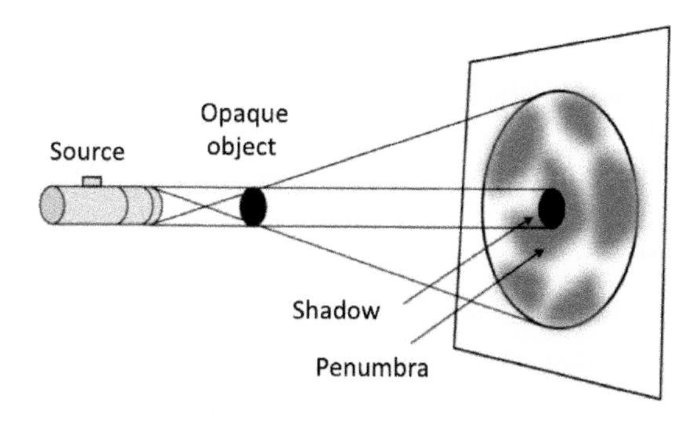

Fig. 7.9: Penumbra effect, degrading photolithography resolution. The effect is reduced by approaching the mask close to the wafer.

$$\delta \approx 3\sqrt{\lambda h},$$

where λ is the light wavelength and h is the thickness of the polymer film.[4] Inserting typical values, we obtain that δ is on the order of 2–3 μm. This steers researchers to work with resists of limited thicknesses and short wavelengths.

An example of a protocol for the fabrication of an SU8 mold

Suppose that the resist, the optical aligner, spincoater, and a silicon or glass substrate are all available. The substrate must be cleaned and dried, which is achieved in two steps. The first step consists of immensing the wafer in acetone (to dissolve organic residue), in an ultrasound bath, and then drying. The second step consists of performing the same operation but with alcohol. The substrate must then be placed on a hotplate at 120°C for 5 min for dehydration. It is necessary to have a substrate with no traces of contaminant whatsoever. Then the resist is spread on the wafer with a spincoater. The system is then heated to 65°C for 60 s, then at 95°C for 180 s, which serves both to eliminate the violent nature of the residual solvents and also as a first hardening step. It is after the exposure step that the baking step known as *post bake* (PBM) takes place. To encourage the progressive rearrangement of material during thermal deformation, heating is achieved in two steps: the wafer is heated to 65°C for 60 s, and then to 95°C for 120 s. The system is now ready for the development step. The system is dipped in a developing solution for 180s. The SU8 is a negative resist, so the exposed motives stay, and the non-exposed motives are dissolved. The whole system is then rinsed with alcohol and dried with nitrogen gas. To finish the protocol and obtain a stable mold, the whole system is heated to 200°C for 2 h.

[4] Mask aligners generally have two positions: contact (the mask is in contact with the resist) and proximity (the mask is 10 μm above the resist). The second mode is more often favourable because the first carries the risk of the resist, which is still in a sticky state, adhering to the mask.

7.4.5 Commonly used resists and the development step

We now properly discuss resists. Resists used for photolithography must possess the following characteristics:

1. a large contrast between the solubilities of the exposed and unexposed parts;
2. high photosensitivity; and
3. high resistance to certain classes of chemical agents.

The solutions used for photolithography are composed of a photosensitive resist, a solvent (that reduces the viscosity of the solution, allowing the spreading on a spin-coater), and an additive (in order to control the kinetics of the photoreaction). For positive resists, the light breaks or weakens the internal bonds of the resist, and induces a rearrangement of the molecule to a more soluble form. For negative resists, the action of the light is to induce the formation of covalent bonds between principal or secondary chains, making them soluble by a process we will not describe here. Examples of 'positive' resists include DQN (a mix of diazoquinone with the resist phenoline novolak), and AZ. These resists are soluble in highly basic solutions such as KOH, TMAH, ketones, and acetates. As an example we cite the negative resists KTFR (a polyisoprene elastomer produced by Kodak), and most importantly, SU8, which we already mentioned. SU8 can be used as a mould, but can also form structures with a high aspect ratio.

With regard to microfabrication, the choice between positive and negative resists is not trivial: negative resists adhere better to the substrate than positive resists, and tend to be more chemically resistant. However, the contrast of the photosolubility of negative resists is weaker than those of positive resists. These characteristics are all factors in the refinement of the microfabrication process. After the insolation step comes the development step. This step consists of immersing the system in a solvent. This step must be carried out at a carefully controlled temperature. It is necessary to remember that the relevant processes during the development phase have a high temperature dependence. Then, either the exposed part (for positive resists) or the unexposed part (for negative resists) is eliminated. The photolithographic process concludes by the reinforcing of the cross-linking of the resist, incubating the system well above the glass transition temperature typically for tens of minutes.

7.5 Direct writing or maskless photolithography

Direct writing (also known as maskless lithography) refers to any technique or process capable of altering the chemistry, depositing, removing, dispensing, or processing various types of materials over different surfaces following a predetermined layout or pattern. There are various means to achieve the desired patterns and they can be broadly classified into additive and subtractive techniques. Additive techniques such as ink jet printing, dip pen nano-lithography (DPN), and micropens add material to a substrate based on a CAD layout. Subtractive techniques such as focused-ion beam

(FIB), and laser micromachining selectively remove material from a substrate using an ion beam and laser source, respectively.

Fig. 7.10 shows an example of direct writing technology, called laser direct writing (LDW).

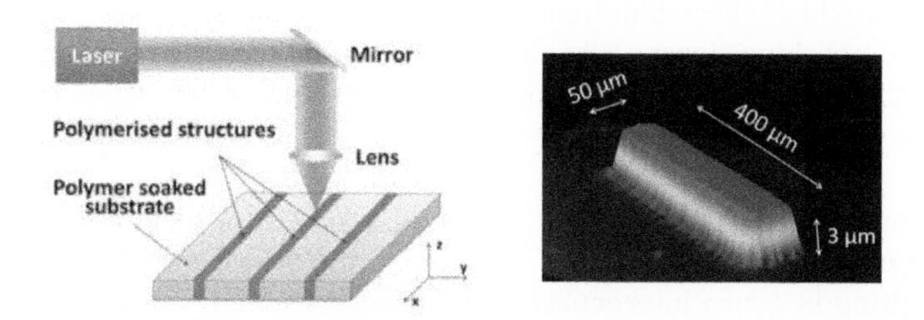

Fig. 7.10: (Left) Principle of direct writing: a spot is illuminated, translated so as the whole surface is exposed. (Right) Example of the direct writing of a photosensitive hydrogel structure [12]. It was obtained with Dilase 650 Kloe, using a 266 nm wavelength and 50 mW output-power source.

In this case, a laser beam illuminates a photo-sensitive polymer layer, which, in a same manner as negative photo-resists, cross-links under illumination. After washing, a structure in form of parallel rails is formed. The figure shown on the left shows an example of a structure obtained with a photo-curable PNIPAM hydrogel. Dimensions indicated on the figure illustrates the fact that the resolution of LDW is comparable to standard photolithography, the limit being again linked to diffraction. LDW is more flexible than standard lithography, but slower, owing to the fact that, in the former case, one small spot is exposed one after the other, while in the latter case, the whole object is illuminated. In the former case, the process is serial, and in the other, parallel.

7.6 Microfabrication methods for silicon and glass devices

7.6.1 Silicon as a microfabrication material

The use of silicon is justified for the following reasons:

- the ready availability of the material, and the considerable documentation of its properties;
- the large number of well-established microfabrication processes, which allow for the fabrication of devices with greater than submicrometric precision;
- the possibility of integration with electronic circuits;
- its anisotropy, advantageous for microfabrication; and
- its physico-chemistry, compatible with many processes.

Germanium is not a viable alternative to silicon because the native dioxide GeO_2 is soluble in water, making it inconvenient for photolithography. Silicon is delivered in the form of wafers that constitute a monocrystal. Silicon is produced by crystalline growth, process carried out in a very clean environment (around class 1 or 10). The growth process, developed by Czochralski, consists of slowly pulling a crystal from an ultrapure bath of silicon while the whole ensemble rotates. A cylindrical crystal is obtained, from which slices hundreds of micrometres thick are cut. This is followed by the atomic polishing phase. All this leads to the fabrication of a wafer, atomically smooth on the polished surface, whose crystallographic orientation is identified by marks on the wafers.

From a crystallographic point of view, silicon is a cubic crystal, whose structure is the same as that of diamond (Fig. 7.11). We recognize two cubic networks with overlapping centres, resulting in a two interpenetrating face-centered cubic structure. The side of one cubic face is 5.43 Å. Each tetravalent silicon atom belonging to a given network is at the centre of a tetrahedron, whose points are made of atoms of the partner network, as shown in Fig. 7.11. The highest-density planes are the (111) planes, which form an angle of 54.74° with respect to the (100) planes.[5] We will later see that these planes play an important role in the wet-etching process.

Table 7.1 displays a number of mechanical, electrical, and thermal properties of silicon.

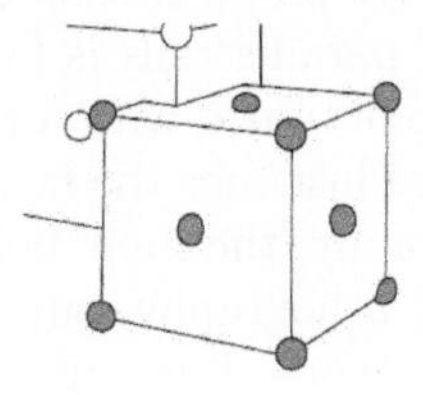
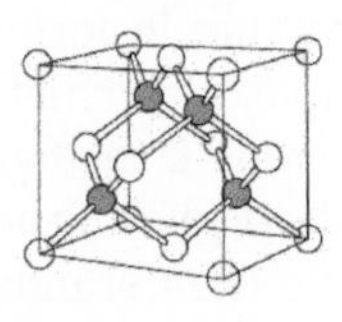

7.11: Silicon structure. Two cubic networks with overlapping centers, resulting in a two interpenetrating face-centered cubic structure. The side of one cubic face is 5.43 Å.

The Young's modulus of silicon is comparable to stainless steel. A similar statement can be made for polysilicon (composed of silicon atoms in a structurally disordered state) (160 MPa).[6] Silicon is a fragile material, as is glass. This signifies that there is no plastic zone. The stress rupture is on the order of 7 GPa, with a maximal strain of 3.5 percent. Silicon is an insulator at ambient temperature. It is an excellent piezoresistor, a poor piezoelectrics, and an excellent thermal conductor (comparable to a metal). Thermal expansion of silicon is small: in the range of temperatures around 300 K, the corresponding coefficient is equal to 2.33×10^{-6}, a value comparable to glass. The coefficient decreases with temperature, and becomes negative around 100 K. Native silicon dioxide SiO_2 adheres well to the substrate, is thermally stable, has an excellent chemical resistance and is insoluble in water (unlike germanium dioxide).

[5] Examining the elementary cell, this angle is seen to be arctan $\sqrt{2}$.

[6] Polysilicon is the structure adopted by silicon, when deposited on a substrate to create microstructures.

Physical size	Value	Units
Rupture strength	7	GPa
Young's modulus	190	GPa
Density	2.33	g/cm^3
Thermal conductivity	2.33	W/cm K
Thermal dilation	2.33×10^{-6}	K^{-1}
Electrical permittivity	11.9	F/m
Electrical field breakdown	3×10^5	V/cm
Electrical resistivity	2.3×10^5	Ω cm
Thermal diffusivity	0.9	cm^2/s
Specific heat (at constant pressure)	0.7	J/g K
Fusion temperature	1415	K

Table 7.1 Physical characteristics of silicon.

Currently, silicon wafers are delivered with one or both sides polished, the location of the crystallographic orientation (defined by two planes) marked, and with different diameters and thicknesses. To get an idea of typical sizes, the 4-inch wafer is frequently used because current machines in the field of MEMS have adopted this format, though the trend is going towards increasing these sizes. The reason for this is practica: the larger the wafer, the more microsystems can be created in a single batch. There is a disparity between microelectronics, which often uses 8-inch and 12-inch wafers, and microfluidics, which uses smaller wafers. A typical wafer thickness is 500 μm. It is interesting to mention SOI wafers (*silicon on insulator*), which possess three layers, two made from silicon and one made from silicon dioxide. They facilitate the fabrication of a certain number of devices, notably those using sacrificial layers.

7.6.2 Wet etching of silicon and glass

Wet etching consists of exposing an object, on which certain parts have been protected by a mask, to a chemical reagent, in a liquid, whose action leads to material removal. This process was in use as early as in the fifteenth century to decorate armours. To accomplish this, the armour was covered with wax, the parts destined for exposure to reagents were cut out, and then it was all dipped into a reactive bath for a determined period of time. When the armour was taken out of the liquid bath and the wax was eliminated by heating, an etched or hollowed-out pattern was obtained. The discovery of photosensitive materials by Niépce, the inventor of photography, represents an important milestone in the history of this technique. Later, the microelectronic revolution of the 1950s caused an extraordinary refinement of wet etching. We now have a large amount of data on etching velocities in different conditions, which allow us to control the geometry of the etched object to just a few micrometres.

An important distinction must be made between isotropic and anisotropic etching. Isotropic etching is carried out equally in the three spatial directions, and can be used for forming structures such as spherical cavities; examples include fluoric acid in glass and EDP in silicon. Anisotropic etching consists of a chemical attack that is carried out preferentially along one of the crystallographic planes. It is possible to create cavities that have facets, which can be useful. This type of etching is not possible in glass, since it is an amorphous solid. Figure 7.12 illustrates these two types of etching for silicon. In the first case (left-hand figure), EDP is used for isotropic etching, while in the second case (right-hand figure), the base KOH is used to etch anisotropically, which is carried out along the planes (111).

Fig. 7.12: (Left) Isotropic wet etching using EDP in glass. (Right) anisotropic wet etching using KOH in silicon.

Isotropic etching of silicon and glass. Isotropic etching generally implies the use of acids: a commonly used system for silicon is composed of HNA, which is a mix of $HF/HNO_3/CH_3COOH$. For glass, a commonly used system is fluoric acid. In these two cases, the etching is carried out at ambient temperature, with etching velocities of tens of micrometres per minute. The attack is accomplished in four steps: the transport of the reactant towards the surface, the local development of chemical reactions, the formation of a cavity, then the evacuation of the product. The transport steps are controlled by the mixing of the reactants in the bath in which the material is submerged. As the etching velocities are high (in comparison to the velocities of the evolution of the diffusion fronts), it is important to mix well to ensure the homogeneity of the concentrations of reactive chemicals in the volume of the system. Chemically speaking, the attack of the reactant is in fact a subtle process. For example, the action of HNA on silicon makes use of the following global reaction:

$$\mathrm{Si} + \mathrm{HNO_3} + 6\mathrm{HF} \longrightarrow \mathrm{H_2SiF_6} + \mathrm{HNO_2} + \mathrm{H_2O} + \mathrm{H_2}\ (\text{gas}).$$

Simplifying the reaction model, we can consider the action of HNA to essentially consist of breaking the covalent bonds of the crystal, and producing a fluoride compound of silicon (H_2SiF_6) that is soluble in fluoric acid. The reader can refer to the work of Marc Madou [2] for a detailed presentation of this process.

Since we are dealing with a chemical reaction, the temperature of the bath must be maintained constant in order to control the depth of the etching. Typically, regulation to a few tenths of a degree is necessary to guarantee precision on the order of 10 percent of the depth. The state of the surface is also an important issue: physically, to obtain smooth surfaces, it is useful to place the system in the regime where diffusive processes (which tend to make reactive concentrations uniform) are quicker than the chemical kinetics. This regime is chosen by controlling the temperature of the system. The corresponding conditions are now known, and are just one of several parameters defining microfabrication protocols. It is useful here to refer to the specialized literature. In an ideal situation, the mask is not attacked by the reactant. In practice, however, there is a chemical attack of rates on the order of a few per cent of those in the intended process itself. For example, silicon is used as a mask for HNA, but it itself is also attacked by HNA, although at a rate 80 times slower than that of silicon. It is thus important to work with thick masks (hundreds of nanometres) to ensure that the masked part remains covered during the entire etching step. However, it has also been shown that the thicker the mask, the less geometric precision is achieved when the pattern is transferred from the lithographic mask to the photosensitive resist. A compromise must thus be reached. As a final point, it is important to note that the chemical reactant etches not only the volume of the exposed part, but also under the mask. This phenomenon can be elegantly used to form original structures, such as planes fixed on rounded supports, or to make tips for atomic force microscopy (AFM).

Anisotropic etching of silicon. Anisotropic wet etching of silicon is a very common technique in the field of MEMS because it allows the formation of flat surfaces, such as wells or canals with flat walls (but with a trapezoidal cross-section, as we will later see). Here, the reactants that are used are alkanes at high concentrations (several moles per litre); unlike isotropic etching that uses acids, the medium in which the object to be etched is immersed is strongly basic.

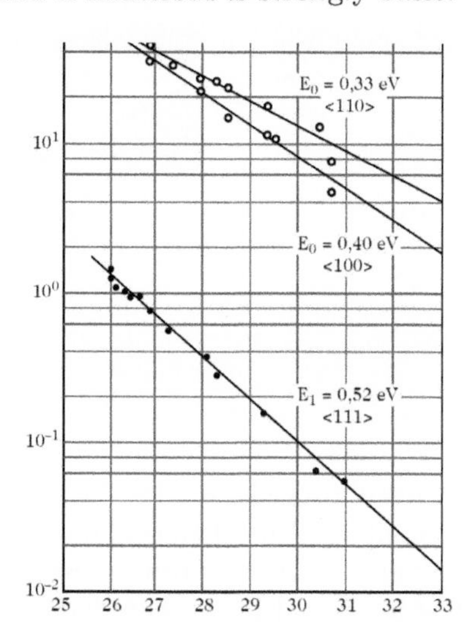

7.13: The evolution of the etching speed with respect to temperature for different crystallographic orientations (anisotropic etching). The lowest velocities seem to correspond to the densest planes. The contrast between slow and fast orientations is one order of magnitude [2].

The principle of anisotropic etching is based on the diagram shown in Fig. 7.13.

The etching rate depends on the crystallographic orientation that the chemical attack follows. The velocity of etching is slow along the planes (111) (on the order of 13 μm per hour), and faster along other orientations. The corresponding ratio of etching velocities is on the order of 60. The consequence of such a contrast is that the silicon crystal immersed in base will make the cut-out forms appear spontaneously along the planes where the etching is the slowest, i.e. along the planes (111). An analogous process occurs during the phenomenon of faceted growth, where the visible facets are those associated with the slowest growth velocities, which also happen to be the highest-density planes. It is this process that causes cubic crystals such as salt to have an orthorhombic structure (constructed from the planes (111)). Unlike isotropic etching, the process of anisotropic etching is not completely understood in all its detail today. There are several concurrent models that describe the various mechanisms in play. The reader can refer to specialized literature for a description of these models. As for the state of the surface, anisotropic etching theoretically allows for the production of an atomically smooth surface formed from terraces. In fact, in the common case of silicon etching by KOH, the production of hydrogen bubbles introduces local fluctuations in concentration of the chemical reactants, inducing inhomogeneities in the etching velocities, and thus leading to more roughness in the surfaces. These perturbations can be reduced by using moderately elevated temperatures (thus reducing the flux of bubbles, which raises the temperature), and thorough mixing of the bath. In a standard well-tuned process, the surfaces have a local roughness on the order of 20 nm.

It is not always simple to guess the shape produced by anisotropic etching. Figure 7.14 illustrates this point.

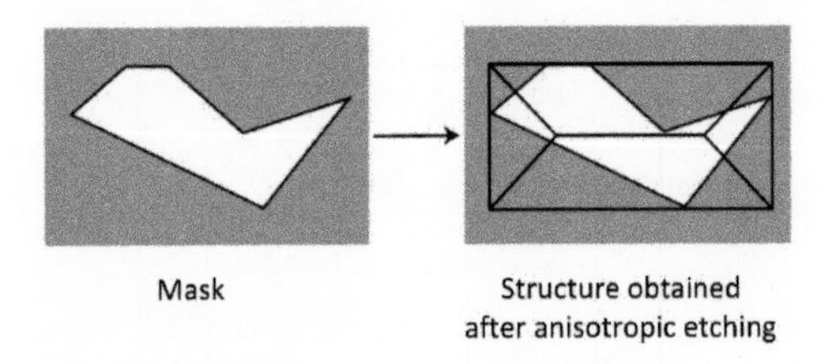

7.14: Structure (black line) obtained with anisotropic etching [2].

7.6.3 Dry etching of silicon

Dry etching is the removal of a substrate exposed to ionic species contained in a gaseous or plasma phase. Depending on the conditions, the shapes obtained can be anisotropic or isotropic. Here, we will limit ourselves to a very brief presentation of the methods and principles of dry etching. The interested reader can find a detailed presentation of the technique in Ref. [2].

We begin by a general classification: Fig. 7.16 represents the four most common types of dry etching.

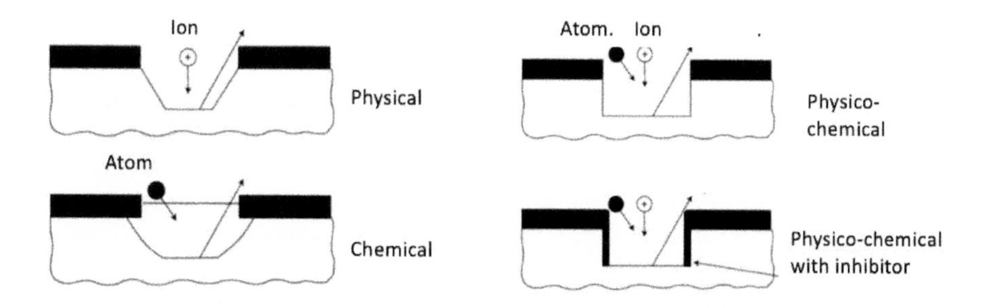

Fig. 7.15: Different types of dry etching for silicon (from top to bottom): physical, chemical, physico-chemical, and physico-chemical with inhibitor [2].

- **Physical etching:** here, ions are accelerated in an electric field and bombard the surface of a target (the object to be etched). An etching effect is produced by the physical action of the flux of incident ions. Physical etching is anisotropic and not selective.
- **Chemical etching:** chemical species migrate towards the surface of the target under the action of an electric field, and chemical reactions occur on the surface, thus producing volatile species and holes.
- **Physico-chemical etching:** the two preceding actions are combined. This type of etching is commonly used.
- **Physico-chemical etching with inhibitor:** the process consists of depositing a protective layer along the sides of the etched cavities during the process. This technique allows to create structures with high aspect ratios (e.g. 1:30).

We describe in this section four important techniques used in microfluidics.

Physical dry etching:. The action of a plasma on a target placed at the cathode (Fig. 7.16) depends on the kinetic energy of the ions accelerated toward the target. For kinetic energies less than 10 eV, nothing happens: the ions are not sufficiently energetic to eject material. For energies between 10 eV and 5000 eV, ejection of material occurs, resulting in physical etching. For energies between 10 and 20 keV, ions penetrate the target in depth. This enables ionic implantation, used in semiconductor technology.

Physical dry etching is carried out at low pressures (typically a few mTorr). using ballistic ions. The simplest configuration (Fig. 7.16) is known as *sputtering*.

Reactive ion etching (RIE). In reactive ion etching (RIE), the object to be etched is in contact with reactive components contained in a gas or a plasma. The principle of this method can be decomposed into six steps:

Chemical etching consists of the following steps.

1. Generation of chemical reagents;

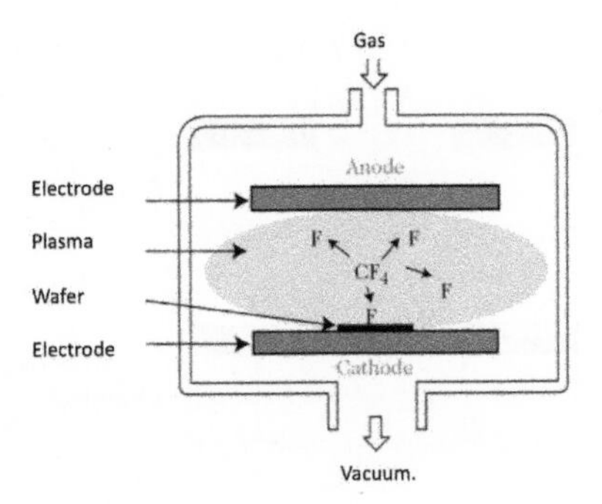

7.16: Plasma etching of silicon.

2. Diffusion of reactive species towards the target;
3. Adsorption;
4. Reaction with the target and the formation of a volatile component;
5. Desorption of the component; and
6. Diffusion in the gas or plasma.

An example is CF_4. This component does not react with silicon, while CF_3 radicals and F^- ions, created by the plasma, do. Several phenomena, not discussed here, are involved in the chemical etching (charge effect of charge, 'bullseye' effect, etc.). For details, the reader can refer to Ref. [2].

Physico-chemical etching with inhibitor: deep reactive ion etching (DRIE). RIE can be used with an inhibitor, whose function is to form a polymerized film that protects the vertical walls from etching.

Fig. 7.17: Two microfabricated structures obtained by DRIE. (Left) Forest of pillars (Reprinted from [11], with permission from Wiley and Sons. Copyright 2022.). (Right) Spring attached to a small mass.

The technique is known as deep RIE (DRIE). It allows to make structures with large aspect ratios (e.g. 30:1). Two spectacular examples are given in Fig. 7.17.

7.6.4 Techniques of deposition onto silicon and glass

Deposition is a process that plays a crucial role in microfabrication. Today, a large variety of techniques allows to deposit all sorts of materials: metals, insulators, semiconductors, polymers, or proteins. We will present here the principles without delving into the detail. Most vapour deposition techniques fall into two categories:

1. PVD: *physical vapour deposition.* The object (or target) is exposed to a vapour containing molecules to be deposited. These molecules adsorb on the substrate and form a layer.
2. CVD: *chemical vapour deposition.* Vaporized species, in contact with the target, react with the surface, forming components chemically bonded to it.

Thermal evaporation. In this case, the material to be deposited faces the target, in a chamber maintained at a low pressure. Next, the material is heated and sublimated. The molecular flux created in this way adsorbs on the target surface. The working pressures are on the order of 10^{-8} Torr). Due to its simplicity, thermal deposition is commonly used in laboratories. Speeds of deposition are slow (a few Å per second); the method is mostly used for metallic deposits.

Sputtering. Here, a cold plasma is created. A cathode is subjected to a ionic flux (for instance, Ar^+), which ejects the anodic material. The ejected material then deposits on the target. In practice, ionic kinetic energies are in the range of 0.3 – 2keV. Sputtering is interesting because it allows to deposit a large variety of materials with a strong adhesion.

CVD: chemical vapor deposition. The substrate is exposed to a gas containing reactive species. There are two possibilities:

- the reaction takes place in the gas and the products of the reaction are adsorbed onto the target;
- the reaction takes place at the surface of the target.

Most CVD equipments are based on heterogeneous reactions, since the adherence of the film produced is superior to that of the homogeneous case. Figure 7.18 roughly represents the mechanism in the heterogeneous case.

CVD allows for the deposition of insulators (for masks or electric insulation), and of polysilicon (for surface micromachining). The deposition rates in all cases are slow (on the order of μm per hour), and this is a problem for the microfabrication of thick structures. Because of this slow deposition speed, many polysilicon structures (e.g. motors, beams) reproduced in the literature are only a few micrometres high.

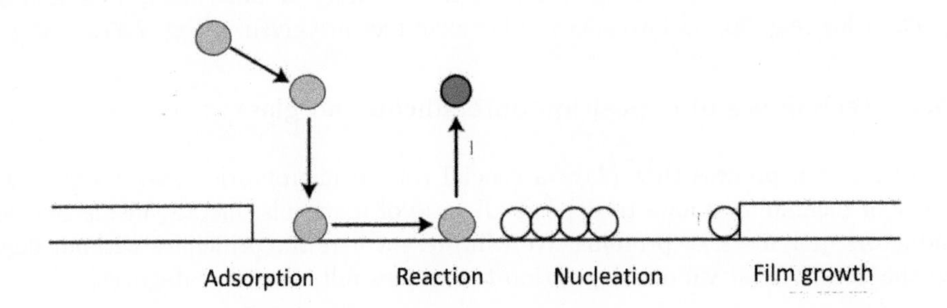

Fig. 7.18: Succession of steps involved in CVD: production of species in the plasma, adsorption, surface reaction, diffusion, aggregation, desorption of volatile components, and film formation.

Here the CVD techniques most commonly used for MEMS and microfluidics are described briefly. There are basically three types of techniques:

- **LPCVD (Low-pressure chemical vapor deposition):** deposition takes place at low pressure (1 Torr). This is used for polysilicon, a fundamental material for MEMS. Deposition speeds are in the order of 1μm/hour.
- **APCVD (Atmospheric-pressure chemical vapor deposition):** in this case, deposition takes place at atmospheric pressure.
- **PECVD (Plasma-enhanced chemical vapor deposition):** this technique deals with systems for which thermal activation is not sufficient to allow heterogeneous chemical reactions. A plasma is thus used, whose role is to activate chemical reactions at the target surface using ionic bombardment.

Evaporating a material on a pre-existing relief does not necessarily lead to uniformly thick layers. We may have conformal and non-conformal deposition. The two cases are shown in Fig. (7.19):

- Conformal deposition: the deposited film has a constant thickness. This is achieved, typically, by working with highly energetic particles.[7]
- Non–conformal deposition: the film thickness is not uniform. This situation is generally avoided, although, in some cases, it can be useful for closing submicrometric channels.

7.6.5 Bonding

Once the etching step is completed, we have grooves carved onto a wafer, and must now cover them in a watertight manner, in order to create functional microfluidic devices. Sealing is not an easy procedure. If it is poorly done, leaks obviously appear.

[7]These particles reach all parts of the target before forming chemical bonds with the substrate

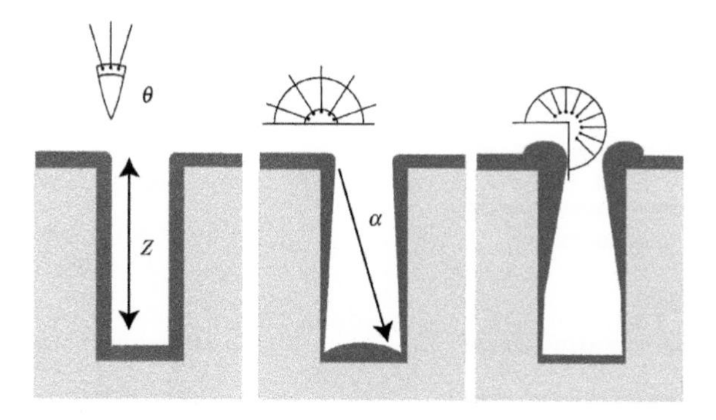

Fig. 7.19: Conformal and non-conformal deposition. Conformal deposition (left) takes the form of the substrate, while non-conformal deposition (right) develops forms different from the substrate (From [2])

To make seals between glass-silicon, glass-glass, or silicon-silicon, fusion techniques are used, with or without an electric field. We describe these techniques below.

Anodic bonding. Anodic bonding allows for the sealing of glass onto a metal or semi-metal such as silicon, in the presence of an electric field. The electric field induces the migration of Na^+ ions in the glass to the interface. Trapped at the interface, they produce an intense local electric field. Sealing is assured by the interpenetration of atoms at the interface. For glass–silicon, this mix of atoms is made possible by an elevated temperature (400°C) to which the system is raised, and by the intense local electric field reduced the sodium trapped for the interface.Typical voltages used to ensure anodic sealing are on the order of 1 kV.

Fusion Bonding. It is possible to seal two pieces of silicon or glass by fusion bonding. Physically, thermal agitation causes the reorganization of silicon atoms at the interface, establishing a small amount of interpenetration that ensures bonding. Depending on the material, temperatures used a range between 600 and 1100°C.

7.6.6 Application: fabrication of a silicon membrane

To apply the previous ideas, the steps leading up to achieving a membrane on a silicon wafer are presented in Fig. 7.20.

The first step consists of depositing a silicon dioxide layer (SiO_2) on a silicon wafer. For that, an oven is used. Then, a layer of photosensitive positive resist is spin-coated. After this comes the photolithographic step, which creates a hole in the resist layer. In the hole, the SiO_2 layer is etched with RIE technique. Further, by using wet etching, a cavity, with oblique walls, is formed in the wafer, with a detph close to its thickness In this way, a silicon membrane is created.

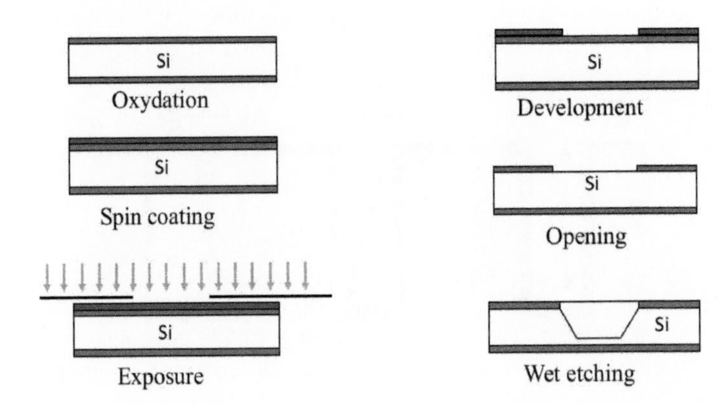

Fig. 7.20: Process of fabrication of a silicon membrane, involving oxydation, spin coating, photolithography, RIE, and wet etching. For achieving the task, seven steps and one mask are needed.

7.6.7 Micropump for insulin injection

The insulin pump, made by Debiotech, is shown in Fig. 7.21 [13]. This micropump can be held on the tip of a finger as shown in Fig. 7.21 (right). The structure includes a polysilicon membrane, called a 'pump diaphragm', a few micrometres thick, made by deposition. Note that silicon is not piezo-electrical but polysilicon is. This membrane is actuated by an electrical field provided by a circuit (not shown in the figure).

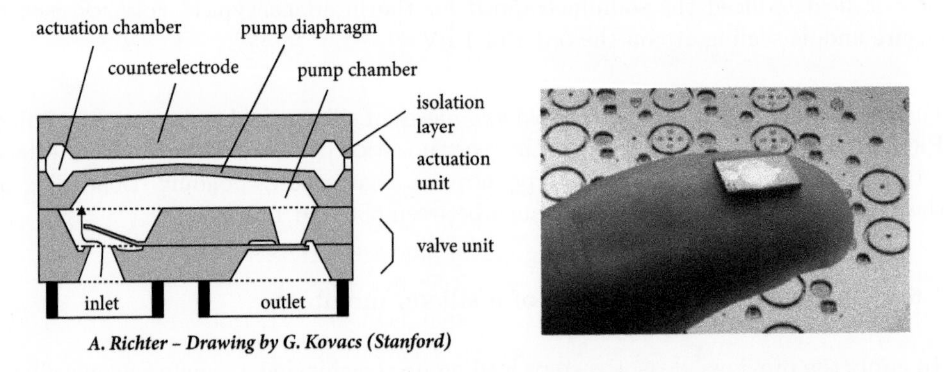

Fig. 7.21: Micropump made by Debiotech [13], which benefited from a large echo in the 1990s.

When the diaphragm raises, fluid is pulled, and when it come back to the rest, it pushes the same fluids out of the chamber. Pumping is possible with the action of two valves, which open and close, in function of the movement of the diaphragm, preventing the captured liquid from flowing out of the chamber when the membrane returns to its rest position. With this system, flow rates in the range of microlitres per minute are obtained. The pump adapts well to the problem of insulin delivery, with its small size, along with the flow rate. In practice, the pump is placed on the skin, at the liver level, and actuated with a battery. It is often called 'skin pump'. The fabrication is delicate, the most critical part being the valves, and it can be achieved using the technological tools shown above.

7.6.8 Electron beam lithography - Nanoguitar

A nanoguitar is not exactly a microfluidic object, but it illustrates, in an amazing manner, the capabilities of MEMS technology, although extended to the nanometric scale. The silicon guitar shown in Fig. 7.22 was made in 1997, in Cornell Laboratory, by the group led by Craighead, using technological ideas presented in the chapter. Its size is ten μm, it includes six chords, as in standard guitars. There sizes are submicrometric.

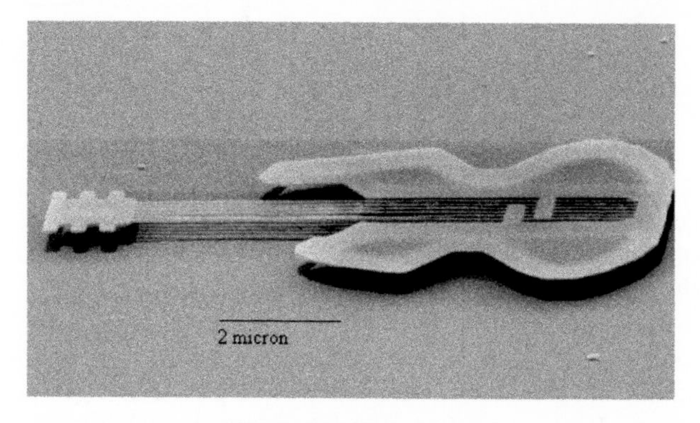

Fig. 7.22: The world's smallest guitar is 10 μm long, with six strings each about 50 nm wide (i.e. 100 atoms). Made by Cornell University researchers from crystalline silicon, it announced, at that time (in 1997) a new generation of electromechanical devices.

The guitar was made by using electro beam lithography, which uses the same ideas as photolithography, but with an access to much smaller scales, owing to the use of electrons (whose wavelengths are in the amstrong range) and not photons (for which 200 – 300 nm wavelengths are typically used, as we saw in the chapter). Difficult to play the Albidoni concerto with this guitar. Should it be possible, the sound would be difficult to hear. Considering that the vibration frequency is inversely proportional to the length, we would need ears able to hear sounds in the MHz range, i.e. ultra-sounds, while the acoustic range of human ears do not exceed 20 kHz.

7.7 PDMS-based moulding – soft lithography

7.7.1 Historical note

Two papers, published in 1998, represent landmarks in the field. The first one [1] introduced the concept of 'soft lithography', i.e. microfabrication based on lithography and soft materials. The second one [15], titled Rapid Prototyping of Microfluidic Systems in Poly(dimethylsiloxane)', made it possible ' to design and fabricate' microfluidic systems in an elastomeric material – poly(dimethylsiloxane) (PDMS)– in less than 24 h'. The work was preceded by two precursor papers reporting the fabrication of PDMS channels and stamps, with a sensibly different approach [16, 17]. Two figures of Ref. [15] are shown in Fig. 7.23.

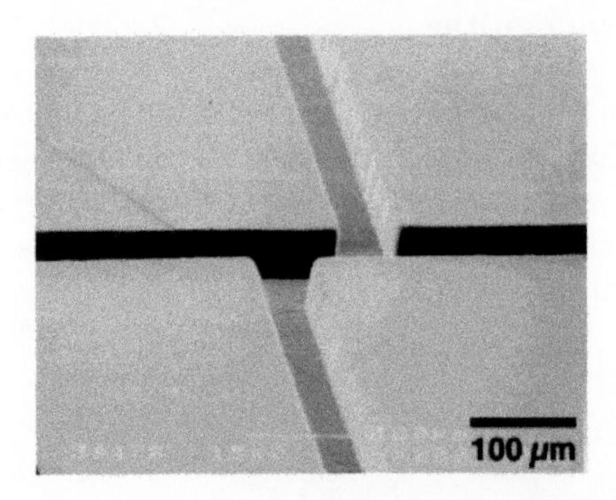

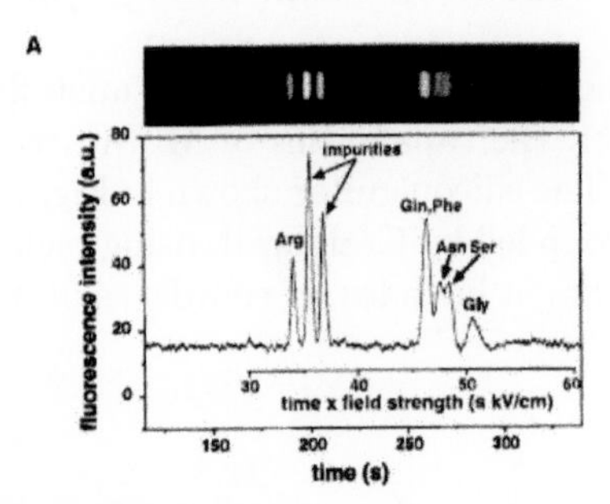

Fig. 7.23: (Left) One of the first PDMS microfluidic devices, made in 1998, and dedicated to separate nucleotides. [15]. (Right) Electropherogram demonstrating the separation of nucleotides in a PDMS device. (Reprinted from [15], with permission of the American Chemical Society. Copyright 2022.)

The left–hand part of Fig. 7.23 represents a PDMS device designed to separate ionic species.[8] Widths and heights are on the order of 50 μm. The right-hand figure shows an electropherogram. Ions are separated in few minutes, with a resolution comparable that of early silicon devices [18] (see shapter 6). Although chromatography is not the most important application of PDMS, the work demonstrated that performing chromatographic studies with this material is possible. At that time, this appeared quite as a surprise: as PDMS surface chemistry is poorly controlled, the efficiency of the separation process could have been compromised.

The message taken away by many scientists was not about chromatography. It was that PDMS technology was an easy-to-use microfabrication method capable of producing a wide range of interesting devices A master student can learn it in one day. The simplicity of the technological process gave rise to a surge of creativity. One could compare it to the transition from centralized informatics to lab–top computers. Many laboratories, interested in microfluidics, but hindered by the complexity of silicon and glass technology, or with little access to clean rooms, adopted PDMS technology and developed new directions that paved the way for outstanding innovations.

[8]See Chapter 5

7.7.2 PDMS Characteristics

Polydimethylsiloxane (PDMS) belongs to a large family of polymers. Low molecular weight PDMS is a liquid used in lubricants, antifoaming agents, and hydraulic fluids. At higher molecular weights, PDMS is a soft, compliant rubber or resin. Among the commercially available forms, Sylgard 184 (Dow Corning) and RTV 615 (GE Bayer Silicones) are the most widely used.

The formula of PDMS is (–Si$(CH_3)_2$O)–. The semi-structural fomula is shown in Fig. 7.24. The siloxane chains are substantially separated by hydrophobic methyl groups (CH_3), leading to high flexibility of the material. The same methyl groups are responsible for the hydrophobicity of the material.

PDMS mixed with a cross-linker, and heated, forms an elastomer whose principal properties are shown in Table 7.2 (from [14]).

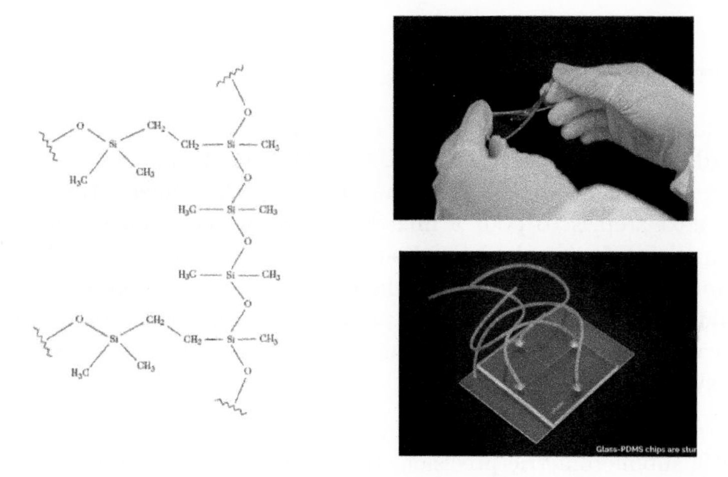

Fig. 7.24: Left: Semi-structural fomula of PDMS; Right: Manipulation of a PDMS cross-linked material. (Courtesy of uFluidix.).

Most, if not all, properties listed in Table 7.2 are extremely important: optical transparency is obviously critical for imaging, electrical insulation for the establishment of electric fields, low Young moduli for valving, gas permeability for liquid filling, absence of reactivity for chemistry and biology. In practice, not all chains are cross-linked. In the material, free short chains migrate to the surface and accumulate. This process is responsible for the time-dependency of the surface properties of the material.

PDMS is unique. There have been several attempts to explore other elastomers for microfluidic applications, with the goal of finding materials that could offer advantages over PDMS. All failed. For instance, in the case of polyurethane, degassing was difficult, due to a high surface tension, and demoulding was challenging due to a stronger

Property	Characteristics
Density	Around 0.9 kg/m^3
Optical	Transparency between 300 nm and 2200 nm
Electrical	Insulating with breaking field 20 kV/cm
Mechanical	Elastomeric with Young's modulus $\sim$ 350–800 kPa
Thermal	Insulating; thermal conductivity $\sim$ 0.2 W/m/K
Interfacial	Low surface energy ($\sim$ 20 mN/m)
Permeability	Permeable to gas, apolar organic solvents, small permeability to water (pervaporation speeds in the nm/s range)
Reactivity	Inert, oxidizable by a plasma
Toxicity	Non-toxic

Table 7.2 Cross-linked PDMS properties, using a cross-linker in the 1:10–1:5 concentration range

adhesion [19]. To these constraints, one must add the difficulty of realizing connections without leakage.

7.7.3 How to make PDMS channels ?

To create a microchannel, four steps are needed (see Fig. 7.25). To start, we need a mould made of a hard material which, often, is SU8 or plastics. The mould fabrication is made in a clean room. When carried out in an ordinary environment, the risk of failure is high. The next step is to pour a mix of PDMS and cross-linker onto the mould. Then, the temperature is raised at about 70°C. During this phase, PDMS cross-links, and becomes solid. After peeling off the PDMS, an object representing the negative of the mould is obtained. This object is placed onto a substrate, often a glass slide, on which it weakly adheres. Most often, this adhesion is reinforced by exposing the surfaces to an oxygen plasma.

The precision of the structures obtained this way is surprisingly high. Whitesides et al. [6] demonstrated a submicrometric precision. In practice, owing to various uncertainties entering the process, the microchannel transverse dimensions lie between 2 and 200μm.

7.7.4 Remarks on PDMS

The properties of PDMS deserve a number of remarks:

- Transparency in the visible spectrum enables flow visualization, which is often crucial in microfluidic studies. Nonetheless, its opacity to hard UV (below 300 nm) leads, in some case, to limitations. For instance, it is difficult to cure materials embedded in PDMS structures.
- The Young's modulus of reticulated PDMS lies in the range 300–850 kPa. It increases with the cross–linker concentration, the largest values leading to 'hard PDMS', which is no longer deformable. The deformability of PDMS allows valves

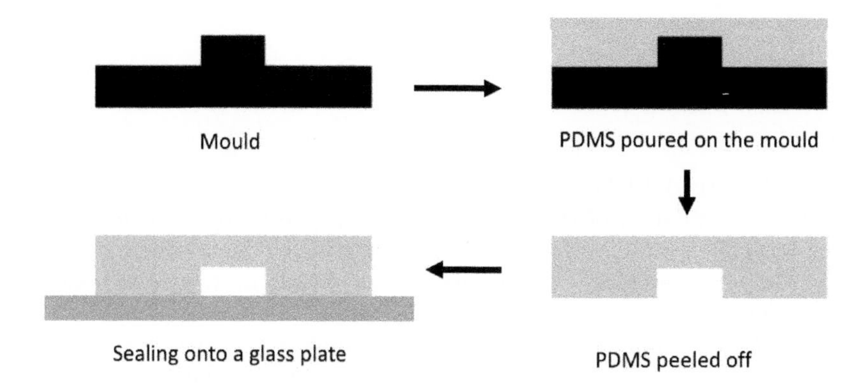

Fig. 7.25: Fabrication of a PDMS microchannel.

to be made, as will be shown later. Under pressure, microchannels deform, and, consequently, their hydrodynamic resistance decreases. An example is shown in Fig. 7.26 [23].

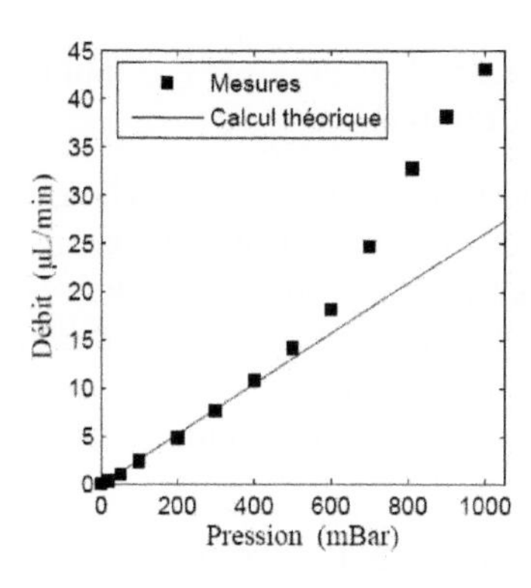

7.26: Flow rate (Q) evolution in a channel (227 μm wide, 24.2 μm high, 2.8 cm long),as a function of pressure ΔP. The straight line is the formula $\Delta P = \frac{12\mu L}{h^3 w} Q$ in which μ is the viscosity, and L, h, and w are the channel length, height, and width [20] (Courtesy of J. Goulpeau.)

In the experiment of Ref. [20], at 1 bar, the channel cross-section increases by 30% and the resistance by 100%. Consequently, the flow rate increases by a factor of two.

An important phenomenon, which is the opposite of the previous one, is the 'roof collapse', i.e. the fact that the roof of a microchannel can collapse and stick to the floor, obstructing the channel. A criterion, calculated by Huang et al [21], for the case of shallow channels, reads:

$$\frac{w\gamma}{Eh^2} > 0.13$$

in which w is the width, h is the height, E is the Young modulus, and γ is the surface tension. In practice, it is wise to restrict ourselves to aspect ratios w/h below 10.

- The elastomeric character of PDMS enables the watertightness of microfluidic connections, which is crucial. The elasticity of the material also allows the fabrication of valves and pumps using membranes, as noted above.
- Untreated (native) PDMS is hydrophobic, and becomes hydrophilic (temporarily) after the oxidation of the surface by oxygen plasma or after immersion in a strong base. The plasma breaks Si–CH_3 bonds, and replaces the methyl groups by hydroxyl groups, which, immersed in water, builds up hydrogen bonds. With this mechanism, the surface, initially hydrophobic, becomes hydrophilic. On the other hand, the formation of reactive silane groups at the surface allows covalent bonding of the PDMS with the silanol groups of other oxidated surfaces. PDMS surfaces can thus strongly bond to different materials, such as glass, silicon, or plastics. Being able to seal PDMS onto substrates represents an extremely important feature, which is instrumental for the creation of functional devices.
- Its permeability to gas facilitates filling in open and blind channels.[9] PDMS permeability is also interesting for cell culturing. On the other hand, its solubility to non-polar organic solvents makes PDMS unsuitable for many chemistry applications.
- Its weak surface energy facilitates the process of peeling off the mold from the substrate. This property plays a key role in the technology.
- The properties of PDMS evolve with respect to time, owing to the diffusion of small free chains in the materials, migrating to the surface. Typically, an oxidized surface remains hydrophilic for thirty minutes, and then evolves towards hydrophobicity. In general, because of the difficulty of controlling surface chemistry, patterning PDMS surfaces remains challenging. Efforts conducted during the past decade in this area have turned out, despite a number of temporary successes, to be unconvincing.
- It is not possible to deposit metallic electrodes on PDMS because of poor adherence and the elevated temperatures involved in the process. PDMS structures can nevertheless be placed on a glass or silicon wafer, on which electrodes were deposited.

7.7.5 Flow lithography

Flow lithography is one of the many applications of soft lithography. The principle is shown in Fig. 7.27 (A), and the presentation of the technology is taken directly from Ref. [30]. A stream of a photosensitive initiator is passed through the rectangular, PDMS microfluidic device shown in the figure. Particle arrays of mask-defined shapes are formed by exposing the flowing oligomer to controlled pulses of ultraviolet (UV)

[9] Air bubbles trapped during filling escape by migrating through PDMS, which is porous to gas.

light. Rapid polymerization kinetics permits the particles to form quickly ($\sim$ 0.1 s), and oxygen-aided inhibition near the PDMS surfaces allowed for particle flow without wall attachment and clogging (oxygen, which diffuses through the PDMS surfaces leaves a non-polymerized lubricating layer near the PDMS walls).

An extension of the technique was made in [31], leading the production of multifunctional particles bearing more than a million unique codes (see Fig. 7.27 (B)).

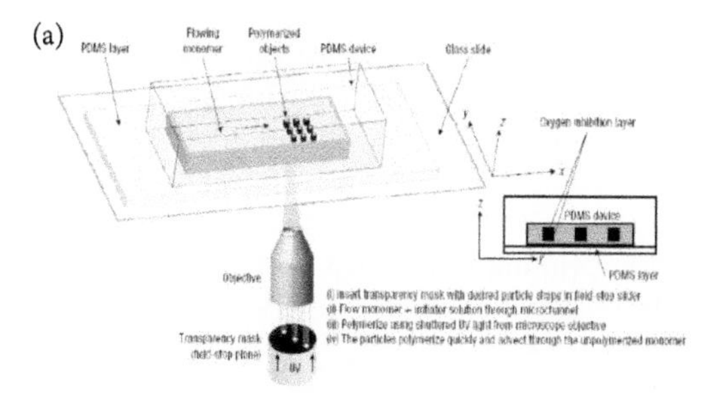

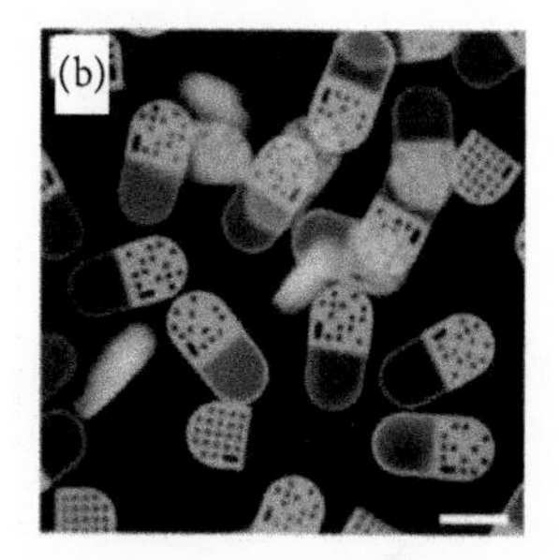

Fig. 7.27: A - Schematic depicting the experimental setup used in flow lithography [30] (Caption taken from the reference). A mask containing the desired features is inserted in the field-stop plane of the microscope. The monomer stream flows through the all-PDMS device in the direction of the horizontal arrow. Particles are polymerized, by a mask-defined UV light beam emanating from the objective, and then advect within the unpolymerized monomer stream. The side-view of the polymerized particles can be seen in the inset shown on the right (Reprinted from Ref. [30] with permission from Springer Nature. Copyright 2022). B Typical image of multifunctional particles made with the technology. The scale bar is 100 μm (Reprinted from Ref. [31] with permission of the American Association for the Advancement of Science. Copyright 2022.)

7.7.6 Microcontact printing

This technology represents another application of soft photolithography. The principle is represented in Fig. 7.28 (Left).

We start with a PDMS stamp. The stamp is the negative of a mould made in hard material, as we explained in this chapter. Inking of the stamp is performed by immersion. The PDMS stamp absorbs the ink. After drying out, the stamp is brought into contact with a hydrophilic substrate, and immersed in a buffer. The molecules captured by the stamp migrate towards the substrate, and form a hydrophobic self-assembling monolayer (SAM). This technology allows surfaces to be patterned with all sorts of molecules. An example is shown in Fig. 7.28 (Right). The figure shows a L pattern of fibronectin, deposited onto a glass surface. The system is immersed in a suspension of Hela cells, whose membranes bind to fibronectin [32]. This experiment allowed to visualize the behaviour of these cells in the presence of constraints imposed by the patterning. The work enlightened the question of the influence of the extracellular matrix on the cell structure, in various situations, e.g. during cell division.

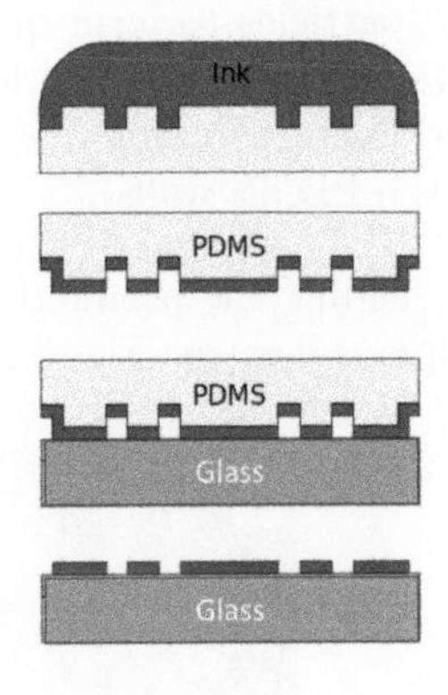

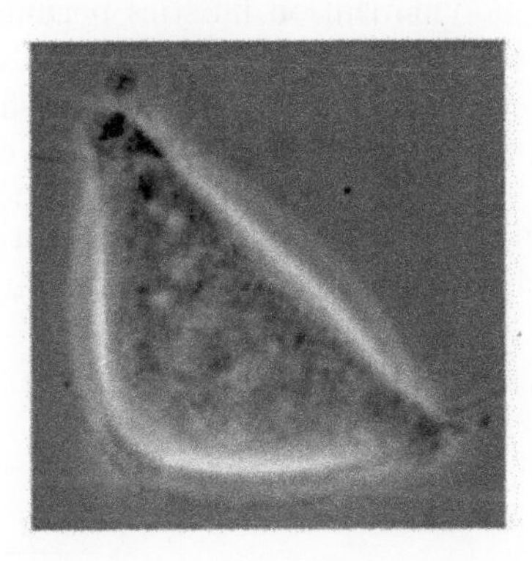

Fig. 7.28: (Left) Principle of microcontact printing technology. From top to bottom, inking, drying, contact, and separation; (Right) Hela cell deformed by a fibronectin patch in form of a L, created by contact photolitography. (Reprinted from Ref. [32], with permission from Springer Nature; Copyright 2022.)

Microcontact printing is used industrially, to carry out high content screening (HCS) of pharmaceutical drugs.

7.7.7 PDMS valving and pumping

As mentioned above, the elasticity of PDMS makes possible the creation of microvalves. The idea developed by S. Quake [22] is shown in Fig. 7.29.

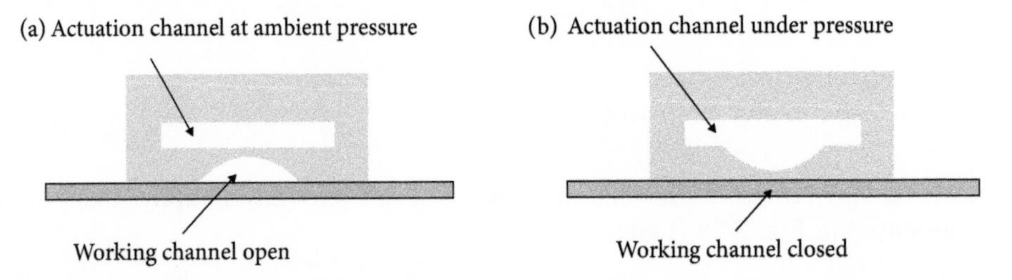

Fig. 7.29: Principle of functioning of PDMS valves, often called 'Quake's valves', shown in the case of a 'push-down' valve (see below). (A) Actuation channel at ambient pressure: the working channel (below) is open.(B) Actuation channel pressurized: the working channel is closed. Owing to the round shape of the working channel, leakage is extremely weak.

The figure can be misleading because the two channels shown on the figure are normal with respect to each other, not parallel. Nonetheless, we will use it for explaining how valves work. The valve of Fig. 7.29 is thus composed of two layers, one above the other. This superposition of layers refers to 'multilayer soft lithography'. In the upper layer, a microchannel, called 'actuation channel', is filled with a liquid called 'actuation fluid'. Below it, there is a second channel, called 'working channel', through which the working fluid flows. In between, the separating wall is thin enough to be deformed under a fraction of bar. This thin wall forms a membrane of typically 5-10 μm in thickness. Thus, as the actuation channel is pressurized, it deflects the membrane and obstructs the working channel. Owing to the round shape of the working channel, the valve is tightly sealed.[10] In early versions, air was used to pressurize the actuation channel. However, due to PDMS permeability, bubbles formed. This option was abandoned, and, even though some presentations still mention that gas is the actuation fluid, gases are no longer used. Two types of valves exist (see Fig. 7.30):

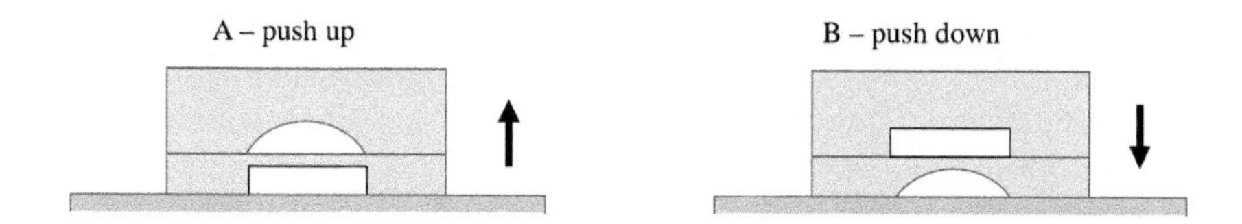

Fig. 7.30: (A) Push-up valve. (B) Push-down valve.

One is push-up and the other push-down; in the former, pressure is applied on the lower channel, to push the membrane up, while in the latter, pressure is applied on the upper channel. The characteristics of the valves have been analysed in several papers [23, 24, 26]. In an optimized design, the dead-volume is indistinguishable from noise.Valves sustain pressures up to several bars. Dynamically, Quake's valves behave as low pass band filters, with a characteristic frequency on the order of 100 Hz.[11] This frequency, which decreases as the system size increases, is large enough for biological applications. The pressures needed for closing the valve are on the order of a fraction of bar. Attempts to miniaturize the valves have been made[12]. Thus far, valves smaller than 6 x 6 μm^2 have not yet been reported [26].

The push-down valve was modeled by J. Goulpeau et al. [23]. This led to the circuit shown in Fig. 7.31.

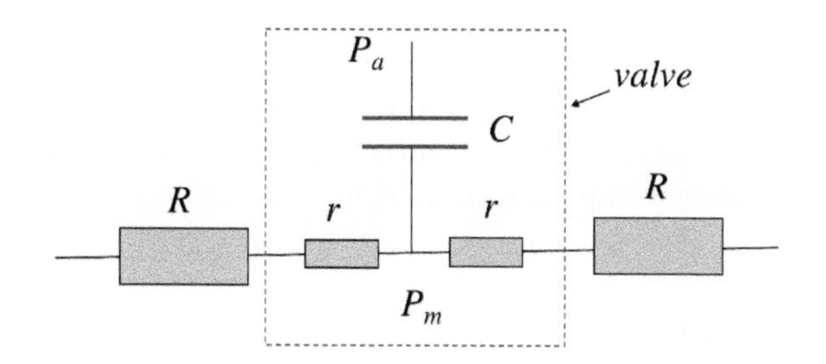

7.31: Electrical circuit proposed in [23] for modelling PDMS valves. C is the capacitance characterizing membrane deformability, r a resistance depending on the valve state (open or closed), and R is the resistance of the external circuit.

[10]The round shape is obtained by using a positive photo-resist, for instance, AZ100, which, when heated in a certain range of temperatures, rounds off.

[11]There is another characteristic frequency, much higher, linked to the channel deformability, which we do not discuss here [23].

[12]In this approach, a Moore's law has been suggested [25, 26]

The capacity expresses the deformability of the membrane, as discussed in Chapter 4. The resistance R is the circuit resistance and r, the valve resistance, depends on the actuation pressure P_a. Its expression is:

$$r = \frac{r_0}{1 - ((P_a - P_m)/P_c)^3}$$

in which P_a is the actuation pressure, P_m is the pressure in the channel, P_c is the closure pressure, i.e. the pressure at which the valve is closed, and r_0 is the value of r when $P_a = P_m$, i.e. when the membrane is flat. The exponent 3 in the formula is obtained by using the Reynolds model for a piston approaching a surface, in the framework of lubrication approximation. When $P_a - P_m = P_c$, r is infinite, and the valve is closed. The theory agrees well with the experiment [23].

With at least two valves actuated in a periodic manner, one realizes a peristaltic pump. A three-valves pump is shown in Fig. 7.32.

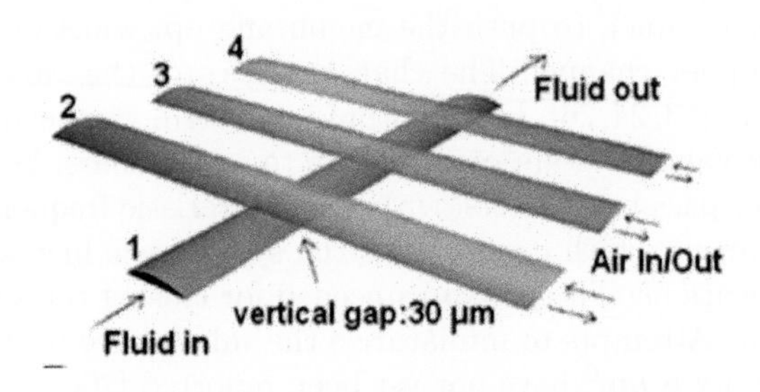

7.32: By actuating three valves sequentially, a net flow along the working channel can be obtained.

These peristaltic pumps provide flows on the order of one μL/min, and sustain counter-pressures on the order of one bar or so. The maximum flow rate obtained in Ref. [23] is:

$$Q_{max} \approx 0.4 \frac{P_c}{R}$$

This maximum was found close to 7.5 μL/min, for a 44 μm micropump. These figures represent well the highest performances one can obtain with these pumps. Technology–wise, these miniaturized pumps represent a leap, in term of simplicity and cost, compared to hard material pumps.

Thus, due to the simplicity of the design, their small sizes, and their natural integrability, thousands of valves, along with many pumps, can be integrated on a device. This opened a route towards the implementation of complex functions on microfluidic devices, suggesting an equivalence between MOS–FETs and valves, i.e. microelectronics and microfluidics. The concept has represented a conceptual jump of the field, giving rise to an important number of innovations.

Illustration 1: Controlling droplet formation with PDMS valves

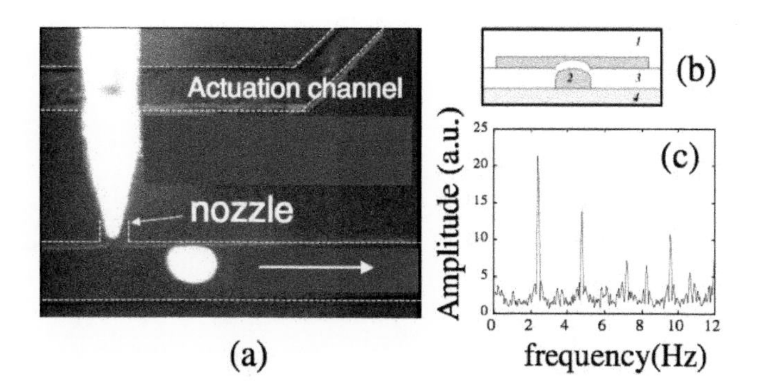

Fig. 7.33: (A) Top view of the experiment of Ref. [27]. Tetradecane (black) and water with fluorescein (white) meet in the T–junction, emitting droplets.(B) Sketch of the actuation system; (1) working channel; (2) actuation channel; (3) PDMS; (4) glass substrate.(C) Fourier power spectrum for a T–junction producing droplets without modulation, i.e. with the valve permanently open.

An experiment [27] performed in 2006 showed that PDMS valves can control droplet emitters. In Fig. 7.33 (A) and (B), a push-up valve has been inserted in the entry channel of a T–junction, in which oil (black) and water (white) are introduced, for the purpose of producing water-in-oil droplets. The valve is periodically opened and closed, thus modulating the flow rate of the dispersed phase. Droplet production is a nonlinear process characterized by an internal frequency shown in Fig. 7.33 (C). The coupling between this frequency and the modulation frequency of the valve give rise to complex behaviour. In the experiment, Arnold tongues were observed. In practice, by working in resonant tongues, i.e. in conditions where the internal frequency resonates with the modulation frequency, the emission frequency and the droplet size can be finely tuned.

Illustration 2: Automatized nanoliter DNA purification

Figure 7.34 shows a biological application of PDMS valve technology [28].

The sample is introduced and mixed with reagents in the ring micromixer. After mixing is completed (a few seconds), the mixture is brought in contact with functionalized beads, which capture DNA. After rinsing, and elution, purified DNA is retrieved. The entire process is carried out automatically.

Illustration 3: Large scale integration

An influential contribution [29], appealing from a conceptual viewpoint, is shown in Figure 7.35.

The device contains 2056 microvalves. Two different reagents can be separately loaded, mixed pairwise, and selectively recovered, making it possible to perform distinct assays in 256 subnanolitre reaction chambers and then recover a reagent. The microchannel layout consists of four central columns in the flow layer consisting of 64 chambers per column, with each chamber containing 750 pL of liquid after compartmentalization

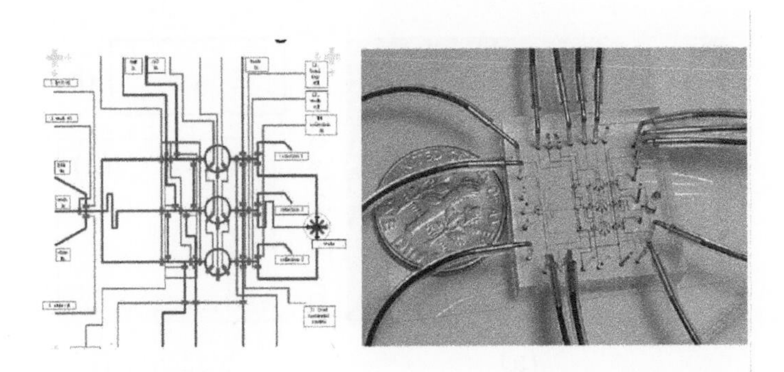

Fig. 7.34: Cell isolation, cell lysis, DNA purification, and recovery, are carried out on the microfluidic chip in nanolitre volumes without any pre– or postsample treatment. Measurable amounts of DNA are extracted in an automated fashion from as little as a single mammalian cell and recovered from the chip. The right-hand figure is an image of the device, using coloured fluids for visualization, in a mutliplexed version (three systems in parallel). (Reprinted from Ref. [28], with permission from Springer Nature. Copyright 2022.)

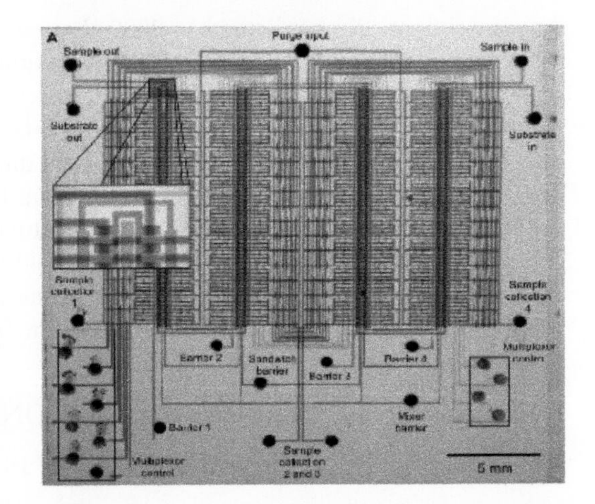

7.35: The device, whose channels are visualized with dyes, contains 2056 microvalves. The system performs distinct assays in 256 nl subreaction chambers. (Reprinted from [29], with permission from the American Association for Advancement of Science. Copyright 2022.)

and mixing. Liquid is loaded into these columns through two separate inputs under low external pressure (20 kPa), filling up the array in a serpentine fashion. Barrier valves on the control layer function to isolate the sample fluids from each other and from channel networks on the flow layer used to recover the contents of each individual chamber. These networks function under the control of a multiplexor and several other control valves. Each of the 256 chambers on the chip can be individually addressed and its respective contents recovered for future analysis using only 18 connections to the outside world.

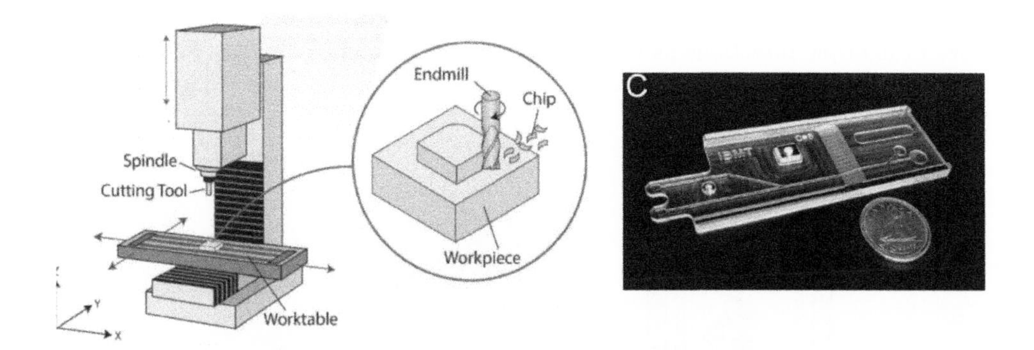

Fig. 7.36: CNC (Computer Numerical Control) micromilling. On the left, a drawing showing the principle of the technology, and, on the right, an example of a microchannel made with it (Reprinted from Ref. [36], with permission from Royal Society of Chemistry. Copyright 2022).

7.8 Computer Numerical Control (CNC) Micromilling

Computer numerical control (CNC) micromilling is an extension of milling (see Fig. 7.36). [13] The technique creates microscale features via cutting tools removing the material. It is mostly used for creating micro and millichannels in plastic materials (COC, PMMA,...). Endmills – the most common cutting tool for milling – are available in many profiles, in a variety of materials (titanium nitride, titanium carbonitride,titanium aluminium-nitride), and with different coatings [36]. The diameter of the cutting tool can be as small as 25 μm. An example of a device is shown in Fig. 7.36. A trained machinist can realize the device in thirty minutes. There exist limitations. With micromilling, it is difficult to create a microchannel smaller than 100 μm width. The surface roughness is quite large (on the order of five hundred nanometres - in practice, tool traces are visible on the surfaces -). Still, micromilling is extremely useful. It complements well the techniques based on photolithography, described in the chapter. Today, micromilling is widespread in microfluidic platforms.

7.9 3D Printing or Additive Manufacturing (AM)

Over the last two decades, three dimensional printing (also called Additive Manufacturing (AM)), has undergone significant growth and diversification in terms of applications, materials, and capabilities.This gave rise to a multitude of acronyms, sometimes puzzling the reader. The presentation given here is extremely brief, and we refer the reader to Refs. [39, 40] for detailed information.

Digital light processing stereolithography (DLP-SLA)

[13]Milling was invented in 1818 [37]

DLP-SLA (Digital Light Processing Stereolithography) technique is illustrated in Fig. 7.37. A 365 nm light is projected onto a build platform, i.e. a flat surface that can move in multiple directions as commanded by the printer's software. Following each exposure, the build platform is raised then lowered back into the resin bath for the next 50 μm layer [40].

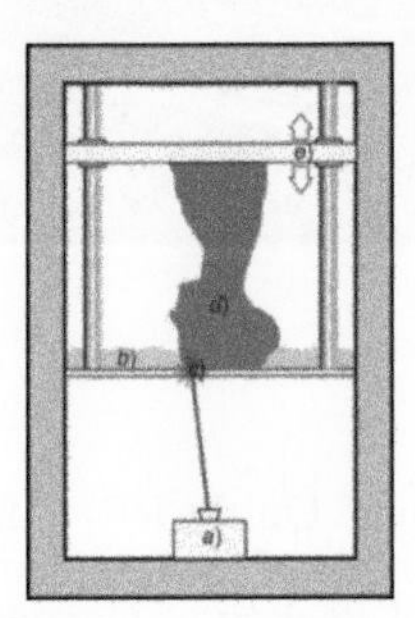

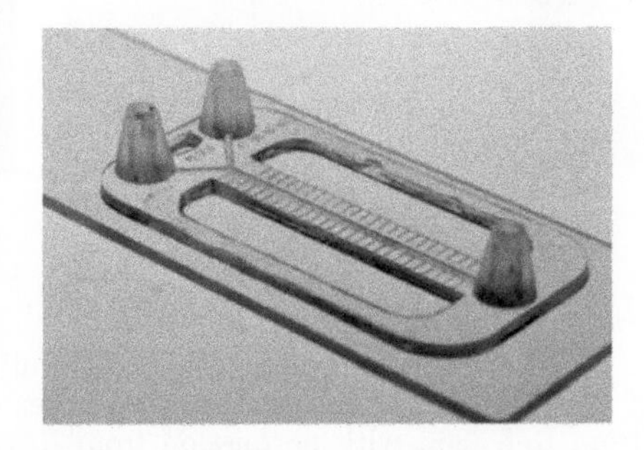

Fig. 7.37: DLP-SLA method, in which 365 nm light was projected onto a build platform, i.e. a flat surface that can move in multiple directions as commanded by the printer's software. Following each exposure, the build platform is raised then lowered back into the resin bath for the next 50 μm layer (Reprinted from Ref. [40] with permission from the American Chemical Society. Copyright 2022.)

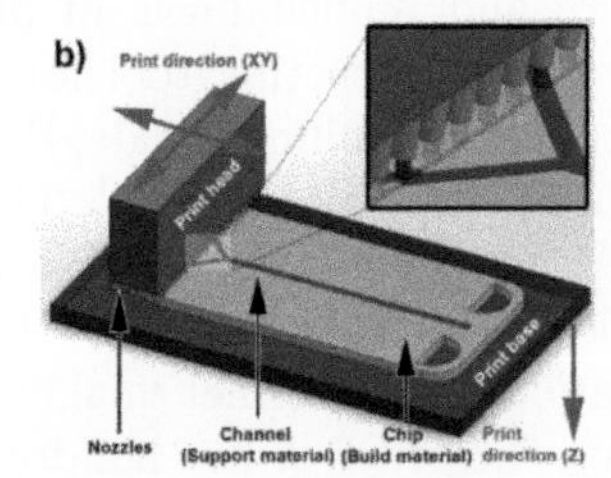

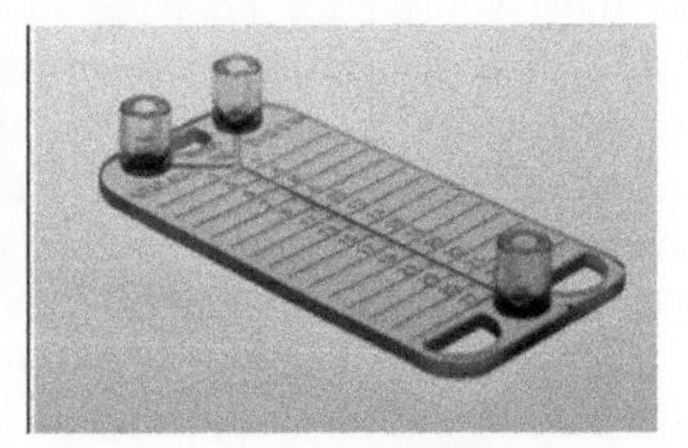

Fig. 7.38: Polyjet method, where two sets of four-micronozzle arrays (build and support material, respectively) are spraying microdroplets of polymer to form the device. Following each pass, UV lamps polymerize the material before the layer is leveled by a roller and scraper (Reprinted from Ref. [40] with permission from the American Chemical Society. Copyright 2022.)

Polyjet (photopolymer inkjet printing)

As sketched in Fig. 7.38, polyjet printing consists in jetting a photopolymer through an array of nozzles, which spray microdroplets onto the build surface, where the material is polymerized using an integrated UV light source.

FFF (Fused filament fabrication) or FDM (fused deposition moulding)

FFF (Fused filament fabrication) or FDM (fused deposition moulding) printing consists in extruding thermal plastic through a heated nozzle, controlled by two precision stepper motors (see Fig. 7.39).

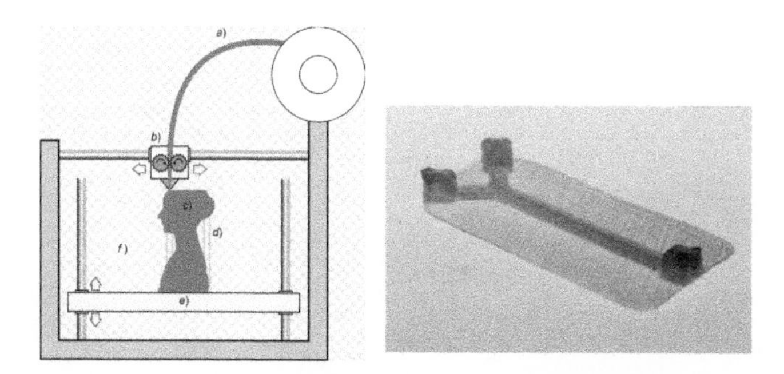

Fig. 7.39: FDM method, in which molten plastic is extruded through a heated nozzle. The features are formed by moving the nozzle in the XY-plane until the current layer is complete, before moving to the next layer. (Reprinted from Ref. [40] with permission from the American Chemical Society. Copyright 2022.)

7.10 Paper microfluidics

7.10.1 Background

Paper microfluidics appeared in the years 2005–2007 [33]. The idea was to replace the standard microfluidic materials (glass, silicon, plastics, or PDMS) by paper. Paper is much cheaper (its cost is a few hundredths of a cent per sheet), is more availalbe, and is burnable (reducing risks of contamination). This material is thus particularly suitable for diagnostic applications in developing countries. Paper has been used for decades for diagnostics, but the novelty brought by G. Whitesides' group was patterning. By using a wax printer [34], it is possible to design channels along which liquids flow in a way similar to microfluidic devices. With this possibility, paper microfluidics appeared as a new branch of microfluidic technology, holding several of its promises, while using a cheaper and more available substrate. Figure 7.40 shows the patterning technique allowing to create a paper microfluidic device.

What about chemistry ? The cellulose fibers, which form the paper, are supramolecular assemblies of polymer chains of glucose [42]. These chains include hydroxyl groups, which are potentially reactive. Although, in many cases, this reactivity does not hinder the functioning of the paper devices, it must be discussed when paper used is as a chemical or biological assay.

7.10.2 Paper microfluidics functionalities

Paper naturally integrates several functionalities [42]:

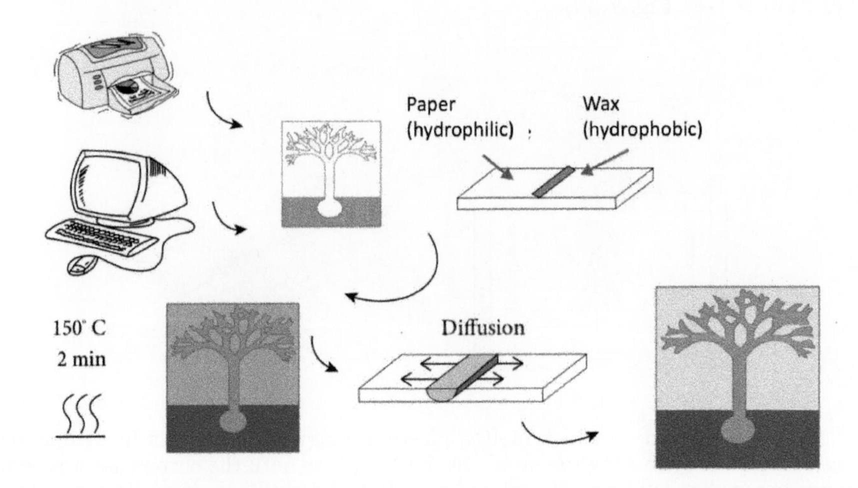

Fig. 7.40: Paper technology using wax printing. The tree is wax–printed. The paper sheet is heated, so as to melt the wax. By melting, the wax diffuses across the sheet and form a physical hydrophobic barrier. Fluorescein is then injected at the bottom of the tree. It spontaneously fills the trunk, the branch and the leaves. (From L. Magro, MMN Lab, IPGG (2015).)

1. Solute transport without pump: as we saw in Chapter 4, Washburn law applies in paper [41], i.e. the flow speed decreases as the inverse of the square root of time. No pump is required. The work is done by capillarity. Note that this is not specific to paper: in microchannels, spontaneous filling is also achieved, thanks to the action of capillarity (see Chapter 4).

2. Filtration: pores are between 5 and 50 μm in size, and therefore they can block objects bigger than these dimensions.

3. Storage: paper can store dried or freeze-dried entities, like blood in dried blood spots (DBS).

4. Valving: . valving can be achieved by folding and unfolding the paper sheet. An example is shown in Fig. 7.41 [43].

5 Multiplexing: in multiplexed assays, several diseases can be diagnosed at the same time, on the same device. An example, using a model of urine, illustrates this point (see Fig. 7.42).

6. Sample to answer devices: with paper technology, is it possible to create fully integrated systems, often called 'sample to answer' devices ? The first example, proposed to commercialization, was realized by Diag4All. It was called 'paper machine' [44]. Recently, several fully integrated devices were proposed [43], some of them motivated by the COVID pandemy [43,45]. Fig. 7.43 shows an integrated device, called 'Covidisc', which performs nucleic acid extraction, followed by loop

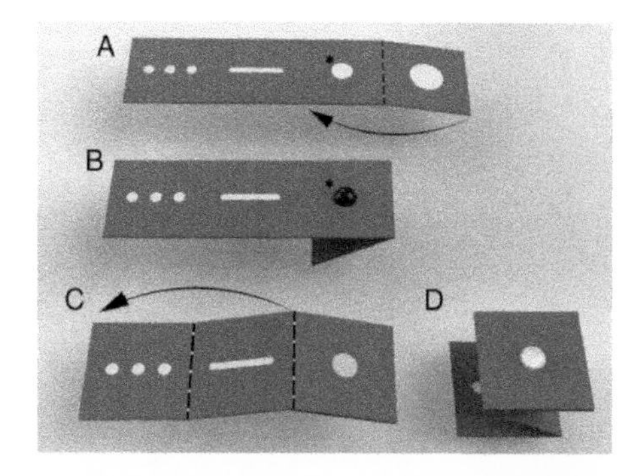

Fig. 7.41: Paper-folding steps for fluidic manipulation. Schematics of how the paper strip in the microfluidic device is folded for each step. The arrows indicate the direction of folding. (A) The layout of the multiplex paper device. (B) The magnetic beads and the DNA in the sample are mixed and added to the sample zone. (C) The second (sample zone) and third panels are then folded to come into contact with each other, forming a distributing channel for DNA elution. (D). The distributing channel is then folded onto the remaining square, which is then placed against the plastic device to cover the three LAMP chambers [43] (Creative Commons License 4.0.)

7.42: Multiplexed analysis of a model sample of urine (glucose solution) [34]. The white is the paper, the brown is the wax, defining channels. Two colorimetric detections are performed in duplicate (reprinted from Ref. [34] with permission of American Chemical Society. Copyright 2022).

amplification (LAMP), with all reagents freeze–dried on it. The system showed performances, in terms of sensitivity and specificity, equivalent to PCR [45].

7.11 Other technologies

7.11.1 Casting

The principle of casting is illustrated in Fig. 7.44 [47].

It includes three steps: heating, embossing, and de–moulding. The most commonly used material for this technology is polymethylmethacrylate (PMMA). When raised to a temperature on the order of 170 °C and pressurized at several tens of bars, it become deformable. In this way, it is possible to create structures by using a press and a mould, as indicated in Fig. 7.44. In general, the mould is made of silicon or metal. The materials used in this technology are not limited to PMMA. They include

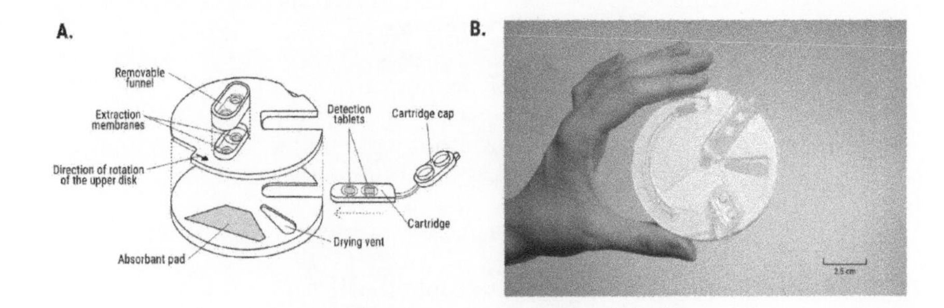

Fig. 7.43: Description (A) and photograph (B) of the COVIDISC. The device, based on paper microfluidics, performs a molecular detection of pathogens, i.e. a detection based on their nucleic acids. It includes an extraction membrane, and a section enabling LAMP amplification. To pass from the extraction to the amplification zone, the upper disk is rotated. Clinical test carried out on 99 COVID samples, with charges distributed in a broad range (from a fraction to hundreds of copies per millilitre), has led to 97 % sensitivity and a 100 % specificity [45, 46].

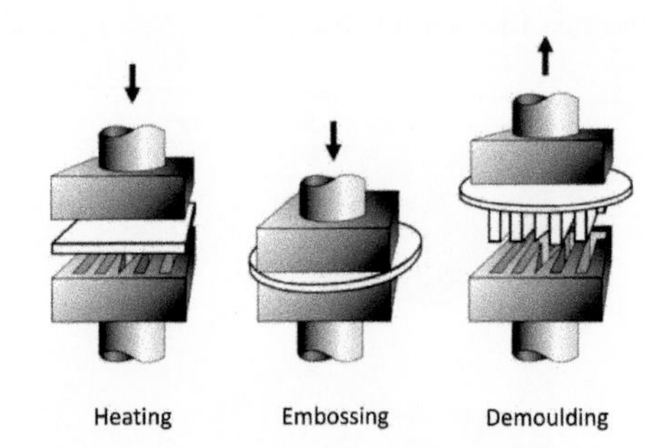

7.44: Casting technology. The process consists in heating, embossing, and demoulding.

plastics such as PC, PI, PE, PET, PVC, or PEEK. The absolute precision of casting is impressive: it is on the order of tens of nanometres [47].

7.11.2 Microinjection

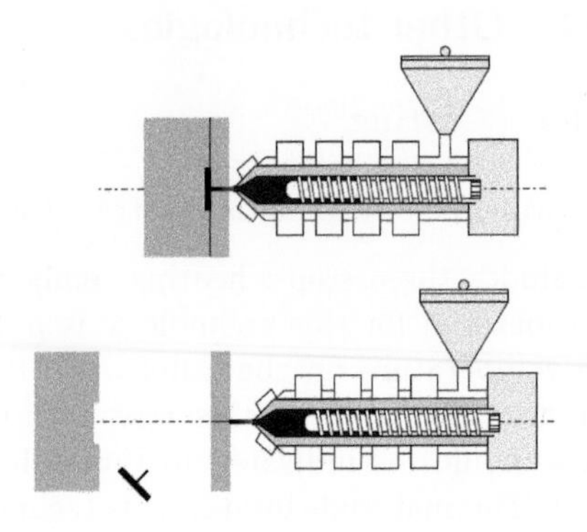

7.45: Plastic moulding injection technology. A plastic material is injected at high temperatures in a mould It is further cooled down to solidify the object

Microinjection is a technique issued from plasturgy. It is shown schematically in Fig. 7.45.

The method can be broken down into three phases:

- The liquid plastic material is injected into a mold under a vacuum and pressure, at a temperature above the glass transition temperature T_g of the plastic.
- The system is cooled down below the glass transition temperature.
- After taking off the mould, a structure is obtained.

The moulds are delicate to make and expensive but, once they are realized, they can give rise to considerable throughputs (milions of device per year are currently made). The technique is interesting for the production of microfluidic devices at low cost.

7.11.3 Laser ablation

Here, plastic is sublimated by the application of an intense laser beam. Examples are shown in Fig. 7.46.

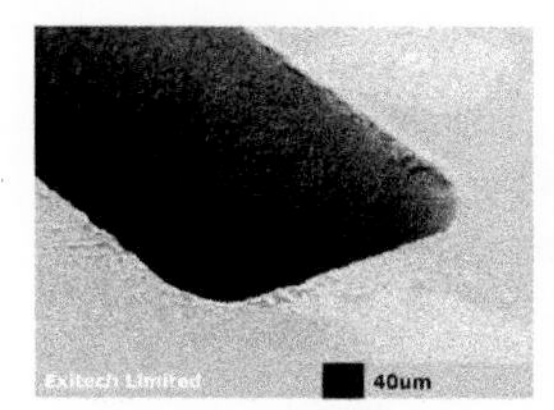

7.46: Micro-trenches made by laser ablation.

UV and femtosecond lasers are used for this technique. The micromachining precision is mediocre (a few micrometres), and the roughness of the surfaces is on the order of 200 to 500 nm. The technique is extremely useful for digging small holes in glass, so as to inject fluids in microfluidic devices.

7.11.4 Sticker technology

A technology, called 'sticker', exploiting the permeability of PDMS to control cross-linking, was developed in 2008 [38]. The process is shown in Fig. 7.47.

A PDMS mould is applied on a photopolymerisable acrylate droplet. After UV illumination, the droplet is cross-linked, but not entirely (see Fig. 7.47 (A)). As oxygen diffuses through the mould, cross-linking is inhibited close to the walls, which enables demoulding. The residual liquid layer allows to seal the system to a glass substrate (see Fig. 7.47 (B)). Owing the large Young modulus of the cross-linked material, the device can sustain high pressures. One limitation is the reactivity of acrylates, which may raise biocompatibility issues.

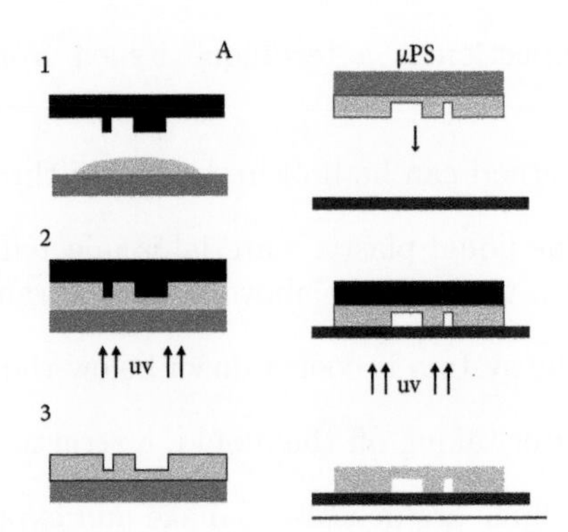

7.47: (A) Structuration of a microfluidic device: a PDMS mould is placed on an photocuring droplet. After cross-linking, the structure is removed from the mould. (B) Sealing of the structure onto a glass slide, using the residual liquid layer to achieve bonding [38].

7.11.5 Actuators based on responsive hydrogels

In the vast family of polymers, some are responsive to stimuli, such as temperature, pH or light. This is the case of PNIPAM (Poly(N-isopropylacrylamide)).[14] When cross-linked with an acrylamide, it forms an hydrogel, which is hydrophilic below a temperature called 'low solubility critical temperature' (LCST) and hydrophobic above. For PNIPAM, the transition temperature is 32°C. Why is it so? This is explained in Fig. 7.48 (taken from [48]).

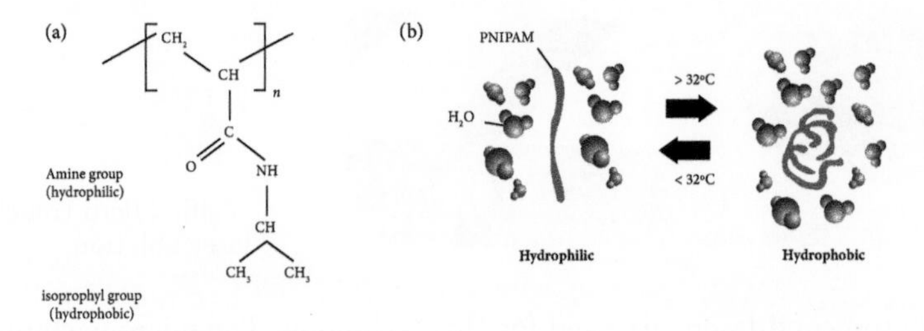

Fig. 7.48: (A) Developed formula of PNIPAM, showing the amine group (hydrophilic) and the isopropyl group (hydrophobic). (B) At low temperature (i.e. below 32 °), energy prevails, and the molecule adopts a linear structure, to expose the hydrophilic group to water and decrease energy. This is the swollen state. At larger temperature, entropy prevails and the molecule collapses to protect the amine group from water. Only hydophobic groups are exposed, and, consequently, the molecule becomes hydrophobic.

At low temperature (i.e. below 32°C), energy prevails, and the molecule adopts a linear structure to expose the hydrophilic group to water and, consequently, decrease the energy of the system.This is the swelling state, in which the molecule is hydrophilic. At larger temperature, entropy prevails and the molecule forms a compact structure to protect the amine group from water. This is the collapsed state. As only hydrophobic groups are exposed, the molecule is hydrophobic. Exploiting this property, the community has elaborated microfluidic actuators. The first demonstration was made in 2000 [49] (see Fig. 7.49).

[14]The molecule is quite recent: it was synthesized in 1968.

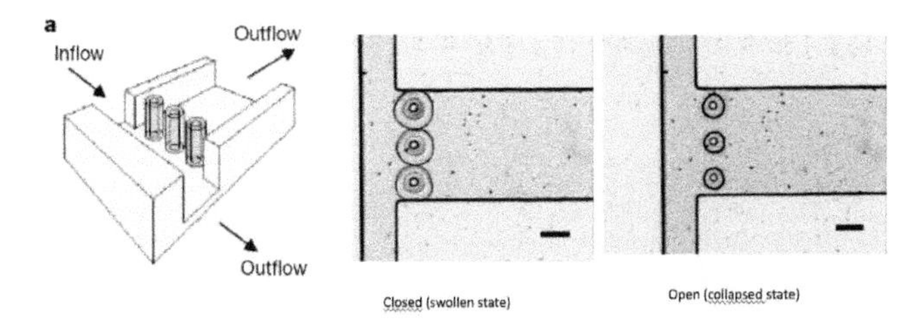

Fig. 7.49: First hydrogel microvalves, based on in-situ photocuring of PNIPAm around pillars (Reprinted from [49] with permission of Springer Nature. Copyright 2022).

Here the excitation which provokes the swelling and the collapse of the hydrogel is not the temperature, but the pH. In Fig. 7.49, the polymer is cross-linked around three pillars, forming annular layers, which swell or collapse, thus close and open the channel, depending on the pH of the solution. The technology had a number of limitations, concerning the control, the presence of cross-linked residues in the channel, and the slow response time, so that it did not spread out in the community. Over the years, progress has been made [50]. Recently, a breakthrough was made [12, 51]. Fig. 7.50 shows a system of 25 cages, that can open and close, in less than one second, under excellent control.

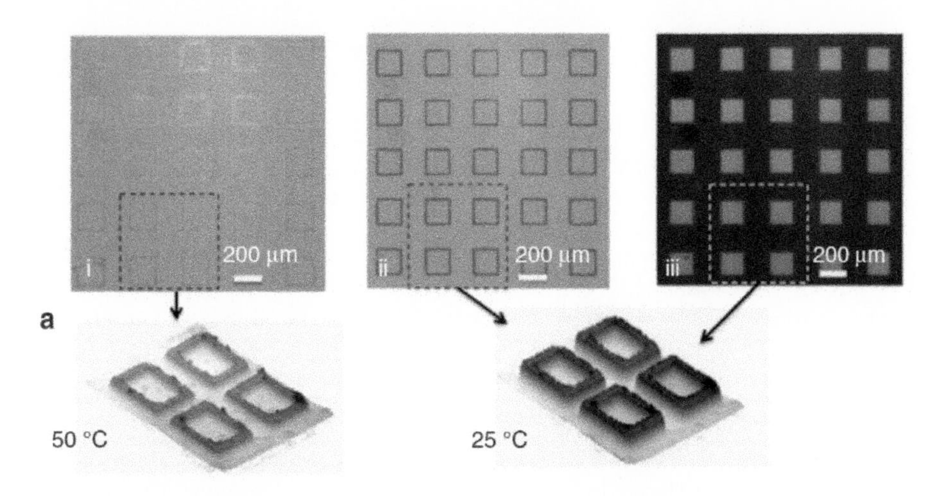

Fig. 7.50: Sequence of fluid manipulations for a system of 25 hydrogel cages, 200 x 200 μm in size with 10 μm thick walls, in a microfluidic chamber. (i) the system is initially filled with a fluorescein solution at 50°C, i.e. with the cages open; (ii) then, temperature is decreased to 25°C. The hydrogel swells and the cages close. (iii) the system is further rinsed with DI water, leaving fluorescein captured in the cages [12].

The device of Fig. 7.50 is initially filled with a fluorescein solution at 50°C, i.e. with the cages open. Then, temperature is decreased to 25°C. The hydrogel swells and the cages close. Finally, the system is rinsed with DI water, leaving fluorescein captured in the cages [12]. The technology allows to trap cells, analyse and process them.

References

[1] Y. Xia, G. M. Whitesides, *Angew. Chem. Int. Ed.*, **37**, 550 (1998).

[2] M. Madou, *Fundamentals of Microfabrication*, CRC Press (2000).

[3] R. Choudhury, *Handbook of Microlithography, Micromachining and Microfabrication*, Vols 1 and 2, SPIE Press, Bellingham, WA (1997).

[4] N. T. Guyen, S. T. Wereley, S. A. M. Saegh *Fundamental and Applications of Microfluidics*, Artech House (2019).

[5] T. Kawai, K. Sawada, Y. Takeuchi, *Proc. MEMS 2001*, Interlaken, **22**, 25 (2001).

[6] D. J. Beebe, G. A. Mensing, G. M. Walker, *Ann. Rev. Biomed. Eng.*, **4**, 261 (2002).

[7] A. Tan, K. Rodgers, J-P. Purrihy, C. O' Mathuna, J. D. Glennon, *Lab on a Chip*, **1**, 7 (2001).

[8] T. Jues, R. Prunieres, E. Brunet, P. Tabeling (2002).

[9] 'History of Lithography', http://www.whatislithoprinting.com/history.html.

[10] J. Danglad-Flores, S. Eickelmann, H. Riegler, *Chemical Engineering Science*, **179**, 257 (2018).

[11] Y. Liu, P. F. Eng, O. J. Guy, K. Roberts, H. Ashraf, N. Knight, *IET Nanobiotechnol.*, **7**, 59 (2013).

[12] L. d'Eramo, B. Chollet, M. Leman, E. Martwong, M. Li, H. Geisler, J. Dupire, M. Kerdraon, C. Vergne, F. Monti, Y. Tran, P. Tabeling *Microsystems & Nanoengineering*, **4**, 1 (2018).

[13] D. Maillefer, S. Gamper, B. Frehner, P. Balmer, H. van Lintel, P. Renaud, *Proc. MEMS 2001*, Interlaken, 414–417 (2001).

[14] J. McDonald, G. Whitesides, *Science*, **276**, 779 (1997).

[15] D. C. Duffy, J. C. McDonald, O. J. A. Schueller, G. M. Whitesides, *Anal. Chem.*, **70**, 4974 (1998).

[16] E. Delamarche, H. Schmid, B. Michel, H. Biebuyck, *Adv. Mater.*, **9**, 741(1997).

[17] C. S. Effenhauser, G. J. M. Bruin, A. Paulus, M. Ehrat, *Anal. Chem.*, **69**, 3451 (1997).

[18] J. Harisson, A. Manz, Z. Fan, H. Ludi, H. M. Widmer, *Anal. Chem.*, **64**, 1926 (1992).

[19] K. Domanskya, D. C. Lesliea, J. Mc Kinney, J. P. Fraser, J. D. Sliza, T. Hamkins-Indika, G. A. Hamiltona, A. Bahinskia, D. E. Ingber, *Lab Chip*, **13**, 3956 (2013).

[20] J. Goulpeau, *Intégration de puce à ADN dans un microsysteme fluidique*, PhD thesis (2005).

[21] Y. Y. Huang, W. Zhou, K. J. Hsia, E. Menard, I. J.–U. Park, J. A. Rogers, A. G. Alleyne *Langmuir*, **21**, 8058 (2005).

[22] M. Unger, H. P. Chou, T. Thorsen, A. Scherre, S. Quake, *Science*,**288**, 113 (2000).

[23] J. Goulpeau, D. Trouchet, A. Ajdari, P. Tabeling, *J. Appl. Phys.*, **98**, 044914 (2005).
[24] J. Liu, C. Hansen, S. Quake, *Anal. Chem.*, **75**, 4718 (2003).
[25] J. W. Hong, S. R. Quake, *Nature Biotech.*,**21**, 1179 (2003).
[26] I. E. Araci, S. T. Quake, *Lab. on a Chip*, **12**, 2803 (2012).
[27] H. Willaime, V. Barbier, L. Kloul, S. Maine, P. Tabeling, *Phys. Rev. Lett.*, **96**, 504501 (2006).
[28] J. Wook Hong, V. Studer, G. Hang, W. French Anderson, S. R. Quake, *Nat. Biotech.*, **22**, 235 (2004).
[29] T. Thorsen, S. J. Maerkl, S. R. Quake, *Science*,**298**, 580 (2002).
[30] A. Dendukuri, D. C. Pregibon, J. Collins, T. A. Hatton, P. S. Doyle, *Nat. Mat*, **5**, 366 (2006).
[31] D. C. Pregibon, M. Toner, P. S. Doyle, *Science*, **315**, 1393 (2007).
[32] M. Thery, V. Racine, A. Pepin. M. Piel, Y. Chen, J. B. Sibarita, M. Bornens, *Nature Cell Biol.*, **7**, 947 (2005).
[33] A. Martinez, S. Phillips, L. Butte, G. M. Whitesides, *Angewandte Chemie*,**46**, 1318 (2007).
[34] A. Martinez, S. Phillips, L. Butte, G. M. Whitesides, *Anal. Chem.*, **82**, 3 (2010).
[35] J. Credou, T. Berthelot, *Journal of Materials Chemistry*, **2**, 4767 (2014).
[36] D. J. Guckenberger, T. E. de Groot, A. M. D. Wan, D. J. Beebe, E. W. K. Young, *Lab. Chip*, **15**, 2364 (2015).
[37] R. S. Woodbury, S. Kobayashi, *Studies in the history of machine tools*, 110 (1974).
[38] D. Bartolo, G. Degre, P. Nghe, V. Studer, *Lab Chip*, **8**, 274 (2008).
[39] A. K. Au, W. Huynh, L. F. Horowitz, A. Folch, *Angew. Chem. Int. Ed.*, **55**, 3862 (2016).
[40] N. P. Macdonald, J. M. Cabot, P. Smejkal, R. M. Guijt, B. Paull, M. C. Breadmore, *Anal. Chem.*, **89**, 3858 (2017).
[41] E. Fu, S. A Ramsey, P. Kauffman, B. Lutz, P. Yager, *Microfluidics and nanofluidics*, **10**, 29 (2011).
[42] L. Magro, C. Escadafal, P. Garneret, B. Jacquelin, A. Kwasiborski, J-C. Manuguerra, F. Monti, A. Sakuntabhai, J. Vanhomwegen, P. Lafaye, P. Tabeling, *Lab on a Chip*, **17**, 2347 (2017).
[43] J. Reboud, G.Xub, A. Garretta, M. Adrikoc, Z.Yanga, E. M. Tukahebwac, C. Rowellc, J. M. Cooper, *PNAS*, **116**, 4834 (2019).
[44] J. T. Connelly, J. P. Rolland, G. M. Whitesides, *Anal. Chem*, **87**, 7595 (2015).
[45] P. Garneret, E. Coz, E. Martin, J-C. Manuguerra, E. Brient-Litzler, V. Enouf, D. F. González Obando, J. C. Olivo-Marin, F. Monti, S. van der Werf, J. Vanhomwegen, P. Tabeling, *Plos One*, **16**, e0243712 (2021).
[46] E. Coz, P. Garneret, E. Martin, D. Freitas do Nascimento, A. Vilquin, D. Hoinard, M. Feher, Q. Grassin, J. Vanhomwegen, J-C. Manuguerra, S. Mukherjee, J-C. Olivo-Marin, E. Brient-Litzler, M. Merzoug, E. Collin, P. Tabeling, B. Rossi, MedRxiv : https://doi.org/10.1101/2021.10.03.21264480, (2021).
[47] Y. Chen, A. Pepin, *Electrophoresis*, **22**, 187 (2001).
[48] W. He et al., *Journal of Physics: Conference Series*, **1676**, 012063 (2020).

[49] D. J. Beebe, J. S. Moore, J. M. Bauer, Q. Yu, R. H. Liu, C. Devadoss, B-H. Jo, *Nature*, **404**, 588 (2000).
[50] A. Richter, G. Paschew, S. Klatt, J. Lienig, K. F. Arndt, H.-J. P. Adler, *Sensors*, **8**, 561 (2008).
[51] B. Chollet, L. Deramo, E. Martwong, M. Li, J. Macron, T. Q. Mai, P. Tabeling, Y. Tran,*ACS Applied Mat. Int.*, **8**, 24870 (2016).

Index